GRIBOV-80
MEMORIAL VOLUME
QUANTUM CHROMODYNAMICS AND BEYOND

GRIBOV-80
MEMORIAL VOLUME

QUANTUM CHROMODYNAMICS AND BEYOND

Proceedings of the Memorial Workshop Devoted
to the 80th Birthday of V N Gribov

The Abdus Salam International Centre for Theoretical Physics,
Trieste, Italy 26 – 28 May 2010

EDITORS

YU L DOKSHITZER

LPTHE, Université Paris 6, Paris, France

P LÉVAI & J NYÍRI

KFKI Research Institute for Particle and Nuclear Physics,
Budapest, Hungary

World Scientific

NEW JERSEY • LONDON • SINGAPORE • BEIJING • SHANGHAI • HONG KONG • TAIPEI • CHENNAI

Published by

World Scientific Publishing Co. Pte. Ltd.
5 Toh Tuck Link, Singapore 596224
USA office: 27 Warren Street, Suite 401-402, Hackensack, NJ 07601
UK office: 57 Shelton Street, Covent Garden, London WC2H 9HE

British Library Cataloguing-in-Publication Data
A catalogue record for this book is available from the British Library.

GRIBOV-80 MEMORIAL VOLUME
Quantum Chromodynamics and Beyond
Proceedings of the Memorial Workshop Devoted to the 80th Birthday of V N Gribov

ISBN-13 978-981-4350-18-1
ISBN-10 981-4350-18-4

Printed in Singapore by B & Jo Enterprise Pte Ltd

PREFACE

The quest for understanding the laws of Nature saw the creation of elementary particle physics in the second half of the 20th century. Among the theorists who built it, the name of Vladimir Naumovich (Volodya) Gribov occupies an exceptional place.

Gribov belonged to the generation of physicists, now almost extinct, who did not take quantum field theory for granted but participated in its creation. Fundamental problems were addressed, "first principles" questioned, genuine ideas and techniques developed to later become the basic art of the trade.

The great majority of "elementary particles" are *hadrons* whose spectrum and interactions — *strong interactions* — posed, and keep posing, the most challenging problems in the theoretical physics history. By the 1960 two things have become clear: 1) that the relativistic nature (crossing symmetry) poses severe restrictions on the buildup of the theory of interacting particles and 2) that at high energies certain simplifications should occur that one may hope to exploit to construct a consistent theoretical description of a multitude of complex phenomena. The quest for constructing the relativistic theory of high energy particle interactions had been formulated.

Gribov was the first to apply the idea of Regge poles to relativistic particle physics. He developed the relativistic theory of complex angular momenta, introduced the notion of reggeons and constructed their effective interaction theory both in its mathematical and physical aspects: from the prediction of shrinkage of the diffractive cone to the space–time parton picture of high energy interactions, to relativistic multiple scattering theory, physics of inelastic diffraction, the multiplicity fluctuation pattern in multi-hadron production.

Gribov's "vacuum pole" (Pomeron) became the core object in the description of total, elastic and multi-particle production cross sections of high energy hadron collisions; his ideas have laid the basis for the concept of partons and to neutrino oscillations, and marked the breakthrough in understanding of second order phase transitions.

Quantum Chromodynamics has inherited many a notion of the general pre-microscopic theory. The start of the exciting Large Hadron Collider operation should help to resolve many outstanding questions that were left unanswered by the "old theory" of high energy hadron interactions.

The Gribov-80 Workshop was organized at the International Centre for Theoretical Physics (ITCP) in Trieste with the invaluable support from its scientists Luciano Bertocchi and Daniele Treleani. The Workshop ran between 26 and 28 May 2010 and brought together theorists of all ages from all over the world. It provided a great opportunity to meet, share new ideas and discuss recent developments in a friendly informal atmosphere of a "Russian style seminar" where the notion of formal politeness towards the speaker is considered secondary to the wish of getting to the heart of the physical issue being discussed.

This book contains the collection of the presented talks and contributed papers, honoring the memory of a great scientist of the twentieth century.

We thank all the participants for having accepted our invitation to Trieste. We are grateful to Prof. Bertocchi and Prof. Treleani for the kind hospitality of the ICTP, and we thank Nadia for her kindness and patience to help the Workshop with her secretary work.

The financial support of the ICTP, the MTA KFKI RMKI, and the Hungarian OTKA Grant NK77816 is gratefully acknowledged.

We hope to meet again at the next Gribov Memorial Workshop.

Yuri Dokshitzer, Péter Lévai, Julia Nyíri

ICTP, Trieste, 2010

CONTENTS

RECENT RESULTS IN GENERAL FIELD THEORY AND GRAVITY

REMEMBERING VOLODYA

MESSAGE TO THE GRIBOV-80 WORKSHOP

ANDRÉ MARTIN

Theory Division, CERN, 1211 Geneva 23, Switzerland
andre.martin@cern.ch

Dear Julia, dear friends

I am sorry not to be able to attend the celebration of the 80th birthday of Volodya Gribov.

I consider Volodya as one of the very best theoreticians of his generation, and I am very sad that he is not anymore with us.

My first contact with Volodya was not physical. It was through the communication he made at the 1960 Rochester conference, which was read, in absentia, by Francis Low. In this work Volodya showed that the common belief that, at high energies the scattering amplitude behaves like $sf(t)$ is inconsistent with continued unitarity in the t channel. It is Léon Van Hove who drew my attention on this work. Later I gave a talk at Imperial College about this. Paul Matthews and Abdus Salam were completely bewildered. Nobody expected that. It was the beginning of a revolution in high energy particle physics. Later Volodya made other very important contributions, but I believe that it is this one which drew the attention of Western physicists on Volodya.

We met for the first time in Geneva during the 1962 conference which was so important. I remember that, with my friend Jurko Glaser we had passionate discussions with him and his inseparable friend Pomeranchuk. The next meeting was in Erevan, in Armenia where a small group of western physicists (among whom Gell-Mann, Lederman, Schwarz) had been invited. This was my first visit to USSR. We were beginning to become closer and closer friends. Then we met in Rochester in 1967, where Volodya managed to come by bargaining with the Russian authorities: he would not agree to give lectures in India (with whom USSR wanted to make friends) if he was not allowed to go to Rochester. Then there was Kiev, Leningrad, Alushta. I should also mention that Thierry, our second son, visited Julia and Volodya in Moscow, where they spoke of the planned expedition of Volodya's son to Pamir, which ended up in a tragedy. Our last meeting was in Budapest (where I had been already in 1977 and met Julia for the first time) in 1996. It was obvious that Julia was already very worried about Volodya's health, but we (Schu, my wife and I) could not imagine that Volodya would leave us one year later. Our only consolation is that his major contributions to theoretical physics will remain.

Budapest, 1996.

I miss Volodya too. He was a wonderful combination of qualities to have as a friend: full of enthusiasm and sensitivities about life and the world in general; a huge talent and shining brilliance as a theoretical physicist with whom to discuss and argue about our field and where it was going; and overall a great pleasure to be with, whether in Moscow, Stanford, or New York.

Sidney Drell
SLAC, Palo Alto
Stanford, CA, USA

Stanford, 1989.

IN MEMORY OF A FRIEND

SEMEN S. GERSHTEIN

Institute of High Energy Physics, State Research Center
Protvino, Moscow Region, Russia

It is really hard to write about Volodya. There are no words that can express my admiration for his talent and the charm of his outstanding personality. Time can not soothe the pain of his untimely death. It only makes it worse, especially for the scientific world.

The achievements which make Gribov a remarkable person are very well described in the previous works of Yu. Dokshitzer and others. So, I will dwell more on our personal contacts and impressions.

The first time I saw Volodya in 1957 or 1958 at the seminar hosted by Landau, who was my post graduate supervisor at the time. I remember a young man, whom I had not seen before, stand up during the talk and start firing questions at the speaker. And the latter found it difficult to satisfy the young man with his answers. Landau took the young man's side and agreed with his objections to the speaker. This inquisitive young man was Volodya.

I got to know him closer when in 1958 after doing my doctorate degree I came to work at Leningrad Physico-Technical Institute (LPHTI). The atmosphere at the Theoretical Department was really pleasant. The Head of the Theoretical Department Ilya Mironovich Shmushkevich, the friend of I.A. Pomeranchuk, was sticking to Landau's school and followed its traditions.

One of the important elements of the activity of the Theoretical Department was a seminar. Shmushkevich tried to clarify every aspect of the issue in depth. Therefore the seminars went on for hours and involved lots of argument. Gradually Gribov took the leading role and was the one who could clearly define the problems and on most occasions find solution to them.

The seminar used to attract not only those who worked in the Theoretical Department. Ludwig Faddeev was one who attended them and used to give a comprehensive explanation to arising mathematical problems. People in the Department worked in a very warm atmosphere of understanding. We were young and tied by interest in science and close friendship. I remember our amicable feasts. Sometimes we were invited by Shmushkevich to his place. I still remember those friends who are alive and those who passed away. Besides Volodya there were V. Shekhter, A. Anselm, I. Dyatlov, S. Maleev, Yu. Petrov and others.

We made friends with Volodya at once. Neither of our scientific career was smooth. Upon graduation from the University we had a hard time finding a job in a scientific institution. Volodya started as a teacher at an evening school, and I became a teacher in a village school 100 km from Moscow. After Stalin's death the situation gradually changed. L.D. Landau (who was the one to examine me on his theoretical minimum) was able to take me to a postgraduate school, and K.A. Ter-Martirosyan has demonstrated his strong will by making Volodya accepted by LPHTI.

Apart from our professional problems we had a similar family situation. We both had young sons and enjoyed exchanging our experiences.

Volodya had a very rare talent for evaluating new works. He showed interest in other topics, not always connected with his work. He would go deep into the matter, find good qualities of the work (if there were any) and its weak points. His remarks, if the author listened to them, often led to new turns in research. Kind-hearted, he at the same time was intolerant both towards scolism and unreasonable results. In this respect he took after Landau. A few years later when Volodya started working with I.Ya. Pomeranchuk, the latter said: "You can hardly imagine the pleasure of working with Volodya. He very much reminds me of Landau".

Volodya himself set high standards to his works. He never wanted to publish them as fast as possible, searching for additional proofs of his findings. I remember when in 1958 I arrived in Leningrad and met Volodya, he was discussing the representation that determined analytical properties of the scattering amplitude over two variables, i.e. energy and momentum transfer, and was looking for convincing arguments in favour of the one. What he told me (and it was not published) coincided in essence with the Mandelstam representation that appeared the same year. Hence Volodya was well prepared to use it and obtained the results, considered nowadays classical, on the asymptotical behaviour of hadron scattering amplitudes at high energies. Yu. Dokshitzer and L. Frankfurt wrote about it in the Preface to the book *V.N. Gribov: Gauge Theories and Quark Confinement.*

In 1960 I started to work in the Joint Institute for Nuclear Research in

Dubna. Several experiments on weak interactions that were rather interesting for me, were being prepared there. However, Volodya and I kept in touch. He often visited Moscow and stayed there for long periods of time.

The results obtained by Volodya greatly impressed Landau and Pomeranchuk. Landau thought that Volodya's approach was a way out of the tight corner for the quantum field theory, where the latter found itself after the discovery of the "zero-charge". From his point of view unobservable quantities should have been taken away from the theory, among them were field operators ψ, and consequently, the Hamiltonian, which could be constructed only from the field operators. Following Landau's ideas the theory should be based on scattering amplitudes with their properties: analyticity (causality), unitarity and crossing symmetry (relativism). Volodya's results quite obviously pointed to the fruitfulness of such an approach and gave extremely interesting predictions (for example, the one pointed to the shrinkage of the diffractive cone in hadron scattering, i.e. to the growth of hadron interaction radius with energy).

Landau highly appreciated Volodya's talent, his possession and devotion to science. I remember him saying many times that he together with Volodya would continue writing his course on theoretical physics. It happened when Landau gained consciousness in the hospital after the accident.

Everyone who knows Volodya, marks his terrific intuition. There is a lot of examples: the physical interpretation of the instanton, his remarks that in non-abelian theories anti-screening was quite possible, his space-time picture of interactions, that was the predecessor of the quark-parton model, etc. Not all of such remarks were timely appreciated by Volodya's colleagues.

In 1965 S. Alliluev, A. Logunov and I managed to explain the behaviour of hadron scattering at large angles, which was established by Orire, with the help of the model of scattering on Gaussian potential. In this model perturbation theory approximations grow at first, but then they decrease. Summing them up we have got the required law. When in Leningrad I asked Volodya if it was possible to obtain the required result in the ℓ-plane. "No problem" — was the answer. "These are branchings, they should be able to explain this phenomenon perfectly well."

Our conversation took place in a small room right before a theoretical seminar. There were a lot of people, but Volodya's remark did not seem to interest anybody. It was only in a few years, that A. Anselm treated this question in his doctorate thesis and derived the Orire law summing branching in the ℓ-plane. (He did not even remember Volodya's remark.)

Volodya not only had an amazing intuition in physics, he created also the necessary mathematical apparatus.

In 1968 my ten-year old son and I together with Volodya, his wife Lilya and their son went to the Caucasus on a guided tour. Our way was to the Chegemskoe clove not far from Nal'chik. Volodya took a tiny wireless with him. And one night we heard over the radio the news about American astronauts on the Moon. We were so happy and proud of this achievement. That over-the-top news was followed by alarming news – clouds started to gather over the "Prague spring" and people's hope for the "socialism with a human face" began to melt. Our march was over at the other end of Chegemskoe clove at the foot of the main Caucasian range. I sent my son back to Moscow by plane, where my mother was to meet him. I myself started for a mountaineers' camp near the Adyr-Su, a tributary of the Boksan. We started for the neighbouring Adyl-Su clove to begin our ascend. (Between these two cloves there is the famous Boksan neutrino observatory, where experiments on solar neutrino detection were later carried out.)

On my way back I dropped in the post office on Boksan and was given a cable. "We are in Mestya. Volodya has had a heart attack." I was just shocked. The cable was sent about a fortnight ago. I decided to start for Mestya immediately. It would have taken too much time to get to the place by any means of transport. That meant to go by peripheral roads. I returned to our camp to take my papers and the same night I left the camp to pass Adyr-Su passage. I did not have any mountaineer's equipment, so I had to literally crawl over shabby ice bridges across clefts. Finally I passed the cleft, but had to walk along a big ice-flow Leksor. It was only late in the evening that I reached the path leading to Mestya. In the suburbs of Mestya (the capital of Svanetya, Georgian region) I met with Yura Petrov. It was quite unexpected. On learning from a mountaineers group who left the camp a few hours earlier, that I was on my way to the town, he decided to meet me.

I would like to tell a few words about Yura Petrov. He was one of the most faithful friends of Volodya. In the winter of 1941–1942 his parents starved to death in the city under siege, and the twelve-year boy was evacuated from Leningrad together with other children of the orphanage to the Caucasus. In the summer of 1942 the Caucasus was occupied by the Nazi, and the boy ran away and became a homeless child. After the war he attended an industrial school, then it was the time at a technical college, and university. He was hired in memory of his father-physicist by the Physico-Technical Institute, but the level of his knowledge was not sufficient. He was in charge of making numeric calculations of a nuclear reactor, which was going to be constructed. However, in a short time Yura brilliantly mastered physics of reactors and in the end became one of the most qualified specialists. In the 1970s together with Konoplev K. he designed the "Pik" reactor with a unique $10^{15}\mathrm{cm}^{-2}\mathrm{sec}^{-1}$ neutrino flux. Now this reactor is going to be launched.

Learning about Volodya's disease, Yura immediately took a plane from Leningrad. When I arrived, Volodya felt quite well. He was given a separate room in a tourist camp. People around him treated him very warmly. The woman-doctor who visited him was an excellent specialist, and Volodya said that the pain in his heart ceased when she came. Svanetya used to be an isolated area behind the remote cols at that time. But when enlightment reached it, many people were carried away with it. (For example, there was a worker in the camp whose name was Edison, and his brother's name was Newton). Volodya and I continued discussing a recently published work by Veneziano, which we began to do in Chegem.

We kept discussing politics too. It seemed that everything was going to the better. But there was the first warning bell... I returned to Moscow on the 21st of August and got a shock when I saw a newspaper. It said that the countries of Warsaw Pact moved troops into Czechoslovakia.

In March 1980 I was going to visit Leningrad to celebrate Volodya's 50th anniversary, but I heard that he was in Dubna. I came to Dubna and saw a completely happy, a bit embarrassed Volodya and learnt about the changes in his life. I was very glad about his happiness and the fact that he was able to win it. I had known Julia for a long time, meeting her in Hungary at the neutrino conferences and I was glad about their union. On the 25th of March, Volodya's birthday, besides me there was Leva Okun, Arkady Migdal and his wife Tanya, Volodya's sister Inna and Julia's friend Livia. We happily celebrated Volodya's 50th birthday.

I am sure that it was Julia who was able to save Volodya from a terrible shock connected with the death of his son, Lyonia. Volodya loved him very much and Lyonia himself had been doing very well and had carried out a brilliant work in theoretical high energy physics shortly before his death.

When Volodya and Julia settled in Moscow, I often visited them when they came back from lectures in Moscow Physico-Technical Institute. Volodya's head was crowded with new ideas. We spoke a lot about his quark confinement model, the role of massless quarks, the importance of non-abelian symmetry and the nature of the chiral anomaly. I think it was easy for him to discuss these things with me because I never keep to any doctrine and always try to understand a physical idea. Besides, Volodya knew about our work with Jakov Borisovich Zeldovich on supercritical nuclear charges. And appealed to it. To explain the nature of the supercritical vacuum he developed an analogy with a remarkable effect known in condensed matter physics: the Andreev reflection. But he did not hurry to write practically ready works, thinking them over and over thoroughly again. He liked a phrase said, I think, by Bergman: "My film is practically ready. All I have to do is to shoot it". Unfortunately, Volodya did not have time "to shoot" a lot of things, and Julia and Yu. Dokshitzer had to do a heroic work: to write his articles according to his notes.

I liked to visit Gribovs' hospitable house, discuss different questions about life, talk to growing Pal, who was so loved by Volodya.

I remember our meeting at the university in Minnesota, where Volodya came earlier than he had planned because he learnt that I had to go away. He had just recovered from his illness, but as usual he smoked a lot (lighter cigarettes, though). In the morning we went to the workshop where Volodya showed his typical features of character. He could not be indifferent as far as science was concerned. He hated pseudo-science and was able to criticize a speaker severely. The way the question was put and hence the received result seemed to him absurd. He felt so nervous that he left the workshop.

That is how I remember him: talented, full of passion and wise.

Minneapolis, 1995

PROGRESS IN QCD

GRIBOV REGGEON CALCULUS IN QCD, SUSY AND GRAVITY

L.N. LIPATOV

Petersburg Nuclear Physics Institute
Gatchina, 188300, St.Petersburg, Russia
lipatov@thd.pnpi.spb.ru

We remind the Gribov approach to the hadron high energy scattering. It was based on the effective field theory for the Pomeron interactions. In QCD and in gravity the gluons and gravitons are reggeized and therefore at high energies one should reformulate these theories in terms of the reggeons and their interactions. We review the basic ideas of the BFKL approach in QCD and in supersymmetric models and generalize them in the framework of the gauge-invariant effective theory for the reggeized gluon interactions. The similar generally covariant action for the reggeized graviton interactions is formulated in terms of the effective currents satisfying a non-linear evolution equation.

Keywords: BFKL equation; effective field theory; quantum gravity.

1. Regge poles and Gribov Pomeron calculus

In the Regge kinematics

$$s = 4E^2 >> -t = \vec{q}^{\,2} \tag{1}$$

the t channel partial wave expansion for the scattering amplitude with the definite signature $p = \pm 1$ has the form

$$A^p(s,t) = s \int_{a-i\infty}^{a+i\infty} \frac{d\omega}{2\pi i} \left((-s)^\omega - p s^\omega\right) f^p_\omega(t)\,. \tag{2}$$

In the framework of the Regge pole hypothesis we have

$$f^p_\omega(t) = \frac{\gamma^2(t)}{\omega - \omega_p(t)}\,, \tag{3}$$

where ω_p is the Regge trajectory assumed to be linear

$$\omega_p(t) = \Delta - \alpha' \vec{q}^{\,2}\,. \tag{4}$$

In this case the amplitude has the Regge form[1]

$$A^p_{Regge}(s,t) = \xi_p(t)\, s^{1+\omega_p(t)}\, \gamma^2(t)\,, \;\; \xi_p(t) = e^{-i\pi\omega_p(t)} - p\,. \tag{5}$$

S. Mandelstam demonstrated, that the Regge poles generate the cut singularities in the ω-plane.[2] For example, in the case of the two reggeon exchange we have for the contribution of the Mandelstam cut the following expression

$$A^p_{Mand}(s,t) = \xi_p\, s \int \frac{d^2k}{(2\pi)^2} \Phi^2(k, q-k) s^{\omega_{p1}(-k^2)} s^{\omega_{p2}(-(q-k)^2)}\,. \tag{6}$$

According to the Gribov signature conservation rule one obtains

$$p = p_1\, p_2\,. \tag{7}$$

To describe an approximately constant behaviour of the total cross-sections σ and the amplitudes related with it by the optical theorem a special reggeon - Pomeron with vacuum quantum numbers is introduced

$$\sigma = \frac{1}{s}\, \Im A(s.0)\,, \;\; A(s,t) \approx is\, s^{\Delta_P - \alpha'_P q^2}\,, \tag{8}$$

where the intercept Δ_p is small.

V. Gribov, I. Pomeranchuk and K. Ter-Martirosyan proved that the Mandelstam cuts appear in an analytic continuation of the multi-particle t-channel unitarity relation[3]

$$\Im_t f_\omega(t) \sim \sum_n \int d\Omega_n |f^{(n)}_\omega|^2\,. \tag{9}$$

The produced and virtual particles are unified in their rapidities in clusters, which are strongly separated each from others in relative rapidities

$$0 < y_1 \ll y_2 \ll \ldots \ll y_k \ll \ln s\,. \tag{10}$$

A cluster of particles in the Gribov picture interacts with other clusters by the exchange of Pomerons. The Pomeron Green function in a linear approximation for its Regge trajectory has a non-relativistic form

$$G_0 = \frac{1}{E + \Delta - \frac{k^2}{2m}}\,, \;\; E = -\omega\,, \;\; \alpha' = \frac{1}{2m}\,. \tag{11}$$

It leads to an effective 2+1 dimensional reggeon field theory described by the Gribov action[4]

$$S = \int dy\, d^2\rho \left(\phi^* (\partial_y - \Delta)\phi + \frac{1}{2m}\, |\partial_\mu \phi|^2) + \lambda \phi^* \phi^2 + \ldots \right)\,. \tag{12}$$

2. Gluon reggeization and multi-Regge processes in QCD

The elastic Born amplitude in the Regge limit of QCD has the form

$$M_{AB}^{A'B'}(s,t)|_{Born} = g\,T^c_{A'A}\,\delta_{\lambda_{A'}\lambda_A}\,\frac{2s}{t}\,g\,T^c_{B'B}\,\delta_{\lambda_{B'}\lambda_B}\,, \tag{13}$$

corresponding to the conservation of helicities λ of the colliding particles A and B. In the leading logarithmic approximation (LLA)

$$\alpha_s \ln s \sim 1\,,\;\; \alpha_s = \frac{g^2}{4\pi} \ll 1 \tag{14}$$

it is multiplied by the Regge factor

$$M_{AB}^{A'B'}(s,t) = M_{AB}^{A'B'}(s,t)|_{Born}\, s^{\omega(t)}\,, \tag{15}$$

where the gluon Regge trajectory in LLA is given below

$$\omega(-|q|^2) = -\frac{\alpha_s N_c}{4\pi^2}\int d^2k\,\frac{|q|^2}{|k|^2|q-k|^2} \approx -\frac{\alpha_s N_c}{2\pi}\ln\frac{|q^2|}{\lambda^2}\,. \tag{16}$$

In the multi-Regge kinematics, where the pair energies $\sqrt{s_k}$ of the produced gluons are large in comparison with momentum transfers $|q_i|$ and transverse components $k_{r\perp}$ of the particle momenta the production amplitude has the factorized form[5]

$$M_{2\to 2+n} \sim \frac{s_1^{\omega_1}}{|q_1|^2}\,gT^{d_1}_{c_2c_1}C(q_2,q_1)\,\frac{s_2^{\omega_2}}{|q_2|^2}...gT^{d_n}_{c_{n+1}c_n}\,C(q_{n+1},q_n)\,\frac{s_{n+1}^{\omega_{n+1}}}{|q_{n+1}|^2}\,, \tag{17}$$

where ω_i are the Regge trajectories and $C(q_i,q_{i-1})$ are the Reggeon-Reggeon-gluon vertices. In LLA they are

$$\omega_r = -\frac{\alpha_s N_c}{2\pi}\left(\ln\frac{|q_r^2|}{\mu^2} - \frac{1}{\epsilon}\right)\,,\;\; C(q_2,q_1) = \frac{q_2\,q_1^*}{q_2^* - q_1^*}\,, \tag{18}$$

where the produced gluons are assumed to have definite helicities. The total cross-section can be calculated in terms of squares of these amplitudes

$$\sigma_t = \sum_n \int d\Gamma_n\,|M_{2\to 2+n}|^2\,. \tag{19}$$

The production amplitudes should satisfy the Steinmann relations, claiming that the amplitudes do not have simultaneous singularities in the overlapping channels

$$\Delta_{s_r}\Delta_{s_{r+1}} M_{2\to 2+n} = 0\,. \tag{20}$$

In an accordance with these relations we have the following representation for the one particle production amplitude $M_{2\to 3}$ in the Regge ansatz

$$M_{2\to 3} = c_1(-s)^{j(t_2)}(-s_1)^{j(t_1)-j(t_2)} + c_2(-s)^{j(t_1)}(-s_2)^{j(t_2)-j(t_1)}\,. \tag{21}$$

In the case of two particle production the corresponding expression for the multi-Regge amplitude contains five terms

$$M_{2\to 4} = d_1(-s)^{j_3}(-s_{012})^{j_2-j_3}(-s_1)^{j_1-j_2} + d_2(-s)^{j_1}(-s_{123})^{j_2-j_1}(-s_3)^{j_3-j_2}$$

$$+ d_3(-s)^{j_3}(-s_{012})^{j_1-j_3}(-s_2)^{j_2-j_1} + d_4(-s)^{j_1}(-s_{123})^{j_3-j_1}(-s_2)^{j_2-j_3}$$

$$+ d_5(-s)^{j_2}(-s_1)^{j_1-j_2}(-s_3)^{j_3-j_2}\,, \; j_r = j(t_r)\,. \tag{22}$$

These formulas can be considered as dispersion relations in the multi-Regge kinematics, because the amplitudes contain real and imaginary contributions. The imaginary parts can be constructed with the use of the unitarity conditions and real parts are obtained from above relations. In the leading logarithmic approximation the imaginary contributions to the production amplitudes $M_{m\to n}$ are small. Therefore the unitarity conditions give a possibility to construct a closed non-linear equations for the amplitudes, which are called the bootstrap relations. Symbolically they have the form[5]

$$\pi\omega(t_1)\, M_{2\to 2+n} = \sum_r \Im_{s_{0r}} M_{2\to 2+n} = \sum_t M_{2\to 2+t} M_{2+t\to 2+n}\,. \tag{23}$$

The knowledge of $M_{2\to 2+n}$ gives a possibility to construct the total cross-section and the scattering amplitude with color singlet quantum numbers in the crossing channel. This amplitude can be expressed in terms of the Pomeron wave function satisfying the BFKL equation[5]

$$E\,\Psi(\vec{\rho}_1,\vec{\rho}_2) = H_{12}\,\Psi(\vec{\rho}_1,\vec{\rho}_2)\,,\; \sigma_t \sim s^{\Delta}\,,\; \Delta = -\frac{\alpha_s N_c}{2\pi}\,E_0\,. \tag{24}$$

Here H_{12} is the BFKL Hamiltonian and Δ is the Pomeron intercept. In the coordinate (impact parameter) representation the hamiltonian has the form[6]

$$H_{12} = \frac{1}{p_1 p_2^*}(\ln|\rho_{12}|^2)p_1p_2^* + \frac{1}{p_1^* p_2}(\ln|\rho_{12}|^2)p_1^*p_2 + \ln|p_1p_2|^2 - 4\psi(1)\,, \tag{25}$$

where

$$\rho_{12} = \rho_1 - \rho_2\,,\; \rho_r = x_r + iy_r\,,\; \Delta = 4\alpha N_c \ln 2\,/\pi\,. \tag{26}$$

This equation is invariant under the Möbius transformations[7]

$$\rho_k \to \frac{a\rho_k + b}{c\rho_k + d} \tag{27}$$

and, as a result, the eigenfunctions of H_{12} are classified by the conformal weights

$$m = \gamma + n/2\,,\; \widetilde{m} = \gamma - n/2\,, \tag{28}$$

expressed in terms of the anomalous dimension $\gamma = 1/2 + i\nu$ of twist-2 operators and integer conformal spin n. The corresponding energies have the property of the holomorphic separability and are expressed in terms of the logarithmic derivative $\psi(x)$ of the gamma function

$$E = 2\chi(n,\gamma) = \psi(m) + \psi(1-m) + \psi(\widetilde{m}) + \psi(1-\widetilde{m}) - 4\psi(1)\,. \qquad (29)$$

These properties lead to the integrability of the BFKL dynamics.[8,9]

3. Pomeron in $N = 4$ SUSY

The next-to-leading corrections to the BFKL kernel was calculated in Ref. 10. The corresponding two-loop expression for its eigenvalue has the form

$$\omega = 4\,\hat{a}\;\chi(n,\gamma) + 4\;\hat{a}^2\,\Delta(n,\gamma)\,,\;\hat{a} = g^2 N_c/(16\pi^2)\,, \qquad (30)$$

where $\chi(n,\gamma)$ is the Born expression. This result contains non-analytic terms proportional to $\delta_{n,0}$ and $\delta_{n,2}$. But in the case of the $N = 4$ extended supersymmetric gauge theory these Kroniker symbols are cancelled and the two loop result has the property of the hermitian separability[11]

$$\Delta(n,\gamma) = \phi(M) + \phi(M^*) - \frac{\rho(M) + \rho(M^*)}{2\hat{a}/\omega}\,,\; M = \gamma + \frac{|n|}{2}\,, \qquad (31)$$

where

$$\rho(M) = \beta'(M) + \frac{1}{2}\zeta(2)\,,\;\beta'(z) = \frac{1}{4}\left[\Psi'\Big(\frac{z+1}{2}\Big) - \Psi'\Big(\frac{z}{2}\Big)\right] \qquad (32)$$

and

$$\phi(M) = 3\zeta(3) + \Psi''(M) - 2\Phi(M) + 2\beta'(M)\Big(\Psi(1) - \Psi(M)\Big), \qquad (33)$$

$$\Phi(M) = \sum_{k=0}^{\infty}\frac{(-1)^k}{k+M}\left(\Psi'(k+1) - \frac{\Psi(k+1) - \Psi(1)}{k+M}\right). \qquad (34)$$

It means, in particular, that the eigenvalue in $N = 4$ has a remarkable property of the maximal transcendentality.[12] This property allowed to calculate the anomalous dimensions of twist-two operators up to three loops in $N = 4$ SUSY.[13] The $N = 4$ four-dimensional conformal field theory due to the Maldacena guess is equivalent to the superstrings living on the anti-de-Sitter 10-dimensional space.[14–16] Therefore the Pomeron in $N = 4$ SUSY is equivalent to the reggeized graviton in the superstring theory.

Let us write the BFKL equation in a diffusion approximation

$$j = 2 - \Delta - D\,\nu^2\,,\;\gamma = 1 + \frac{j-2}{2} + i\nu\,, \qquad (35)$$

where the parameters Δ and D depend on the coupling constant. According to the energy-momentum conservation we have $D = \Delta$ and therefore the anomalous dimension of twist-2 operators is

$$\gamma = (j-2)\left(\frac{1}{2} - \frac{1/\Delta}{1+\sqrt{1+(j-2)/\Delta}}\right). \tag{36}$$

On the other hand, due to the AdS/CFT correspondence the diffusion approximation for the BFKL equation corresponds to the linear graviton Regge trajectory

$$j = 2 + \frac{\alpha'}{2}\,t\,,\ t = E^2/R^2\,,\ \alpha' = \frac{R^2}{2}\,\Delta\,. \tag{37}$$

Because the behavior of the anomalous dimension at large coupling constants is known

$$\gamma = -\sqrt{2\pi(j-2)}\,\hat{a}^{1/4}\,, \tag{38}$$

one can calculate the Pomeron intercept in the strong coupling limit[13]

$$j = 2 - \Delta\,,\ \Delta = \frac{1}{2\pi}\,\hat{a}^{-1/2}\,. \tag{39}$$

The duality between the BFKL Pomeron and reggeized graviton means, that the Gribov Pomeron calculus could be described in the framework of the approach based on the effective action for the reggeized gravitons. It is the reason, why we consider below the high energy scattering in gravity.

4. Effective action approach in QCD

To begin with, we consider the effective field theory for reggeised gluons.[17] The corresponding effective action is local in the rapidity space

$$y = \frac{1}{2}\ln\frac{\epsilon_k + |k|}{\epsilon_k - |k|}\,,\ |y - y_0| < \eta\,,\ \eta << \ln s\,. \tag{40}$$

We can introduce the anti-hermitian fields v_μ describing the usual gluons and the gauge invariant fields $A_\pm$ describing the production and annihilation of the reggeized gluons

$$v_\mu(x) = -iT^a v_\mu^a(x)\,,\ A_\pm(x) = -iT^a A_\pm^a(x)\,,\ \delta A_\pm(x) = 0\,. \tag{41}$$

The fields $A_\pm$ satisfy the kinematical constraints

$$\partial_- A_+ = \partial_+ A_- = 0\,, \tag{42}$$

meaning, that the Sudakov variables in the clusters of particles with approximately equal rapidities have the Sudakov parameters of the cluster momenta strongly ordered in the quasi-multi-Regge kinematics.

The effective action for particles inside a cluster has the form

$$S = \int d^4x \left(L_{QCD} + Tr(V_+ \partial_\mu^2 A_- + V_- \partial_\mu^2 A_+) \right) , \tag{43}$$

where L_{QCD} is the usual QCD action and the effective currents $V_\pm$ are given below

$$V_+ = -\frac{1}{g}\partial_+ P \exp\left(-\frac{g}{2}\int_{-\infty}^{x^+} v_+(x')d(x')^+ \right) = v_+ - g v_+ \frac{1}{\partial_+} v_+ + \cdots . \tag{44}$$

The Feynman rules for this action are derived in Ref. 18. The effective action approach gives a possibility to construct various reggeon vertices needed to calculate NLO and NNLO corrections to the BFKL kernel.

Another application of this effective field theory is a verification of the BDS ansatz[19] for the inelastic amplitudes in the planar approximation for $N = 4$ SUSY. It was shown[20,21] with the use of the BFKL equation for the color octet states, that the BDS amplitude $M^{2\to 4}$ should be multiplied in the multi-Regge kinematics, where $s, s_2 > 0; s_1, s_3 < 0$, by the factor

$$M^{2\to 4} = (1 + i\Delta_{2\to 4})\, M^{BDS}_{2\to 4} , \tag{45}$$

where in LLA

$$\Delta_{2\to 4} = \frac{a}{2}\sum_{n=-\infty}^{\infty} (-1)^n \int_{-\infty}^{\infty} \frac{d\nu}{\nu^2 + \frac{n^2}{4}} (V^*)^{i\nu - \frac{n}{2}} V^{i\nu + \frac{n}{2}} \left(s_2^{\delta(\nu,n)} - 1 \right) \tag{46}$$

and $\delta(\nu, n)$ is the eigenvalue of the Hamiltonian. Further, the quantities V depend on transverse components of produced particle momenta k_1, k_2 and momentum transfers q_1, q_2, q_3. Using the duality transformation,[9] we can present them in the simple form

$$V = \frac{q_3 k_1}{k_2 q_1} \to \frac{z_{03} z_{0'1}}{z_{0'3} z_{01}} . \tag{47}$$

Moreover, the correction $\Delta_{2\to 4}$ turns out to be a function of 4-dimensional anharmonic ratios

$$i\Delta_{2\to 4} = \frac{a^2}{4} Li_2(\chi) \ln \frac{\chi t_2 s_{13}}{s_3 t_1} \ln \frac{\chi t_2 s_{02}}{t_3 s_1} + \ldots , \; \chi = 1 - \frac{s s_2}{s_{012} s_{123}} . \tag{48}$$

In the two-loop approximation this expression can be derived also from analytical properties and factorization[23] or directly from recently obtained exact result for $M_{2\to 4}$.[24] The effective action allows one to construct the

Mandelstam cut contributions corresponding to the composite states of an arbitrary number n of reggeized gluons in the adjoint representation.[22] The holomorphic hamiltonian for such state has the form

$$h = \ln(Z_1^2 \partial_1) - 2\psi(1) + \ln \partial_{n-1} + \sum_{k=1}^{n-2} h_{k,k+1}\,,\; p_k = Z_{k-1,k}\,. \tag{49}$$

The pair hamiltonian corresponds to the interaction of neighbors in the spin chain considered in Ref. 8

$$h_{1,2} = \ln(Z_{12}^2 \partial_1) + \ln(Z_{12}^2 \partial_2) - 2\ln Z_{12} - 2\psi(1)\,. \tag{50}$$

The total Hamiltonian h coincides with the hamiltonian for the integrable open spin chain.[22] The corresponding monodromy matrix has the form

$$t(u) = L_1(u) L_2(u) ... L_{n-1}(u) = \begin{pmatrix} A(u) & B(u) \\ C(u) & D(u) \end{pmatrix}, \tag{51}$$

where L_i are L-operators constructed from the Möbius group generators. The monodromy matrix contains the integrals of motion and allows to derive the equation for the Baxter function $Q(u)$

$$[D(u), h] = 0\,,\; D(u)Q(u) = (u-i)^{n-1} Q(u-i)\,. \tag{52}$$

5. Effective action for gravity

To construct the interaction of the reggeised gravitons with the usual gravitons we use again the locality in the rapidity space

$$y = \frac{1}{2} \ln \frac{\epsilon_k + |k|}{\epsilon_k - |k|}\,,\; |y - y_0| < \eta\,,\; \eta << \ln s\,. \tag{53}$$

The various effective graviton vertices and the graviton Regge trajectory in LLA for gravitons were calculated many years ago[25] In particular, it was found that the Regge trajectory is ultravioletly finite in $N = 4$ SUSY.[26] The effective field theory for multi-Regge processes was derived in Ref. 27. Here we construct the generally covariant effective action for quasi-multi-Regge processes. In the Einstein general relativity the graviton field is described by the metric tensor $g_{\mu\nu}$

$$d^2S = \sum_{\mu\nu} g_{\mu\nu}\, dx^\mu\, dx^\nu\,,\; \delta g_{\mu\nu} = D_\mu \chi_\nu + D_\mu \chi_\nu\,, \tag{54}$$

where D_μ are covariant derivatives and χ_ν are infinitesimal parameters of general covariant transformations. The reggeon fields A^{++} and A^{--} describing production and annihilation of the reggeized gravitons are assumed to be invariant under the general covariant transformation with the parameter $\chi_\nu(x) \to 0$ at $x \to \infty$. They satisfy also the kinematical constraints

$$\partial_+ A^{++} = \partial_- A^{--} = 0 \tag{55}$$

The effective action for the interactions of these fields local in rapidity is given below

$$S = \int d^4x \left(-\frac{\sqrt{-g}}{2\kappa} R + j_{++} A^{++} + j_{--} A^{--} \right), \tag{56}$$

where R is the Riemann curvature. The currents j_{++} and j_{--} are completely defined by the invariance of the action under the general coordinate transformations. They can be constructed in the perturbation theory. For example, the perturbative expansion of j_{++} has the form

$$j_{++} = h_{++} - X^2_{\sigma+} + X_{\rho+} X_{\sigma+} h_{\rho\sigma} - X_{\sigma+} \frac{\partial_\sigma}{\partial_+} X^2_{\rho+} + \ldots, \tag{57}$$

where

$$g_{\mu\nu} = \eta_{\mu\nu} + h_{\mu\nu}\,, \quad X_{\sigma+} = h_{\sigma+} - \frac{1}{2}\frac{\partial_\sigma}{\partial_+} h_{++} \tag{58}$$

and the Minkovsky metric $\eta_{\mu\nu}$ has the signature $(-,+,+,+)$. If we shall introduce the new current j^- according to the definition

$$j_{++} = \partial_+ j_+\,, \quad j^- = j_+\,, \tag{59}$$

it satisfies the "fan" equation

$$g^{\rho -} D_\rho j^- = -g^{--} - \frac{1}{4} \left(D^\rho j^- \right) \left(D_\rho j^- \right). \tag{60}$$

Here D_ρ are covariant derivatives acting on the world vector j^-, which is transformed under the infinitesimal coordinate transformation as follows

$$\delta j^- = 2 g^{\rho -} \chi_\rho + \chi^\rho D_\rho j^-. \tag{61}$$

From the above action one can reproduce the results of Refs. 25, 26.

Acknowledgments

I thank the Hamburg University for the hospitality and my collaborators J. Bartels and A. Sabio Vera for helpful discussions. This work was supported by the grant RFFI-10-02-01338-a.

References

1. V. N. Gribov, *Sov. Phys. JETP* **14**, 478 (1962).
2. S. Mandelstam, *Nuovo Cim.* **30**, 1148 (1963).
3. V. N. Gribov, I. Ya. Pomeranchuk and K. A. Ter-Martirosyan, *Phys. Rev.* **B139**, 184 (1965).
4. V. N. Gribov, *Sov. Phys. JETP* **26**, 414 (1968).
5. L. N. Lipatov, *Sov. J. Nucl. Phys.* **23**, 338 (1976);
V. S. Fadin, E. A. Kuraev and L. N. Lipatov, *Phys. Lett.* **B60**, 50 (1975);
E. A. Kuraev, L. N. Lipatov and V. S. Fadin, *Sov. Phys. JETP* **44**, 443 (1976).
6. L. N. Lipatov, *Phys. Lett.* **B309**, 394 (1993).
7. L. N. Lipatov, *Sov. Phys. JETP* **63**, 904 (1986).
8. L. N. Lipatov *High energy asymptotics of multi-colour QCD and exactly solvable lattice models*, hep-th/9311037, unpublished.
9. L. N. Lipatov, *Nucl. Phys.* **B548**, 328 (1999).
10. V. S. Fadin and L. N. Lipatov, *Phys. Lett.* **B429**, 127 (1998);
M. Ciafaloni and G. Camici, *Phys. Lett.* **B430**, 349 (1998).
11. A. V. Kotikov and L. N. Lipatov, *Nucl. Phys.* **B582**, 19 (2000).
12. A. V. Kotikov and L. N. Lipatov, *Nucl. Phys.* **B661**, 19 (2003).
13. A. V. Kotikov, L. N. Lipatov, A. I. Onishchenko and V. N. Velizhanin, *Phys. Lett.* **B595**, 521 (2004); [Erratum-ibid. **B632**, 754 (2006)].
14. J. M. Maldacena, *Adv. Theor. Math. Phys.* **2**, 231 (1998).
15. S. S. Gubser, I. R. Klebanov and A. M. Polyakov, *Phys. Lett.* **B428**, 105 (1998).
16. E. Witten, *Adv. Theor. Math. Phys.* **2**, 253 (1998).
17. L. N. Lipatov, *Nucl. Phys.* **B452**, 369 (1995); *Phys. Rept.* **286**, 131 (1997).
18. E. N. Antonov, L. N. Lipatov, E. A. Kuraev and I. O. Cherednikov, *Nucl. Phys.* **B721**, 111 (2005).
19. Z. Bern, L. J. Dixon and V. A. Smirnov, *Phys. Rev.* **D72**, 085001 (2005).
20. J. Bartels, L. N. Lipatov and A. Sabio Vera, *Phys. Rev.* **D80**, 045002 (2009).
21. J. Bartels, L. N. Lipatov and A. Sabio Vera, *Phys. Rev.* **D80**, 045002 (2009).
22. L. N. Lipatov, *J. Phys.* **A42**, 304020 (2009).
23. L. N. Lipatov, preprint, hep-th/1008.1015.
24. L. N. Lipatov and A. Prygarin, preprint, hep-th/1008.1016.
25. L. N. Lipatov, *JETP* **82**, 991 (1982).
26. L. N. Lipatov, *Phys. Lett.* **B116**, 411 (1982).
27. L. N. Lipatov, *Nucl. Phys.* **B365**, 614 (1991).

THE QCD ROTATOR

P. HASENFRATZ

Institute for Theoretical Physics
University of Bern
Sidlerstrasse 5, CH-3012 Bern, Switzerland

The low lying spectrum of QCD in a special environment (δ-regime) can be obtained in chiral perturbation theory (ChPT). The NNL order result predicts the spectrum in terms of the low energy constants of ChPT: F, Λ_1, Λ_2 and a mass M. Since the size of the spatial box in the δ-regime is a few fermi, only numerical *simulations* can create the QCD data. The low lying stable spectrum in the box is a relatively simple numerical problem. Comparing the data with the predictions of ChPT will be a precise way to determine three of the low energy constants of chiral perturbation theory.

1. Introduction and summary

We consider here an analytic result on the low lying *spectrum* of QCD in a special environment.This environment has a spatial box $L_s \times L_s \times L_s$, where $L_s = O(\text{fm})$, while the 4th (euclidean) extension is infinite: $L_t \to \infty$. The spectrum is created by the 'would be Goldstone bosons' of QCD. The energies of these low lying excitations are much smaller than those of the standard excitations with non-zero impulses E $\approx 2\pi/L_s$. Actually, in leading order, the low lying spectrum is a simple quantum mechanical rotator.[1] Similar observations were done in condensed matter physics context also: in a finite box at zero temperature the lowest excitations in the spontaneously broken phase are related to the slow precession of the order parameter described by a rotator.[2,3]

The finite box creates an infrared-safe environment. This allows to study the chiral limit first and switching on the symmetry breaking terms later. In two-flavor QCD, in the leading order (L) of chiral perturbation, theory the $SU(2) \times SU(2) \sim O(4)$ rotator has an inertia Θ proportional to the size of the spatial box: $\Theta = F^2 L_s^3$. Here F is the pion decay constant in the chiral limit. In the next-to-leading order (NL) the inertia is corrected $\Theta = F^2 L_s^3 (1+ \propto 1/F^2 L_s^2)$,[4] where $1/F^2 L_s^2$ is the small expansion parameter.

The logarithms and the additional low energy constants l_1 and l_2 enter only in the NNL order.

It took a rather long time until recently two works presented the NNL order result in the chiral limit.[5,6] These works follow different techniques, among others the regularizations are also different: dimensional regularization in Ref. 5 vs. lattice regularization in Ref. 6. A full comparison of the two final results request to solve a general problem which is interesting in its self. Of course, the physical content of the calculation of a quantum field theory is independent of the regularization. The matching can be obtained exactly in a 1-loop calculation.[7] In our case, however we have an *effective* quantum field theory (namely, chiral perturbation theory). This, as far as we know, is an untouched problem.

In this work we follow[5] using dimensional regularization. The NNL symmetry breaking corrections has been completed recently,[8] where also dimensional regularization was used.

Almost certainly, this spectrum will never be measured in a real experiment. On the other hand, it will be certainly measured in numerical experiments in QCD (stochastic, 'Monte Carlo' approaches). The works[9,10] are interesting steps in this way.

The analytic result comes from chiral perturbation theory. The result is expressed in terms of a small number of low-energy constants.

Comparing the results from numerical experiments with that of the chiral perturbation theory results gives a test for QCD. Or, we might assume that QCD is the theory of the strong interactions and use the analytic results vs. numerical experiments to determine more precisely some of the low energy constants of chiral perturbation theory.

In the environment of the δ-regime it is natural to divide the degrees of freedom into $fast$ and *slow* modes. The fast modes can be treated in perturbation theory, while the slow modes build the rotator, whose energy excitations are much smaller than those of the standard Goldstone boson excitations which carry finite momenta.

Let us summarize the final results before going over to some of the details. We quote the result for $N = 4$, which corresponds to 2-flavor QCD. Up to NNL order the rotator spectrum in the chiral limit has the form*

$$E_l = \frac{1}{2\Theta} l(l+2)\,, \ l = 0, 1, 2, \ldots \,. \tag{1}$$

The corrections from the perturbative expansion enter in the inertia Θ

*Terms with other Casimir forms are expected to enter beyond NNL order.

which, up to NNL order in the chiral limit, reads

$$\Theta = F^2 L_s^3 \left\{ 1 - \frac{2}{F^2 L_s^2} \bar{G}^* + \frac{1}{(F^2 L_s^2)^2} \left[0.088431628 + d0d0\bar{G}^* \frac{1}{3\pi^2} \left(\frac{1}{4} \ln(\Lambda_1 L_s)^2 + \ln(\Lambda_2 L_s)^2 \right) \right] \right\} . \tag{2}$$

Here Λ_1, Λ_2 are the standard scales related to the bare low energy constants l_1, l_2 in the chiral Lagrangian.

The constants $\bar{G}^*$ and $d0d0\bar{G}^*$ are related to the constrained Green's function $\bar{D}^*(x)$ and its second time derivative $\partial_0\partial_0\bar{D}^*(x)$, which enter the perturbation theory:

$$\bar{D}^*(0) = \frac{1}{L_s^2}\bar{G}^* \,, \quad \bar{G}^* = -0.2257849591 \,. \tag{3}$$

$$\partial_0\partial_0\bar{D}^*(0) = \frac{1}{L_s^4} d0d0\bar{G}^* \,, \quad d0d0\bar{G}^* = -0.8375369106 \,. \tag{4}$$

In the Green's function $\bar{D}^*(x)$ the non-perturbative slow modes (rotator modes) are missing (notation D^*) and the UV singularity is also subtracted (notation $\bar{D}$). The precision of the numerical numbers are estimated to be 10^{-9}, or better.

2. The leading rotator

We consider the low energy limit of QCD with two massless quarks $m_u = m_d = 0$. The global symmetry is $SU(2) \times SU(2) \sim O(4)$. It is somewhat more intuitive to use the 'O(4) language'. We can imagine a magnet described by a field with 4 components in the internal space. We have an effective non-linear $O(N = 4)$ σ-model with

$$\mathbf{S}(x) = [S(x)_0, S(x)_1, S(x)_2, S(x)_3], \quad \mathbf{S}^2(x) = 1, \tag{5}$$

and x lives in $d = 4 = (d-1)+1$, where $(d-1)$ is for the 'space'. We consider an $L_s \times L_s \times L_s \times L_t$ box with spatial volume $V_s = L_s^3$, while the (Euclidean) time extension is taken very large, $L_t \to \infty$. In the $L_s \to \infty$ limit the system has a net magnetization and massless Goldstone bosons.

We consider a cylinder geometry (δ-regime), where L_s is finite, but sufficiently large so that the (would be) Goldstone bosons dominate the finite size effects. Due to the microscopic magnetic moments, the $L_s \times L_s \times L_s$

spatial box has a net magnetization on each time slices. Since L_s is finite, this net magnetization is moving around as the function of the time t.

In leading order (L) we ignore the fluctuations of the small (microscopic) magnets in the box: they are parallel. The length of the total magnetization remains constant, but it's direction is slowly changing:

$$A_{\rm rot} = \frac{F^2 V_s}{2} \int dt\, \dot{\mathbf{e}}(t)\dot{\mathbf{e}}(t)\,. \tag{6}$$

Here $\mathbf{e}(t)$ is the direction of the total magnetization in the internal $O(4)$ space at the time t:

$$\mathbf{e}(t) = [e(t)_0, e(t)_1, e(t)_2, e(t)_3],\ \ \mathbf{e}(t)^2 = 1\,. \tag{7}$$

Equation (6) describes a quantum mechanical $O(4)$ rotator, the QCD rotator in leading order,[1] with a discrete energy spectrum above the ground state:

$$\mathbf{H} = \frac{\mathbf{L^2}}{2\Theta}\,,\ \ \Theta = F^2 V_s\,,\ \ E_l = \frac{1}{2\Theta}\, l(l+2)\,,\ \ l = 0\,,1\,,2\,,\ldots\,. \tag{8}$$

In Eq. (8) $\mathbf{H}$ is the Hamiltonian, $\mathbf{L}$ is the $O(4)$ angular momentum and Θ is the inertia of the rotator. Our aim is to determine the corrections up to NNL order, where the (bare) low energy constants l_1, l_2 first enter.

3. Summary of the NNL results and discussions

The Lagrangian, up to 1-loop level in the chiral limit, has the form $L_{\rm eff} = L^{(2)}_{\rm eff} + L^{(4)}_{\rm eff}$, where,

$$\begin{aligned} L^{(2)}_{\rm eff} &= \frac{F^2}{2} \partial_\mu \mathbf{S}\, \partial_\mu \mathbf{S}\ , \\ L^{(4)}_{\rm eff} &= -l_1\, (\partial_\mu \mathbf{S}\, \partial_\mu \mathbf{S})(\partial_\nu \mathbf{S}\, \partial_\nu \mathbf{S}) - l_2\, (\partial_\mu \mathbf{S}\, \partial_\nu \mathbf{S})(\partial_\mu \mathbf{S}\, \partial_\nu \mathbf{S})\,. \end{aligned} \tag{9}$$

Here F, l_1, l_2 are the bare low energy constants. We use dimensional regularization. In the renormalization process the bare l_1, l_2 are replaced by the finite Λ_1 and Λ_2 whith dimension of a mass.

The final result for the low lying rotator spectrum in NNL order in the two flavour ($N = 4$) case is

$$E_l = \frac{1}{2\Theta}\, j(j+2)\,,\ \ j = 0\,,1\,,2\,,\ldots\,, \tag{10}$$

where

$$\Theta = F^2 L_s^3 \left\{ 1 - \frac{2}{F^2 L_s^2}(-0.22578495) + \frac{1}{(F^2 L_s^2)^2} \left[0.08841628 + (-0.83753691)\frac{1}{3\pi^2} \left(\frac{1}{4} \ln(\Lambda_1 L_s)^2 + \ln(\Lambda_2 L_s)^2 \right) \right] \right\} . \quad (11)$$

The result in Eq. (11) requests a rather technical calculation. We make a few words on that in Section 4. Note the simple Casimir form $j(j+2)$ in Eq. (10).[5] Beyond NNL order we expect other Casimir forms also.

We believe, that the δ-regime gives a good possibility to constrain the low energy constants of ChPT theory. First, up to NNL order in the δ-regime, the number of low energy constants is the same as in NL order in the p-regime (see Eq. (11)). A further good feature is the following. We want to compare the results of chiral perturbation theory with the numerical simulations of QCD. This simulation requires to calculate the low lying spectrum in a box. Measuring the low lying spectrum is among the simplest and cleanest numerical problems.

A difficulty is, that for precise results the box size might need to reach 3 fm, or larger. This number is suggested by the NL correction, which is approximately 50, 30 and 20 procent of the leading (L) order result, if L_s is 2, 2.5, and 3 fermi, respectively. However, the NNL order correction is unexpectedly small, it is only 2 procent at 3 fm. Anyhow, the lattice community is close to reaching 3 fm today and will be there tomorrow.

4. Separating the slow and fast modes

The direction of the magnetization $\mathbf{e}(t)$ moves much slower than the single microscopic magnets. The procedure is to integrate out the fast modes and obtain an action which depends only on the slow modes $\mathbf{e}(t)$. This is then a (generalized) rotator, i.e. a problem in quantum mechanics which can be solved easily. The task of separating the fast and slow modes accours in the regimes also, where $L_t = O(L_s)$.[11–14] The difficulty in the δ-regime is that the number of slow modes is infinite (as opposed to the case where $L_t = O(L_s)$ and we have a few slow modes only). We refer to Ref. 5 concerning the solution of this technical problem.

Acknowledgments

I thank Julia Nyíri for inviting me to this exciting meeting. I am indebted for discussions with G. Colangelo, A. Frenkel, H. Leutwyler, F. Niedermayer, M. Weingart and Ch. Weyermann. This work is supported in part by the Schweizerischer Nationalfonds. The author acknowledges support by DFG project SFB/TR-55. The "Albert Einstein Center for Fundamental Physics" at Bern University is supported by the "Innovations-und Kooperationsprojekt C-13" of the Schweizerischer Nationalfonds.

References

1. H. Leutwyler, *Phys. Lett.* **B189**, 197 (1987).
2. M. E. Fisher and V. Privman, *Phys. Rev.* **B32**, 447 (1985).
3. E. Brezen and J. Zinn-Justin, *Nucl. Phys.* **B257**, 867 (1985).
4. P. Hasenfratz and F. Niedermayer, *Z. Phys.* **B92**, 91 (1993).
5. P. Hasenfratz *Nucl. Phys.* **B828**, 201 (2010).
6. F. Niedermayer and C. Weiermann, *Nucl. Phys.* **B842**, 248 (2011).
7. A. Hasenfratz and P. Hasenfratz, *Phys. Lett.* **B93**, 165 (1980).
8. M. Weingart, arXiv:1006.5076 [hep-lat].
9. W. Bietenholz, arXiv:1007.0285 [hep-lat].
10. W. Bietenholz et al., *Phys. Lett.* **B687**, 410 (2010).
11. J. Gasser and H. Leutwyler, *Phys. Lett.* **B184**, 83 (1987).
12. J. Gasser and H. Leutwyler, *Phys. Lett.* **B188**, 477 (1987).
13. J. Gasser and H. Leutwyler, *Nucl. Phys.* **B307**, 763 (1988).
14. P. Hasenfratz and H. Leutwyler, *Nucl. Phys.* B343, 241 (1990).

CONFINEMENT IN YANG–MILLS THEORIES: ELEMENTS OF A BIG PICTURE

MIKHAIL SHIFMAN

*William I. Fine Theoretical Physics Institute, University of Minnesota, Minneapolis, MN 55455, USA**
and
Jefferson Physical Laboratory, Harvard University, Cambridge, MA 02138, USA

I give a brief review of advances in Yang–Mills theory at strong coupling. Analytical theory of quark (color) confinement on which Gribov worked non-stop in the last twenty years of his life seems closer now than ever before.

1. Introduction

Fig. 1. Young Volodya Gribov.

In the early 1970s, when quantum chromodynamics (QCD) was born, BH (that's how Gribov's colleagues would address him) was already a

*Permanent address.

theoretical physicist with the world-wide reputation. Among us — graduate students at ITEP — Vladimir Naumovich had the reputation of the deepest thinker, with the broad knowledge of everything in quantum physics, with a passion to strong interactions, and the toughest seminar warrior, whose appearance could instantaneously instill fear even in the best speakers and lecturers. Later, in 1979, I myself had an occasion to learn this the hard way. Everybody knows the famous story of David Politzer's seminar in Gatchina, where at that time (early 1980s) BH was the head of the Theory Division. Only 5 minutes after the beginning of David's talk, he was interrupted by Gribov, who requested to make a number of explanatory comments. The audience got agitated, everyone wanted to speak simultaneously with Gribov, and in this noise and tumult they forgot about the speaker. An hour later or so, when Gribov explained to the audience all intricacies of the seminar topic, they found David Politzer in one of the adjacent rooms, sitting on a windowsill and playing a little horn.

With the advent of QCD Gribov delved into its depths. All his efforts in the next 25 years were devoted to the mysteries of quark (color) confinement, and from these depths he brought to us a few precious gems of understanding. He was the first to suggest the tunneling interpretation of instantons, and, shorly after, he discovered the celebrated Grobov copies and Gribov horizons, let alone his seminal contributions in perturbative QCD in which he created the Gribov school of leading, next-to-leading, next-to-next-to-leading, and *ad infinitum*, logarithms — the school which is equal to none. Unfortunately, BH did not live up to the breakthrough advances of the present day. Mostly current advances are due to penetration of supersymmetry-based ideas in non-Abelian Yang–Mills theories. I am sure he would be fascinated and deeply involved in these developments.

Almost 40 years after my first encounter with BH, I am honored and proud to be invited to this Conference celebrating Gribov's 80th birthday, to give a talk on what's happening in QCD now.

The problem of strong interactions turned out to be extremely difficult (despite the fact that the underlying Lagrangian is firmly established) and the progress is slow and painful. This is a usual story with the strong coupling regime: whenever theorists find themselves at strong coupling, they are in trouble. Yang–Mills theories are no exception.

My task today is to outline some contours of the strong interaction theory which gradually emerged from obscurity during the last 15 years or so. Yes, the theory is incomplete, but those parts which are already in existence are beautiful, and continue to grow.

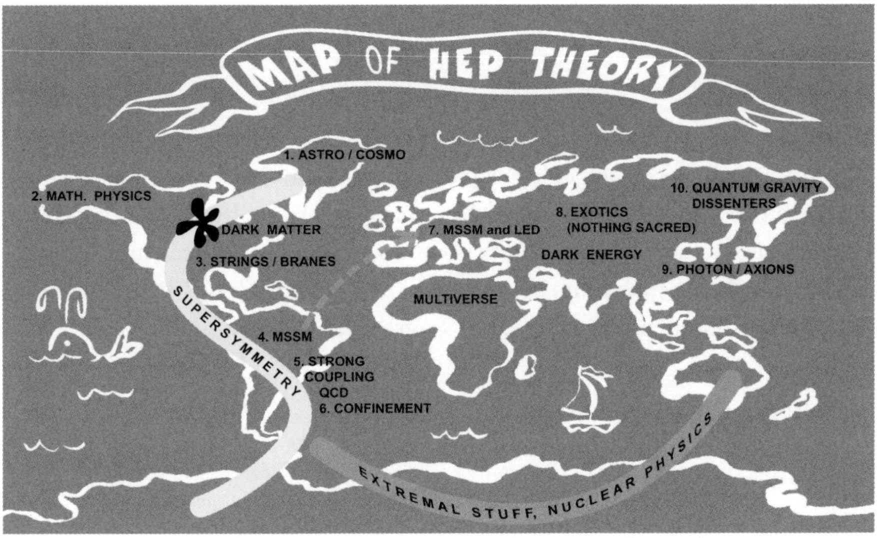

Fig. 2. Map of the HEP theory.

First, to give a general idea of the role which this theory plays in high energy physics (HEP) I would like to chart approximate contours of main areas that are under intense development in the theoretical community of today. To this end I display a symbolic map in Fig. 2. The linkage of the HEP theory map to the earth's geography is arbitrary and does not mean anything. You should pay attention only to interconnections of various areas of HEP. You see that QCD and strongly coupled gauge theories at large, occupy a vast area. The advances there, and in the neighboring areas, crucially depend on the exchange of ideas between them, as is demonstrated by solutions of supersymmetric Yang–Mills models, discovery of the strongly-interacting quark-gluon plasma, discovery of superconductivity in high-density QCD, and so on. This list could have been continued.

2. QCD

In all processes with hadron participation strong interactions play a role. All matter surrounding us is made of protons and neutrons the most common representatives of the class of hadrons. Even if you do not see them explicitly, they will necessarily show up at a certain stage or in loops. The fundamental Lagrangian governing strong interactions is

$$\mathcal{L} = \sum_f \bar{\psi}_f \left(i\not{D} - m_f\right) \psi^f - \frac{1}{4} G^a_{\mu\nu}\, G^{\mu\nu\, a}\,. \tag{1}$$

The first term describes color-triplet quarks and their coupling to color-octet gluons. The second Yang–Mills term describes the gluon dynamics. Both terms taken together comprise the Lagrangian of quantum chromodynamics (QCD). Much in the same way as the Schrödinger equation codes all of quantum chemistry, the QCD Lagrangian codes all of

- nuclear physics;
- Regge behavior;
- neutron stars;
- chiral physics;
- light & heavy quarkonia;
- glueballs & exotics;
- exclusive & inclusive hadronic scattering at large momentum transfer;
- interplay between strong forces & weak interactions,
- quark-gluon plasma;

and much more.

Although the underlying Lagrangian (1), and asymptotic freedom it implies at short distances,[1] are established beyond any doubt, the road from this starting point to theoretical control over the large-distance hadronic world is long and difficult. The journey which started 40 years ago is not yet completed. *En route*, many beautiful theoretical constructions were developed allowing one to understand various corners of the hadronic world. Here I am unable even to list them, let alone discuss in a comprehensible way. Therefore, I will focus only on one – albeit absolutely global – aspect defining the hadronic world: the confinement phenomenon.

3. Confinement in Non-Abelian Gauge Theories: the Dual Meissner Effect

The most salient feature of pure Yang-Mills theory is linear confinement. If one takes a heavy probe quark and an antiquark separated by a large distance, the force between them does not fall off with distance, while the potential energy grows linearly. This is the explanation of the empiric fact that quarks and gluons (the microscopic degrees of freedom in QCD) never appear as asymptotic states. The physically observed spectrum consists of color-singlet mesons and baryons. The phenomenon got the name color confinement, or, in a more narrow sense, quark confinement. In the early days of QCD it was also referred to as infrared slavery.

Quantum chromodynamics (QCD), and Yang–Mills theories at strong coupling at large, are not yet analytically solved. Therefore, it is reasonable to ask:

Are there physical phenomena in which interaction energy between two interacting bodies grows with distance at large distances? Do we understand the underlying mechanism?

The answer to these questions is positive. The phenomenon of linearly growing potential was predicted by Abrikosov[2] in the superconductors of the second type. The corresponding set up is shown in Fig. 3. In the middle of this figure we see a superconducting sample, with two very long magnets attached to it. The superconducting medium does not tolerate the magnetic field. On the other hand, the flux of the magnetic field must be conserved. Therefore, the magnetic field lines emanating from the N pole of one magnet find their way to the S pole of another magnet, through the medium, by virtue of a flux tube formation. Inside the flux tube the Cooper pair condensate vanishes and superconductivity is destroyed. The flux tube has a fixed tension, implying a constant force between the magnetic poles as long as they are inside the superconducting sample. The phenomenon described above is sometimes referred to as the Meissner effect.

Of course, the Meissner effect of the Abrikosov type occurs in the Abelian theory, QED. The flux tube that forms in this case is Abelian. In Yang–Mills theories we are interested in non-Abelian analogs of the Abrikosov vortices. Moreover, while in the Abrikosov case the flux tube is that of the magnetic field, in QCD and QCD-like theories the confined

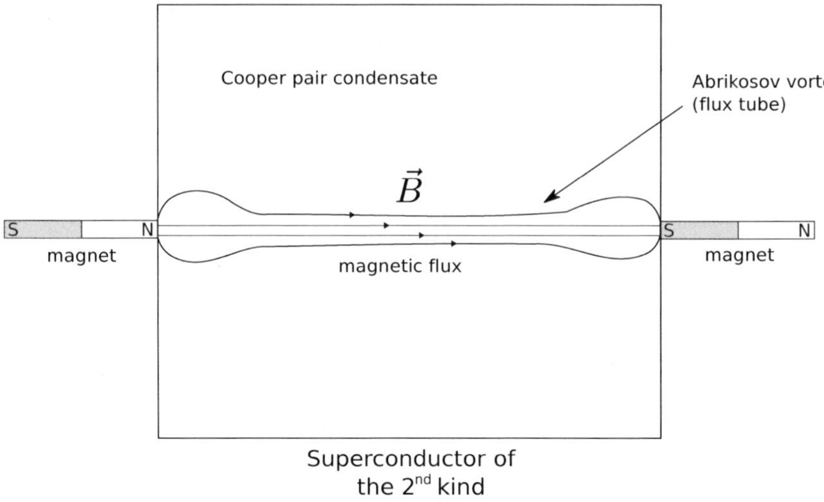

Fig. 3. The Meissner effect in QED.

objects are quarks; therefore, the flux tubes must be "chromoelectric" rather than chromomagnetic. In the mid-1970s Nambu, 't Hooft, and Mandelstam (independently) put forward an idea[3] of a "dual Meissner effect" as the underlying mechanism for color confinement. Within their conjecture, in chromoelectric theories "monopoles" condense leading to formation of "non-Abelian flux tubes" between the probe quarks. At this time the Nambu–'t Hooft–Mandelstam paradigm was not even a physical scenario, rather a dream, since people had no clue as to the main building blocks such as non-Abelian flux tubes. After the Nambu–'t Hooft–Mandelstam conjecture had been formulated many works were published on this subject, to no avail.

A long-awaited breakthrough discovery came 20 years later: the Seiberg–Witten solution[4] of $\mathcal{N}=2$ super-Yang–Mills theory slightly deformed by a superpotential breaking $\mathcal{N}=2$ down to $\mathcal{N}=1$. In the $\mathcal{N}=2$ limit, the theory has a moduli space. If the gauge group is SU(2), on the moduli space, $SU(2)_{\rm gauge}$ is spontaneously broken down to U(1). Therefore, the theory possesses the 't Hooft–Polyakov monopoles[5] in the quasiclassical regime. Of course, in this regime they are very heavy and play no role in dynamics. Using the power of $\mathcal{N}=2$ supersymmetry, two special points on the moduli space were found[4] at strong coupling, (the monopole and dyon points), in which the monopoles (dyons) become massless. In these points the scale of the gauge symmetry breaking

$$\mathrm{SU}(2) \to \mathrm{U}(1) \tag{2}$$

is determined by Λ, the dynamical scale parameter of $\mathcal{N} = 2$ super-Yang–Mills theory.

All physical states can be classified with regards to the unbroken U(1). It is natural to refer to the U(1) gauge boson as to the photon. In addition to the photon, all its superpartners, being neutral, remain massless at this stage, while all other states, with nonvanishing "electric" charges, acquire masses of the order of Λ. In particular, two gauge bosons corresponding to SU(2)/U(1) – it is natural to call them $W^{\pm}$ – have masses $\sim \Lambda$. All such states are "heavy" and can be integrated out.

In the low-energy limit, near the monopole and dyon points, one deals with electrodynamics of massless monopoles. One can formulate an effective local theory describing interaction of the light states by dualizing the original phton. This is a U(1) gauge theory in which the (magnetically) charged matter fields M, $\tilde{M}$ are those of monopoles while the U(1) gauge field that couples to M, $\tilde{M}$ is *dual* with respect to the photon of the original theory. The $\mathcal{N} = 2$ preserving superpotential has the form $\mathcal{W} = \mathcal{A}M\tilde{M}$, where $\mathcal{A}$ is the $\mathcal{N} = 2$ superpartner of the dual photon/photino fields.

Now, if one switches on a small $\mathcal{N} = 2$ breaking superpotential, the only change in the low-energy theory is the emergence of the extra $m^2\mathcal{A}$ term in the superpotential. Its impact is crucial: it triggers the monopole condensation, $\langle M \rangle = \langle \tilde{M} \rangle = m$, which implies, in turn, that the dual U(1) symmetry is spontaneously broken, and the dual photon acquires a mass $\sim m$. As a consequence, the Abrikosov flux tubes are formed. Viewed inside the dual theory, they carry fluxes of the magnetic field. With regards to the original microscopic theory these are the electric field fluxes.

Thus, Seiberg and Witten demonstrated, for the first time ever, the existence of the dual Meissner effect in a judiciously chosen non-Abelian gauge field theory. If one "injects" a probe (very heavy) quark and antiquark in this theory, a flux tube forms between them, with necessity, leading to linear confinement.

The flux tubes in the Seiberg–Witten solution were investigated in detail.[6] These flux tubes are Abelian, and so is confinement caused by their formation. What does that mean? At the scale of distances at which the flux tube is formed (the inverse mass of the Higgsed U(1) photon) the gauge group that is operative is Abelian. In the Seiberg–Witten analysis this is the dual U(1). The off-diagonal (charged) gauge bosons are very heavy in this scale and play no direct role in the flux tube formation and confinement that ensues. Naturally, the spectrum of composite objects in this case turns out to be richer than that in QCD and similar theories with non-Abelian confinement. By non-Abelian confinement I mean such dynamical regime in which at distances of the flux tube formation all gauge bosons are equally important.

Moreover, the string topological stability is based on $\pi_1(\mathrm{U}(1)) = \mathbb{Z}$. Therefore, N strings do not annihilate as they should in SU(N) QCD-like theories.

The two-stage symmetry breaking pattern, with SU(2) $\to$ U(1) occurring at a high scale while the dual U(1) $\to$ nothing at a much lower scale, has no place in QCD-like theories, as we know from experiment. In such theories, presumably, all non-Abelian gauge degrees of freedom take part in the string formation, and are operative at the scale at which the strings are formed. The strings in the Seiberg–Witten solution are believed to belong to the same universality class as those in QCD-like theories. However, in the limit of large-m deformations, when a non-Abelian regime presumably sets in and non-Abelian strings develop in the model considered by Seiberg and Witten, theoretical control is completely lost. Thus, the status of the statement of the same universality class is conjectural.

4. Non-Abelian strings

In a bid to better understand string-based confinement mechanism in Yang–Mills theories that might be more closely related to QCD people continued searches for models supporting non-Abelian strings. If a model in which non-Abelian strings develop in a fully controllable manner, i.e. at weak coupling, could be found and the passage from Abelian to non-Abelian strings explored, this would provide us with evidence that no phase transition occurs between the two regimes in the Seiberg–Witten solution.

In the technical sense, what does one mean when one speaks of non-Abelian flux tubes? Apparently, the orientation of the magnetic field in the tube interior must be free to strongly fluctuate inside the SU(N) group. There is no such freedom in the Abelian string of the Abrikosov type. In other words, in addition to translational moduli, the theory on the string world sheet must acquire orientational moduli (Fig. 4).

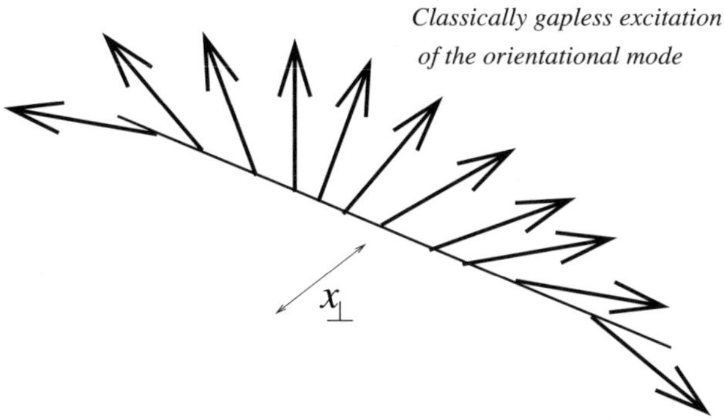

Fig. 4. Orientational moduli on the string world sheet.

If one thinks there is a kind of string theory behind QCD confining dynamics, such behavior is natural. Indeed, string theory is formulated in higher dimensions. Bringing it to $D = 4$ requires compactification of some dimensions. If (some of) compact dimensions have isometries, the corresponding sigma model on the 4D string world sheet will have classically massless internal degrees of freedom. For instance, compactification on S_2 gives rise to CP(1) sigma model on the string world sheet. Two-dimensional infrared dynamics will then generate a mass gap for these orientational degrees of freedom.

That's why searches for non-Abelian flux tubes and non-Abelian monopoles in the bulk Yang–Mills theories continued, with a decisive breakthrough in 2003-04.[7,8] By that time the program of finding field-theoretical analogs of all basic constructions of string/D-brane theory was in full swing. BPS domain walls, analogs of D branes, had been identified in supersymmetric Yang–Mills theory.[9] It had been demonstrated that such walls support gauge fields localized on them. BPS saturated string-wall junctions had been constructed.[10] Topological stability of the non-Abelian strings under consideration is due to the fact that

$$\pi_1\left(\frac{\mathrm{SU}(2)\times\mathrm{U}(1)}{Z_2}\right)\to \text{nontrivial}\,. \tag{3}$$

5. Basic Bulk Theory: setting the stage

Non-Abelian strings were first found in $\mathcal{N}=2$ super-Yang–Mills theories with $\mathrm{U}(2)_{\mathrm{gauge}}$ and two matter hypermultiplets.[7,8] The $\mathcal{N}=2$ vector multiplet consists of the U(1) gauge field A_μ and the SU(2) gauge field A^a_μ, (here $a=1,2,3$), and their Weyl fermion superpartners $(\lambda^1,\,\lambda^2)$ and $(\lambda^{1a},\,\lambda^{2a})$, plus complex scalar fields a, and a^a. The global $\mathrm{SU}(2)_R$ symmetry inherent to $\mathcal{N}=2$ models manifests itself through rotations $\lambda^1\leftrightarrow\lambda^2$.

The quark multiplets consist of the complex scalar fields q^{kA} and $\tilde{q}_{Ak}$ (squarks) and the Weyl fermions ψ^{kA} and $\tilde{\psi}_{Ak}$, all in the fundamental representation of the SU(2) gauge group ($k=1,2$ is the color index while A is the flavor index, $A=1,2$). The scalars q^{kA} and $\bar{\tilde{q}}^{kA}$ form a doublet under the action of the global $\mathrm{SU}(2)_R$ group. The quarks and squarks have a U(1) charge too.

If one introduces a non-vanishing Fayet–Iliopoulos parameter ξ the theory develops isolated quark vacua, in which the gauge symmetry is fully Higgsed, and all elementary excitations are massive. In the general case, two matter mass terms allowed by $\mathcal{N}=2$ are unequal, $m_1\neq m_2$. There are free parameters whose interplay determines dynamics of the theory: the Fayet–Iliopoulos parameter ξ, the mass difference Δm and a dynamical scale parameter Λ, an analog of the QCD scale Λ_{QCD} (Fig. 5). Extended supersymmetry guarantees that some crucial dependences are holomorphic, and there is no phase transition.

Both the gauge and flavor symmetries of the model are broken by the squark condensation. All gauge bosons acquire the same masses (which are of the order of inverse string thickness). A global diagonal combination of color and flavor groups, $\mathrm{SU}(2)_{C+F}$, survives the breaking (the subscript $C+F$ means a combination of global color and flavor groups).

While $\mathrm{SU}(2)_{C+F}$ is the symmetry of the vacuum, the flux tube solutions break it spontaneously. This gives rise to orientational moduli on the string world sheet.

The bulk theory is characterized by three parameters of dimension of mass: ξ, Δm, and Λ. As various parameters vary, the theory under consideration evolves in a very graphic way, see Fig. 5. At $\xi = 0$ but $\Delta m \neq 0$ (and $\Delta m \gg \Lambda$) it presents a very clear-cut example of a model with the standard 't Hooft–Polyakov monopole. This is due to the fact that the relevant part of the bosonic sector is nothing but the Georgi–Glashow model. The monopole is unconfined — the flux tubes are not yet formed.

Switching on $\xi \neq 0$ traps the magnetic fields inside the flux tubes, which are weak as long as $\xi \ll \Delta m$. The flux tubes change the shape of the monopole far away from its core, leaving the core essentially intact. Orientation of the chromomagnetic field inside the flux tube is essentially fixed. This is due to the fact that all off-diagonal gauge bosons (W bosons) are heavy in this limit. Thus, the flux tubes supported in this limit are *Abelian.* (They are commonly referred to as the Z_N strings.)

With $|\Delta m|$ decreasing, fluctuations in the orientation of the chromomagnetic field inside the flux tubes grow. Simultaneously, the monopole which no loner resembles the 't Hooft–Polyakov monopole, is seen as a *string junction.*

Finally, in the limit $\Delta m \to 0$ the transformation is complete. A global SU(2) symmetry restores in the bulk. All three gauge bosons have identical masses. Orientational (exact, classically massless) moduli develop on the string world sheet making it non-Abelian. The string world sheet theory is CP(1) (CP($N-1$) for generic values of N). Two-dimensional CP($N-1$) models with four supercharges are asymptotically free. They have N distinct vacuum states.

Each vacuum state of the worldsheet CP($N-1$) theory presents a distinct string from the standpoint of the bulk theory. There are N species of such strings; they have degenerate tensions $T_{\rm st} = 2\pi\xi$. The ANO string tension is N times larger.

Two different strings can form a stable junction. Figure 6 shows this junction in the limit

$$\Lambda_{\rm CP(1)} \ll |\Delta m| \ll \sqrt{\xi} \tag{4}$$

corresponding to the lower left corner in Fig. 5. The magnetic fluxes of the U(1) and SU(2) gauge groups are oriented along the z axis. In the limit (4) the SU(2) flux is oriented along the third axis in the internal space. However, as $|\Delta m|$ decreases, fluctuations of B_z^a in the internal space grow, and at $\Delta m \to 0$ it has no particular orientation in SU(2) (the lower right

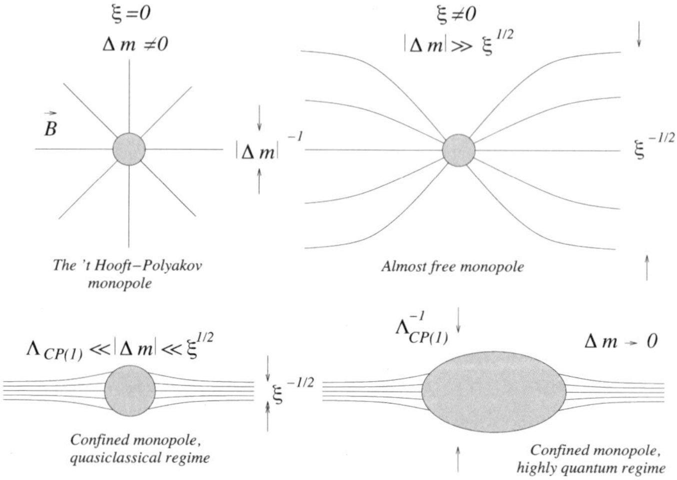

Fig. 5. Various regimes for monopoles and strings.

corner of Fig. 5). In the language of the world-sheet theory this phenomenon is due to restoration of the O(3) symmetry in the quantum vacuum of the CP(1) model.

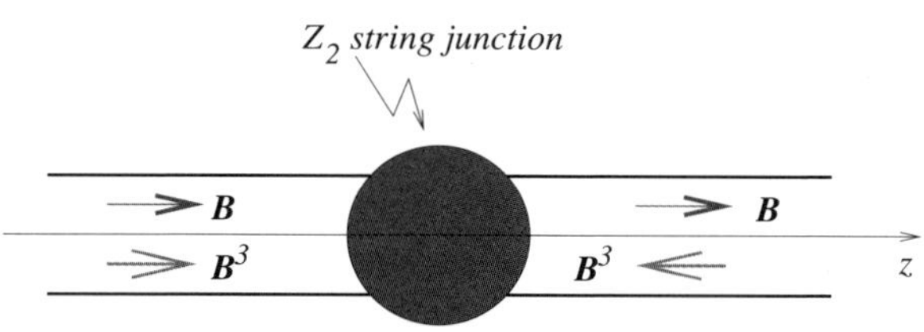

Fig. 6. Z_2 string junction.

Evolution from the upper right corner in Fig. 5 to the lower right corner is in fact the transformation of the Abelian string into non-Abelian. $\mathcal{N} = 2$ supersymmetry guarantees that it is smooth, with no phase transition.

The junctions of degenerate strings present what remains of the monopoles in this highly quantum regime.[11,12] It is remarkable that, despite the fact we are deep inside the highly quantum regime, holomorphy allows one to exactly calculate the mass of these monopoles. This mass is given by the expectation value of the kink central charge in the worldsheet CP($N-1$) model (including the anomaly term), $M_M \sim N^{-1} \langle R\,\psi_L^\dagger\,\psi_R\rangle$.

6. Towards $\mathcal{N}=1$

The "unwanted" feature of $\mathcal{N}=2$ theory, making it less similar to QCD than one would desire, is the presence of the adjoint chiral superfields $\mathcal{A}$ and $\mathcal{A}^a$. One can get rid of them making them heavy. To this end we can endow the adjoint superfield with a mass term of the type $\mu\mathcal{A}^2$, through the $\mathcal{N}=1$ preserving superpotential

$$\mathcal{W} = \frac{\mu}{2}\left[\mathcal{A}^2 + (\mathcal{A}^a)^2\right]. \tag{5}$$

Now, supersymmetry of the bulk model becomes $\mathcal{N}=1$. At large μ the adjoint fields decouple.

With the deformation superpotential (5) the 1/2 BPS classical flux tube solution stays the same as in the absence of this superpotential.[13] Moreover, the number of the boson and fermion zero modes, which become moduli fields on the string world sheet, does not change either. For the fermion zero modes this statement follows from an index theorem proved in Ref. 14. If the string solution and the number of zero modes remain the same, what can one say about the string world-sheet theory?

The bulk deformation (5) leads to a remarkable, *heterotic* deformation of the CP(1) model on the world sheet, with $\mathcal{N}=(0,2)$ supersymmetry. The discovery of non-Abelian strings in $\mathcal{N}=1$ bulk theories is a crucial step on the way to the desired $\mathcal{N}=0$ theories. Moreover, the heterotically deformed CP(1) model is very rich by itself exhibiting a number of distinct dynamical scenarios unknown previously.

To understand the emergence of $\mathcal{N}=(0,2)$ supersymmetry in the world-sheet Lagrangian recall that $\mathcal{N}=2$ Yang–Mills theories which support non-Abelian flux tubes have eight supercharges. The flux tube solutions are 1/2 BPS-saturated. Hence, the effective low-energy theory of the moduli fields on the string world sheet must have four supercharges. The bosonic moduli consist of two groups: two translational moduli $(x_0)_{1,2}$ corresponding to translations in the plane perpendicular to the string axis, and two orientational moduli whose interaction is described by CP(1) (see Fig. 4). The fermion moduli also split in two groups: four supertranslational moduli

ζ_L, $\zeta_L^\dagger$, ζ_R, $\zeta_R^\dagger$ plus four superorientational moduli. $\mathcal{N} = 2$ supersymmetry in the bulk and on the worldsheet guarantees that $(x_0)_{1,2}$ and $\zeta_{L,R}$ form a free field theory on the worldsheet completely decoupling from (super)orientational moduli, which in turn form $\mathcal{N} = (2,2)$ supersymmetric CP(1) model.

When one deforms the bulk theory to break $\mathcal{N} = 2$ down to $\mathcal{N} = 1$, one has four supercharges in the bulk and expects two supercharges on the world sheet. Two out of four supertranslational modes, ζ_R and $\zeta_R^\dagger$, get coupled to two superorientational modes ψ_R and $\psi_R^\dagger$.[16] At the same time, ζ_L and $\zeta_L^\dagger$ remain protected. Thus, the right- and left-moving fermions acquire different interactions; hence, the flux tube becomes heterotic!

This breaks two out of four supercharges on the world sheet. Edalati and Tong outlined[16] a general structure of the chiral $\mathcal{N} = (0,2)$ generalization of CP(1). Derivation of the heterotic CP(1) model from the bulk theory was carried out in Ref. 15. The derivation was extended to generic N in Ref. 17.

7. Heterotic non-Abelian string

The Lagrangian of the heterotic CP($N-1$) model can be written as[15]

$$
\begin{aligned}
L_{\text{heterotic}} = {} & \zeta_R^\dagger \, i\partial_L \, \zeta_R + \left[\gamma \, g_0^2 \, \zeta_R \, G_{i\bar{j}} \left(i\,\partial_L \phi^{\dagger\,\bar{j}}\right)\psi_R^i + \text{H.c.}\right] \\
& - g_0^4 \, |\gamma|^2 \left(\zeta_R^\dagger \, \zeta_R\right)\left(G_{i\bar{j}} \, \psi_L^{\dagger\,\bar{j}} \psi_L^i\right) \\
& + G_{i\bar{j}} \left[\partial_\mu \phi^{\dagger\,\bar{j}} \, \partial_\mu \phi^i + i\bar{\psi}^{\bar{j}} \gamma^\mu D_\mu \psi^i\right] \\
& - \frac{g_0^2}{2}\left(G_{i\bar{j}} \psi_R^{\dagger\,\bar{j}} \, \psi_R^i\right)\left(G_{k\bar{m}} \psi_L^{\dagger\,\bar{m}} \, \psi_L^k\right) \\
& + \frac{g_0^2}{2}\left(1 - 2g_0^2|\gamma|^2\right)\left(G_{i\bar{j}} \psi_R^{\dagger\,\bar{j}} \, \psi_L^i\right)\left(G_{k\bar{m}} \psi_L^{\dagger\,\bar{m}} \, \psi_R^k\right). \qquad (6)
\end{aligned}
$$

The constant γ in Eq. (6) is the parameter which determines the "strength" of the heterotic deformation, and the left-right asymmetry in the fermion sector. It is related to the parameter μ in Eq. (5) (e.g. $\gamma \propto \mu$ at small μ). The third, fourth and fifth lines in Eq. (6) are the same as in the conventional $\mathcal{N} = (2,2)$ CP($N-1$) model, except the last coefficient.

Introduction of a seemingly rather insignificant heterotic deformation drastically changes dynamics of the CP(1) model, leading to spontaneous supersymmetry breaking. At small μ (small γ) the field ζ_R represents a massless Goldstino, with the residue $\langle R\,\psi_R^\dagger\,\psi_L\rangle$. As well known, a nonvanishing bifermion condensate $\langle R\,\psi_R^\dagger\,\psi_L\rangle$ develops in the undeformed model. Thus, the vacuum energy

$$\mathcal{E}_{\rm vac} = |\gamma|^2\,\left|\langle R\,\psi_R^\dagger\,\psi_L\rangle\right|^2 \neq 0\,. \tag{7}$$

Therefore, upersymmetry is spontaneously broken. A nonvanishing $\mathcal{E}_{\rm vac}$ for arbitrary values of γ in heterotically deformed CP($N-1$) models was obtained in Ref. 18 from the large-N expansion. Spontaneous breaking of SUSY in heterotic $CP(N-1)$ was anticipated in Ref. 19. Very interesting results, such as the infrared fixed point at $g^2\gamma^2 = \frac{1}{2}$, were recently obtained in the perturbative study of this model.[20]

8. Large-N solution of the heterotic CP($N-1$) model

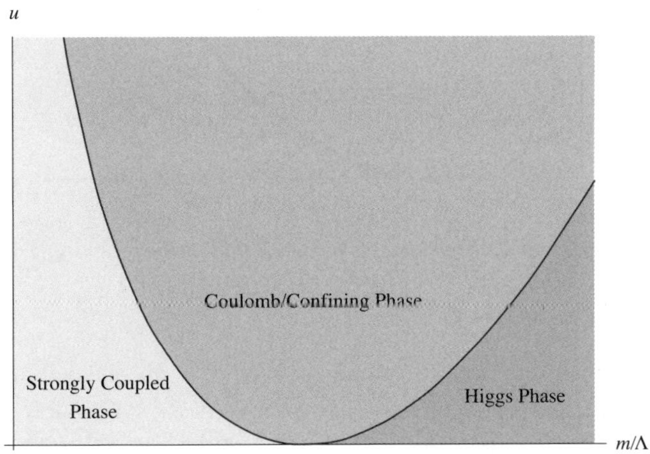

Fig. 7. The phase diagram of the twisted-mass deformed heterotic CP($N-1$) theory. The parameter u denotes the amount of deformation and is related to γ.

To reveal a rich dynamical structure of the heterotically deformed CP($N-1$) models it is instructive to add twisted masses which correspond to $\Delta m \neq 0$ introduced above. Moreover, the most convenient choice of the twisted masses is that preserving the Z_N symmetry of the model which exists at $\Delta m = 0$,

$$m_k = m \exp\left(i\frac{2\pi\, k}{N}\right), \qquad k = 0, 1, 2, \ldots, N-1\,. \tag{8}$$

where m is a complex parameter setting the scale of the twisted masses. For simplicity I will take it real. Now we have two variable parameters, m and γ, the strength of the heterotic deformation. The breaking vs. nonbreaking of the above Z_N determines the phase diagram. This model can be solved at large N using the $1/N$ expansion.[21] I will present here just two plots exhibiting main features of the solution.

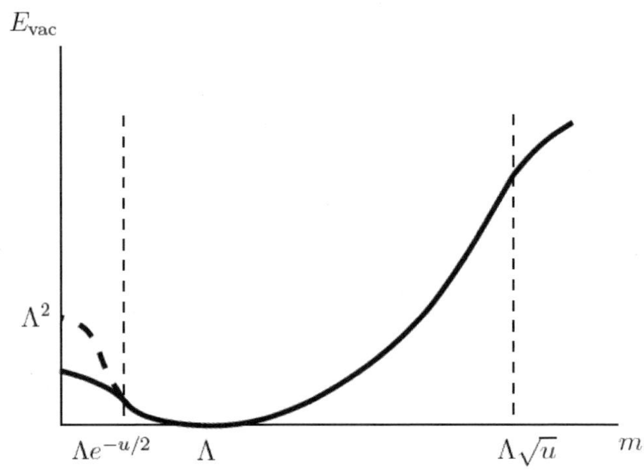

Fig. 8. Vacuum energy density vs. m. The dashed line shows an unstable quasivacuum.

Figure 7 displays three distinct regimes and two phase transition lines. Two phases with the spontaneously broken Z_N on the left and on the right are separated by a phase with unbroken Z_N. This latter phase is characterized by a unique vacuum and confinement of all U(1) charged fields ("quarks"). In the broken phases (one of them is at strong coupling) there are N degenerate vacua and no confinement.

Figure 8 shows the vacuum energy density at a fixed value of γ. It demonstrates that supersymmetry is spontaneously broken everywhere except a circle $|m| = \Lambda$ in the Z_N-unbroken phase. The first phase transition occurs at strong coupling (small $|m|$) while the second phase transition is at weak coupling (large $|m|$). Both phase transitions between the three distinct phases are of the second kind.

One must be able to translate this rich world sheet-dynamics into (presumably) highly nontrivial statements regarding the bulk theory at strong coupling.

9. Instead of conclusion

The progress in understanding dynamics of non-Abelian theories at strong coupling was painfully slow. But what a progress it is! To properly appreciate the scale of achievements, please, look back in the 1970's and compare what was known then about strong interactions to what we know now. Just open old reviews or textbooks devoted to this subject, in parallel with fresh publications. Of course, a pessimist might say that the full analytical theory is still elusive. Will it ever be created? And what does it mean, "the full analytical theory," in the case when we are at strong coupling? The richness of the hadronic world is enormous. Unlike QED we will never be able to analytically calculate all physical observables with arbitrary precision. But do we really need this? To my mind, what is really needed is the completion of the overall qualitative picture of confinement in non-supersymmetric theories, supplemented by a variety of approximate quantitative tools custom-designed to treat particular applications. A large number of such tools are already available.

Acknowledgments

I am very grateful to A. Yung with whom I shared the pleasure of working on the issues discussed in this talk. This write-up was completed during my sabbatical at Harvard University. This work is supported in part by the DOE grant DE-FG02-94ER408.

References

1. D. J. Gross and F. Wilczek, *Phys. Rev. Lett.* **30**, 1343 (1973); H. D. Politzer, *Phys. Rev. Lett.* **30**, 1346 (1973).
2. A. Abrikosov, *Sov. Phys. JETP* **32**, 1442 (1957) [Reprinted in *Solitons and Particles*, Eds. C. Rebbi and G. Soliani (World Scientific, Singapore, 1984), p. 356];

H. Nielsen and P. Olesen, *Nucl. Phys.* **B61**, 45 (1973) [Reprinted in *Solitons and Particles*, Eds. C. Rebbi and G. Soliani (World Scientific, Singapore, 1984), p. 365].

3. Y. Nambu, *Phys. Rev.* **D10**, 4262 (1974);
G. 't Hooft, *Gauge theories with unified weak, electromagnetic and strong interactions,* in Proc. of the E.P.S. Int. Conf. on High Energy Physics, Palermo, 23-28 June, 1975 ed. A. Zichichi (Editrice Compositori, Bologna, 1976);
S. Mandelstam, *Phys. Rept.* **23**, 245 (1976).
4. N. Seiberg and E. Witten, *Nucl. Phys.* **B426**, 19 (1994), (E) **B430**, 485 (1994), arXiv:hep-th/9407087; *Nucl. Phys.* **B431**, 484 (1994), arXiv:hep-th/9408099.
5. G. 't Hooft, *Nucl. Phys.* **B79**, 276 (1974);
A. M. Polyakov, *JETP Lett.* **20**, 194 (1974).
6. M. R. Douglas and S. H. Shenker, *Nucl. Phys.* **B447**, 271 (1995), arXiv:hep-th/9503163; A. Hanany, M. J. Strassler and A. Zaffaroni, *Nucl. Phys.* **B513**, 87 (1998), arXiv:hep-th/9707244.
7. A. Hanany and D. Tong, *JHEP* **0307**, 037 (2003), arXiv:hep-th/0306150.
8. R. Auzzi, S. Bolognesi, J. Evslin, K. Konishi and A. Yung, *Nucl. Phys.* **B673**, 187 (2003), arXiv:hep-th/0307287.
9. G. R. Dvali and M. A. Shifman, *Phys. Lett.* **B396**, 64 (1997), (E) **B407**, 452 (1997), arXiv:hep-th/9612128.
10. M. Shifman and A. Yung, *Phys. Rev.* **D67**, 125007 (2003), arXiv:hep-th/0212293.
11. M. Shifman and A. Yung, *Phys. Rev.* **D70**, 045004 (2004), arXiv:hep-th/0403149.
12. A. Hanany and D. Tong, *JHEP* **0404**, 066 (2004), arXiv:hep-th/0403158.
13. M. Shifman and A. Yung, *Phys. Rev.* **D72**, 085017 (2005), arXiv:hep-th/0501211.
14. A. Gorsky, M. Shifman and A. Yung, *Phys. Rev.* **D75**, 065032 (2007), arXiv:hep-th/0701040.
15. M. Shifman and A. Yung, *Phys. Rev.* **D77**, 125016 (2008), (E) **D79**, 049901 (2009), arXiv:0803.0158 [hep-th].
16. M. Edalati and D. Tong, *JHEP* **0705**, 005 (2007), arXiv:hep-th/0703045.
17. P. A. Bolokhov, M. Shifman and A. Yung, *Phys. Rev.* **D79**, 085015 (2009), arXiv:0901.4603 [hep-th]; qit Phys. Rev. **D79**, 106001 (2009), arXiv:0903.1089 [hep-th].
18. M. Shifman and A. Yung, *Phys. Rev.* **D77**, 125017 (2008), D(E) **81**, 089906 (2010), arXiv:0803.0698 [hep-th].
19. D. Tong, *JHEP* **0709**, 022 (2007), arXiv:hep-th/0703235.
20. X. Cui and M. Shifman, *Perturbative Aspects of Heterotically Deformed CP(N-1) Sigma Model*, arXiv:1009.4421 [hep-th].
21. P. A. Bolokhov, M. Shifman and A. Yung, *Phys. Rev.* **D82**, 025011 (2010), arXiv:1001.1757 [hep-th].

GLUON CONFINEMENT AND QUANTUM CENSORSHIP

JÁNOS POLÓNYI

Strasbourg University, High Energy Theory Group, CNRS-IPHC,
23 rue du Loess, BP28 67037 Strasbourg Cedex 2 France
polonyi@ires.in2p3.fr

The dynamical Maxwell-cut, a degeneracy is shown to be a precursor of condensate in the ϕ^4 and the sine-Gordon models. The difference of the way the Maxwell-cut is obtained is pointed out and quantum censorship, the generation of semiclassically looking phenomenon by loop-corrections is conjectured in the sine-Gordon model. It is argued that quantum censorship and gluon confinement exclude each other.

Keywords: Renormalization group; condensate; confinement.

1. Introduction

The peculiarity of a condensate arises from its macroscopic occupation. The order of magnitude of the typical quantum fluctuations is $\mathcal{O}(\hbar^{1/2})$ in a trivial, perturbative vacuum. When a condensate is present then the field is supposed to display an $\hbar$-independent, $\mathcal{O}(\hbar^0)$ expectation value. The expectation value of the square of the field in a state with n particles is $\hbar\mathcal{O}(n)$ thus an occupation number $\mathcal{O}(\hbar^{-1})$ is needed to make up the condensate.

The macroscopically high occupation number is possible in the absence of strong repulsive forces acting between the particles only. In other words, the presence of a condensate suggests a high degree of degeneracy in the vacuum, a necessary condition of the semiclassical approximation. A dynamical generalization of the Maxwell-cut was found by inspecting the semiclassical contributions to the functional renormalization group equations in theories with condensate.[1] It turned out that the loop contributions may change this picture and quantum censorship was proposed as a mechanism which reduces a semiclassical degeneracy to an approximate degeneracy driven by quantum fluctuations.[2,3]

The high degree of degeneracy indicating the emerging condensate influences the asymptotic sector of a theory in a fundamental manner. When created semiclassically then the degeneracy consists of localized modes. In the quantum censorship scenario the loop contributions are handled in the extended plane wave basis and their way of realizing the degeneracy is based on extended modes. The Yang-Mills vacuum contains a condensate, as well, and the structure of its soft is obviously important for confinement. It is argued below that gluon confinement excludes the building up of the quantum censorship.

2. Tree-level condensates

We consider condensate in a weakly coupled Euclidean field theory for a real scalar field $\phi(x)$ where the sharp UV cutoff in momentum space will be lowered by the recursive equation

$$e^{-\frac{1}{\hbar}S_{k-\Delta k}[\phi]} = \int D[\phi']e^{-\frac{1}{\hbar}S_{k-\Delta k}[\phi+\phi']}. \tag{1}$$

Here $\phi(x)$ and $\phi'(x)$ denote fields with support in $0 < p < k - \Delta k$ and $k - dk < p < k$, respectively. We follow for simplicity the local potential approximation where the form

$$S_k[\phi] = \int dx \left[\frac{1}{2}(\partial\phi)^2 + U_k(\phi)\right] \tag{2}$$

is assumed. Most of the examples mentioned below correspond to the initial condition $U_\Lambda(\phi) = m_B^2\phi^2/2 + g_B\phi^4/4!$ in four space-time dimensions. The functional integral will be evaluated in the loop-expansion and the Wegner-Houghton equation[4]

$$\partial_k U_k(\phi) = -\frac{\hbar k^3}{16\pi^2}\ln[k^2 + U_k''(\phi)], \tag{3}$$

follows by assuming trivial saddle points. This is an exact equation because the small parameter of the loop-expansion is actually $\hbar\Delta k/k$ because each loop integral is restricted to a shell of thickness Δk in momentum space. But what happens when nontrivial saddle points are encountered?

The simplest strategy to find an answer is to ignore the loop corrections altogether and consider the tree-level evolution only.[1] It turns out that the nontrivial saddle point of the blocking relation (1) is the hallmark of spontaneous symmetry breaking, a condensate in the vacuum. There is no evolution on the tree-level as long as the action reaches its minimum at $\phi'(x) = 0$ within the shell $k - dk < p < k$. But if the potential has negative curvature at the initial condition at $k = \Lambda$ say at $\phi = 0$ then the gradual decrease of the gliding cutoff k leads to nontrivial saddle points at some finite scale $k = k_{cr}$. The evolution from this scale was followed numerically by setting up an iterative process where the cutoff was decreased by a small but finite amount, $k \to k - \Delta k$ and for each value of the homogeneous field configuration Φ a saddle point ϕ'_{cl} was sought within the family of plane waves with momentum k and the new potential $VU_{k-\Delta k}(\Phi) = S_k[\Phi + \phi'_{cl}]$ was calculated. The potential evolved in an interval around zero, $|\phi| < \Phi_k$ where Φ_k increases from zero as k decreases from k_{cr} and the form

$$V_k^{inst}(\phi) = -\frac{k^2}{2}\phi^2 \tag{4}$$

was found as shown in Fig. 1.

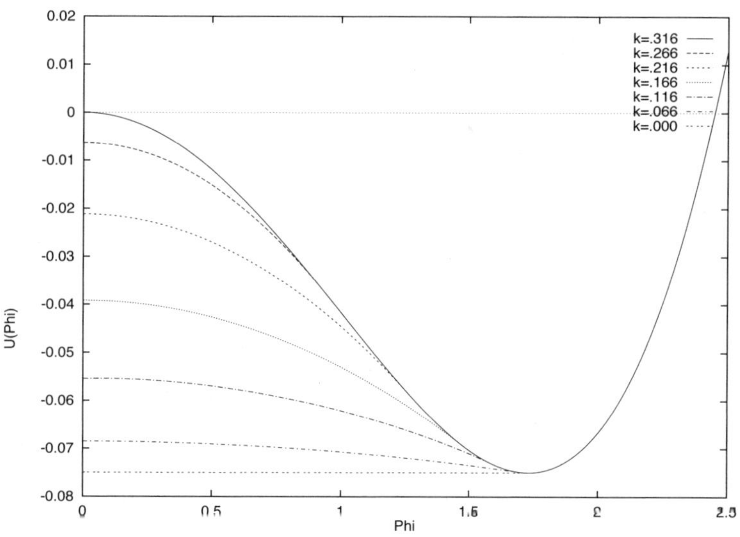

Fig. 1. The evolution of the potential[1] in the unstable region $|\Phi| < \Phi_k$ for $m_B^2 = -0.1$, $g_B = 0.2$ in units of $k_{init} = \Lambda = 1$.

Such a potential makes the action (2) degenerate at the scale $p = k$ by canceling the kinetic term in. What happened numerically was that the action lost convexity at $k = k_{cr}$ and in the next step, $k \to k_{cr} - \Delta k$ a shallow, $\mathcal{O}(\Delta k/k)$ minimum was found in the action as the function of the plane wave amplitude. This saddle point made the action $S_{k_{cr}-\Delta k}[\phi]$ degenerate for modes with $p = k - \Delta k$ with a slightly larger amplitude.

This scenario repeated itself during the iteration, namely the saddle point, found in a shallow minimum made the new action degenerate again. The best description of the dynamics of a mode with a given momentum p is provided by $S_k[\phi]$ with $k = p$, hence the degeneracy of the blocked actions at their cutoff can be interpreted as the indication within the renormalization group scheme that the dynamics is degenerate for all modes with small enough amplitude and momentum $p < k_{cr}$. This is a dynamical Maxwell cut, a precursor of the condensate, and leads to the traditional flattening of the effective potential at $k = 0$, $U_{eff}(\Phi) = V_{k=0}(\Phi)$ between the vacuum expectation values $\langle 0|\phi(x)|0\rangle = \pm\Phi_{vac}$. It is natural to interpret the plane wave saddle points as domain walls. There is a saddle point for each scale $p < k_{cr}$ and the functional integration over their soft zero modes, arising from the their breaking of the space-time symmetries makes the system strongly coupled as expected by analogies with the mixed phase of first order phase transitions.

Though these results are simple and reasonable, they raise some doubt and concern. The first question concerns the role of loop-correction. Can they alter the conclusion in a qualitative manner? The procedure is based on the sharp cutoff which prevents us from including higher orders of the gradient expansion in the action (2). Will the general picture remain valid when these terms are taken into account in some manner? The idea of the renormalization group is the construction of the renormalized, full action by integrating a differential equation. The emergence of a saddle point usually signals the appearance of non-analytic terms. Will the evolution equation remain integrable in k at k_{cr}?

As of the trouble spots, the first is related to the shallowness of the minimum of the action around the saddle point because it removes the factor $\Delta k/k$ from the small parameter of the loop expansion and we fall back on the $\hbar$-expansion. To make things worse, we even loose the $\hbar$-expansion in the degenerate limit, $\Delta k \to 0$, having a flat integrand. The evolution equation (3) requires the availability of the loop expansion even if the higher orders are resummed in the differential equation limit $\Delta k \to 0$. In general, we have neither analytical nor numerical method at hand to tackle constant integrands, the ultimate strong coupling limit.

Another troubling aspect of the soft zero mode dynamics of the domain walls is that it renders the physics of the mixed phase, $|\langle 0|\phi(x)|0\rangle| < \Phi_{vac}$ non-perturbative and opens the possibility of having new free, relevant parameter in the scalar model, such as the value of the condensate.[13] The true vacuum lies at the common point of the stable region $\langle 0|\phi(x)|0\rangle > \Phi_{vac}$ and the mixed phase therefore "half" of the quantum fluctuations which tend to decrease $\phi(x)$ might well be non-perturbative.

3. Loop-corrections

The first of the previous questions is taken up in this section only, the eventual modification of the dynamical Maxwell-cut by including the loop corrections within the ansatz (2). The potential is non-analytical according to the tree-level analysis therefore the numerical integration of Eq. (3) must be done without assuming a polynomial representation. This is possible by using the spline representation for the potential.[2] It is interesting to follow the evolution of the dimensionless curvature of the action (2) for a mode $p = k$, $P_k(\Phi) = 1 + \partial_\phi^2 U_k(\Phi)/k^2$, depicted in Fig. 2. It shows a sudden drop at a finite scale followed by the stop of the program. The numerical algorithm makes dynamical adjustment of the precision needed during the integration and stops because the system of linear equations for the spline coefficients becomes nearly singular and requires extreme precision. The value of k where this happens depends slightly on the accuracy but seemed to be not moving significantly up to our limit, several thousand splines. The potential approaches the form (4) at the last steps as can clearly be seen from Fig. 3. The lesson of such a failed attempt to reach the infrared end point suggests a full or nearly complete dynamical Maxwell-cut.

The possible problems of the loop-expansion for a nearly degenerate action is avoided when the evolution of the effective action is followed,[5–7] an algorithm with $\Delta k/k$ as the only small parameter. Similar, sudden drop of the curvature of the action was observed at finite scale in this scheme when smooth cutoff was used,[2] leaving room for the eventual inclusion of higher orders of the gradient expansion in the ansatz (2). Naturally the difficulties of the degenerate bare action remain in this scheme in the disguise of the problem of justifying any usable ansatz for the effective action.

Can we decide whether the action is exactly or only nearly degenerate? The analytical efforts presented so far[8–11] leave room for reaching true singularities during integrating the evolution equation in the form of degenerate action involving non-analytical terms.[3] Numerical methods, based on finite amount of computer power can give no satisfactory answer neither. I believe that the following options are left open:

- The evolution equation driven by loop-contributions leads to a truly degenerate action as in the tree-level case. We have no analytical or numerical methods to tackle such models.

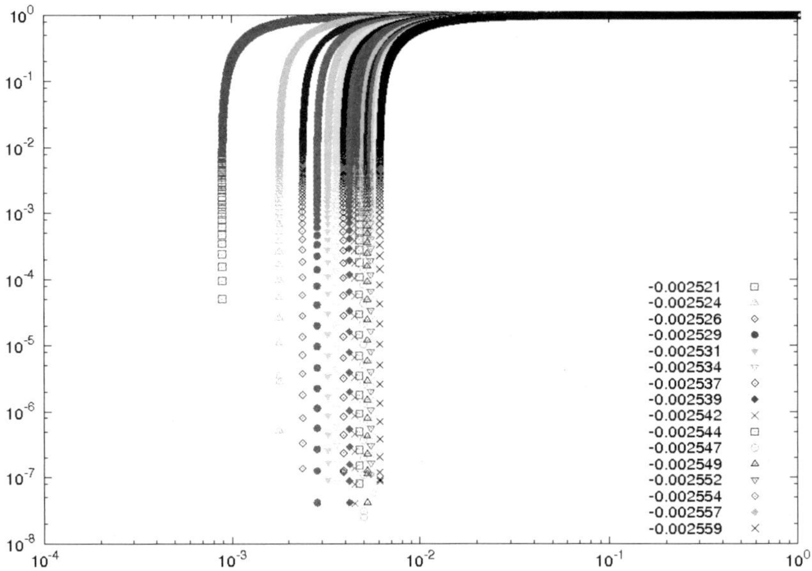

Fig. 2. The dependence of the dimensionless degeneracy for vanishing field amplitude, $P_k(0)$ of the action on the scale k for $g_B = 0.8$, the bare mass square values are shown in the figure.[2]

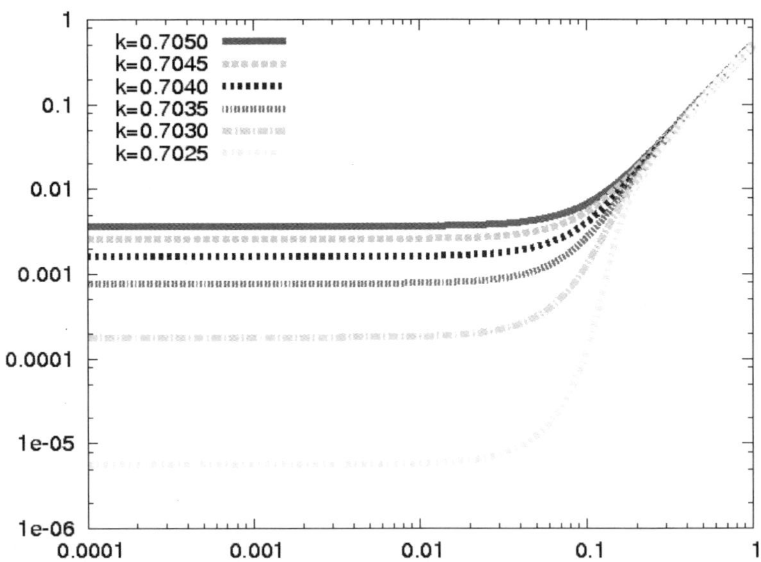

Fig. 3. The evolution[2] of $P_k(\Phi)$ as the function of the field Φ for $m_B^2 = -0.5$ and $g_B = 0.5$.

- The loop-contributions manage to keep the action regular and generate an approximative dynamical Maxwell-cut. Such a shielding of the semiclassical singularity by quantum fluctuations which make up a similar effect is called Quantum Censorship.

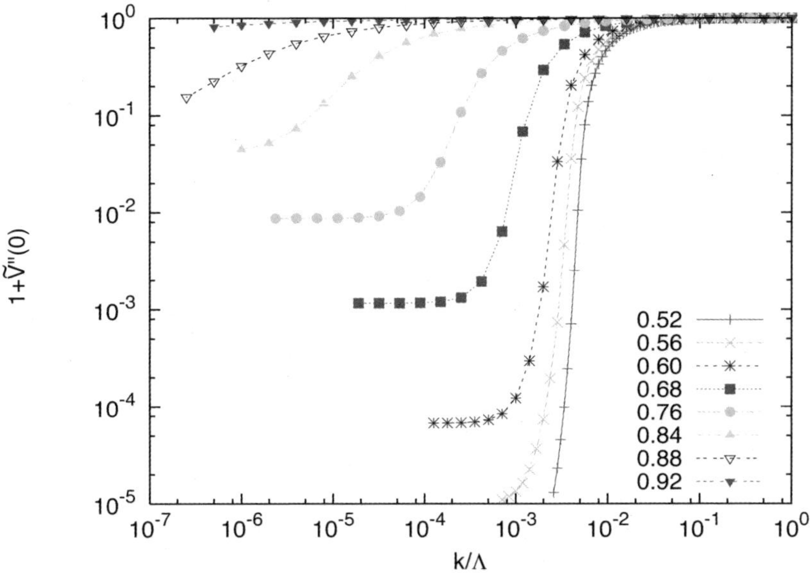

Fig. 4. The evolution[3] of $P_k(0)$ in the sine-Gordon model for different values of β_r, indicated in the figure.for $m_B^2 = -0.5$ and $g_B = 0.5$.

Quantum Censorship seems to be realized in the two-dimensional sine-Gordon model[3] defined by the bare dimensionless potential $\tilde{U}_{k=\Lambda}(\phi) = k^{\ 2}U_B(\phi) = u_B \cos(\sqrt{8\pi}\beta_r\phi)$. The effective potential must be constant, being the only convex periodic function, but this naturally does prevent the model to display highly nontrivial dynamics in the infrared. The tree-level evolution gives saddle points and produces degenerate action[12] for $k < k_{cr} \neq 0$ in the phase $\beta_r < 1$. When the loop-contributions are added then we observe a sudden drop of the curvature of the action, characteristic of the dynamical Maxwell-cut followed by a surprising stabilization of the curvature at very small values.[3]

4. Yang-Mills theories

We now leave the territory of well established results and make an attempt to interpret the difference of Figs. 3 and 4 and to conjecture about its relevance for Yang-Mills theories.

It seems reasonable to assume that the increase of the accuracy of the numerical algorithm would find a plateau in Fig. 4 for $\beta_r < 0.6$, too. When this state of affairs is compared with Fig. 3 which follows the evolution of models from nearly at the critical point to deep into the symmetry broken phase with the same numerical accuracy then one has the impression that Quantum Censorship is not observed in the non-periodic ϕ^4 model.

Accepting this interpretation one wonders about the possible source of this difference between the two models. I believe that it is to be found in the non-propagating nature of the excitations in the mixed phase of the ϕ^4 model. The domain walls formed in a vacuum with $|\langle 0|\phi(x)|0\rangle| < \Phi_{vac}$ can be deformed with small energy investment. As mentioned before, such a deformation represent the Goldstone modes of the broken space-time symmetries. These modes restore the homogeneity of the vacuum like in a liquid but leave an important imprint on the dynamics by generating dissipation. In the language of solid state physics the sound wave of the mixed phase is damped by the reflection and the traverse of a domain walls and the velocity of sound is reduced to zero. In particle physics this situation is interpreted as having no asymptotic particle states. The domain walls in the small β_r phase of two dimensional sine-Gordon model where the periodic symmetry of the theory is dynamically broken[12] are kinks, stable propagating particles. As a result, elementary plane wave excitations have more chance to propagate and the long distance structure of the theory can be reconstructed by means of plane waves. Once this is possible the gradual turning on the plane waves during the evolution may lead us to the correct vacuum.

Let us no turn to Yang-Mills theories. The vacuum of the scalar ϕ^4 model lies at the boundary of the non-perturbative mixed phase and can safely be approached from the stable region. This is not the case in Yang-Mills theories where the vacuum is within the mixed phase. In fact, the strong chromo-magnetic attraction among gluons is supposed to generated a liquid vacuum filled with condensate[14] which is inhomogeneous,[15,16] an analogy of the mixed phase of the ϕ^4 model. The conjecture that the integration of the evolution equation of Yang-Mills theories in the plane wave basis would run into a "naked singularity" seems reasonable when gluons are confinement. In fact, otherwise there should be colored asymptotic states.

Though confinement excludes strictly particle-like asymptotic gluon states only, non-particle like asymptotic states are not expected to exists neither according to the general view. In other words, color confinement and Quantum Censorship are exclusive properties. The verification of this conjecture is a challenging question because I think that we lack some technical elements related to gauge symmetry, Wick rotation and the ansatz for the effective action.

There have been impressive advances achieved in applying the functional renormalization group for Yang-Mills theories[17] by relying on gauge fixing and modified Slavnov-Taylor identities. A different method which avoids gauge fixing altogether or is explicitly independent of it would be useful to assure that the singularities arising from the degeneracy of the unphysical sector are properly separated.

Another issue awaiting for careful consideration is the return to real time and Minkowski space-time. The renormalization group studies of the Euclidean theory can shed light on the way the contributions of the off-shell modes pile up as the long distance physics is approached. But confinement of color is beyond this issue, it concerns the dynamics of modes on the mas-shell. A basic element of the renormalization group idea is the successive dealing with the degrees of freedom. The order of their elimination is in principle arbitrary but it is advised to start with simple, perturbative modes and finish with soft, large amplitude fluctuations. In fact, the piling up the informations gained during the elimination process makes the effective dynamics better prepared to deal with the non-perturbative modes at a later stage of the elimination process. Recall the renormalization group approach to fermions at finite density where the blocking zooms into the Fermi sphere in the Brioullin zone instead of the zero momentum point as for bosonic particles. In order to address the confinement problem in Yang-Mills theory with the renormalization group method we have to zoom into the mass-shell which is possible in Minkowski space-time only. There is reason to suspect that mass-shell singularities are stronger in the Yang-Mills vacuum than for non-confining models. For instance the perturbative collinear divergences are stronger for non-Abelian gauge theories and the view of confinement as an Anderson localization in space-time[18] suggests that pinch-singularities may arise, too. The liquid models of the vacuum indicate the presence of an unusually large number of soft modes which enhance the dressing, as well.

Finally, the truncation of the ansatz used in solving the evolution equation may be critical for models with condensate. The (approximate) dynamical Maxwell-cut can not be obtained in the scalar ϕ^4 theory or the sine-Gordon model when the loop corrections are collected in a truncated power or Fourier series representation of the local potential. In a similar manner one expects the need of more flexible ansatz than those what has been used so far in Yang-Mills models to address the issue of degeneracy.

5. Summary

Seemingly disparate points are related in this work. The precursor of the formation of a condensate, a large degree of degeneracy when the vacuum is constructed by the successive turning on the modes in the plane wave basis is claimed to be related to gluon confinement. Even if correct, this view does not add much to our understanding of color confinement, it rather orients our attention to some difficulties waiting us along the road. Gauge fixing is an ever returning problem for non-perturbative methods for Yang-Mills models. The Wick rotation and the choice of the ansatz are important problems of the functional renormalization group method independently of their possible role in Yang-Mills theories. Their improvement would be a gain for other domains, too.

Finally, another subjective remark about the functional renormalization group. I think that it is a promising method whose limitation is not yet in sight. It can conveniently interpolate between numerical and more intuitive, analytically based schemes to solve strongly coupled theories. As such, it should ultimately integrate into itself the experiences gained in lattice gauge theory and use them where this latter is not reliable, in real time scattering processes and finite particle density.

Acknowledgments

I thank Jean Alexandre and Vincent Pangon for the opportunity to work with them on the nice problems[1–3,13] covered in this paper.

References

1. J. Alexandre, V. Branchina and J. Polónyi, *Phys. Lett.* **B445**, 351 (1999).
2. V. Pangon, S. Nagy, J. Polónyi and K. Sailer, *Symmetry breaking and the functional RG scheme*, arXiv:0907.0144, submitted to Phys. Rev.
3. V. Pangon, S. Nagy, J. Polónyi and K. Sailer, *Quantum censorship in two dimensions*, arXiv:0907.0496, to appear in Phys. Lett.
4. F. J. Wegner and A. Houghton, *Phys. Rev.* **A8**, 401 (1973).
5. J. F. Nicoll and T. S. Chang, *Phys. Lett.* **A62**, 287 (1977).

6. C. Wetterich, *Phys. Lett.* **B301**, 90 (1993).
7. T. R. Morris, *Int. J. Mod. Phys.* **A9**, 2411 (1994).
8. A. Ringwald and C. Wetterich, *Nucl. Phys.* **B334**, 506 (1990).
9. N. Tetradis and C. Wetterich, *Nucl. Phys.* **B383**, 197 (1992).
10. C. Wetterich, **Z. Phys. C57**, 451 (1993).
11. D. Litim, J. Pawlowski and L. Vergara, *Convexity of the effective action from functional flows*, arXiv:hep-th/0602140.
12. S. Nagy, I. Nandori, J. Polónyi and K. Sailer, *Phys. Lett.* **B647**, 152 (2007).
13. J. Alexandre, V. Branchina and J. Polónyi, *Phys. Rev.* **D58**, 16002 (1998).
14. G. K. Savvidy, *Phys. Lett.* **B71**, 133 (1977).
15. H. B. Nielsen and P. Olesen, *Nucl. Phys.* **B160**, 380 (1979).
16. T. H. Hansson, K. Johnson and C. Peterson, *Phys. Rev.* **D26**, 2069 (1982).
17. J. M. Pawloswki, *Ann. Phys.* **322**, 2831 (2007).
18. K. Johnson, L. Lellouch and J. Polónyi, *Nucl. Phys.* **B367**, 675 (1991).

FROM CONFINEMENT TO SUPERFLUIDITY?

V.I. ZAKHAROV

Institute of Theoretical and Experimental Physics,
117125, B. Cheremushkinskaya, 25, Moscow, Russia
vzakharov@itep.ru

We describe a unified picture of confining and deconfined phases of Yang-Mills theories in terms of nonperturbative vacuum defects. The confinement is related to condensation of (magnetic) strings. The phase transition at $T = T_c$ is viewed as change of dimensions, $4d \to 3d$. Namely, all the defects become time oriented. As a result, percolation of strings becomes percolation of 3d trajectories or, in field theoretic language, condensation of a 3d scalar field. The condensation, in turn, might signal superfluidity of the quark-gluon plasma. The notes are mostly a mini-review. A remark on entanglement and confinement is added.

Keywords: Confinement; quark-gluon plasma; nonperturbative QCD.

1. Introduction

The problem of confinement occupied Vladimir Naumovich Gribov during the last years of his scientific career.[1] Thus, it seems to be the most natural choice to talk about confinement at the conference Gribov-80. I do not think that we have actually advanced much in understanding the problem compared to the Gribov's time but, probably, we see it in a somewhat different way. To put it short, the confinement in non-Abelian case reduces to a problem in string theory, and is difficult for this reason. While in Abelian case it is a problem in field theory.

There exist various approaches to confinement. I will talk about the approach which unifies the languages of the lattice and continuum-theory. In both cases, –lattice and continuum,– we will discuss "vacuum defects". The terminology is not yet commonly used although examples of defects are known to everyone. For example, the instanton in this language is called a point-like defect. The terminology might look misleading because for the instanton of size ρ positioned at x_0 the gluon field strength squared is equal to

$$(G^a_{\mu\nu}G^a_{\mu\nu})_{instanton} \sim \frac{1}{g^2}\frac{\rho^4}{(\rho^2+(x-x_0)^2)}, \quad (1)$$

Although the instanton position, x_0 is indeed a point in four dimensions (4d), its size is typically of order $\rho \sim \Lambda^{-1}_{QCD}$ and (1) is a smeared distribution. The usage of the "point-like defect" with respect to the instanton becomes actually justified in the dual formulations of the Yang-Mills theories. According to this approach, defects live in extra dimensions. One of the extra coordinates, z is conjugate to the size in 4d. Thus, (1) corresponds to a point-like defect in 4d with coordinate $z \sim \rho^{-1}$.[a]

Moreover, in confining theories there is a horizon in the z-coordinate so that

$$z < z_H, \quad z_H \sim \Lambda^{-1}_{QCD} \quad (2)$$

Thus, one may say that instantons (1) live near the horizon. The "near-horizon" physics is infrared physics and the dual models are to replace the field theoretic approach just in this region.

In field theoretic language, instantons practically exhaust the list of the quasiclassical fluctuations, or "defects" in our language. In dual formulations, things are very different. There is a kind of a zoo of defects, all of them living near the horizon. I am not aware of a review that would list all possible defects, for a partial list see Ref. 2. Also in these notes, we should make selection, which defects we discuss in some detail.

We will concentrate on the so called magnetic strings, see Ref. 3 and references therein. The definition of the magnetic strings is that they are closed in the vacuum and can be open on the 't Hooft line. The choice of the magnetic defects is rather unique for a number of reasons. First, it is practically for sure that the magnetic strings are relevant to the confinement, for review see Refs. 4, 5. The very idea that condensation of dual, in our case magnetic degrees of freedom, is responsible for confinement of "direct" (in our case, color) charges goes back to about 40 years, see in particular.[6] The stringy realization of this idea is more recent.

Moreover, the magnetic strings are singled out by the fact that they are both predicted by the continuum theory (dual models) and seen on the lattice.[3] We need this double check since the evidence in either approach alone is not so strong. Indeed, the dual models for QCD are not well developed

[a]Actually, the instanton is a D0 brane wrapped around an extra compact coordinate θ, see also below.

(yet). One can only claim that the universality class is known reliably.[7,8] No details of metric in the extra coordinates are known. In view of this, we decide to rely only on the topological properties of the existing models which may survive large deformations of the metrics. The lattice results, on the other hand, are much more precise. However, the interpretation is difficult since the lattice results are formulated in a specific language, foreign to the continuum-theory community. Confronting the lattice data with the dual-models predictions helps to decipher the data.

The outline of the notes as follows. In Sect. 2 we give some details on the geometry of extra dimensions relevant to the dual formulation at zero temperature and at temperatures above the deconfining phase transition. In Sect. 3 we overview the lattice language in the context of the defects. The point is that comparison of the lattice data and continuum theory predictions involves actually a subtle issue of modification of the soft fields ("living near the horizon") by the process of measurement inherent to the lattice simulations.[9] In this section we also summarize the lattice evidence for the magnetic strings. In Sect. 4 we discuss possible implications for the quark-gluon plasma.

It is worth emphasizing that the notes are mostly a mini-review. The basic original papers are Refs. 2, 10, 11.

2. Geometry in extra coordinates

2.1. *Geometry at small temperatures*

Duality assumes that the Yang-Mills theories, poorly defined perturbatively at large distances are completed in the infrared as string theories in extra dimensions. The simplest motivation is to recollect that introducing strings with tension σ which can be open on the Wilson line immediately allows to reproduce the confining potential at large distances,

$$\lim_{R\to\infty} V_{Q\bar{Q}}(R) \;=\; \sigma R \;. \tag{3}$$

Note that the confinement which is a quantum effect in the field theoretic language emerges at the classical level in the string language. This would be the main advantage of the dual formulation, if it is actually found.

We could be more ambitious and attempt on reproducing by strings the whole of the Cornell potential,

$$V_{Q\bar{Q}}(R) \;=\; -\frac{const}{R} + \sigma R \;. \tag{4}$$

Postulate to this end that the string tension runs as function of it length, $\sigma(l)$. Clearly, at short distances one should have then $\sigma(l) \sim l^{-2}$ while $\sigma(l) \sim const$ at large l. The problem now is that the length l is not the only characteristic of the shape of a string. An ingenious step is to postulate that the string ends on our 4d space but is stretched otherwise into an extra, z direction with a non-trivial metric. Then minimization of the area of the string, with account of the non-trivial metric, determines the potential uniquely in terms of the running tension.

Phenomenologically, one can reverse the problem and construct a metric which reproduces the Cornell potential (4). In particular, the choice[12]

$$ds^2 = \frac{const}{z^2} \exp(cz^2/2)\big(dz^2 + dx_\mu^2\big), \tag{5}$$

with the parameter $c = 0.9\ GeV^2$, does reproduce (4) with reasonable accuracy. The position of the horizon is

$$z_H = \sqrt{\frac{c}{2}}\,,$$

and this is an example of a "soft wall".[b]

There is one more crucial ingredient in the construction discussed. Namely, there exists another, compact θ-direction.[7] Wrapping n_θ times around this circle implies a non-trivial topological charge associated with the defect,

$$Q_{top} = \pm n_\theta\,,$$

where the sign depends on the direction of the wrapping. Moreover, the geometry in the θ, z coordinates is not trivial and the radius of the θ-circle vanishes on the horizon:

$$R_\theta(z = z_H) = 0\,. \tag{6}$$

At finite temperature, and in the Euclidean space-time the time coordinate is also cyclic, as usual. Its radius, however, does not depend on z:

$$R_\tau(z) = \frac{1}{T}\,. \tag{7}$$

There could be further extra dimensions but these are not relevant for our purposes here.

---

[b]The metric (5) is given for illustration and is not explicitly used below.

2.2. *Defects*

According to the rules of the string theory, there exist D-branes which are defects of $D+1$ dimensions. One can consider

$$D0,\ D2,\ D4$$

branes and we are free to choose their orientation in the six dimensional space we are considering. The probability to find a brane as an excitation is

$$W_{defect} \sim e^{-S_{defect}} .$$

For our purposes, it is important only that the classical action of any brane wrapped around the θ-direction vanishes at the horizon because of (6):

$$T < T_c: \quad S_{(D2n)}(z = z_H) \; = \; 0, \quad n = 0, 2, 4 \ wrapping\ n_\theta = 1, 2, \ldots . \qquad (8)$$

The vanishing of the classical action (8) implies that the density of such defects is determined by their interaction, (a well known story in case of instantons, which are $(n = 0, n_\theta = \pm 1)$ defects in the classification (8)). Known approximations do not allow to evaluate the resulting density.[c]

2.3. *Phase transition*

The deconfinement phase transition is viewed now as a change of geometry, the so called Hawking-Page transition.[14] Namely, at $T > T_c$ the geometry in (τ, z) coordinates becomes that of a cigar:

$$R_\tau(z = z_H) \; = \; 0 \; , \quad T > T_c \; , \qquad (9)$$

where R_τ is the radius of the compact Euclidean time directions. Respectively, the geometry in the (θ, z) coordinates is that of a cylinder:

$$R_\theta(z) \; = \; const \; , \quad T > T_c \; . \qquad (10)$$

The change of geometry implies, for example, that the instantons, or D0 branes wrapped around the θ-direction become suppressed at $T > T_c$,[15] as is well known since long from other considerations.

To the contrary, all the defects wrapped around the τ-direction become tensionless in the infrared:[2]

$$T > T_c: \quad S_{(D2n)}(z = z_H) \; = \; 0, \ n = 0, 2, 4 \ wrapping\ n_\tau = 1, 2, \ldots . \qquad (11)$$

---

[c] In case of SUSY YM theories with large N_c there is a melting of lower- dimension D-branes onto higher dimensional ones.[13] In the pure YM theory such a melting could well be relevant to the phenomenology.[2] Let us also note that the vanishing of the action (8) mixes up the whole of the large-N_c counting.

Thus, all the defects look now in the time direction. Their time dependence is therefore trivial and they depend now only on the 3d coordinates.

To summarize, the phase transition is now the loss of one dimension,

$$4d \rightarrow 3d \ \ at\ T = T_c \tag{12}$$

for nonperturbative physics.

3. Lattice

3.1. *From quasiclassics to fine tuning*

Another source of information about the defects are lattice simulations. However, the interpretation of the data is not straightforward because the image of the defects seen on the lattice is determined not only by our continuum-theory expressions but by the lattice spacing a as well. Recognition of the image of a quasiclassical object[d] in high-resolution measurements is a pure theoretical problem, not so much specific for the lattice. But somehow the problem was not appreciated thoroughly enough and our understanding of the lattice data is somewhat limited.

The problem is how, say, the instanton (1) looks on the lattice. Measurements on the lattice are performed with a certain value of the lattice spacing a, $(a\rho_{instanton}) \rightarrow 0$. The lattice spacing can be viewed as space resolution in measuring the Yang-Mills fields on the lattice. In particular, by virtue of the uncertainty principle huge zero-point fluctuations are induced by the measurement:

$$< (G^a_{\mu\nu} G^a_{\mu\nu})_{zero-point} > \ = \ \frac{f(\alpha_s(a))}{a^4} \ , \tag{13}$$

where the function $f(\alpha_s)$ is calculable perturbatively.

How the instanton field is seen on the background of the zero-point fluctuations? The "default answer" to this question is:

$$(G^a_{\mu\nu} G^a_{\mu\nu})_{instanton} \ \approx \ \frac{f(\alpha_s(a))}{a^4} + \frac{const}{g^2} \frac{\rho^4}{(\rho^2 + (x - x_0)^2)} \qquad (wrong!)$$

and this is a wrong answer. The reality, for review and further references see Ref. 16, is that the instanton is squeezed by the singular zero-point fluctuations and becomes a singular field itself. In more detail, the 4d volume

---

[d]Quasiclassics itself corresponds to a very poor resolution, $a \gg \Lambda^{-1}_{QCD}$.

occupied by the instanton[e] shrinks to zero with $a \to a$:

$$V_{4d}(instanton) \sim (a \cdot \Lambda_{QCD})^r \Lambda_{QCD}^{-4} , \tag{14}$$

where the exponent $r \neq 0$ and its measured value lies in the interval $r = (2-4)$.

Moreover, the instanton field becomes singular so that the topological charge associated with the instanton remains unchanged,

$$(G^a_{\mu\nu}\tilde{G}^a_{\mu\nu})_{instanton} \sim (a \cdot \Lambda_{QCD})^{-r} \Lambda^4_{QCD} \tag{15}$$

Indeed, the total number of the zero modes cannot change as a function of the resolution a since it determines a physical matrix element:

$$< Q^2_{top} > = < (n_+ - n_-)^2 > \sim \Lambda^4_{QCD} V_{lattice}, \tag{16}$$

where $V_{lattice}$ is the 4d volume of the lattice, $n_\pm$ are the numbers of chiral zero modes and (16) is nothing else but the famous Veneziano-Witten relation.

It is important to realize that the shrinking of the quasiclassical configurations as a result of performing measurements with high resolution is not a lattice artefact or curiosity but a manifestation of a general phenomenon. In a number of examples, one can trace such a shrinking theoretically.[9] Probably the pioneering example of this type is the Kapitza pendulum.

Another example is provided by the polymer representation of field theory. Imagine that we have a Higgs field with a vacuum expectation value

$$< \phi_H >^2 \sim m^2_H$$

where m_H is a fixed mass scale. Our standard visualization of the condensate is that it is "everywhere the same". Imagine, however, that we would perform measurements with high resolution of lattice spacing a, $m_H \cdot a \to 0$. Then the condensate would correspond to an infinite cluster of trajectories. This cluster would be very dilute so that probability of a given link to belong to the the infinite cluster in 4d case is of order

$$\theta^{link}_{inf.\ cluster} \sim (m_H \cdot a)^2 \tag{17}$$

and tends to zero in the continuum limit as a power of the lattice spacing. On the other hand, the action associated with the trajectory is divergent

[e] By the volume occupied by the instanton we understand in fact the volume occupied by fermionic zero modes which are easier to measure on the lattice. Because of the index theorem it is reasonable to assume that the distribution of the density of a zero mode follows distribution of the density of topological charge.

in the continuum limit:

$$S_{inf.\ cluster} \approx \ln 7\frac{L}{a} \quad . \tag{18}$$

where L is the total length of trajectory while $\ln 7$ refers to the case of a 4d rectangular lattice. Eqs. (17), (18) demonstrate that, indeed, quasiclassics in measurements with high resolution looks as a fine tuning of a singular action and small probability to encounter such a singular field. In this case the fine tuning is well understood theoretically.

3.2. *Magnetic strings at* $T = 0$

The high-resolution image of strings is actually not known theoretically. On the other hand there is rich lattice data, for review see Refs. 4, 5. The magnetic strings, on the lattices with small lattice spacing a have total area in physical units

$$(Area)_{tot} \sim \Lambda^2_{QCD} \cdot V_{lattice} \ . \tag{19}$$

Eq. (19) corresponds to a vanishing probability of a given plaquette to belong to the magnetic strings,

$$\theta^{strings}_{plasimquette} \sim (\Lambda_{QCD} \cdot a)^2 \ .$$

The action of the strings is singular, in analogy with (18). The magnetic-strings are associated with non-trivial density of the topological charge, as predicted by theory outlined in Sect. 2.

3.3. *Magnetic strings at* $T > T_c$

It is most remarkable that the magnetic strings become time oriented at the phase transition. The effect is known since long on the lattice, see, in particular Ref. 17. However, it is only recently that it was realized[20] that this phenomenon is exactly what is predicted by the dual models, see (12). As far as the magnetic strings get time oriented their intersection with a time slice, or a 3d volume becomes a percolating infinite cluster. As is mentioned in the preceding subsection, an infinite cluster in any number of dimensions corresponds to a scalar field condensed. Thus, there is strong lattice evidence that

$$\langle \phi_{3d} \rangle \neq 0 \ , \tag{20}$$

and this might have important consequences as we discuss in the conclusions.

3.4. *Digression on entanglement and confinement*

We did not explain yet relation of the magnetic strings to the confinement. Magnetic strings, by definition, are closed in the vacuum and can be open on the external 't Hooft line (which is nothing else but the trajectory of heavy monopoles). According to the famous 't Hooft criterion in non-Abelian theories either quarks or magnetic monopoles are confined. Since in our world it is the quarks which are confined, the prediction fir the magnetic-string tension is:

$$\sigma_M \;=\; 0\;. \tag{21}$$

This prediction agrees perfectly with the lattice observation that the magnetic strings percolate in the vacuum, or form an infinite cluster.

In a more model-dependent way, the strings are related to the confinement through the stochastic model, see, e.g., Refs. 4, 5 for detail. Imagine that the minimum-area (A_{min}) surface is spanned on a Wilson line, $W(C)$. Introduce then the probability p for a magnetic string to penetrate a plaquette belonging to this minimal-area surface. Then the effect of strings can be estimated by assuming that each penetration of the Wilson line by the magnetic string results in a minus sign and is stochastic in nature. This is like linking number. In this way, one comes to the area law for the Wilson line:

$$< W(C) > \;\approx\; exp(-\rho_{surface}A_{min})\;\;, \tag{22}$$

where $\rho_{surface}$ is the density of the magnetic strings, defined in terms of Eq. (19). Numerically, Eq. (22) works well.[4]

Now, we would like to jump from Eq. (22) to discussion of another point. Namely, Eq. (22) assumes actually long-distance communication. Indeed, there is no other way to ensure the area law (22) but to have a massless exchange. This is clearly seen either in the vortex picture we are discussing or in the monopole picture of confinement. In the latter case we also need a massless particle exchange in the vacuum to derive confinement.

Thus, we expect to have for the vacuum correlation length

$$l^{corr}_{vac} \;\rightarrow \infty \tag{23}$$

On the other hand, in the hadronic world we have

$$l^{corr}_{excitations} \;\approx\; \frac{1}{m_{glueball}}\;, \tag{24}$$

where $m_{glueball}$ is the lightest glueball mass. Which is in contrast with (23).

So far we are lacking, however, more precise definition, in terms of observables, of the "vacuum correlation length" (23). Our guess is that the correlation length (23) is manifested as the correlation length in the entanglement entropy.[f] Thus, our guess is that the confinement is a combination of two lengths

$$l_{entanglement} = \infty, \qquad l_{correlator} = \frac{1}{m_{glueball}}. \tag{25}$$

The prediction (25) can be checked on the lattice, see Refs. 19, 20. The evidence which is available now[19,20] is inconclusive, to our mind.

4. Conclusions. Towards Superfluidity

One of our central points comes as conclusions. It is quite well known that having a condensed complex 3d field might well result in superfluidity. Indeed, condensation of 3d complex field $< \Phi(\mathbf{r}) > \neq \mathbf{0}$ results in a massless real scalar field. And it is well known that the phase of the condensate wave function does represent an extra light degree of freedom in case of the superfluidity. Further details and references can be found in Ref. 10.

Thus, paradoxically enough, if we start in the non-Abelian case from $T = 0$ we have confinement but in complicated terms of string theory which are not well understood yet. If we go to the temperatures $T > T_c$ the same stringy defects which are responsible for the confinement become objects of a 3d field theory. We come to superfluidity, which is a kind of the same simple as the theory of the confinement in 4d and in the Abelian case.

Acknowledgments

I would like to acknowledge, with gratitude and always-alive memory the deep influence which Vladimir Naumovich Gribov had on our attitude to physics and on our way of thinking in physics. I am thankful to Julia Nyíri for the invitation to attend the conference Gribov-80 and write material for the proceedings.

References

1. V. N. Gribov, *Eur. Phys. J.* **C10**, 91 (1999), arXiv:hep-ph/9902279; *Orsay lectures on confinement (III)*, arXiv:hep-ph/9905285.
2. A. S. Gorsky, V. I. Zakharov and A. R. Zhinitsky, *Phys. Rev.* **D79**, 106003 (2009), arXiv:0902.1842 [hep-ph].

---

[f] Let us mention that in the literature[18] one finds argumentation which leads to conclusions opposite to what we are suggesting.

3. A. Gorsky and V. I. Zakharov, *Phys. Rev.* **D77**, 045017 (2008), arXiv:0707.1284 [hep-th].
4. J. Greensite, *Prog. Part. Nucl. Phys.* **51**, (2003) 1, arXiv:hep-lat/0301023.
5. V. I. Zakharov, *Phys. Atom. Nucl.* **68**, 573 (2005), *Yad. Fiz.* **68**, 603 (2005), arXiv:hep-ph/0410034.
6. A. M. Polyakov, *Phys. Lett.* **B59**, 82 (1975).
7. E. Witten, *Adv. Theor. Math. Phys.* **2**, 505 (1998), arXiv:hep-th/9803131; *Phys. Rev. Lett.* **81**, 2862 (1998), arXiv:hep-th/9807109.
8. T. Sakai and Sh. Sugimoto, *Prog. Theor. Phys.* **114**, 1083 (2005), arXiv: hep-th/0507073.
9. V. I. Zakharov, *in* "Sense of Beauty in Physics" Miniconference in Honor of Adriano Di Giacomo on his 70th Birthday, Pisa, Italy, 26-27 Jan 2006. arXiv:hep-ph/0602141.
10. M. N. Chernodub, H. Verschelde and V. I. Zakharov, arXiv:1007.1879 [hep-ph], arXiv:0905.2520 [hep-ph].
11. M. N. Chernodub, A. Nakamura and V. I. Zakharov, arXiv:0904.0946 [hep-ph].
12. O. Andreev and V. I. Zakharov, *Phys. Rev.* **D74**, 025023 (2006), arXiv: hep-ph/0604204.
13. E. Gava, K. S. Narain and M. H. Sarmadi, *Nucl. Phys.* **B504**, 214 (1997), arXiv:hep-th/9704006.
14. S. W. Hawking and D. N Page, *Commun. Math. Phys.* **87**, 577 (1983).
15. O. Bergman and G. Lifschytz, *JHEP* **0704**, 043 (2007), arXiv:hep-th/0612289.
16. F. V. Gubarev *et al.*, *JETP Lett.* **82**, 343 (2005), arXiv:hep-lat/0505016; V. I. Zakharov, arXiv:hep-ph/0612341; E.-M. Ilgenfritz *et al.*, arXiv:0912.2281 [hep-lat].
17. M. Engelhardt, K. Langfeld, H. Reinhardt and O. Tennert, *Phys. Rev.* **D61**, 054504 (2000), arXiv:hep-lat/9904004.
18. Sh. Ryu and T. Takayanagi, *Phys. Rev. Lett.* **96**, 181602 (2006), arXiv:hep-th/0603001; I. R. Klebanov, D. Kutasov and A. Murugan, *Nucl. Phys.* **B796**, 274 (2008), arXiv:0709.2140 [hep-th].
19. P. V. Buividovich and M. I. Polikarpov, *Nucl. Phys.* **B802**, 458 (2008), arXiv:0802.4247 [hep-lat].
20. Y. Nakagawa, A. Nakamura, S. Motoki and V. I. Zakharov, arXiv:0911.2596 [hep-lat].

SIGMA-MESON AND CONFINEMENT SINGULARITY

V.V. ANISOVICH*, V.A. NIKONOV and T.O. VULFS

Petersburg Nuclear Physics Institute, Gatchina, 188300, Russia
** anisovic@thd.pnpi.spb.ru*

We investigate a hypothesis according to which the sigma-meson is a remnant of the confinement singularity in the white channel.[1] Introducing the confinement singularity $1/s^2$ into the $\pi\pi$-interaction block, we observe a splitting of the white component of the confinement singularity ($1/s^2$) into two poles which are diving on the second sheet of the complex-s plane. The poles correspond to states which are mixtures of gluonic, flavour-singlet $q\bar{q}$, and $\pi\pi$ components. The low-lying pole is located at the complex-M region $(690 \pm 160) - i(400 \pm 150)$ MeV, the second one goes in the region of large masses: ($\mathrm{Re}\, M \gtrsim 1400$, $-\mathrm{Im}\, M \gtrsim 500$) MeV.

1. Introduction

Presently we have no reliable evidence for the existence of the σ meson – a resonance in the region 280–900 MeV. This resonance, provided it really exists, should reveal itself in the $\pi\pi$ channel as a pole in the complex-M plane, in the ($IJ^{PC} = 00^{++}$) partial wave.

Numerous allocations of such a pole gave values distributed over all the low-energy interval $\sqrt{s} \equiv M \lesssim 900$ MeV, with various widths from 200 MeV up to 1000 MeV. Such a situation emerged in the nineties[2] and it did not change till now (see recent studies: $(552^{+84}_{-106}) - i(232^{+81}_{-72})$ MeV[3] and $(484 \pm 17) - i(255 \pm 10)$ MeV[4]). The reason for uncertainties is the strong dependence of the results of fits on assumptions made in the analysis.

In the nineties, we have fitted meson spectra using standard K-matrix techniques.[5] The main advantages of such a technique are:
(i) an opportunity to fit simultaneously to several reactions, thus to increase the number of experimental points that is essential for the reconstruction of the analytical amplitude,
(ii) the K matrix correctly takes into account the threshold singularities in all investigated channels (such as $\pi\pi, K\bar{K}, \eta\eta$, etc),

(iii) also, the unitarity is accounted for correctly (for the $\hat{K}$ matrix the requirement $\hat{K} = \hat{K}^{+}\hat{K}^{+}$ is fulfilled in the physical region, at $\sqrt{s} > 2\mu_\pi$).

However, in the K-matrix amplitude the left-hand cut owing to crossing channels is ambiguously determined (remind that t and u channel meson exchanges depend on couplings and form factors, which are not well known). It is just the impossibility to write down precisely the contribution of left-hand cuts that leads to a freedom in the interpretation of the $\pi\pi \to \pi\pi$ amplitude in the region $\sqrt{s} < 900$ MeV. In our K-matrix analyses,[5] we model the contribution of the left-hand cut at $s < 0$ by several pole parameters which were objects of the fit. Describing the partial amplitude in the region $280 \le \sqrt{s} \le 1900$ MeV, usually we did not see the pole which can be interpreted as the σ-meson. However, in some solutions (not the best ones) such poles appear. Under this factual background, in Ref. 6 we fit the amplitude 00^{++} in the region $280 \le \sqrt{s} \le 900$ MeV separately within the dispersion relation approach sewing the low-energy N/D-solution with the K-matrix one at $450 \le \sqrt{s} \le 1900$ MeV.

As a result, the best fit, accounting for the left-hand cut contribution, contained the σ-meson pole at $M_\sigma = (430 \pm 150) - i(320 \pm 130)$ MeV[6] which is precisely on the border of the applicability region of the determined K-matrix amplitude.

One would think that the ambiguity problem may be solved with the help of the investigation of the $\pi\pi$ scattering in all three (u, d, s) channels. However, this requires the analytical continuation of pole terms into the regions which are that far from the pole mass and because of that one needs to know not only both resonance form factors but energy dependence of resonance widths as well. The high spin states lead to the divergence of amplitude in crossing channels. It is only the summing over all set of states that resolve these divergences resulting finally in the Regge behavior – this way requires long and not always reliable calculations.

In the K-matrix analysis not only the masses and full widths of resonances but also the couplings of resonances to different channels are calculated. The coupling is a criterium that we deal just with a particle, though unstable. Besides, the coupling interrelations allow one to define the quark content of a particle, provided this is a $q\bar{q}$ state. In this way, the states found in the K-matrix analysis can be classified as members of nonets: $(n\bar{n}\cos\phi + s\bar{s}\sin\phi, -n\bar{n}\sin\phi + s\bar{s}\cos\phi)$. It was found that $[f_0(980), f_0(1300)]_{n=1}$, $[f_0(1500), f_0(1750)]_{n=2}$, where n is the radial quantum number while the broad state $f_0(1200-1600)$ turns out to be the glueball descendant – it acquired a large width because of the accumulation of widths of neighboring states (such a phenomenon had been studied both in nuclear physics[7–9] and in meson physics[10]).

So, in the K-matrix systematics there is no room for the $q\bar{q}$ state; in addition, K-matrix analysis gives a candidate for the low-lying scalar glueball. Still, in Ref. 11 the possibility of two gluonium-rich states in the low energy region is argued.

Here we would like to turn the attention of the reader to some specific mechanism of the σ-meson formation. Within the frame of this mechanism, the σ-meson is the remnant of the confinement singularity. By its content, the σ-meson is the mixture of $\pi\pi$, gluonium and flavour-singlet $q\bar{q}$ state thus being a gluonium-rich state – it is in agreement with statement of Ref. 11.

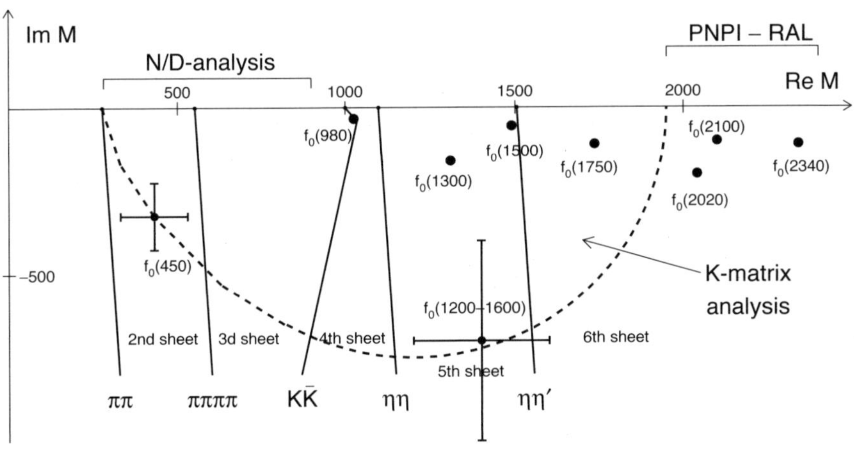

Fig. 1. Complex-M plane for the ($IJ^{PC} = 00^{++}$) mesons. The dashed line encircles the part of the plane where the K-matrix analysis[12] reconstructs the analytical K-matrix amplitude: in this area the poles corresponding to resonances $f_0(980)$, $f_0(1300)$, $f_0(1500)$, $f_0(1750)$ and the broad state $f_0(1200-1600)$ are located. Beyond this area, in the low-mass region, the pole of the light σ-meson is located (the position of pole, $M = (430 - i320)$ MeV, corresponds to the result of N/D analysis[6]). In the high-mass region one has resonances $f_0(2030)$, $f_0(2100)$, $f_0(2340)$ (see Ref. 1). Solid lines stand for the cuts related to the thresholds $\pi\pi, \pi\pi\pi\pi, K\bar{K}, \eta\eta, \eta\eta'$.

The analytical amplitude should give the σ-meson as a pole in the lower part of the complex-$M_{\pi\pi}$ plane. Such a complex plane for isoscalar/scalar states is shown in Fig. 1.

We think that one of the most reliable methods for studying the $\pi\pi$ amplitude is the analysis of the diffractive $\pi\pi$ production in high-energy

πN collisions. The dominance of the π-meson exchange at $|t| \sim 0.1$ GeV^2 allowed to extract $\pi\pi$-interaction amplitudes (see Ref. 5 and references therein), but a more precise information is obtained by the use of the t-channel reggeon exchange.[12] The K-matrix description of the diffractive $\pi N \to \pi\pi + N$ reaction both using the hypothesis about π-exchange dominance at small t and within the framework of reggeon exchange expansion give rather stable characteristics of the studied resonances in the $J^P{=}0^+, 1^-, 2^+, 3^-, 4^+$ waves; positions of the resonances in the wave $IJ^{PC}{=}00^{++}$ found in Ref. 12 are shown in Fig. 1.

2. Confinement interaction in the $q\bar{q}$ sector

Description of mesons of the $q\bar{q}$ sector provide us with information about quark confinement interactions. The interactions contain t-channel singularities of scalar and vector type. The t-channel exchange interaction can be both in white and colour states, $\mathbf{c} = \mathbf{1}{+}\mathbf{8}$ though, of course, the colour-octet interaction plays a dominant role in meson formation.

The observed linearity of the $q\bar{q}$-meson trajectories in the (n, M^2) planes,[13] where n is the radial quantum number of the $q\bar{q}$-meson with mass M, provides us with the t-channel singularity $V_{conf} \sim 1/q^4$ or, in coordinate representation, $V_{conf} \sim r$. In the coordinate representation the confinement interaction can be written in the following potential form:[1,14]

$$V_{conf} = (I \otimes I)\, b_S\, r + (\gamma_\mu \otimes \gamma_\mu)\, b_V\, r\ , \qquad (1)$$
$$b_S \simeq -b_V \simeq 0.15\ \text{GeV}^{-2}\ .$$

The first term in (1) refers to scalar interaction $(I \otimes I)$, the second one to vector $(\gamma_\mu \otimes \gamma_\mu)$ - in the $q\bar{q}$ sector the scalar and vector forces are approximately equal.

The spectral integral equation for the meson-$q\bar{q}$ vertex (or for the $q\bar{q}$ wave function of the meson) was solved by introducing a cut-off into the interaction (1): $r \to re^{-\mu r}$. The cut-off parameter is small: $\mu \sim 1-10$ MeV; if μ is changing in this interval, the $q\bar{q}$-levels with $n \leq 7$ remain practically the same.

In Refs. 14, 15, the spectral integral equations were solved using the barrier interaction. The leading term reads:

$$re^{-\mu r} \to 8\pi \left(\frac{4\mu^2}{(\mu^2 - t_\perp)^3} - \frac{1}{(\mu^2 - t_\perp)^2} \right). \qquad (2)$$

In general, having in mind that in the framework of spectral integration (as in dispersion technics) the total energy is not conserved, we have to write for the momentum transfer squared:

$$t_\perp = (k_1^{\perp p} - k_1^{'\perp p'})_\nu(-k_2^{\perp p} + k_2^{'\perp p'})_\nu\ , \tag{3}$$

where k_1 and k_2 are the momenta of the initial quarks ($p = k_1 + k_2$, $p^2 = s$, $k_a^{\perp p} = k_a - p(k_a p)/p^2$), while k_1' and k_2' are those after the interaction ($p' = k_1' + k_2'$, $p'^2 = s'$, $k_a'^{\perp p'} = k_a' - p'(k_a' p')/p'^2$ and $s \neq s'$). If the energies in the initial and final states are equal ($s = s'$), we have $t^\perp = -\vec{q}^{\,2}$, and the t-channel singular term in the right-hand side of (2) is equivalent to the interaction $r\,e^{-\mu r}$ after the Fourier transform. In the general case we can write

$$r^N e^{-\mu r} = \int \frac{d^3 q}{(2\pi)^3} e^{-i\vec{q}\vec{r}} I_N(t_\perp)\ , \tag{4}$$

$$I_N(t_\perp) = \frac{4\pi(N+1)!}{(\mu^2 - t_\perp)^{N+2}} \sum_{n=0}^{N+1} \left(\mu + \sqrt{t_\perp}\right)^{N+1-n} \left(\mu - \sqrt{t_\perp}\right)^n$$

2.1. *White remnants of the confinement singularities*

We have serious reasons to suspect that the confinement singularities (the t-channel singularities in the scalar and vector states) have a complicated structure. In the colour space they are octet states but, may be, they contain also white components. The octet exchange interaction contains quark-antiquark and gluonic blocks. Therefore the question is the following: does the $V^{(1)}_{confinement}(q^2)$ have the same singular behavior as $V^{(8)}_{confinement}(q^2)$?

The observed linearity of the (n, M^2)-trajectories, up to the large-mass region, $M \sim 2000 - 2500$ MeV,[13] favors the idea of the universality in the behavior of the potentials $V^{(1)}_{confinement}$ and $V^{(8)}_{confinement}$ at large r, or small q. To see that let us consider, as an example, the process $\gamma^* \to q\bar{q}$, Fig. 2a. We discuss the colour neutralization mechanism of outgoing quarks as a breaking of the gluonic string by newly born $q\bar{q}$-pairs. At large distances, which correspond to the formation of states with large masses, several new $q\bar{q}$-pairs should be formed. It is natural to suggest that a convolution of the quark–gluon combs governs the interaction forces of quarks at large distances, see Fig. 2b. The mechanism of the formation of new $q\bar{q}$-pairs to neutralize colour charges does not have a selected colour component. In this case all colour components $3 \otimes \bar{3} = 1 + 8$ behave similarly, that is, at small q^2 the singlet and octet components of the potential are uniformly singular, $V^{(1)}_{confinement}(q^2) \sim V^{(8)}_{confinement}(q^2) \sim 1/q^4$.

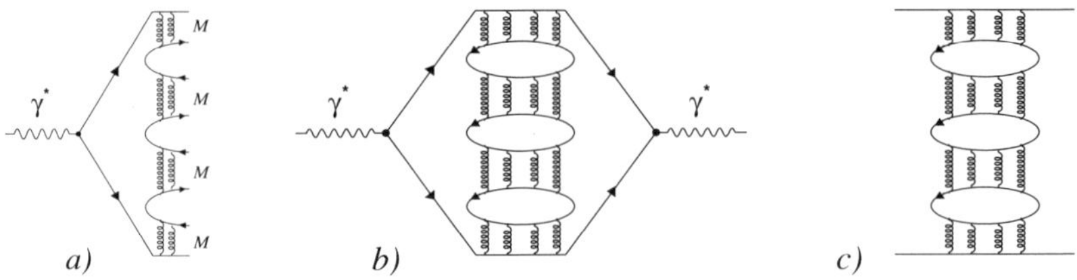

Fig. 2. a) Quark–gluonic comb produced by breaking a string by quarks flowing out in the process $e^+e^- \to \gamma^* \to q\bar{q} \to mesons$ (denoted as M); the t-channel quarks are responsible for colour neutralization of the comb.[16] b) Convolution of the quark–gluonic combs. c) Example of diagrams describing interaction forces in the $q\bar{q}$ systems.

If the confinement singularities have, indeed, white constituents, this raises immediately the following questions:
(i) How do these constituents reveal themselves in white channels?
(ii) Can they be identified?

In the scalar channel we face the problem of the σ meson ($IJ^{PC} = 00^{++}$): its existence is quite probable, although there are no reliable data for it. If the white scalar confinement singularity exists, it would be reasonable to consider it as the σ meson revealing itself: because of the transitions into the $\pi\pi$ state, the confinement singularity could move to the second sheet. If so, the σ meson can certainly do not reveal itself as a lonely amplitude singularity $1/t^2$ but a standard Breit-Wigner pole or a group of poles.

A similar scenario may be valid also for the vector confinement singularity in the $\pi\pi\pi$ ($IJ^{PC} = 01^{--}$) channel. In this case it is natural to assume that the white confinement singularity couples with the channel $\rho\pi$, splits and dives into the complex-$M_{\pi\pi\pi}$ plane.

3. Confinement singularity and dispersion relation solution

The partial pion–pion scattering amplitude being a function of the invariant energy squared, $s = M^2$, can be represented as a ratio $N(s)/D(s)$[17]), where $N(s)$ has a left-hand cut due to the "forces" (the interactions due to t- and u-channel exchanges), and the function $D(s)$ is determined by the rescatterings in the s-channel. $D(s)$ is given by the dispersion integral along the right-hand cut in the complex-s plane. The inclusion of the strong singularity into the $\pi\pi$ scattering block leads to easy modifications in the method.

3.1. *Inclusion of the confinement singularity, $1/s^2$, into the N/D-amplitude*

The $\pi\pi$ scattering block related to the s-channel confinement singularity reads:

$$G(s)\frac{1}{s^2}G(s). \tag{5}$$

The s-channel rescatterings give a set:

$$\begin{aligned} A(s) &= G(s)\frac{1}{s^2}G(s) + G(s)\frac{1}{s^2}\Pi(s)\frac{1}{s^2}G(s) + \cdots \\ &= \frac{G^2(s)}{s^2 - \Pi(s)} = G^2(s)\left[s^2 - \int\limits_{4\mu_\pi^2}^{\infty}\frac{ds'}{\pi}\frac{G^2(s')\rho(s')}{s'-s}\right]^{-1} \end{aligned} \tag{6}$$

Here $\rho(s)$ is the invariant $\pi\pi$ phase space,

$$\rho(s) = (16\pi)^{-1}\sqrt{(s-4\mu_\pi^2)/s}. \tag{7}$$

In the N/D representation the amplitude (6) reads:

$$A(s) = \frac{N(s)}{D(s)}, \quad D(s) = s^2 - \int\limits_{4\mu_\pi^2}^{\infty}\frac{ds'}{\pi}\frac{\rho(s')N(s')}{s'-s-i0}.$$

The N-function, being determined by the left-hand singularities caused by forces due to t-channel and u-channel meson exchanges, is written as an integral along the left cut as follows:

$$N(s) = \int\limits_{-\infty}^{s_L}\frac{ds'}{\pi}\frac{L(s')}{s'-s}, \tag{8}$$

where the value s_L marks the beginning of the left-hand cut. For example, for the one-meson exchange diagram $g^2/(m^2-t)$ the left-hand cut starts at $s_L = 4\mu_\pi^2 - m^2$, and the N-function in this point has a logarithmic singularity; for the two-pion exchange, $s_L = 0$.

3.2. *Approximation for the $\pi\pi$ amplitude in the region $0 \le \sqrt{s} \le 1$ GeV*

The left-hand integral for $N(s)$, Eq. (8), we replace by the following sum:

$$N(s) = \int\limits_{-\infty}^{s_L}\frac{ds'}{\pi}\frac{L(s')}{s'-s} \to 16\pi\sqrt{s}\sum_n\frac{L_n}{s_n - s}, \tag{9}$$

where L_n and s_n are "force parameters", where $-\infty < s_n < s_L$.

The pole approximation *ansatz* (9) allows us to calculate easily the rescattering block which, we believe, can be responsible for the region $0 \le \sqrt{s} \le 1$ GeV:

$$\int\limits_{4\mu_\pi^2}^{\infty} \frac{ds'}{\pi} \frac{\rho(s')N(s')}{s'-s-i0} = \sum_n \frac{L_n\sqrt{4\mu_\pi^2 - s_n}}{s-s_n} + i\sqrt{s-4\mu_\pi^2}\sum_n \frac{L_n}{s-s_n} \tag{10}$$

For the scattering amplitude we have in the physical region at $s > 4\mu_\pi^2$ the following expression:

$$\exp\left(i\delta_0^0(s)\right)\sin\delta_0^0(s) = \frac{\sqrt{s-4\mu_\pi^2}\sum\limits_n L_n(s-s_n)^{-1}}{s^2 - \sum\limits_n\left(\sqrt{4\mu_\pi^2 - s_n} + i\sqrt{s-4\mu_\pi^2}\right)L_n(s-s_n)^{-1}} \tag{11}$$

and on the first sheet at $s < 4\mu_\pi^2$:

$$\sqrt{s-4\mu_\pi^2} \to i\sqrt{4\mu_\pi^2 - s} \tag{12}$$

3.3. *Results of the fit*

Here we present formulae for the fit and the corresponding results.

Writing the unitarity condition for the scattering amplitude, we have a freedom in choosing the phase space factor. Let us give an example:

$$Im\,A(s) = \rho(s)|A(s)|^2 \to Im\,fA(s) = \frac{\rho(s)}{f}|fA(s)|^2 \tag{13}$$

$$Im\,A_f(s) = \rho_f(s)|A_f(s)|^2 \quad A_f(s) = fA(s), \quad \rho_f(s) = \frac{\rho(s)}{f}$$

Below we use

$$\rho_f(s) = \sqrt{s-4\mu^2} \equiv \rho_{thr}(s). \tag{14}$$

Such a notation is used because $\rho_{thr}(s) = \sqrt{s-4\mu^2}$ includes the threshold singularity only.

The dispersion relation gives:

$$A_{thr}(s) = \frac{g^2}{s^2 - \int\limits_{4\mu^2}^{\infty}\frac{ds'}{\pi}\frac{g^2\sqrt{s'-4\mu^2}}{s'-s-i0}}, \qquad e^{2i\delta_0^0(s)} = \frac{A_{thr}(s)}{A^*_{thr}(s)}. \tag{15}$$

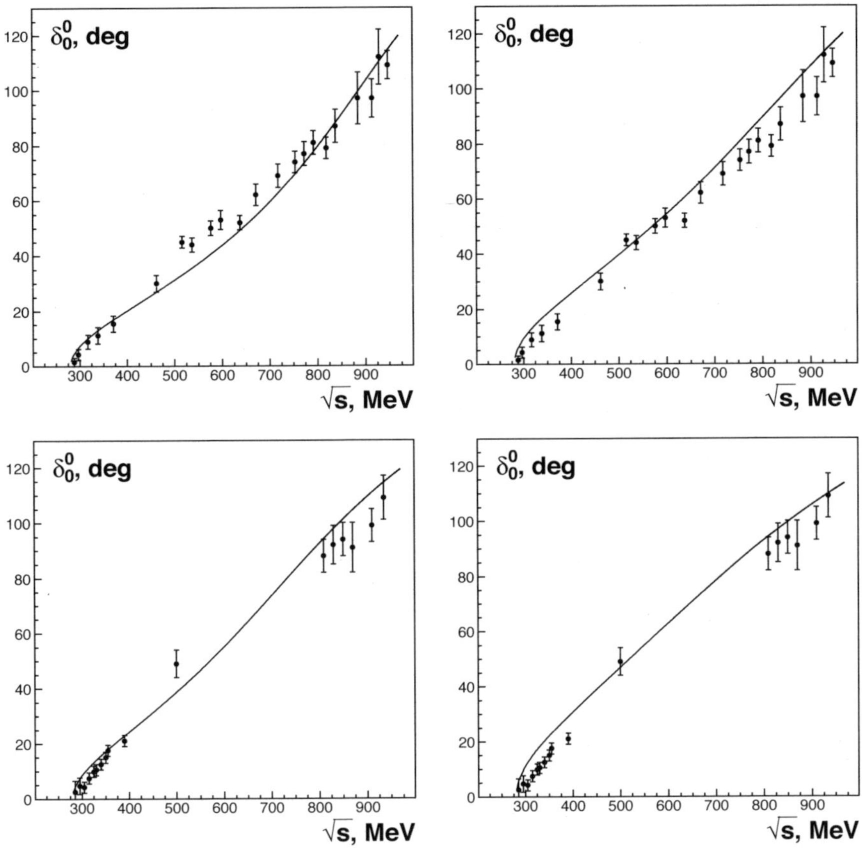

Fig. 3. Solutions for $A^{II}_{thr}(s)$ (parameters are given in (18) in μ_π units):
Solution I: a=3.1, b=1.0, c=7.7, d=9.0, a_0^0=0.30, (M_{pole-1} = 896 − i274) MeV, (M_{pole-2} = 2164 − i2511) MeV; data are from Ref. 5.
Solution II: a=2.8, b=1.5, c=7.7, d=4.5, a_0^0=0.40, (M_{pole-1} = 829 − i391) MeV, (M_{pole-2} = 2169 − i1252) MeV; data are from Ref. 5.
Solution III: a=2.4, b=1.3, c=10, d=2; a_0^0=0.380, (M_{pole-1} = 726 − i344), (M_{pole-2} = 2810 − i560) MeV; data are from Refs. 4, 18.
Solution IV: a=2.1, b=1.8, c=10, d=2, a_0^0=0.500, (M_{pole-1} = 637 − i467) MeV, (M_{pole-2} = 2810 − i560) MeV, data are from Refs. 4, 18.

One subtraction in the pion loop diagrams results in

$$A^I_{thr}(s) = \frac{g^2}{s^2 - \left(a + s \int\limits_{4\mu^2}^{\infty} \frac{ds'}{\pi} \frac{g^2\sqrt{s'-4\mu^2}}{s'(s'-s-i0)}\right)} = \frac{g^2}{s^2 - \left(a_I + ig^2\sqrt{s-4\mu^2}\right)} \tag{16}$$

The amplitude with two subtractions reads

$$A^{II}_{thr}(s) = \frac{g^2}{s^2 - \left(a_{II} + b_{II}s + ig^2\sqrt{s-4\mu^2}\right)} \tag{17}$$

Below we use the representation (17) for the estimation of the low-energy $\delta^0_0(s)$. We can write:

$$e^{2i\delta^0_0(s)} = \frac{D^{II}(s)}{D^{II*}(s)} \tag{18}$$
$$D^{II}(s) = [k-(a-ib)][k-(-a-ib)][k-(c-id)][k-(-c-id)]\,.$$

With parameters (a, b, c, b) we fit the data for $\delta^0_0(s)$ in the energy interval $280 \le \sqrt{s} \le 950$ MeV, see Fig. 3.

4. Conclusion

Introducing confinement singularity $1/s^2$ into the $\pi\pi$-interaction block, we restore the $(IJ^{PC} = 00^{++})$-amplitude at $\sqrt{s} \lesssim 900$ MeV applying the low-energy dispersion relation method. With a good database description at $280 \le \sqrt{s} \le 900$ MeV, we observe the splitting of the white component of the confinement singularity $(1/s^2)$ into two poles which are diving on the second sheet of the complex-s plane. Poles correspond to states which are mixtures of gluonic, flavour-singlet $q\bar{q}$, and $\pi\pi$ components. In the solutions shown in Fig. 3, the low-lying pole is located in the complex-M region:

$$M_\sigma = (690 \pm 160) - i(400 \pm 150)\text{MeV}, \tag{19}$$

the second one goes in the region ($\text{Re}\, M \gtrsim 1400$, $-\text{Im}\, M \gtrsim 500$) MeV.

Acknowledgement

We thank L.G. Dakhno, M.A. Matveev and A.V. Sarantsev for helpful discussions.

References

1. A.V. Anisovich, V.V. Anisovich, M.A. Matveev, V.A. Nikonov, J. Nyíri and A.V. Sarantsev, *Mesons and Baryons*, World Scientific, Singapore, 2008.
2. W.-M. Yao, *et al.*, PDG, *J. Phys.* **G33**, 1 (2006).
3. M. Ablikim *et al.* (BES Collab.) *Phys. Lett.* **B645**, 19 (2007).
4. R.Garcia-Martin, J.R. Pelaez and F.J. Yndurian, *Phys. Rev.* **D76**, 074034 (2007).

5. V.V. Anisovich and A.V. Sarantsev, *Eur. Phys. J.* **A16**, 229 (2003); V.V. Anisovich, A.A. Kondashov, Yu.D. Prokoshkin, S.A. Sadovsky and A.V. Sarantsev, *Yad. Fiz.* **60**, 1489 (2000) [*Phys. Atom. Nucl.* **60**, 1410 (2000)]; V.V. Anisovich and A.V. Sarantsev, *Phys. Lett.* **B382**, 429 (1996).
6. V.V. Anisovich and V.A. Nikonov, *Eur. Phys. J.* **A8**, 401 (2000).
7. I.S. Shapiro, *Nucl. Phys.* **A122**, 645 (1968).
8. I.Yu. Kobzarev, N.N. Nikolaev and L.B. Okun, *Sov. J. Nucl. Phys.* **10**, 499 (1970).
9. L. Stodolsky, *Phys. Rev.* **D1**, 2683 (1970).
10. V.V. Anisovich, D.V. Bugg and A.V. Sarantsev, *Phys. Rev.* **D58**, 111503 (1998).
11. S. Narison, *Nucl. Phys.* **B509**, 312 (1998).
12. V.V. Anisovich and A.V. Sarantsev, *Int. J. Mod. Phys.* **A24**, 2481 (2009); *Yad. Fiz.* **72**, 1950 (2009) [*Phys. Atom. Nucl.* **72**, 1889 (2007)]; *Yad. Fiz.* **72**, 1981 (2009) [*Phys. Atom. Nucl.* **72**, 1920 (2009)].
13. A.V. Anisovich, V.V. Anisovich and A.V. Sarantsev, *Phys. Rev.* **D62**, 051502(R) (2000).
14. V.V. Anisovich, L.G. Dakhno, M.A. Matveev, V.A. Nikonov and A.V. Sarantsev, *Yad. Fiz.* **70**, 480 (2007) [*Phys. Atom. Nucl.* **70**, 450 (2007)].
15. V.V. Anisovich, L.G. Dakhno, M.A. Matveev, V.A. Nikonov and A.V. Sarantsev, *Yad. Fiz.* **70**, 68 (2007) [*Phys. Atom. Nucl.* **70**, 63 (2007)]; *Yad. Fiz.* **70**, 392 (2007) [*Phys. Atom. Nucl.* **70**, 364 (2007)].
16. V.N. Gribov, '*The Gribov Theory of Quark Confinement*', World Scientific, Singapore (2001).
17. G.F. Chew and S. Mandelstam, *Phys. Rev.* **119**, 467 (1960); G.F. Chew, *The Analytic S-Matrix*, W.A. Benjamin, New York, 1966.
18. J.R. Pelaez and F.J. Yndurain, *Phys. Rev.* **D71**, 074016 (2005).

QCD JETS AND PARTON SHOWERS

B. R. WEBBER

University of Cambridge, Cavendish Laboratory,
J.J. Thomson Avenue, Cambridge CB3 0HE, UK
brw1@cam.ac.uk

I discuss the calculation of QCD jet rates in e^+e^- annihilation as a testing ground for parton shower simulations and jet finding algorithms.

Keywords: QCD; jets; Monte Carlo simulations.

1. Introduction

The production of jets of hadrons in all kinds of high-energy collisions is dramatic evidence of the pointlike substructure of matter. QCD predictions of the rates of production of different numbers of jets are well confirmed and provide good measurements of the fundamental coupling α_S. The latest triumph in this respect is the calculation of the 5-jet rate in e^+e^- annihilation to next-to-leading order, i.e. $\mathcal{O}(\alpha_S^4)$.[1] Figure 1 shows that calculation compared to data from the ALEPH experiment at LEP.[2] The observable shown is $L_{45} \equiv -\ln(y_{45})$, where y_{45} is the value of the jet resolution parameter at which five jets are just resolved using the k_t-jet algorithm.[3] There is good agreement over the range shown, and the uncertainty in the prediction is remarkably small considering this quantity is $\mathcal{O}(\alpha_S^3)$ at leading order. The value of the strong coupling obtained from the NLO fit to the region $L_{45} < 6$ is

$$\alpha_S(M_Z) = 0.1156^{+0.0041}_{-0.0034}\,, \tag{1}$$

which is in good agreement with the world average value obtained from other observables.

However, looking at a wider range of y_{45} values, Fig. 2, we see that the region used in the NLO fit represents only a small part of the full distribution. Most events have $L_{45} > 6$, with a distribution that turns over at $L_{45} \sim 8$, whereas the fixed-order prediction continues to rise more and

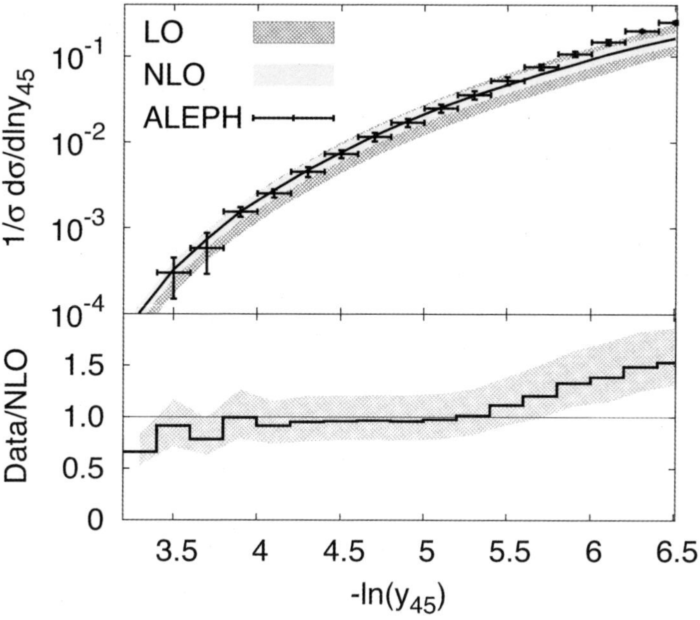

Fig. 1. ALEPH data[2] on the differential 5-jet rate, with the NLO prediction from Ref. 1.

more rapidly with increasing L_{45} (note the logarithmic vertical scale in Fig. 1).

What this means physically is that most events have a two-jet structure that can only be resolved into five jets by using a high-resolution jet algorithm. However, it is important to understand this internal structure of the jets as well as possible, for example to search for highly-boosted new particles whose decays might look like or be hiding inside QCD jets. To achieve better understanding we need progress on two fronts:

(1) Calculations of jet substructure in the region beyond the reach of fixed-order perturbation theory;
(2) Jet algorithms that probe jets in a way that reveals their substructure in informative ways.

Although the era of LEP physics is past, e^+e^- annihilation can still serve as a good testing ground for ideas on both these topics, as I hope to illustrate in the following sections.

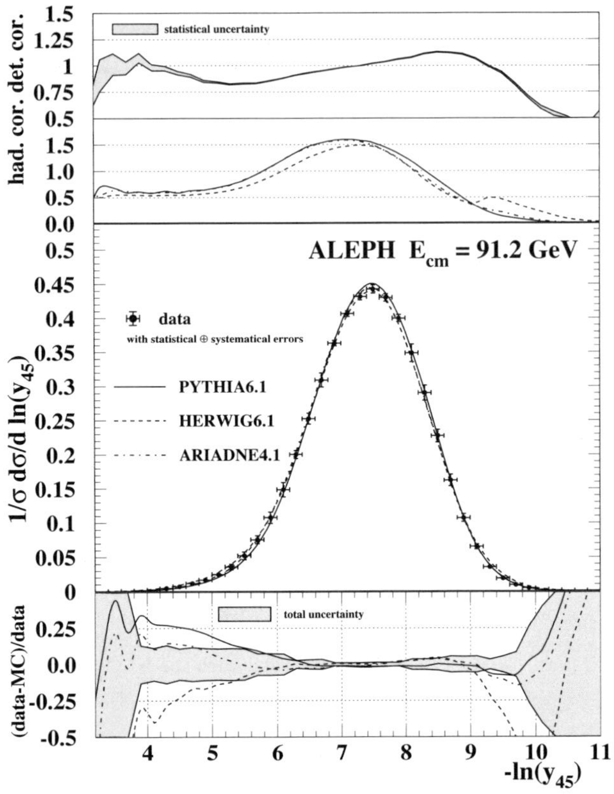

Fig. 2. ALEPH data[2] on the differential 5-jet rate, with event generator predictions.

2. Parton showers

The reason for the breakdown of fixed-order predictions at high L_{45}, where most of the data lie, is that QCD matrix elements have soft and collinear singularities that give rise to logarithmic enhancement of higher-order contributions. In fact there are up to two factors of L_{45} for every extra power of α_S, so if the coefficient were unity we would expect a breakdown at $L_{45} \sim 1/\sqrt{\alpha_S} \sim 3$. As we shall see, in fact the coefficient is more like $2/3\pi$, which does indeed imply a breakdown at $L_{45} \sim 6$. Ideally we would like to be able to sum these enhanced terms to all orders in a closed form that would exhibit the turnover in the distribution, as is the case for several other e^+e^- observables.

Table 1. Jet fractions in $e^+e^- \to$ hadrons to NLL order in $L = \ln(1/y_{\rm cut})$, expanded to third order in $a = \alpha_S/\pi$.

$$
\begin{aligned}
R_2 &= 1 + a(R_{21}L + R_{22}L^2) + a^2(R_{23}L^3 + R_{24}L^4) + a^3(R_{25}L^5 + R_{26}L^6) + \ldots \\
R_{21} &= 3C_F/2 \\
R_{22} &= -C_F/2 \\
R_{23} &= -3C_F^2/4 - 11C_FC_A/36 + C_FN_f/18 \\
R_{24} &= C_F^2/8 \\
R_{25} &= 3C_F^3/16 + 11C_F^2C_A/72 - C_F^2N_f/36 \\
R_{26} &= -C_F^3/48
\end{aligned}
$$

$$
\begin{aligned}
R_3 &= a(R_{31}L + R_{32}L^2) + a^2(R_{33}L^3 + R_{34}L^4) + a^3(R_{35}L^5 + R_{36}L^6) + \ldots \\
R_{31} &= -3C_F/2 \\
R_{32} &= C_F/2 \\
R_{33} &= 3C_F^2/2 + 7C_FC_A/12 - C_FN_f/12 \\
R_{34} &= -C_F^2/4 - C_FC_A/48 \\
R_{35} &= -9C_F^3/16 - 137C_F^2C_A/288 - 7C_A^2C_F/160 + 5C_F^2N_f/72 + C_FC_AN_f/160 \\
R_{36} &= C_F^3/16 + C_F^2C_A/96 + C_FC_A^2/960
\end{aligned}
$$

$$
\begin{aligned}
R_4 &= a^2(R_{43}L^3 + R_{44}L^4) + a^3(R_{45}L^5 + R_{46}L^6) + \ldots \\
R_{43} &= -3C_F^2/4 - 5C_FC_A/18 + C_FN_f/36 \\
R_{44} &= C_F^2/8 + C_FC_A/48 \\
R_{45} &= 9C_F^3/16 + 71C_F^2C_A/144 + 217C_FC_A^2/2880 - 41C_F^2N_f/720 - C_FC_AN_f/120 \\
R_{46} &= -C_F^3/16 - C_F^2C_A/48 - 7C_FC_A^2/2880
\end{aligned}
$$

$$
\begin{aligned}
R_5 &= a^3(R_{55}L^5 + R_{56}L^6) + \ldots \\
R_{55} &= -3C_F^3/16 - 49C_F^2C_A/288 - 91C_FC_A^2/2880 + 11C_F^2N_f/720 + C_FC_AN_f/480 \\
R_{56} &= C_F^3/48 + C_F^2C_A/96 + C_FC_A^2/720
\end{aligned}
$$

In Ref. 3 we wrote down integral equations for generating functions that can be used to compute the leading and next-to-leading logarithms (NLL) in jet cross sections to any order. Table 1 shows the results up to $\mathcal{O}(\alpha_S^3)$. These equations are for the jet fraction $R_n(y_{\rm cut})$, which is the fraction of events that have precisely n jets at resolution $y_{\rm cut}$. The differential jet rates, like the one in Figs. 1 and 2, are obtained from them by differentiating:

$$
\frac{1}{\sigma_{\rm tot}}\frac{d\sigma}{dy_{k-1,k}} = -\sum_{n=k}^{\infty} \left.\frac{dR_n}{dy_{\rm cut}}\right|_{y_{\rm cut}=y_{k-1,k}} . \tag{2}
$$

Thus to NLL accuracy, in the notation of table 1,

$$
\frac{1}{\sigma_{\rm tot}}\frac{d\sigma}{dy_{45}} = \frac{a^3}{y_{45}}(6R_{56}L_{45}^5 + 5R_{55}L_{45}^4) + \mathcal{O}(\alpha_S^4) . \tag{3}
$$

However, such fixed-order NLL predictions are not much use as they are invalid when L_{45} is not large and need to be resummed when it is large.

Indeed, since $6R_{56} = 197/270 = 0.73$ while (for $n_f = 5$ flavours) $5R_{55} = -7.77$, the prediction (3) is actually negative for $L_{45} < 10$.

The leading double-logarithmic 'abelian' terms, i.e. those proportional to $(aC_F L^2)^{n-2}$, resum to an exponential form:

$$R^{(\mathrm{ab})}_{n+2} \sim \frac{1}{n!}\left(\frac{1}{2}aC_F L^2\right)^n \exp\left(-\frac{1}{2}aC_F L^2\right) \tag{4}$$

This gives the correct qualitative features of the differential distribution (2) at large L, but the numerical values are wrong, e.g. the turn-over occurs at $L_{45} \sim 10$. This is not surprising in view of the comparable non-abelian terms and large NLL corrections.

The easiest way to resum the enhanced terms more completely is to encode them in a parton shower simulation. By this I mean a sequential $1 \to 2$ parton branching process with branching probabilities of the form

$$dP(a \to bc) = \frac{\alpha_S(q')}{\pi}\frac{dq}{q}P_{ba}(z)dz \tag{5}$$

where q is an ordered evolution variable, z measures the energy fraction in the branching, P_{ba} is the corresponding DGLAP splitting function and the argument q' of α_S is a function of q and z in general. The integral equations of Ref. 3 are equivalent to such a process with the following simple properties: the evolution variable is the angle of branching and q' is the relative transverse momentum.

The HERWIG[4] event generator results shown in Fig. 2 are based on a parton shower with precisely these properties. PYTHIA[5] also has a parton shower which, although organized in a different way, ought to be equivalent. ARIADNE[6] is based on a different approach involving colour dipoles rather than partons.[a] All the generators correctly reproduce the main features of the distribution, in particular the turn-over at $L_{45} \sim 8$.

It should be said that the event generators include a lot of additional refinements, such as matching to fixed-order matrix elements at low L_{45} and modelling of hadronization. In particular the latter has quite a strong effect at LEP energies and introduces free parameters which can be tuned to the data. Nevertheless a parton shower, or equivalent, with the correct features is an essential component for reliable extrapolation to the higher energies and different processes encountered at the LHC.

Angular ordering is not the most convenient organization of the parton shower: physical quantities such as transverse momenta and jet masses have

[a]I should emphasise that the discussion in this paper concerning alternative evolution variables and the colour structure of the shower refer only to parton showers as defined by Eq. (5) and not to dipole showers.

to be reconstructed from the shower variables. It would also be preferable to generate the hardest (highest transverse momentum) branchings first, which would make matching to fixed-order matrix elements[7] and NLO improvements[8,9] simpler. These considerations lead us to look at what happens if we order the shower in relative transverse momentum (p_t) rather than angle.

Unfortunately with simple p_t-ordering things start to go wrong even at the leading-log level as soon as gluon branching is involved. Instead of the results in table 1 for the LL coefficients in the 4-jet and 5-jet fractions,

$$\begin{aligned} R_{44} &= C_F^2/8 + C_F C_A/48 \,, \\ R_{56} &= C_F^3/48 + C_F^2 C_A/96 + C_F C_A^2/720 \,, \end{aligned} \tag{6}$$

we get[b]

$$\begin{aligned} R_{44}^{(p_t)} &= C_F^2/8 + C_F C_A/24 \,, \\ R_{56}^{(p_t)} &= C_F^3/48 + C_F^2 C_A/48 + 13 C_F C_A^2/2880 \,. \end{aligned} \tag{7}$$

We could try to fix things up by ordering in p_t and rejecting branchings that are disordered in angle. For the 4-jet rate this cures the problem with gluon branching, Fig. 3(c), but spoils the result for sequential quark branching, Fig. 3(b), while for the 5-jet fraction everything is wrong:

$$\begin{aligned} R_{44}^{(p_t,\theta)} &= 5C_F^2/48 + C_F C_A/48 \,, \\ R_{56}^{(p_t,\theta)} &= 7C_F^3/576 + 13 C_F^2 C_A/1440 + C_F C_A^2/960 \,. \end{aligned} \tag{8}$$

To see what is going wrong, consider the (z_2, θ_2) integration regions for diagrams 3(b) and (c), depicted in Fig. 4. Here z_1, z_2 and θ_1, θ_2 are the (smaller) gluon energy fractions and opening angles in successive branchings, and $\epsilon = \sqrt{y_{\rm cut}}$. Thus in diagram 3(b), p_t-ordering corresponds to $\epsilon < z_2\theta_2 < z_1\theta_1$, giving the integration region A+B. However, the correct region is the angular-ordered one A+C. If we impose angular ordering after p_t-ordering, we get only A, i.e. a deficit in the coefficient of C_F^2.

Now it happens that for this diagram the inclusion of region B compensates for the loss of C as far as the logarithms are concerned, so in this case p_t-ordering alone gives the same result as angular ordering. I will come back to this point later.

---

[b]Thanks to Mike Seymour for pointing out an error in my original calculation of the coefficient of $C_F C_A^2$.

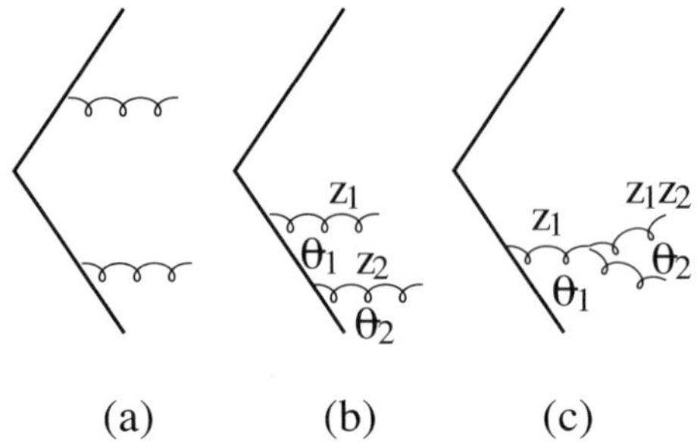

Fig. 3. Leading order diagrams for $e^+e^- \to 4$ jets.

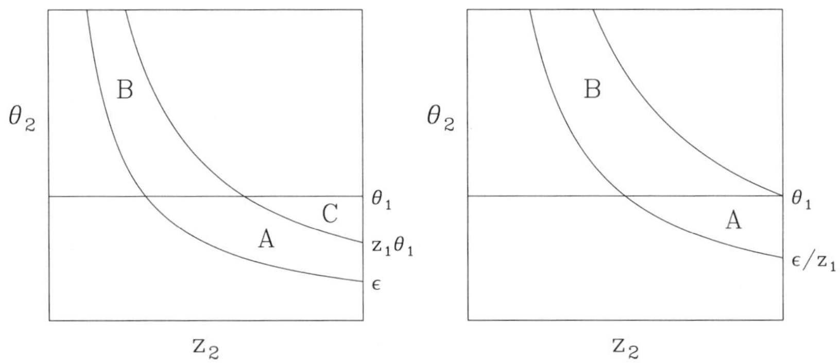

Fig. 4. Integration regions for 4-jet diagrams (b) left and (c) right.

In diagram 3(c), p_t-ordering corresponds to $\epsilon < z_1 z_2 \theta_2 < z_1\theta_1$, i.e. $\epsilon/z_1 < z_2\theta_2 < \theta_1$, as shown on the right in Fig. 4. The region C has disappeared and the p_t-ordered region A+B is just too large, giving an enhanced coefficient of $C_F C_A$. However, because region C is not there, imposing angular ordering after p_t-ordering is equivalent to simply angular ordering, giving the correct region A and hence the correct coefficient of $C_F C_A$.

So perhaps the correct prescription for a p_t-ordered shower is to angular-order only the $g \to gg$ vertices? This corrects the 4-jet rate but in the 5-jet rate the coefficient of $C_F C_A^2$ is too small:

$$
\begin{aligned}
R_{44}^{(p_t,gg)} &= C_F^2/8 + C_F C_A/48 \, , \\
R_{56}^{(p_t,gg)} &= C_F^3/48 + C_F^2 C_A/96 + C_F C_A^2/960 \, . \qquad (9)
\end{aligned}
$$

However, the reason for this is the same as before: if the gluon that branches a second time in Fig. 5(f) is the harder one coming from the first gluon branching, the situation is as on the left in Fig. 4, and we should not angular-order the second gluon branching.

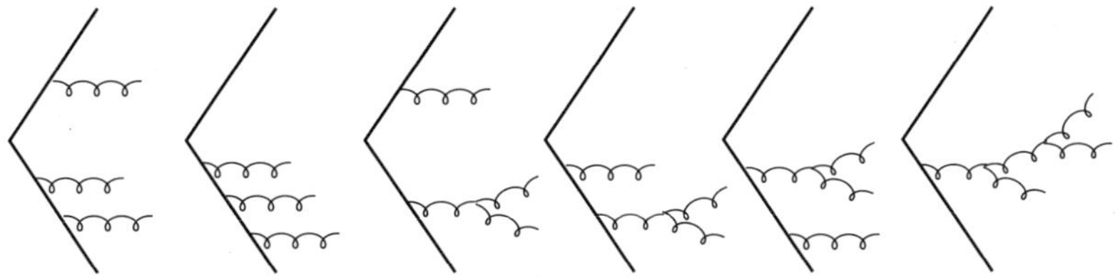

(a) (b) (c) (d) (e) (f)

Fig. 5. Leading order diagrams for $e^+e^- \to 5$ jets.

In summary, the way to get the correct LL (and NLL) jet fractions, to all orders, from a p_t-ordered parton shower is to enforce angular ordering with respect to the branching at which each parton was "created", where this means the branching at which it was the softer of the two produced.[7] More precisely, one should veto branchings that are disordered in angle with respect to their "creation". Technically, a veto means not branching but resetting the p_t scale as if the branching had occurred. This is a common kind of procedure in parton shower generators anyway, for example to correct for flavour thresholds or higher orders in the running coupling.

This looks like a better way to do parton shower event generation. With p_t-ordering one can more easily correct the prediction to NLO, or indeed to any fixed order in α_S in principle. One only has to correct the first few steps in the shower. Unfortunately there is a catch. Everything works fine at the parton level as far as the distribution in phase space is concerned, but the colour structure of the partonic final state is not correct.

Coming back to Fig. 4 (left), we see that, compared to angular ordering, p_t-ordering includes a region of softer, wide-angle gluon emission, B, in place of a region of harder, more collinear emission, C. What this means is that gluon radiation is moved around within the shower, the amount and distribution remaining the same. This is depicted schematically in Fig. 6, where for simplicity we show the large-N_c approximation, as used for hadronization in event generators. In Fig. 6(a), angular ordering assigns a soft, wide-angle gluon, actually emitted coherently by partons b and c, to the parent parton a, which is reasonable because a does have the coherent sum of the colour charges of b and c. In contrast, p_t-ordering assigns this gluon to the harder of b and c, in this case c, as in Fig. 6(b). That is reasonable as far as the momenta are concerned, but it spoils the colour structure by treating c as the colour source and neglecting the coherent contribution of b.

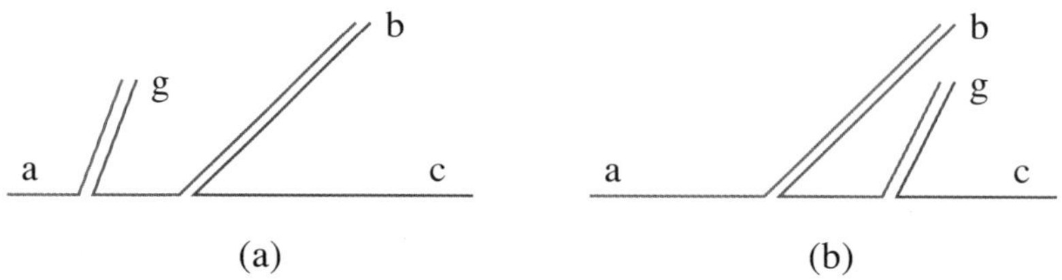

Fig. 6. Large-N_c colour structure of wide-angle gluon emission associated with the parton branching $a \to bc$: (a) angular-ordered shower; (b) p_t-ordered shower.

The colour structure matters when one wants to interface the parton shower to a non-perturbative hadronization model. In the cluster model used by HERWIG, colour-singlet clusters are formed by splitting gluons at the end of the shower into $q\bar{q}$ pairs. Thus in the angular-ordered Fig. 6(a) the clusters connect (gb) and (bc), while in p_t-ordered Fig. 6(b) they connect (bg) and (gc). Similarly in the PYTHIA string hadronization model, the string connects $a - g - b - c$ in Fig. 6(a) but $a - b - g - c$ in Fig. 6(b).

In conclusion, an angular-ordered parton shower sums the LL and NLL enhanced terms and provides partonic final states with colour structure consistent with QCD coherence. This is good for hadronization models but not so convenient for reconstruction of kinematics or for systematic improvement away from the soft and collinear regions. A p_t-ordered shower is better in those respects and, with the right angular veto procedure, can give the correct NLL jet fractions. However the colour structure then needs to be reconfigured according to angular ordering before the partonic final state can be hadronized.

3. Jet algorithms

Recall that the k_t-algorithm for e^+e^- annihilation[3] is defined in terms of the resolution variable

$$y_{ij} = 2\min\{E_i^2, E_j^2\}(1 - \cos\theta_{ij})/Q^2 \,, \tag{10}$$

where $E_{i,j}$ are the energies of final-state objects i and j, θ_{ij} is the angle between their momenta and Q is the centre-of-mass energy. The two objects with the smallest value of y_{ij} are combined into one, this is repeated until all $y_{ij} > y_{\rm cut}$, and the remaining objects are called jets. For the purpose of counting large logarithms of $y_{\rm cut}$, we can write this in the small-angle approximation

$$\epsilon_{ij} = \min\{E_i, E_j\}\theta_{ij}/Q > \epsilon \; , \tag{11}$$

where as before $\epsilon = \sqrt{y_{\rm cut}}$.

As pointed out in Ref. 10, this is just one of a continuum of possible jet algorithms with resolution variable

$$\epsilon_{ij} = \min\{E_i^p, E_j^p\}\theta_{ij}/Q^p \; , \tag{12}$$

where p can be any positive or negative number. In particular $p = -1$ defines the resolution for the e^+e^- analogue of the anti-k_t algorithm,[10] which has the advantage that objects are combined starting with those that have the highest energy rather than the lowest.

When $p < 0$ a supplementary condition is needed, otherwise infinitely soft emissions would be resolved. For anti-k_t we define[c]

$$\begin{aligned} \epsilon_{ij} &= \min\{Q/E_i, Q/E_j\}\theta_{ij} \; , \\ \epsilon_i &= \epsilon Q/E_i \; . \end{aligned} \tag{13}$$

Then if the smallest of the set of $\{\epsilon_{ij}, \epsilon_i\}$ is an ϵ_i, we remove i from the list of objects to be recombined, and if $\epsilon_i < 1$ we call it a jet. Otherwise we just throw it away. Thus every jet has an energy greater than ϵQ and is separated from other jets by an angle greater than ϵ. The resulting LL coefficients in the 4- and 5-jet fractions are

$$\begin{aligned} R_{44}^{\rm anti} &= C_F^2/2 + C_F C_A/8 \; , \\ R_{56}^{\rm anti} &= C_F^3/6 + C_F^2 C_A/8 + C_F C_A^2/48 \; , \end{aligned} \tag{14}$$

where as before the large logarithm is defined as $L = -2\ln\epsilon$. We could introduce an angular resolution δ different from the energy resolution ϵ by multiplying ϵ_{ij} by ϵ/δ. This would just replace $\ln^2\epsilon$ by $\ln\epsilon\ln\delta$.

It is easy to see that leading double-logarithmic abelian terms in the anti-k_t jet rates resum to an exponential form with twice the exponent of the k_t rates (4):

$$R_{n+2}^{\rm (anti,ab)} \sim \frac{1}{n!}\left(aC_F L^2\right)^n \exp\left(-aC_F L^2\right) \tag{15}$$

---

[c]Essentially the same e^+e^- algorithm is described in the FastJet user manual, available at http://www.lpthe.jussieu.fr/∼salam/fastjet/. I thank Gavin Salam for bringing this to my attention.

It should also be possible to resum the non-abelian and NLL terms using techniques like those of Ref. 3.

4. Conclusions

Although the era of high-energy e^+e^- collider experiments is past, at least for a while, it is helpful to study how our tools for analysing hadronic final states perform in the cleaner environment of the annihilation process.

The k_t-jet algorithm has proven useful in all kinds of processes and the e^+e^- jet rates defined in this way are a good place to test alternative resummation methods, particular those involving parton showers ordered in different ways. We have seen that angular-ordered and p_t-ordered showers can both be arranged to resum the leading and next-to-leading logarithms of the k_t-jet resolution $y_{\rm cut}$. The p_t-ordering option is good for matching to fixed-order calculations but causes some difficulties in matching to hadronization models at low scales, owing to its disordered colour structure.

The rather different anti-k_t algorithm has been adopted as the preferred tool for jet finding at the LHC. The analogous e^+e^- algorithm has a simple pattern of leading logarithms, which should be amenable to resummation using techniques similar to those applied to the k_t algorithm.

Acknowledgments

It is a pleasure to recall and acknowledge conversations with Volodya Gribov in many places during the all-too-brief times we spent together. I am also indebted to Stefano Catani, Gavin Salam and Mike Seymour for helpful comments and discussions.

References

1. R. Frederix, S. Frixione, K. Melnikov, G. Zanderighi, arXiv:1008.5313 (2010).
2. A. Heister, *et al.*, *Eur. Phys. J.* **C35**, 457 (2004).
3. S. Catani, Yu.L. Dokshitzer, M. Olsson, G. Turnock and B.R. Webber, *Phys. Letter* **B269**, 432 (1991).
4. G. Corcella, *et al.*, *JHEP* **01**, 010 (2010).
5. T. Sjostrand, S. Mrenna and P.Z. Skands, *JHEP* **05**, 026 (2006).
6. L. Lonnblad, *Comput. Phys. Commun.* **71**, 15 (1992).
7. S. Catani, F. Krauss, R. Kuhn and B.R. Webber, *JHEP* **11**, 063 (2001).
8. S. Frixione and B.R. Webber, *JHEP* **06**, 029 (2002).
9. S. Frixione, P. Nason and C. Oleari, *JHEP* **11**, 070 (2007).
10. M. Cacciari, G.P. Salam and G. Soyez, *JHEP* **04**, 063 (2008).

GLUON EVOLUTION AND SATURATION PROCEEDINGS*

L. D. MCLERRAN

Brookhaven National Laboratory and Riken Brookhaven Center, Physics Dept., Upton, NY 11973, USA

Almost 40 years ago, Gribov and colleagues at the Leningrad Nuclear Physics Institute developed the ideas that led to the Dokhsitzer-Gribov-Lipatov-Altarelli-Parisi and the Balitsky-Fadin-Kuraev-Lipatov equations. These equations describe the evolution of the distributions for quarks and gluon inside a hadron to increased resolution scale of a probe or to smaller values of the fractional momentum of a hadronic constituent. I motivate and discuss the generalization required of these equations needed for high energy processes when the density of constituents is large. This leads to a theory of saturation realized by the Color Glass Condensate.

1. Introduction

About 40 years ago, Gribov and colleagues at the Leningrad Nuclear Physics Institute developed ideas that led to equations that describe the change in quark and gluon distributions in hadrons as a function of both the resolution size of an electromagnetic probe[1–6] and as a function of the fractional momentum of the parton in a hadron.[7,8] These developments led to a revolutionary change in our understanding of strong interaction physics and provided a foundation within QCD for ideas originally developed by Bjorken.[9]

These ideas are manifest in the space-time diagram for hadronic processes developed by Gribov and by Bjorken, as shown in Fig. 1. The red line on this diagram shows a hadronic constituent in the initial state successively radiating and then absorbing quanta in the final state. The position of emission in the figure corresponds to the distance either before or after the collision when the quanta is radiated. The time is of order the distance away from the collision where the radiation takes place, and we are looking

*Presented at the **Gribov Memorial Workshop on Quantum Chromodynamics and Beyond**.

in the laboratory frame where the initial hadron has high energy. Where the initial state and final state hadron meet, the collision takes place. This picture formed the framework of our modern understanding of pp, pA, and AA collisions.[10]

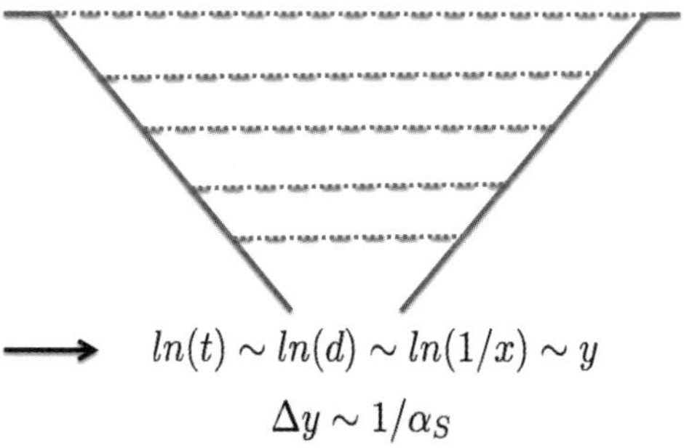

Fig. 1. The space-time diagram that describes the evolution of a hadronic wavefunction.

A difficulty with the evolution equations is that they predict a rapid growth as an inverse power of the fractional momentum of the gluon distribution function. This growth if uncontrolled can lead to violations of unitarity in high energy collisions. This was recognized by Leonid Gribov (Volodya Gribov's son,) Levin and Ryskin,[11] and by Mueller and Qiu[12] who argued that the growth should slow at fixed probe resolution scale. Together with Raju Venugopalan, we argued that to understand such saturation of the gluon density, one needed to replace the idea of a gluon density as an incoherent distribution by the idea of a classical gluon field.[13,14] This idea leads to renormalization group equations that are generalizations of those envisioned by Gribov. These equations can include the effect of the high density of gluons, and do not violate s-channel unitarity.[15–20] The region where the gluon density is large is where the classical field nature of the gluon density is important. The high gluon density provides a natural infrared cutoff in the theory, and makes possible a number of computations of processes that were infrared divergent in the ordinary parton model.

A momentum scale, the saturation momentum, characterizes this matter. The saturation momentum grows at high energies, and the strong coupling constant ultimately becomes small when evaluated at the saturation momentum scale. A novel space-time picture of the early stages of hadron-hadron collisions has emerged.[21–26] The high density matter inside a hadronic wave function is called the Color Glass Condensate, and the high density highly coherent matter produced in hadronic collisions is called the Glasma.

It is the purpose of this lecture to motivate and provide an intuitive motivation for the Color Glass Condensate and the Glasma.

2. Some Qualitative Features of Parton Evolution

We begin by introducing the parton transverse momentum and rapidity distribution df/dyd^2p_T, where $y = ln(1/x)$ and x is the parton momentum as a fraction of the hadron momentum in a frame where a hadron has a very high energy. The parton distribution function is

$$xG(x, Q^2) = \int^{Q^2} d^2p_T \frac{df}{dyd^2p_T} \tag{1}$$

There is need for an upper limit on this integration since the integral would weakly diverge if the limit was taken to infinity. The value of Q^2 can be thought of as a resolution scale and we are counting all partons whose size $r \sim 1/p_T$ is $r < 1/Q$.

The DGLAP evolution equation describes the change in the parton distribution as one changes the resolution scale. Since the integral for $G(x, Q^2)$ is mildly divergent, we see that the hadron has more and more smaller constituents. This means that the density of partons in a hadron $\rho = G/\pi R^2$ times the typical area of a constituent, $1/Q^2$ shrinks to zero $\rho/Q^2 \rightarrow 0$ as $Q^2 \rightarrow \infty$. The evolution in Q^2 takes one into a short distance dilute limit.

The evolution of the gluon density in $y = ln(1/x)$ at fixed Q^2 is given by the BFKL equation. The gluon density grows like $1/x^\delta$ where $\delta \sim 0.2 - 0.3$ at accessible energies. Evolution at fixed Q^2 corresponds to evolution at fixed parton size. Evolution in y takes one to the high parton density limit. For fixed Q^2 we shall soon see that the BFKL evolution equation breaks down, and the rapid growth is tempered. A figure illustrating the behaviour of the gluon distribution function in the $ln(1/x) - ln(Q^2)$ plane is shown in Fig. 2. The red line in the diagram corresponds to the point when evolving in $ln(x)$ at fixed values of $ln(Q^2)$ where the gluon density stops growing rapidly. This line is called the saturation boundary.

The growth of the gluon density is seen experimentally. In Fig. 3, the gluon and quark densities are shown. One sees that gluons dominate the density of particles inside a hadron for $x \leq 10^{-2}$. The density also is rapidly growing with decreasing x. Surely when the density of gluons becomes very large, one can use weak coupling methods to describe the gluons. This is because of asymptotic freedom and that the typical separation between

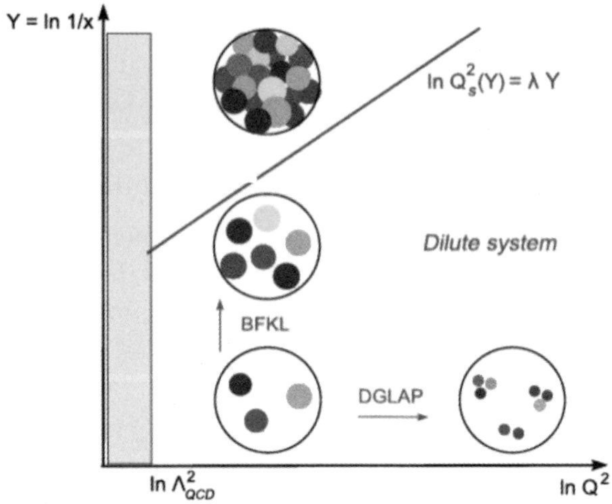

Fig. 2. The gluon density in the $ln(x) - ln(Q^2)$ plane.

gluons is very small. This does not mean the system is perturbative. The high density of gluons can act coherently and generate large interactions. A simple example of coherence turning intrinsically weak interactions into strong forces is gravity. This is because classical fields can add together with the same sign, and because the interaction is long range so that interactions are enhanced due to coherent forces of many nucleons.

3. Saturation of the Gluon Density

To understand how the gluon density might saturate, imagine that the gluons are hard spheres with a size of order $r \sim 1/p_T$. A hadron has a very slowly growing size as energy increases so we will treat the hadron size as fixed. On the other hand, as we go to higher energy, we can probe smaller values of $x_{min} \sim \Lambda_{QCD}/E$. If we start with a low density of gluons of size r_0 at some energy, the hadronic disk begins to become closely packed with gluons of this size as the energy increases. This continues until the gluons are so closely packed that they repel, and begin to act as hard spheres. This is the density $\frac{d\rho}{dyd^2p_Td^2r_T} \sim 1/\alpha_S$. At this density scale their intrinsic weak interaction strength $\sim \alpha_S$ is compensated by their high density $\sim 1/\alpha_S$.

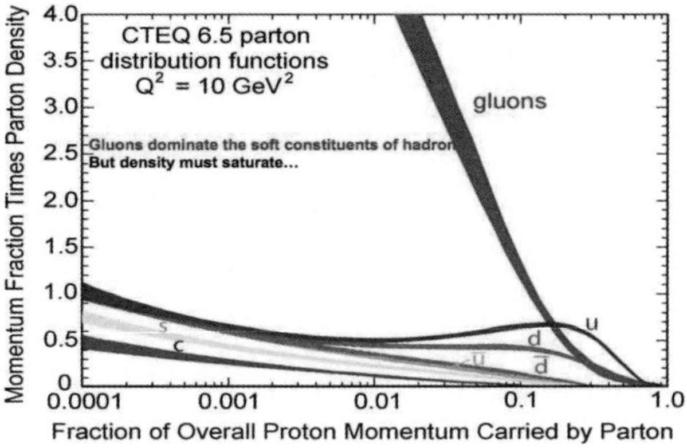

Fig. 3. The distributions of quarks and gluons as a function of x.

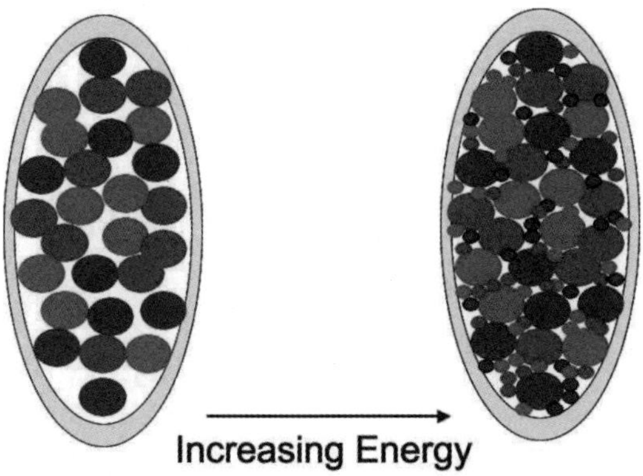

Fig. 4. Saturation of the gluon density.

What happens as we go to yet higher energy? We can still pack in more gluons but they have to have smaller size, $r << r_0$ so that they can fit in the cracks between the gluons of size r_0. This means that the gluon density can grow forever, so long as it is associated with smaller and smaller gluons. There is a characteristic momentum scale which at any energy scale separates the highly coherent gluons from those that are not so coherent. This is the saturation scale Q_{sat}. Our considerations argue that the saturation momentum can grow forever. When it is $Q_{sat} >> \Lambda_{QCD}$ we can use weak coupling methods.

An essential ingredient in this description is Gribov's space-time picture shown in Fig. 5. The smallest x gluon emitted is the lowest on on this figure. Because the phase space density of gluons is so large, we are justified in thinking about this gluon as a classical field. It has however been produced by gluons at much larger $ln(x'/x) \sim 1/\alpha_S$. This gluon has its evolution Lorentz time dilated. This means that the produced gluon field is static. It also has a deeper consequence: The different configurations that yield the gluon field will not quantum mechanically interfere. They are a glass, similar to spin glasses of condensed matter physics

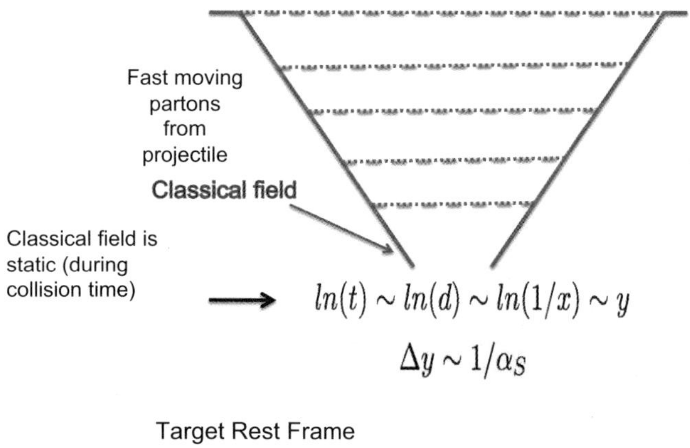

Fig. 5. The space time diagram showing how the glassy nature of the Color Glass Condensate arises.

We therefore call this high density ensemble of gluons the Color Glass Condensate. Color is because it is made of colored gluons. Glass because the field describing it are static and because of its relationship to spin glasses. Condensate because the phase space density of gluons is very high, the gluons are highly coherent, and this high density arises spontaneously.

4. The Renormalization Group Formalism for the CGC

The Color Glass Condensate is described by both gluon fields and sources for them. It has a path integral representation:

$$Z = \int_\Lambda [dA][d\rho]\ exp\{iS[A,\rho] - F[\rho]\} \tag{2}$$

The Yang-Mills action is in the presence of a source,

$$J^\mu = \delta^{\mu +} \rho(x_T, y) \tag{3}$$

The space-time rapidity is $y = ln(x_0^-/x^-)$, where $x^- = (t-z)/\sqrt{2}$. By the uncertainty principle $y \sim ln(1/x)$, where x is the fractional momentum of a constituent. The scale Λ is a separation scale that separates the gluon degrees of freedom between dynamical fields and sources. Its presence leads to renormalization group equations that determine the distribution of gluonic sources, $F[\rho]$.

To leading order at weak coupling, the gluon field is determined by the classical field equations. Once this is done, one then integrates over the incoherent distribution of the sources. Of course there are fluctuations in the gluon field, and if we were to try to compute quantities with longitudinal momentum much less that the cutoff Λ, we would generate large terms proportional to $\alpha_S \, ln(\Lambda/p^+)$. To deal with such quantities, we need to shift the scale Λ to be closer to p^+. This is done by renormalization group methods and leads to the JIMWLK evolution equations.[15–20] When computed on correlation functions, these equations are equivalent to the Balitsky-Kovchegov equations.[27,28] The derivation of the JIMWLK equation requires an analytic computation to all orders in the background classical gluon field for an arbitrary light cone source.

We can understand generic features of the JIMWLK equations. They are for the source functional

$$Z_0 = e^{-F[\rho]} \tag{4}$$

They are of the form of a a Euclidean Hamiltonian evolution equation

$$\frac{d}{dy} Z_0 = -H[d/d\rho, \rho] \; Z_0 \tag{5}$$

For strong and intermediate strength fields, corresponding to the saturation limit, H is second order in $d/d\rho$. The Hamiltonian H has no potential, only a non-linear kinetic energy term. It therefore describes non-linear quantum diffusion. Recall that the ordinary linear diffusion equation is

$$\frac{d}{dt}\psi = -\frac{p^2}{2}\psi \tag{6}$$

It has a solution

$$\psi \sim e^{-x^2/2t} \tag{7}$$

The wavefunction spreads as time goes to infinity, and the exponential behaviour is universal. It is therefore natural to expect that the solution to the JIMWLK equations will allow the saturation momentum to grow forever as $y \to \infty$, and that its solution will be universal. This means that the Color Glass Condensate is universal and is such is fundamental.

5. The Nature of the Color Glass Condensate Fields

At high hadron p^+, x^- is a small co-ordinate and x^+ is big. The components of the gluonic classical fields have the properties that F^{i+} is big, F^{i-} is small, and F^{ij} is of order one. Concentrating on the large components of fields, we conclude therefore that $E \perp B \perp \hat{z}$. The fields are therefore simply Lorentz boosted Coulomb fields. Their distribution in color, polarization and on the two dimensional sheet corresponding to the hadron are what is determined by the theory of the Color Glass Condensate. Note that the density of gluons per unit area up to the saturation momentum is given by dimensional grounds as

$$\frac{1}{\pi R^2}\frac{dN}{dy} \sim \frac{1}{\alpha_S} Q^2_{sat} \tag{8}$$

It is useful to determine the light cone gauge vector potential. We require a vector potential that gives F^{ij} and F^{i-} zero everywhere and F^{i+} a delta function of x^- and an arbitrary function of x_T, independent of x^+. A gauge field that is a two dimensional (in transverse coordinates) gauge transform of vacuum will give zero $F^{\mu\nu}$. If we make the vector potential different gauge transforms of vacuum with a discontinuity at $x^- = 0$, this will give the desired form for $F^{\mu\nu}$. We have therefore

$$A^j = \theta(-x^-)\frac{1}{i}U_1\nabla^j_T U_1^\dagger + \theta(x^-)\frac{1}{i}U_1\nabla^j_T U_1^\dagger \tag{9}$$

(In practice, it is sometimes necessary to spread the source out a bit in x^-, in a manner prescribed by the renormalization group equations.) Notice that although the field strength $F^{\mu\nu}$ is confined to the sheet at $x^- = 0$, the Wigner distribution function corresponding to the gluon distribution function

$$W^{ij}(X,p) = \int dx^- d^2x_T \; e^{ip\cdot x} < A^i(X - x/2)A^j(X + x/2) > \tag{10}$$

is spread out with a distance scale of order $1/p^+$ for components with momentum p^+. The Wigner distribution is however not positive definite, and should not be expected to be so because we are measuring spatial distributions on the size scale of quantum fluctuations.

6. The Glasma

The collision of two sheets of Colored Glass produce color electric and magnetic fields with very different properties than those in the initial sheets. These fields are produced in the time it takes the sheets to pass through one another, which is a very short time $t \sim e^{-\kappa/\alpha_S}/Q_{sat}$, compared to the natural time scale for the classical fields produced after the collision $t \sim 1/Q_{sat}$.

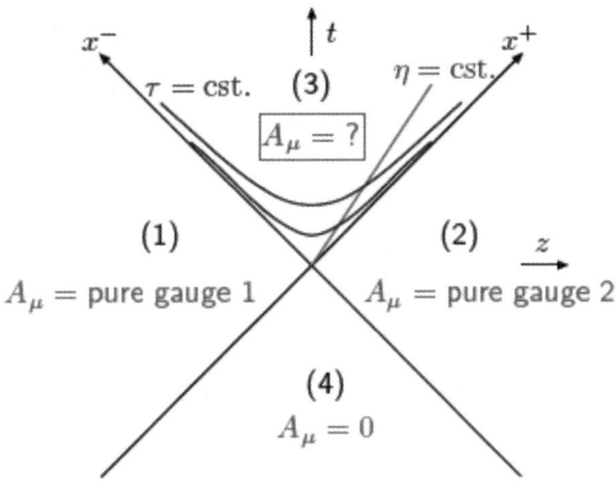

Fig. 6. The space time diagram showing the fields before and after the collision of two sheets of Colored Glass.

In Fig. 6, the fields are shown on a light cone diagram. In the backward light cone there is a gauge we can choose where the fields are zero. On the side lightcones, we choose the fields A_1 and A_2 two be two different pure gauge transforms of vacuum. These choices give the proper field charge density on the sheets at $x^\pm = 0$ for $t < 0$. In the forward light cone, we could try to make the charge density on the sheets be the correct value by choosing the field in the forward light cone to be $A_1 + A_2$. This is in fact correct very close to the forward lightcone as can be seen by inspecting the Yang-Mills field equations. The field $A_1 + A_2$ is however not a gauge transform of vacuum, so the fields will evolve in the forward light cone. One has produced a distribution of colored fields.

The terms in the Yang Mills equations $\vec{A} \cdot \vec{E}$ and $\vec{A} \cdot \vec{B}$ produce sources of colored electric and magnetic charge. There are delta function contributions arising from terms like $A_i \cdot E_j$ and $A_i \cdot B_j$ for $i \neq j$. This means that in the collision the sheets are dusted with equal and opposite charge densities of colored electric and colored magnetic charges. Immediately after the collision, lines of longitudinal color electric and magnetic flux are produced. This is shown in Fig. 7.

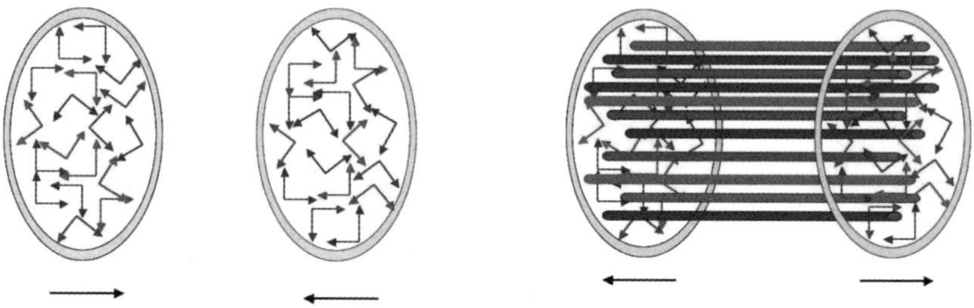

Fig. 7. The formation of longitudinal color electric and color magnetic flux lines during the collision of two sheets of Colored Glass.

The transverse density of these flux tubes is $1/Q_{sat}^2$, and the typical strength of the fields is $A \sim Q_{sat}/g_S$, so they are highly coherent. there is both color electric and color magnetic field, so there is a large topological charge density $\vec{E} \cdot \vec{B}$. The fields evolve and decay according to the classical Yang-Mills equations. Unlike the case where there is pair production needed for the decay of an electric flux tube, because there is both a color electric and a color magnetic field, the Yang-Mills equation allow for the classical decay.

7. Phenomenology of the Color Glass Condensate and the Glasma

There are now a wide variety of phenomenon described by the Color Glass Condensate and the Glasma. The Color Glass Condensate provides a good description of deep inelastic scattering and diffraction in electron-proton collisions at HERA for small values of x. It has provided a good phenomenological description of heavy ion collisions and dAu collisions at the RHIC accelerator. An excellent review of the situation is provided at www.bnl.gov/riken/glasma (to be published in a special edition of Nuclear Physics A). An even better summary of the current situation is provided in the presentation of J. P. Blaizot. Of course at the LHC, the ideas of gluon saturation can and will be tested in pp, pA and AA collisions. At the LHC, x values are very small and the saturation momentum large enough so that computations should become precise. There is also the possibility to test these ideas in the collisions of electrons from nuclei at an electron-heavy ion collider such as eRHIC.

Acknowledgments

It is an honor to speak at a memorial meeting for a truly gifted, creative and absolutely honest physicist: Volodya Gribov. His work has influenced me in many ways including some that I have only come to understand with time.

I thank Luciano Bertocchi, Yuri Dokshitzer, Péter Lévai, Julia Nyíri, and Daniele Treleani for organizing this meeting in honor of Volodya Gribov and the sponsors, ICTP (Trieste), RMKI (Budapest) and INFN. The research of L. McLerran is supported under DOE Contract No. DE-AC02-98CH10886.

References

1. V. N. Gribov and L. N. Lipatov, *Sov. J. Nucl. Phys.* **15**, 438 (1972) [*Yad. Fiz.* **15**, 781 (1972)].
2. V. N. Gribov and L. N. Lipatov, *Sov. J. Nucl. Phys.* **15**, 675 (1972) [*Yad. Fiz.* **15**, 1218 (1972)].
3. Y. L. Dokshitzer, *Sov. Phys. JETP* **46**, 641 (1977) [*Zh. Eksp. Teor. Fiz.* **73**, 1216 (1977)].
4. Y. L. Dokshitzer, D. Diakonov and S. I. Troian, *Phys. Rept.* **58**, 269 (1980).
5. L. N. Lipatov, *Sov. J. Nucl. Phys.* **20**, 94 (1975) [*Yad. Fiz.* **20**, 181 (1974)].
6. G. Altarelli and G. Parisi, *Nucl. Phys.* **B126**, 298 (1977).
7. E. A. Kuraev, L. N. Lipatov and V. S. Fadin, *Sov. Phys. JETP* **45**, 199 (1977) [*Zh. Eksp. Teor. Fiz.* **72**, 377 (1977)].
8. I. I. Balitsky and L. N. Lipatov, *Sov. J. Nucl. Phys.* **28**, 822 (1978) [*Yad. Fiz.* **28**, 1597 (1978)].
9. J. D. Bjorken and E. A. Paschos, *Phys. Rev.* **185**, 1975 (1969)
10. J. D. Bjorken, *Lect. Notes Phys.* **56**, 93 (1976).
11. L. V. Gribov, E. M. Levin and M. G. Ryskin, *Phys. Rept.* **100**, 1 (1983).
12. A. H. Mueller and J. W. Qiu, *Nucl. Phys.* **B268**, 427 (1986).
13. L. D. McLerran and R. Venugopalan, *Phys. Rev.* **D49**, 2233 (1994), arXiv:hep-ph/9309289.
14. L. D. McLerran and R. Venugopalan, *Phys. Rev.* **D49**, 3352 (1994), arXiv:hep-ph/9311205.
15. J. Jalilian-Marian, A. Kovner, L. D. McLerran and H. Weigert, *Phys. Rev.* **D55**, 5414 (1997), arXiv:hep-ph/9606337.
16. J. Jalilian-Marian, A. Kovner, A. Leonidov and H. Weigert, *Nucl. Phys.* **B504**, 415 (1997), arXiv:hep-ph/9701284.
17. J. Jalilian-Marian, A. Kovner, A. Leonidov and H. Weigert, *Phys. Rev.* **D59**, 014014 (1999), arXiv:hep-ph/9706377.

18. E. Iancu, A. Leonidov and L. D. McLerran, *Nucl. Phys.* **A692**, 583 (2001), arXiv:hep-ph/0011241.
19. E. Iancu, A. Leonidov and L. D. McLerran, *Phys. Lett.* **B510**, 133 (2001), arXiv:hep-ph/0102009.
20. E. Ferreiro, E. Iancu, A. Leonidov and L. McLerran, *Nucl. Phys.* **A703**, 489 (2002), arXiv:hep-ph/0109115.
21. A. Kovner, L. D. McLerran and H. Weigert, *Phys. Rev.* **D52**, 6231 (1995), arXiv:hep-ph/9502289.
22. A. Kovner, L. D. McLerran and H. Weigert, *Phys. Rev.* **D52**, 3809 (1995), arXiv:hep-ph/9505320.
23. A. Krasnitz and R. Venugopalan, *Nucl. Phys.* **B557**, 237 (1999), arXiv:hep-ph/9809433.
24. A. Krasnitz and R. Venugopalan, *Phys. Rev. Lett.* **84**, 4309 (2000), arXiv:hep-ph/9909203.
25. T. Lappi, *Phys. Rev.* **C67**, 054903 (2003), arXiv:hep-ph/0303076.
26. T. Lappi and L. McLerran, *Nucl. Phys.* **a772**, 200 (2006), arXiv:hep-ph/0602189.
27. I. Balitsky, *Nucl. Phys.* **B463**, 99 (1996), arXiv:hep-ph/9509348.
28. Y. V. Kovchegov, *Phys. Rev.* **D60**, 034008 (1999), arXiv:hep-ph/9901281.

Minneapolis, 1995

NON-PERTURBATIVE PARTICLE PRODUCTION IN TIME DEPENDENT STRONG NON-ABELIAN FIELDS

P. LÉVAI[a] and V.V. SKOKOV[b]

[a] *KFKI RMKI Research Institute for Particle and Nuclear Physics, 29-33 Konkoly Thege Str., H-1121 Budapest, Hungary*
[b] *GSI Gesellschaft für Schwerionenforschung mbH, Planckstr. 1, D-64291 Darmstadt, Germany*

Non-perturbative production of quark-antiquarks is investigated in the early stage of heavy-ion collisions. The time-dependent study is based on a kinetic description of the fermion-pair production in strong non-Abelian fields. We introduce time-dependent chromo-electric external field with a pulse-like time evolution in order to simulate the overlap of two colliding heavy ions. Our investigation is performed with finite current quark masses and focus on heavy quark-pair production. We have found that the small inverse duration time of the field pulse determines the efficiency of the quark-pair production. Therefore, the expected suppression for heavy quark production, as follows from the Schwinger formula for a constant field, is not seen, but an enhanced heavy quark production appears at ultrarelativistic energies.

1. Introduction

The main aim of ultrarelativistic heavy-ion collisions is to create extreme high energy densities and study the deconfinement of colored quarks and gluons. Experiments at the BNL Relativistic Heavy Ion Collider (RHIC) have been investigated this transition up to $\sqrt{s} = 200$ AGeV center of mass colliding energy and detectors at the CERN Large Hadronic Collider (LHC) are ready to explore the energy range to $\sqrt{s} = 5500$ AGeV.

One way to describe nucleus-nucleus collisions is based on the Glauber-Gribov model: the total cross section is well understood within the probabilistic Glauber model,[1] which only takes into account elastic rescatterings (described by Pomeron exchange) of the initial incoming hadron on various nucleons of the nucleus; in case of increasing energy the coherent interaction of constituents of the incoming hadrons with several nucleons of the nucleus must be summed up following Gribov[2] and shadowing corrections to the superposed proton-proton collisions are resulted in the modified Glauber picture. Interactions between soft partons of different nucleons in the nucleus can be described by multi-Pomeron interactions in the Glauber-Gribov model[3] and can be understood as interactions between strings formed in the collisions. Since the triple-Pomeron vertex is proportional to $A^{1/3}$ in hA collisions, this term becomes very important for hA and AA collisions of very heavy nucleus.[4] Heavy-ion collisions can be described successfully in the Glauber-Gribov model, especially using perturbative QCD based extensions.[5]

On the other hand, at RHIC and especially at LHC energies the colliding nucleus can be considered as two colliding sheets of nucleons owing to huge Lorentz-contraction ($\gamma = 100$ at RHIC and $\gamma = 2750$ at LHC), and their overlap results in a strong chromo-electric and chromo-magnetic field to be built up. Particles, namely gluons and quark-antiquark pairs will be produced from this strong field, similarly to the Schwinger-mechanism in quantum elecrodynamics (QED).[6] The production rate will depend on the field strength, which is varying in time.

In both frameworks, theoretical description of soft particle production can be understood by the introduction of chromo-electric flux tube ('string'), applied in many models.[7–11] String picture is a good example of how to convert the kinetic energy of a collision into field energy. However, at RHIC and LHC energies the string density is expected to be so large that a strong collective gluon field will be formed in the whole available transverse volume. Furthermore, the gluon number will be so high that a classical gluon field as the expectation value of the quantum field can be considered in the reaction volume.[11–13]

Fermion pair production together with boson pair production were investigated by different kinetic models of particle production from strong Abelian[14–20] and non-Abelian[21,22] fields. These calculations concentrated mostly on the bulk properties of the gluon and quark matter, the time evolution of the system, the time dependence of energy and particle number densities, and the appearance of fast thermalization.

We investigated massless fermion and boson production[23,24] and massive fermion-pair production[25,26] in strong Abelian and non-Abelian external electric field. We have realized, that the role of mass becomes important when the collisional energy is increasing and the duration of the pulse becomes comparable to the inverse quark mass.[25] In these papers, the description of the particle production in non-Abelian fields was considered in the framework of the Wigner function formalism. For some special configurations of the external field, it is possible to reduce the problem to the Abelian one. While this reduction is evident for the original Dirac equations, it cannot be easily seen from the kinetic equation for the Wigner function. Here, we demonstrate that the derived kinetic equation, indeed, reduces to Abelian result. Based on this result, we also present an analytic solution for the special pulse-like time dependence of non-Abelian external field. Obtained exact solutions bring new insight in understanding of the production of heavy quark pairs at very high colliding energies.

2. The kinetic equation for the Wigner function

The equation of motion for color Wigner function in the gradient approximation reads:[21,27–29]

$$\partial_t W + \frac{g}{8}\frac{\partial}{\partial k_i}\left(4\{W, F_{0i}\} + 2\left\{F_{i\nu}, [W, \gamma^0\gamma^\nu]\right\} - \left[F_{i\nu}, \{W, \gamma^0\gamma^\nu\}\right]\right)$$
$$= ik_i\{\gamma^0\gamma^i, W\} - im[\gamma^0, W] + ig\left[A_i\,, [\gamma^0\gamma^i, W]\right].$$

Here m denotes the current mass of the fermions, g is the coupling constant, A_μ is the 4-potential of an external space-homogeneous color field and $F_{\mu\nu}$ is the corresponding field tensor

$$F_{\mu\nu} = \partial_\mu A_\nu - \partial_\nu A_\mu - ig[A_\mu, A_\nu]. \tag{1}$$

The color decomposition of the Wigner function with SU(N_c) generators in the fundamental representation is given by

$$W = W^s + W^a t^a, \qquad a = 1, 2, ..., N_c^2 - 1\,, \tag{2}$$

where W^s is the color singlet part and W^a is the color multiplet components. It is also convenient to perform the spinor decomposition separating scalar (a), vector (b_μ), tensor ($c_{\mu\nu}$), axial vector (d_μ) and pseudo-scalar parts (e).

$$W^{s|a} = a^{s|a} + b_\mu^{s|a}\gamma^\mu + c_{\mu\nu}^{s|a}\sigma^{\mu\nu} + d_\mu^{s|a}\gamma^\mu\gamma^5 + ie^{s|a}\gamma^5. \tag{3}$$

The asymmetric tensor components of the Wigner function is convenient to decompose into the axial and polar vectors $c_1^j = c^{j0}$ and $c_2^j = \frac{1}{2}\epsilon^{0\omega\rho j}c_{\omega\rho}$ correspondingly.

3. Kinetic equation in SU(2) with color isotropic external field

After decomposition the equations for the Wigner function in case of pure longitudinal external SU(2) color field with fixed color direction $A_z^a = A_z^\diamond n^a$, where $n^a n^a = 3$ and $\partial_t n^a = 0$,[24] we obtain the following system of equations for the singlet components

$$\partial_t a^s + \frac{3g}{4} E_z^\diamond \frac{\partial}{\partial k_z} a^\diamond = -4\mathbf{k}\mathbf{c}_1^s, \quad (4)$$

$$\partial_t e^s + \frac{3g}{4} E_z^\diamond \frac{\partial}{\partial k_z} e^\diamond = -4\mathbf{k}\mathbf{c}_2^s - 2md_0^s, \quad (5)$$

$$\partial_t b_0^s + \frac{3g}{4} E_z^\diamond \frac{\partial}{\partial k_z} b_0^\diamond = 0, \quad (6)$$

$$\partial_t \mathbf{b}^s + \frac{3g}{4} E_z^\diamond \frac{\partial}{\partial k_z} \mathbf{b}^\diamond = 2[\mathbf{k} \times \mathbf{d}^s] + 4m\mathbf{c}_1^s, \quad (7)$$

$$\partial_t d_0^s + \frac{3g}{4} E_z^\diamond \frac{\partial}{\partial k_z} d_0^\diamond = 2me^s, \quad (8)$$

$$\partial_t \mathbf{d}^s + \frac{3g}{4} E_z^\diamond \frac{\partial}{\partial k_z} \mathbf{d}^\diamond = 2[\mathbf{k} \times \mathbf{b}^\mathbf{s}], \quad (9)$$

$$\partial_t \mathbf{c}_1^s + \frac{3g}{4} E_z^\diamond \frac{\partial}{\partial k_z} \mathbf{c}_1^\diamond = a^s\mathbf{k} - m\mathbf{b}^s, \quad (10)$$

$$\partial_t \mathbf{c}_2^s + \frac{3g}{4} E_z^\diamond \frac{\partial}{\partial k_z} \mathbf{c}_2^\diamond = e^s\mathbf{k}; \quad (11)$$

and the multiplet components

$$\partial_t a^\diamond + gE_z^\diamond \frac{\partial}{\partial k_z} a^s = -4\mathbf{k}\mathbf{c}_1^\diamond, \quad (12)$$

$$\partial_t e^\diamond + gE_z^\diamond \frac{\partial}{\partial k_z} e^s = -4\mathbf{k}\mathbf{c}_2^\diamond - 2md_0^\diamond, \quad (13)$$

$$\partial_t b_0^\diamond + gE_z^\diamond \frac{\partial}{\partial k_z} b_0^s = 0, \quad (14)$$

$$\partial_t \mathbf{b}^\diamond + gE_z^\diamond \frac{\partial}{\partial k_z} \mathbf{b}^s = 2[\mathbf{k} \times \mathbf{d}^\diamond] + 4m\mathbf{c}_1^\diamond, \quad (15)$$

$$\partial_t d_0^\diamond + gE_z^\diamond \frac{\partial}{\partial k_z} d_0^s \delta^{bc} = 2me^\diamond, \quad (16)$$

$$\partial_t \mathbf{d}^\diamond + gE_z^\diamond \frac{\partial}{\partial k_z} \mathbf{d}^s = 2[\mathbf{k} \times \mathbf{b}^\diamond], \quad (17)$$

$$\partial_t \mathbf{c}_1^\diamond + gE_z^\diamond \frac{\partial}{\partial k_z} \mathbf{c}_1^s = a^\diamond\mathbf{k} - m\mathbf{b}^\diamond, \quad (18)$$

$$\partial_t \mathbf{c}_2^\diamond + gE_z^\diamond \frac{\partial}{\partial k_z} \mathbf{c}_2^s = e^c\mathbf{k}. \quad (19)$$

The distribution function for massive fermions is defined by the components a, b:[24]

$$f_q(\mathbf{k}, t) = \frac{ma^s(\mathbf{k}, t) + \mathbf{k}\,\mathbf{b}^s(\mathbf{k}, t)}{\omega(\mathbf{k})} + \frac{1}{2}, \quad (20)$$

where $\omega(\mathbf{k}) = \sqrt{\mathbf{k}^2 + m^2}$. Therefore, to reveal time- and momentum-dependence of the distribution functions, the scalar a, vector b_μ, axial vector d_μ, and axial tensor components of the Wigner function are needed, only.

The initial conditions for the Wigner function in vacuum reads:[21,24]

$$a^s = -\frac{1}{2}\frac{m}{\omega}, \quad \mathbf{b}^s = -\frac{1}{2}\frac{\mathbf{k}}{\omega}. \tag{21}$$

Taking into account the symmetry of the initial conditions and performing vector decomposition,

$$\mathbf{v} = v_z \mathbf{n} + v_\perp \frac{\mathbf{k}_\perp}{k_\perp} + v_x [\mathbf{n} \times \frac{\mathbf{k}_\perp}{k_\perp}] \,, \tag{22}$$

we obtain the following equations for the singlet (let us redefine $\mathbf{c} = \mathbf{c}_1$ to simplify reading):

$$\partial_t a^s + \frac{3g}{4} E_z^\diamond \frac{\partial}{\partial k_z} a^\diamond = -4(k_z c_z^s + k_\perp c_\perp^s), \tag{23}$$

$$\partial_t b_z^s + \frac{3g}{4} E_z^\diamond \frac{\partial}{\partial k_z} b_z^\diamond = 2k_\perp d_x^s + 4mc_z^s, \tag{24}$$

$$\partial_t b_\perp^s + \frac{3g}{4} E_z^\diamond \frac{\partial}{\partial k_z} b_\perp^\diamond = -2k_z d_x^s + 4mc_\perp^s, \tag{25}$$

$$\partial_t d_x^s + \frac{3g}{4} E_z^\diamond \frac{\partial}{\partial k_z} d_x^\diamond = 2(k_z b_\perp^s - k_\perp b_z^s), \tag{26}$$

$$\partial_t c_z^s + \frac{3g}{4} E_z^\diamond \frac{\partial}{\partial k_z} c_z^\diamond = a^s k_z - mb_z^s, \tag{27}$$

$$\partial_t c_\perp^s + \frac{3g}{4} E_z^\diamond \frac{\partial}{\partial k_z} c_\perp^\diamond = a^s k_\perp - mb_\perp^s \;; \tag{28}$$

and for the multiplet:

$$\partial_t a^\diamond + gE_z^\diamond \frac{\partial}{\partial k_z} a^s = -4(k_z c_z^\diamond + k_\perp c_\perp^\diamond), \tag{29}$$

$$\partial_t b_z^\diamond + gE_z^\diamond \frac{\partial}{\partial k_z} b_z^s = 2k_\perp d_x^\diamond + 4mc_z^\diamond, \tag{30}$$

$$\partial_t b_\perp^\diamond + gE_z^\diamond \frac{\partial}{\partial k_z} b_\perp^s = -2k_z d_x^\diamond + 4mc_\perp^\diamond, \tag{31}$$

$$\partial_t d_x^\diamond + gE_z^\diamond \frac{\partial}{\partial k_z} d_x^s = 2(k_z b_\perp^\diamond - k_\perp b_z^\diamond), \tag{32}$$

$$\partial_t c_z^\diamond + gE_z^\diamond \frac{\partial}{\partial k_z} c_z^s = a^\diamond k_z - mb_z^\diamond, \tag{33}$$

$$\partial_t c_\perp^\diamond + gE_z^\diamond \frac{\partial}{\partial k_z} c_\perp^s = a^\diamond k_\perp - mb_\perp^\diamond \,. \tag{34}$$

Here the axial part of the vector b_x and the tensor components c_x, the longitudinal d_z and perpendicular $d_\perp$ parts of the axial vector components do not contribute to the evolution of the distribution function.

4. Numerical solutions

In Ref. 24 we solved the above equations for massless (light) quarks and described their longitudinal and transverse momentum distributions. Here we focus on the integrated particle yields and discuss the obtained results, focusing on massive (heavy) quark production.

In the numerical calculation we use the following parameters: the maximal magnitude of the field strength $E_0 = 0.68$ GeV/fm; the strong coupling constant $g = 2$; the current quark masses $m_{u,d} = 8$ MeV, $m_s = 150$ MeV, $m_c = 1.2$ GeV, $m_b = 4.2$ GeV for light, strange, charm and bottom quarks, correspondingly. The value of maximal magnitude of the field corresponds to the effective string tension $\kappa \sim 1.17$ GeV/fm.

The particle production is ignited by a pulse-like color field simulating a high energy heavy ion collision:[24]

$$E^{\diamond}(t) = E_0 \cdot \left[1 - \tanh^2(t/\tau)\right], \tag{35}$$

where τ is a pulse duration time, E_0 is the maximal magnitude of the field strength at $t = 0$.

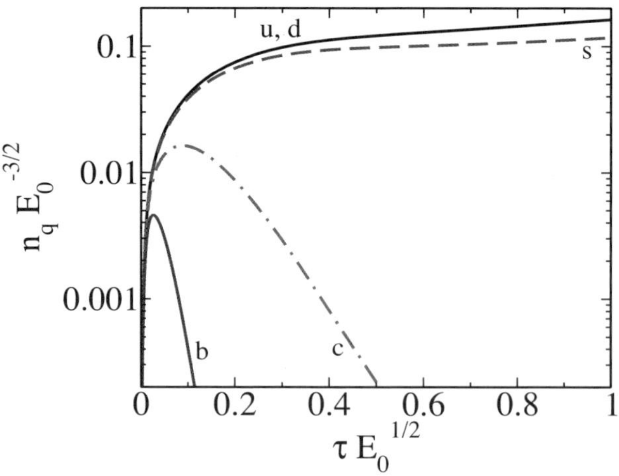

Fig. 1. The dimensionless quark number densities at the asymptotic final state, $n_q(t \gg \tau)E_0^{-3/2}$, as a function of pulse duration time $\tau E_0^{1/2}$.

Here $n_q(t)$ denotes the number density of corresponding quarks given by

$$n_q(t) = 4N_c \int \frac{d^3k}{(2\pi)^3} f_q(\mathbf{k}, t) \ , \tag{36}$$

where q denotes a quark flavour, q=u, d, s, c, b. If the particles are produced in the field (35), both the quark number density and the suppression factor are implicit functions of E_0 and τ.

5. Exact solutions

The kinetic equation for the Wigner function, especially the Eqs. (23)–(34) were solved numerically in Ref. 26. However, after taking into account additional symmetries of the external field we can discover further simplifications and even obtain exact solutions.

To demonstrate this fact we rewrite Eqs. (23)–(34) explicitly for the distribution function $f(\mathbf{k}, t)$:

$$\partial_t f + \frac{3}{4} g E_z^\diamond \frac{\partial}{\partial k_z} f^\diamond = \frac{3}{4} W^\diamond v^\diamond, \tag{37}$$

$$\partial_t v^s + \frac{3}{4} g E_z^\diamond \frac{\partial}{\partial k_z} v^\diamond = -\frac{3}{4} W^\diamond f^\diamond - 2\omega u^s, \tag{38}$$

$$\partial_t u^s + \frac{3}{4} g E_z^\diamond \frac{\partial}{\partial k_z} u^\diamond = 2\omega v^s, \tag{39}$$

$$\partial_t f^\diamond + g E_z^\diamond \frac{\partial}{\partial k_z} f = W^\diamond v^s, \tag{40}$$

$$\partial_t v^\diamond + g E_z^\diamond \frac{\partial}{\partial k_z} v^s = \frac{1}{2} W^\diamond (1 - 2f) - 2\omega u^\diamond, \tag{41}$$

$$\partial_t u^\diamond + g E_z^\diamond \frac{\partial}{\partial k_z} u^s = 2\omega v^\diamond. \tag{42}$$

Here the following new functions were defined

$$f^\diamond = \frac{m a^\diamond + \mathbf{k}\mathbf{b}^\diamond}{\omega}, \tag{43}$$

$$v^{s,\diamond} = \frac{\varepsilon_\perp}{\omega} b_z^{s,\diamond} - \frac{k_z}{\omega} \left(\frac{m a^{s,\diamond} + k_\perp b_\perp^{s,\diamond}}{\varepsilon_\perp} \right), \tag{44}$$

$$u^{s,\diamond} = -\frac{k_\perp}{\varepsilon_\perp} d^{s,\diamond} - 2\frac{m}{\varepsilon_\perp} c_z^{s,\diamond}, \tag{45}$$

$$W^\diamond = \frac{g E \varepsilon_\perp}{\omega^2}. \tag{46}$$

The naming scheme is chosen to be consistent with the U(1) case of our previous work.[23] The transverse one-particle energy in Eq. (1) is defined by $\varepsilon_\perp = \sqrt{k_\perp^2 + m^2}$.

As it follows from Eq. (21) the vacuum state corresponds to zero initial conditions for the functions $f, f^\diamond$, $v^{s,\diamond}$, $u^{s,\diamond}$.

Note, that Eqs. (37)–(42) has the same number of equation for massive and massless particles. Therefore, the massless limit does not lead to any further simplifications. The system (37)–(42) can be transformed to more conventional form, allowing the solution on characteristics. For that we introduce the following new functions

$$F^\pm = f \pm \frac{\sqrt{3}}{2} f^\diamond, \tag{47}$$

$$V^\pm = v^s \pm \frac{\sqrt{3}}{2} v^\diamond, \tag{48}$$

$$U^\pm = u^s \pm \frac{\sqrt{3}}{2} u^\diamond. \tag{49}$$

The equations for these functions read

$$\partial_t F^\pm \pm \frac{\sqrt{3}}{2} g E_z^\diamond \frac{\partial}{\partial k_z} F^\pm = \pm \frac{\sqrt{3}}{2} W^\diamond V^\pm, \tag{50}$$

$$\partial_t V^\pm \pm \frac{\sqrt{3}}{2} g E_z^\diamond \frac{\partial}{\partial k_z} V^\pm = \pm \frac{\sqrt{3} W^\diamond}{4} (1 - 2F^\pm) - 2\omega U^\pm, \tag{51}$$

$$\partial_t U^\pm \pm \frac{\sqrt{3}}{2} g E_z^\diamond \frac{\partial}{\partial k_z} U^\pm = 2\omega V^\pm. \tag{52}$$

The equations for "(+)" and "(−)" functions are completely factorized and can be solved independently. The distribution function f is obtained as

$$f = \frac{F^+ + F^-}{2}. \tag{53}$$

The equations (50)–(52) are very similar to those obtained in the Abelian case (see Eqs. (76)–(77) in Ref. 29 and Eqs. (22)–(24) in Ref. 23). However, only f is physical quantity, while the functions $F^\pm$ carry intermediate information. Nevertheless, from mathematical point of view there is no difference and this analogy allows to exploit U(1) solution to obtain exact analytical results for SU(2)-color case, as we demonstrate below. This is not surprising. For the assumptions we used for the external field it is possible to get the diagonal representation for the color field by performing an appropriate rotation in the color space. In this case the Dirac equations of the basic problem decouple for each color component to U(1) problem.

In the case of a time reversal symmetry of the field strength, $E(t) = E(-t)$, the functions F^+ and F^- are equal to each other. Therefore, the distribution function is defined as $f = F^+ = F^-$. Furthermore, in the Abelian case, it is known that for the field with form of Eq. (35) an analytic solution of Dirac equation exists, as well as analytic solution of the corresponding kinetic equation (see e.g. Ref. 30). Following this analogy we can obtain an analytic solution of Eqs. (50)–(52) for the asymptotic state $t \gg \tau$:

$$f = \frac{\sinh\left[\pi(\theta - \mu^+ + \mu^-)\right] \sinh\left[\pi(\theta + \mu^+ - \mu^-)\right]}{\sinh(2\pi\mu^+)\sinh(2\pi\mu^-)}. \tag{54}$$

Here we introduced the following notations

$$\mu^\pm = \frac{\tau}{2}\sqrt{\left(k_z \pm \frac{E_0}{E_{cr}} m^2 \tau\right)^2 + k_\perp^2 + m^2}, \tag{55}$$

$$\theta = -\frac{E_0}{E_{cr}} m^2 \tau^2, \tag{56}$$

$$E_{cr} = \frac{2m^2}{\sqrt{3}g}. \tag{57}$$

The *Schwinger limit* can be readily obtained from Eq. (54). Indeed, for pulses longer than any scale in the system, $\tau \gg \max\{E_0^{-1/2}, m^{-1/2}\}$, we can use the expansions

$$\theta \pm (\mu^+ - \mu^-) = -\frac{E_0}{E_{cr}} m^2 \tau^2 \mp k_3 \tau + \mathcal{O}\left(\frac{1}{\tau}\right), \tag{58}$$

$$\mu^\pm = \frac{E_0}{2E_{cr}} m^2 \tau^2 \pm k_3 \tau + \frac{2}{\sqrt{3}} \frac{m^2 + k_\perp^2}{4gE_0} + \mathcal{O}\left(\frac{1}{\tau}\right). \tag{59}$$

In the leading order of the expansion parameter τ we obtain the following distribution function

$$f \simeq \exp\left(-\pi \frac{2\varepsilon_\perp}{\sqrt{3}gE_0}\right). \tag{60}$$

After the integration w.r.t. momentum we obtain SU(2) version of the Schwinger formula

$$n = \tau \frac{3(gE_0)^2}{16\pi^2} \exp\left(-\frac{2\pi m^2}{\sqrt{3}gE_0}\right), \tag{61}$$

where we have taken into account the replacement $\int dk_z \to \sqrt{3}gE_0\tau/2$. Thus in SU(2) the suppression factor of heavy particles with mass m_Q to light particles with mass m_q in Schwinger limit is given by

$$\gamma^Q = \exp\left[-\frac{2\pi(m_Q^2 - m_q^2)}{\sqrt{3}gE_0}\right]. \tag{62}$$

The analytical result for the number density of quarks can be also obtained if the following inequality is satisfied

$$\Gamma_K \equiv \frac{E_{cr}}{E}\frac{1}{m\tau} \gg \max\{1, m\tau\}, \tag{63}$$

where we introduce the adiabaticity parameter Γ_K.[a] In this limit the distribution function is given by

$$f \simeq \left[\frac{\sqrt{3}\pi\sqrt{k_\perp^2 + m^2}}{2\omega} gE\tau^2 \mathrm{csch}\,(\pi\omega\tau)\right]^2. \tag{64}$$

The momentum integration of this distribution function can be done analytically in the following cases:

a) *Long pulse duration and undercritical field.* For long pulse duration the constraint $m\tau \gg 1$ in Eq. (63) is satisfied for undercritical field, $E/E_{cr} \ll 1$. After expanding the distribution function in Eq. (64) and performing momentum integration, we obtain the number density

$$n = \frac{3(m\tau)^{3/2}}{4\pi}\tau(gE_0)^2\exp(-2\pi m\tau). \tag{65}$$

The condition of long pulse duration and undercritical field could be realized for the collision of heavy ions in SPS energy range (assuming that the physical picture of classical gluon field is still valid) for light quarks, or for charm and bottom quarks at and below the RHIC energies (the string tension is the order of 1 GeV/fm.)

b) *Short pulse duration.* In the opposite limit, $m\tau \ll 1$, we obtain the number density

$$n \simeq \frac{\tau}{12\pi}(gE_0)^2. \tag{66}$$

In this case the number density of produced particles depends linearly on the duration time, τ, and is independent of the particle mass, m.

[a] In the Abelian case Γ_K is known as the Keldysh adiabaticity parameter. It separates the non-perturbative region $\Gamma_K \ll 1$ from the perturbative multiphoton one $\Gamma_K \gg 1$.

For the parameter set we used in the main part of the manuscript the field magnitude, E_0, is about five times higher than the critical one for the strange quark. The number of produced strange quarks follows the Eq. (66) for pulse duration at least five times less then inverse mass of strange quark $\tau \ll (5m_s)^{-1} \simeq 0.3$ fm/c As we estimated the pulse duration time for RHIC is about 0.1 fm/c, that is only three times less than $(5m_s)^{-1}$. Thus the expression (66) is valid for the strange quark only at higher than RHIC energies.
For heavier particles, e.g. charm quark, the requirement $m_c\tau \ll 1$ leads to smaller values of τ. Indeed, we can rewrite

$$m_c\tau = \frac{m_c}{m_s} \times (m_s\tau) \ll 1. \tag{67}$$

This expression shows that Eq. (66) becomes valid for charm quark on a shorter scale of pulse duration time ($m_c/m_s \simeq 8$).
For light quarks the condition $m_u\tau \ll 1$ is trivially satisfied at RHIC energies. However, now the condition (63) plays more important role. From Eq. (63) we obtain an estimate for the validity of Eq. (66)

$$\tau \ll 4 \cdot 10^{-3}\text{fm}/c. \tag{68}$$

In this limiting case the strange suppression factor tends to unity.

One more analytical result can be obtained from the general solution Eq. (54). If the duration time of the pulse tends to zero, but amplitude is increasing as $E_0 = A_0\tau^{-1}$ (A_0 is a constant), then the distribution function is given by

$$f = \frac{1}{2}\left(1 - \frac{\omega^2 - \frac{3}{4}A_0^2}{\omega_+\omega_-}\right). \tag{69}$$

where $\omega_+ = 2\mu_\pm/\tau$. Since the momentum integral from the above distribution function is divergent, this approximation results in infinitely many new quark-antiquark pairs in unit volume. This is understandable since we pump an infinite amount of energy to the system.

Acknowledgments

P.L. would like to thank Prof. Luciano Bertocchi and Prof. Daniele Treleani for their kind hospitality and the perfect conditions at ICTP for the Gribov-80 Workshop. This work was supported in part by Hungarian OTKA Grants NK062044 and NK077816, MTA-JINR Grant, and RFBR grant No. 03-02-16877.

References

1. R.J. Glauber, in *Lectures in Theoretical Physics*, eds. W.E. Brittin and L.G. Dunham (Inerscience, N.Y., 1959), Vol. 1, p. 315.
2. V.N. Gribov, *Sov. Phys. JETP* **29**, 483 (1969); *ibid.* **30**, 709 (1970).
3. A. Kaidalov, *Nucl. Phys.* **A525**, 39c (1991).
4. K. Tywoniuk, I.C. Arsene, L. Bravina, A.B. Kaidalov and E. Zabrodin, *Eur. Phys. J* **C49**, 193 (2007).
5. G.G. Barnaföldi, P. Lévai, G. Papp, G. Fai, M. Gyulassy, *Eur. Phys. J* **C33**, s609 (2004).
6. J. Schwinger, *Phys. Rev.* **82**, 664 (1951).
7. B. Andersson *et al.*, *Phys. Rep.* **97**, 31 (1983); *Nucl. Phys.* **B281**, 289 (1987).
8. X.N. Wang and M. Gyulassy, *Phys. Rev.* **D44**, 3501 (1991); *Comput. Phys. Commun.* **83**, 307 (1994).
9. H. Sorge, *Phys. Rev.* **C52**, 3291 (1995).
10. V. Topor Pop, M. Gyulassy, J. Barrette, C. Gale, X. N. Wang, and N. Xu, *Phys. Rev.* **C70**, 064906 (2004).
11. V. Topor Pop, M. Gyulassy, J. Barrette, C. Gale, R. Bellwied, and N. Xu, *Phys. Rev.* **C72**, 054901 (2005);
V. Topor Pop, M. Gyulassy, J. Barrette, C. Gale, S. Jeon, and R. Bellwied, *Phys. Rev.* **C75**, 014904 (2007).
12. M. Gyulassy and L. McLerran, *Phys. Rev.* **C56**, 2219 (1997).
13. L. D. McLerran, *Lect. Notes Phys.* **583**, 291 (2002).
14. G. Gatoff, *et al. Phys. Rev.* **D36**, 114 (1987).
15. Y. Kluger *et al.*, *Phys. Rev. Lett.* **67**, 2427 (1991).
16. G. Gatoff and C.Y. Wong, *Phys. Rev.* **D46**, 997 (1992).
17. C.Y. Wong, *et al. Phys. Rev.* **D51**, 3940 (1995).
18. J.M. Eisenberg, *Phys. Rev.* **D51**, 1938 (1995).
19. D.V. Vinnik, *et al.*, *Few-Body Syst.* **32**, 23 (2002).
20. V.N. Pervushin, *et al. Int. J. Mod. Phys.* **A20**, 5689 (2005).
21. A.V. Prozorkevich, S.A. Smolyansky, and S.V. Ilyin, (`hep-ph/0301169`).
22. D.D. Dietrich, *Phys. Rev.* **D68**, 105005 (2003); *ibid.* **D70**, 105009 (2004).
23. V.V. Skokov and P. Levai, *Phys. Rev.* **D71**, 094010 (2005).
24. V. V. Skokov and P. Levai, *Phys. Rev.* **D78**, 054004 (2008).
25. P. Levai and V. Skokov, *J. Phys.* **G36**, 064068 (2009).
26. P. Levai and V. Skokov, *Phys. Rev.* **D82**, 074014 (2010).
27. A. V. Prozorkevich, *et al. Phys. Lett.* **B583**, 103 (2004).
28. H. T. Elze, M. Gyulassy and D. Vasak, *Phys. Lett.* **B177**, 402 (1986); H. T. Elze, M. Gyulassy and D. Vasak, *Nucl. Phys.* **B276**, 706 (1986); S. Ochs and U. Heinz, *Ann. Phys.* **266**, 351 (1998).
29. V. N. Pervushin and V. V. Skokov, *Acta Phys. Polon.* **B37**, 2587 (2006).
30. A.A. Grib, S.G. Mamaev and V.M. Mostepanenko, *Vacuum Quantum Effects in Strong Fields*, (Friedmann Laboratory Publishing, St. Petersburg, 1994); G. V. Dunne, arXiv:hep-th/0406216; K. Fukushima, F. Gelis and T. Lappi, arXiv:0907.4793 [hep-ph].

LEADING TWIST SHADOWING, ONSET OF BLACK DISK REGIME AND POST-SELECTION PHENOMENON

LEONID FRANKFURT

Physics Department, Tel Aviv University Department,
Tel Aviv, Israel
frankfur@tauphy.tau.ac.il

MARK STRIKMAN

104 Davey Lab, Penn State University, University Park, PA 16803, USA
strikman@phys.psu.edu

We summarize results of the Generalization of the Gribov picture of nuclear shadowing in high energy processes off nuclei for calculating shadowing of the parton distributions in nuclei and some properties of the final states in small x processes in DIS are described. It is explained how account of the restrictions due to the energy-momentum conservation allows to choose approximations effective for the description of small x phenomena. Estimates of proximity of the pQCD interaction of small dipoles to the black disk limit are provided and the phenomenon of post-selection suppression of leading parton spectrum (effective fractional energy losses) in the proximity of the black disk regime in DIS is described. We argue that the resulting suppression of the the leading hadron production provides one of the most sensitive signals of the onset of new QCD dynamics. Evidence for importance of the post-selection phenomenon in the production of the leading pions in the $d - Au$ collisions studied at RHIC is presented. We also outline briefly impact of rapid increase of parton distributions within pQCD at small x on the interplay of soft and hard QCD dynamics at the LHC.

Keywords: Leading twist shadowing; black disk regime; forward hadron production.

1. Theory of the leading twist nuclear shadowing

The relationship between nuclear shadowing and diffraction in the total cross section of hadron-nucleus interactions was established a long time ago.[1] Comparison with data shows that accuracy of the theory for the hadron–nucleus interactions is on the level of few % which reflects small admixture of non-nucleonic degrees of freedom in nuclei and small off-shellness of the nucleons in nuclei as compared to the soft strong interaction scale. The Gribov logic was successfully applied for the description of the γA total cross sections. In the limit of a thin nuclear target when only interactions of projectile photon with two nucleons contribute to the shadowing one finds:

$$F_{2A}^{(double)}(x,Q^2) = -2\Re e \int d^2\vec{b} \int_{-\infty}^{\infty} dz_1 \int_{z_1}^{\infty} dz_2 \sum_X A(A-1)$$

$$\left\langle \Gamma_{\gamma^* X}(\vec{b}-\vec{r}_{1\perp})\Gamma_{X\gamma^*}(\vec{b}-\vec{r}_{2\perp})e^{i(z_1-z_2)\Delta_{\gamma^* X}} \right\rangle , \tag{1}$$

where $\sum_X$ denotes the sum over all diffractive intermediate states $A(A-1)$ is the number of the nucleon pairs; $(\vec{r}_{i\perp}, z_i)$ are the transverse and longitudinal (with respect to the direction of the momentum of γ^*, $\vec{q}$) coordinates of the involved nucleons; the integration limit $z_2 \geq z_1$ reflects the underlying space-time evolution of the process; $\Gamma_{\gamma^* X}$ is the $\gamma^* N \to XN$ scattering amplitude in the space of the impact-parameter $\vec{b}$ (to be more precise, the amplitude $\Gamma_{\gamma^* X}$ formally defined for the elastic transition refers to the hadronic component of the virtual photon which already does not contain the small electromagnetic coupling constant); the brackets denote the matrix element between the nuclear ground-states; $\Delta_{\gamma^* X}$ is the longitudinal momentum transfer, or, equivalently, the inverse coherence length for the $\gamma^* \to X$ fluctuation,

$$\Delta_{\gamma^* X} = \frac{M_X^2 + Q^2}{2|\vec{q}|} . \tag{2}$$

It is worth emphasizing that Eq. (1) can be understood as the manifestation of the unitarity of evolution operator $exp(iHt)$ as reflected in the Abramovsky, Gribov, Kancheli (AGK) cutting rules[2] as long as the nucleus is treated as a multinucleon system.

However at large Q^2 an additional step is necessary to calculate the nuclear parton distributions at small x.[3] One has to combine unitarity relations for different cuts of the diagrams corresponding to the different final states for the interaction of the hard probe with a nucleus, with the QCD factorization theorem for hard diffraction.[4] The Q^2 dependence of the obtained results is accounted for in terms of the QCD evolution equation. In the limit of thin target when only double scattering is important one obtains

$$\Delta\left[xf_{j/A}(x,Q^2,b)\right] = xf_{j/N}(x,Q^2,b) - xf_{j/A}(x,Q^2,b)$$
$$= 8\pi A(A-1)\Re e\left(\frac{(1-i\eta)^2}{1+\eta^2}\int_x^{0.1} dx_{I\!P}\beta f_j^{D(4)}(\beta,Q^2,x_{I\!P},t_{\min})\right.$$
$$\left.\int_{-\infty}^{\infty} dz_1 \int_{z_1}^{\infty} dz_2\, \rho_A(\vec{b},z_1)\rho_A(\vec{b},z_2)e^{i(z_1-z_2)x_{I\!P}m_N}\right), \quad (3)$$

where $f_{j/A}(x,Q^2), f_{j/N}(x,Q^2)$ are nuclear and nucleon pdfs, $f_j^{D(4)}(\beta,Q^2,x_{I\!P},t_{\min})$ are diffractive pdfs, $\eta = ReA^{diff}/ImA^{diff} \approx 0.17$ and $\rho_A(r)$ is the nuclear matter density. Here $x_{I\!P} = x(1+M_X^2/Q^2)$ is the light cone fraction transferred from the nucleon to the system X.

Note here that since the Gribov logic allows a direct calculation of F_{2A} using as the input the corresponding diffractive cross section of γ^*N scattering the predictions of all approaches which fit the diffraction data would result in the same shadowing for F_{2A} for the double scattering. However in an approximation which lacks a separation of the leading twist and higher twist contributions, as for example within the some variants of dipole model,[a] it is impossible to extend the results for other processes, for example production of dijets in pA scattering, or $W^{\pm}$-boson in AA scattering.

Numerical studies indicate that dominant contribution to the shadowing originates from the region of relatively large $\beta = Q^2/(M^2+Q^2)$ corresponding to rapidity intervals ≤ 3. For such a restricted interval in rapidity small x approximations like the BFKL approximation is not applicable. (The BFKL approximation predicts energy dependence of the diffractive cross section corresponding to $\alpha_{I\!P} \sim 1.25$ while the HERA experiments find a soft energy dependence corresponding to the effective Pomeron trajectory with $\alpha_{I\!P} \sim 1.11$).

The uncertainties of the predictions are related to the shadowing effects resulting from the interaction of the hard probe with $N \geq 3$ nucleons. Recently we improved the treatment of the multiple interactions, based on the concept of the color fluctuations and were able to account for the presence of both point-like and hadron-like configurations in the virtual photon wave function.[5] The relative contribution of these two components is essentially the only parameter which cannot be directly calculated from the theory. However uncertainty related to this parameter is rather small

[a]Note that calculations in the dipole models usually neglect recoil effects corresponding to the factor $e^{i(z_1-z_2)x_{I\!P}m_N}$ in Eq. (3) as well as the real part of the diffractive amplitude.

as compared to the previous analyses: the difference between two extreme scenarios (models I and II in Fig. 1) is $\leq 20\%$ for $A \sim 200$ and much smaller for light nuclei.

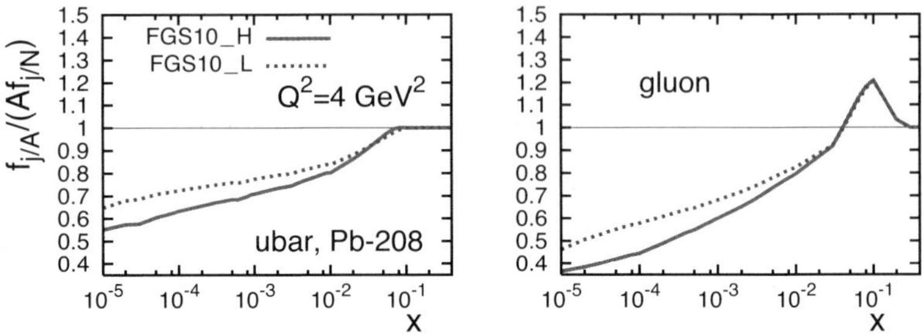

Fig. 1. Predictions of the theory of leading twist shadowing for quark and gluon pdfs. Difference in the predictions of two models of color fluctuation illustrates the range of uncertainties of the predictions.

Using AGK cutting rules it is possible to extend the approach to hard diffractive processes with nuclei and calculate diffractive parton distributions in nuclei. This allows to calculate also the fraction of the cross section which is due to the diffraction in hard quark and gluon induced processes. This fraction could be as large as 30% for the gluon channel at $Q^2 = 4\ GeV^2, x = 10^{-4}$ and it is much more sensitive to the interplay of point-like and hadron-like configurations in a hard probe[3c]. Since the same parameter enters in the calculation of the shadowing for the nuclear pdfs and for diffraction, the simultaneous measurements of the small x inclusive and diffractive cross sections would provide a stringent test of the theory.

Note that first measurements of hard diffraction in the scattering off nuclei will be possible in the process γ + A→ 2 jets +X + gap +A in the near future using ultraperipheral heavy ion collisions at the LHC. Our analysis[6,7] indicates that the rates of the photon – gluon fusion process in the ultraperipheral collisions will be sufficient to observe tens of thousands of dijet events even in the first low lumi run. Hence the measurement of the fraction due to coherent processes which is expected to be on the scale of 10% for $p_t \sim 10$ GeV/c will be possible.

2. Energy-momentum conservation and the applicability of pQCD approximations

It has been understood recently (resummation models) that the natural framework for the theoretical description of small x processes in the kinematics of HERA, and most of the LHC kinematics is the DGLAP approximation, for a recent discussion see Ref. 8. This is because the double logarithmic terms $\alpha_s \ln(Q^2/Q_0^2)\ln(x_0/x)$ give dominant contribution in the small x phenomena. At the same time DGLAP approximation should become inapplicable at extremely small x where $\ln(x_0/x) \gg \ln(Q^2/Q_0^2)$ because of vastly smaller number of allowed steps in $\ln(Q^2/Q_0^2)$ as compared to that in $\ln(x_0/x)$. However corresponding regions are hardly within the reach of the LHC.

Complimentary reasoning which leads to rather similar conclusion follows from the account for the energy-momentum conservation. Really span in rapidity for a high energy process with the radiation of n additional gluons in the multiRegge kinematics is

$$\Delta y = 4 + 3(n+1). \tag{4}$$

Here 4 is the sum of the lengths in rapidity of fragmentation regions in a high energy collisions of two particles, ≥ 3 is the distance in rapidity between adjacent gluon in the parton ladder in the multi Regge kinematics. Rapidity span for pp collisions at LHC is $2\ln(2E/m_N) = 19$. Here E=7 TeV is the maximal energy of the proton beams which can be reached at the LHC. Thus the number of gluons in pp collision which is kinematically allowed in the multi Regge kinematics is ≤ 4. At HERA allowed number of gluons is significantly smaller.[9]

Actual number of gluons produced at the LHC in the multi Regge kinematics is significantly smaller 2 since one should take into account the process of formation of small size dipoles in the fragmentation region. Thus in the most of kinematics of LHC the DGLAP approximation appears to be a legitimate approximation.

Hence in the evaluation of the kinematic boundary of the applicability of pQCD we can use the DGLAP approximation.

3. Where a non-linear regime of QCD evolution sets in?

In the leading log approximation one can derive a relation between the QCD evolution equations and the target rest frame picture of the interaction of small color dipoles of transverse size "d" with targets expressing it through the gluon density in the target, see Ref. 11 and refs. therein:

$$\sigma_{inel}^{\text{``}d\text{''}T} = \frac{\pi^2 F^2}{4}\alpha_S(Q_{eff}^2)x' g_T(x', Q_{eff}^2), \tag{5}$$

where $Q^2 \approx \lambda/d^2, \lambda \sim 4\div 9$; F^2 is the Casimir operator of the color group in the octet and the triplet representation equal to 3 and 4/3 correspondingly.

Matching the behavior of the dipole cross section in the pQCD regime and of large size dipoles in the regime of soft interaction, it is possible to derive interpolation formulae for the dipole–nucleon cross section for all dipole sizes and describe the total cross section of DIS at HERA.

To determine how close is the interaction strength to the maximal allowed by the probability conservation (unitarity) it is necessary to consider the amplitude of the dipole–nucleon interaction in the impact parameter space:

$$\Gamma_d(s,b) = \frac{1}{2is}\frac{1}{(2\pi)^2}\int d^2\vec{q}e^{i\vec{q}\vec{b}}A_{dN}(s,t), \tag{6}$$

where $A_{\text{“}d\text{”}N}(s,t)$ is the elastic amplitude of dipole–nucleon scattering normalized to $ImA_{\text{“}d\text{”}N} = is_{\text{“}d\text{”}N}\sigma_{tot}(\text{“}d\text{”}N)$.

The profile function expresses the probability to scatter at a given impact parameter b in the high energy limit.

It follows from the general principles of quantum mechanics cf.[10] that,

$$\sigma_{inel} = \int d^2b\, \Gamma_{\rm inel} = \int d^2b\, (1 - |1 - \Gamma(b,x,d)|^2), \tag{7}$$

$$\sigma_{tot} = 2\int d^2b\ Re\Gamma(b,x,d), \tag{8}$$

$$\sigma_{el} = \int d^2b\ |\Gamma(b,x,d)|^2\,. \tag{9}$$

Then at high energies, where amplitudes become predominantly imaginary, an important inequality follows:

$$\Gamma_{inel}(b,x,d) \leq 1, \tag{10}$$

which is just a statement of the probability conservation. Thus, the maximal strength of interaction is achieved in the case of complete absorption of a projectile at a given impact parameter b corresponding to

$$\Gamma(b,x,d^2) = \Gamma_{inel}(b,x,d) = 1. \tag{11}$$

Equation (11) is satisfied for $b \leq b_{\max}$. Increase with energy of $b_{\max}$:

$$b_{\max} \propto \ln(x_0/x), \tag{12}$$

is specific and new QCD effect indicating a transition to the new pQCD regime of strong interactions with a small coupling constant. (Actual dependence of b on x is more complicated. Equation (12) is the reasonable numerical fit to the more complicated QCD formulae).

If Eq. (11) is satisfied for $b \leq b_{\max}$ it is legitimate to call this regime as *black disk regime* (BDR). This name is taken from similar behavior of cross section for the scattering of light off a black screen. This is precisely the black-disk behavior that we refer to before.

3.1. *Evaluation of kinematical region of BDR for the scattering of spatially small dipole*

The dependence of the amplitude of dipole–nucleon elastic scattering on momentum transfer t can be extracted from the amplitude of the exclusive vector meson production in the regime where QCD factorization theorem for the exclusive processes allows to express relate the t dependence of the amplitude to the t-dependence of the gluon GPDs. Performing the Fourier transform of this amplitude to the impact parameter space b allows to reconstruct dependence of Γ on b.

Combining this information with the information on the total cross section of the dipole–nucleon interaction allows us to determine $\Gamma_d(s,b)$ as function of the dipole size. A sample of the results for $q\bar{q}$ dipole–proton (nucleus) interaction which represent an update of the analysis of Ref. 12 is presented in Fig. 2. One can see from the figure that for small impact parameters Γ for scattering off a proton and a heavy nucleus are comparable provided one takes into account the leading twist shadowing for the gluons. This reflects an observation that gluon densities in nuclei and proton at $b = 0$ are rather similar. However very few processes with proton are sensitive to $b = 0$ rather than $\langle b \rangle$, while in nuclei $\Gamma(b)$ is practically constant for a broad range of b.

Since the probability of the inelastic interaction, $P_{in}(b) = 1 - |1 - \Gamma(b)|^2$, $P_{in} \geq 3/4$ already for $\Gamma(b) \geq 0.5$. The non-linear relation between $\sigma_{inel}(b,T)$ and $\Gamma(d,b)$ — Eq. (7) combined with the pQCD enhancement of the color octet dipole inelastic interaction by a factor of 9/4 as compared to the color triple dipole indicates that $\Gamma^{octet}(d,b)$ may reach 1/2 already for $\Gamma^{triplet}(d,b) \approx 0.2$.

Hence the results presented in Fig. 2 indicate that gluon induced interactions are close to the BDR for a much larger range of the dipole sizes.

There is an independent evidences for the presence of the strong interaction in the gluon channel. We could estimate the ratio, R_g of diffractive gluon induced cross section and the total gluon induced cross section using correspondingly the gluon diffractive pdfs and gluon pdfs. We found that the current HERA fits lead to $R_g(x = 10^{-4}, Q^2 = 4\mathrm{GeV}^2) \approx 0.3 \div 0.4$ which is pretty close to the BDR value of 0.5. We find that a smaller value

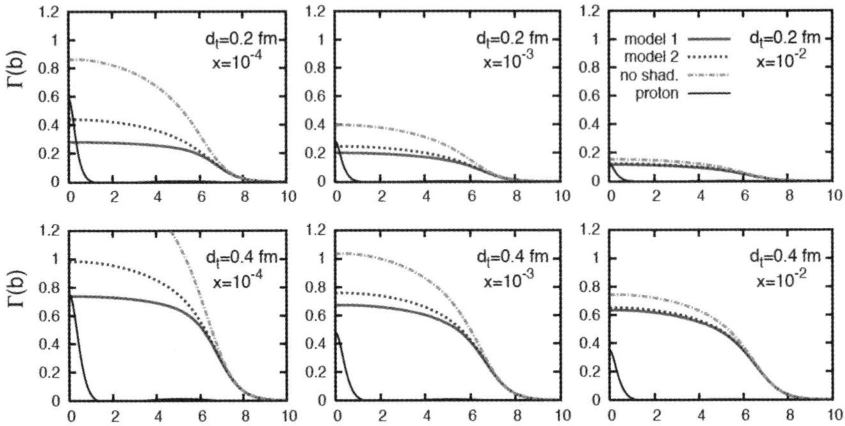

Fig. 2. Impact parameter distribution of $q\bar{q}$ dipole interaction with protons and lead nucleus. The models for the gluon shadowing are the same as in Fig. 1.

of R_g can be explained by the contribution of the scattering at large impact parameters as given by Eq. (13):[13]

$$R_g(x, Q^2) = \frac{\int |\Gamma_{gg}(b)|^2 \, d^2b}{\int 2\Gamma_{gg}(b) d^2b}. \tag{13}$$

It is of interest that the tension of the DGLAP fits with the HERA data starts at similar Q^2, see for example Ref. 8.

4. Small x behavior of structure functions of a nucleon and nucleus

Structure functions of nuclei have been evaluated in the black disc limit by V. Gribov before QCD was recognized as the underlining theory of strong interactions.[15] He assumed that the total cross section of the hadron-hadron collisions tends to a constant at large energies. Account of interplay of soft and hard QCD and rapid increase with energy of hard QCD interactions implies important modifications of V. Gribov formulae.

One of implications of rapid increase with energy of the hard interactions and existence of unitarity restrictions on the value of amplitudes at given impact parameter is the prediction that violation of convergency of pQCD series occurs at given impact parameter b which depend on x, Q^2 — see Eq. (12). As a result structure function can be represented as the sum of two contributions: $F_2 = F_2(central) + F_2(peripheral)$ where the first term corresponds to $b \le b_{max}(x, Q^2)$ and the second to $b \ge b_{max}$ -peripheral. So

$$F_2(central) = cQ^2 \ln(Q^2/Q_0^2) \ln^2(x_0/x), \tag{14}$$

where c is the number calculable in QCD. This formulae is the generalization of the Froissart limit familiar from soft QCD processes but for hard QCD contribution to the structure functions of a nucleon. $\ln(Q^2/Q_0^2)$ is due to polarization operator of the virtual photon. At very large energies $\ln(Q^2/Q_0^2)$ becomes $\ln(x_0/x)$ because of the increase with energy of the parton transverse momenta in the photon wave function.[14] The term $F_2(peripheral)$ is given by the conventional QCD evolution equation.

Thus QCD predicts significantly different behavior of structure functions of a nucleon as compared to the expectations which assumed parton "saturation" at any impact parameters i.e. that $F_2 \approx cQ^2$:

- structure functions of a nucleon will increase with energy forever without slowing down at the energies achievable at accelerators. Predictions of slowing down of increase of structure functions with energy are in variance with nonlinear relations between Green functions in QCD.
- BDR contribution should reveal itself at larger x in the gluon structure function of a nucleon as compared to F_2. This is because gluon dipole build of color octet constituents interacts with target with significantly larger cross section than a quark dipole because the ratio of Casimir operators for octet and triplet representations of color group $SU(3)_c$ is 9/4.
- Since the wave functions of heavy nucleus weakly depends on the impact parameter for $b < R_A$ the energy dependence of $F_{2A}(central) \approx Q^2 \ln(Q^2/Q_0^2)$ which is slower than that for nucleon structure function. At extremely large energies $\ln(Q^2/Q_0^2)$ becomes $\ln(x_0/x)$ as was first argued by V. Gribov[15] within preQCD ideas and confirmed in pQCD as the result of increase of interaction with energy in the form of double logarithmic terms.[14]

5. Post selection effect in BDR — effective energy losses

It was argued in Ref. 16 that in the BDR interactions with the target select configurations in the projectile wave function where the projectile's energy is split between constituents much more efficiently than in the DGLAP regime. The simplest example is inclusive production of the leading hadrons in DIS for $Q \le 2p_t(BDR)$. Interactions with the target are not suppressed up to $p_t \sim p_t(BDR)$, leading to selection of configurations in γ^* where longitudinal fractions carried by quark and antiquark are comparable. The energy splits before the collision — post-selection. As a result to a first approximation the leading hadrons are produced in the independent fragmentation of q and $\bar{q}$:

$$\bar{D}^{\gamma_T^* \to h}(z) = 2 \int_z^1 dy D_q^h(z/y) \frac{3}{4}(1 + (2y-1)^2), \tag{15}$$

leading to a strong suppression of the hadron production at $x_F > 0.3$.

A related effect is a gross violation of the leading twist predictions for the vector meson production in the BDR:

$$\frac{d\sigma^{\gamma_T^* \to VM+A}}{dt} = \frac{M_V^2}{Q^2} \frac{d\sigma^{\gamma_L^* \to VM+A}}{dt} = \frac{(2\pi R_A^2)^2}{16\pi} \frac{3\Gamma_V M_V^3}{\alpha(M_V^2+Q^2)^2} \frac{4|J_1(\sqrt{-t}R_A|^2}{-tR_A^2}. \tag{16}$$

This is a factor of Q^4 weaker Q^2 dependence than in the leading twist.

In the case of a parton of a hadron projectile propagating through the nucleus near BDR effective energy losses were estimated in Ref. 17.

The inelastic collision of a parton produced in a hard high energy collision is described by the imaginary part of the two gluon ladder with the vacuum quantum numbers. By definition, the inelastic cross section is calculable in terms of the probability of inelastic interaction of the parton with a target at a given impact parameter b, [see also Eq. (7)]:

$$\sigma_{\rm inel}(s) = \int d^2b P_{\rm inel}(s, b, Q^2). \tag{17}$$

Since $\sigma_{\rm inel}(s)$ is calculable in QCD, Eq. (17) serves as a useful definition of $P_{\rm inel}(s, b, Q^2)$. Thus, the probability of the inelastic interaction of a quark reads (cf. Eq. (5)):

$$P_{\rm inel}(s, b, Q^2) = \frac{\pi^2}{3} \alpha_s(k_t^2) \frac{\Lambda}{k_t^2} x g_A(x, Q^2, b), \tag{18}$$

where $x \approx 4k_t^2/\hat{s}_{qN}$; $Q^2 \approx 4k_t^2$; $\Lambda \sim 2$ GeV^2; $g_A(x, Q^2, b)$ is impact parameter dependent gluon density in a nucleus. For the gluon dipole, $P_{\rm inel}(s, b, Q^2)$ is $9/4$ times larger. Note also that Eq. (18) is valid only for the onset of the BDR when $P_{\rm inel}(s, b, Q^2) < 1$.

The unitarity limit on $P_{\rm inel}(s, b, Q^2)$ is $P_{\rm inel}(s, b, Q^2) = 1$. When $P_{\rm inel}(s, b, Q^2)$ given by Eq. (18) approaches or exceeds unity, it means that the average number of inelastic interactions, $N(b)$, becomes larger than one. Denoting the gluon density for which the unitarity limit on $P_{\rm inel}(s, b, Q^2)$

is reached as $g_{A,cr}(x, Q^2, b)$, we can evaluate $N(b, x, Q^2)$ as

$$N(x, b, Q^2) = g_A(x, Q^2, b)/g_{A,cr}(x, Q^2, b)\,. \tag{19}$$

As soon as $P_{\rm inel}(s, b, Q^2$ becomes close to one, one can easily evaluate the lower boundary for the energy losses arising from the single inelastic interaction of a parton, which follows from the general properties of the parton ladder. Indeed, the loss of a finite fraction of the incident parton energy, ϵ, arises from the processes of the parton fragmentation into a mass M which does not increase with an increase of the energy. For binary collisions, $M^2 = k_t^2/[\epsilon(1-\epsilon)]$, where k_t is the transverse momentum of the incident parton after the inelastic collision. For the contribution of small ϵ, e.g., $\epsilon \leq 1/4$,

$$\epsilon \approx k_t^2/M^2\,. \tag{20}$$

In the single ladder approximation (NLO DGLAP) approximation the spectrum over the masses M^2 has the following form:

$$\frac{d\tilde{N}}{dM^2} \propto \frac{1}{M^2}(s/M^2)^{\lambda}\theta(M^2 - 4k_t^2)\,, \tag{21}$$

where we have accounted for the high energy behavior of the two-gluon ladder amplitude and effectively taken into account the energy-momentum conservation, i.e., NLO effects. Consequently, the *average* energy loss [for the contribution of relatively small energy losses, $\epsilon \leq \gamma \sim 1/4$, where approximation of Eq. (20) is valid] can be estimated as:

$$\epsilon_N \equiv \langle\epsilon\rangle = \frac{\int_0^{\gamma} d\epsilon\epsilon/\epsilon^{1-\lambda}}{\int_0^{\gamma} d\epsilon/\epsilon^{1-\lambda}} = \gamma\frac{\lambda}{1-\lambda}\,. \tag{22}$$

For the realistic case $\gamma = 1/4$ and $\lambda = 0.2$, this calculation gives the fractional energy loss of 6%. This is a lower limit since we neglected here a significant contribution of larger ϵ.

In the kinematics of the onset of the BDR, the absorption at central impact parameters is due to $N(x, b, Q^2) > 1$ inelastic collisions (interactions with several ladders). The energy of the initial parton is shared before collisions at least between $N(x, b, Q^2)$ constituents in the wave function of the incident parton to satisfy causality and energy-momentum conservation. This quantum field theory effect, which is absent in the framework of eikonal approximation, can be interpreted as an additional energy loss:[16,17]

$$\epsilon_A(b) \approx N(x, b, Q^2)\epsilon_N\,. \tag{23}$$

where ϵ_N is the fraction of energy lost due to the exchange by one ladder, see Eq. (22). Equation (23) is written for the regime when the interaction with the nucleus is close to the BDR, but the interaction with a single nucleon is still far from the BDR. To correct for the onset of the BDR at the single nucleon level, one needs to take into account small probability of secondary interactions. The inclusion of enhanced "Pomeron" diagrams which are suppressed by the energy-momentum conservation would not change our conclusions.

Note also the effective energy losses are somewhat larger for gluons as the $g \to$ gg splitting is more symmetric in the light cone fractions than the qg splitting.

6. Leading hadron production in hadron–nucleus scattering

Production of leading hadrons with $p_t \sim$ few GeV/c in hadron–nucleus scattering at high energies provides a sensitive test of the onset of the BDR. Indeed in this limit pQCD provides a good description the forward single inclusive pion production in pp scattering.[18] At the same time it was found to overestimate grossly the cross section of the pion production in $d - Au$ collisions in the same kinematics. The analysis of Ref. 19 has demonstrated that the dominant mechanism of the single pion production in the NN collisions in the kinematics which was studied at RHIC is scattering of leading quark of the nucleon off the gluon of the target with the median value of x for the gluon to be in the range $x_g \sim 0.01 \div 0.03$ depending on the rapidity of the pion. The nuclear gluon density for such x is known to be close to the incoherent sum of the gluon fields of the individual nucleons since the coherent length in the interaction is rather modest for such distances (cf. Fig. 1). As a result the leading twist nuclear shadowing effects can explain only a very small fraction of the observed suppression[19] and one needs a novel dynamical mechanism to suppress generation of pions in such collisions. It was pointed out in Ref. 19 that the energy fractional energy losses on the scale of $10 \div 15\%$ give a correct magnitude of suppression of the inclusive spectrum due to a steep fall of the cross section with x_F which is consistent with the estimates within the post-selection mechanism.

An important additional information comes from the correlations studies where correlation of the leading pion with the pion produced at the central rapidities x[20,21] which corresponds to the kinematics which receives the dominant contribution from the scattering off gluons with $x_g \sim 0.01 \div 0.02$. The rate of the correlations for pp scattering is consistent with pQCD expectations. An extensive analysis performed in Ref. 17 has demonstrated that the strengths of "hard forward pion" – "hard $\eta \sim 0$ pion" correlations in dAu and in pp scattering are similar. A rather small difference in the pedestal originates from the multiple soft collisions. Smallness of the increase of the soft pedestal as compared to pp collisions unambiguously demonstrates that the dominant source of the leading pions is the dAu scattering at large impact parameters. This conclusion is supported by the experimental observation[22] that the associated multiplicity of soft hadrons in events with forward pion is a factor of two smaller than in the minimal bias dAu events. A factor of two reduction factor is consistent with the estimate of Ref. 17 based on the analysis of the soft component of $\eta = 0$ production for the forward pion trigger. Overall these data indicate that (i) the dominant source of the forward pion production is $2 \to 2$ mechanism, (ii) production is dominated by projectile scattering at large impact parameters, (iii) proportion of small x_g contribution in the inclusive rate is approximately the same for pp and dAu collisions.

A lack of additional suppression of the $x_g \sim 0.01$ contribution to the double inclusive spectrum as compared to the suppression of the inclusive spectrum is explained in the post-selection mechanism as due to relatively small momentum of the gluon in the nucleus rest frame putting it far away from the BDR.

It is difficult to reconcile enumerated features of the forward pion production data with the $2 \to 1$ mechanism[23] inspired by the color glass condensate model. In the scenario of[23] incoherent $2 \to 2$ mechanism is neglected, a strong suppression of the recoil pion production is predicted. Also it leads to a dominance of the central impact parameters and hence a larger multiplicity for the central hadron production in the events with the forward pion trigger. The observed experimental pattern indicates the models[24] which neglect contribution of the $2 \to 2$ mechanism and consider only $2 \to 1$ processes strongly overestimates contribution of the $2 \to 1$ mechanism to the inclusive cross section.

7. Production of two forward pions and double-parton mechanism in *pp* and *dA* scattering

In Ref. 19 we suggested that in order to study the effects of small x gluon fields in the initial state one should study production of two leading pions in nucleon–nucleus collisions. Recently the data were taken on production of two forward pions in dAu. The preliminary results of the studies of the reactions $pp \to \pi^0\pi^0+X$, $dAu \to \pi^0\pi^0+X$, where one leading pion served as a trigger and the second leading pion had somewhat smaller longitudinal and transverse momenta.[25,26] The data indicate a strong suppression of the back to back production of pions in the central $d - Au$ collisions. Also a large fraction of the double inclusive cross section is isotropic in the azimuthal angle $\Delta\varphi$ of the two pions. In order to understand the origin of the suppression and other features of the data we performed a study[27] which is summarized below.

It is instructive to start with the case of pp scattering. In the leading twist mechanism $2 \to 2$ a leading quark from the nucleon and a small x gluon scatter to produce two jets with leading pions. In this kinematics $x_g \leq M^2(\pi\pi)/x_q s_{NN}$. Production of two pions which together carry a large fraction of the nucleon momentum can occur only if x_q is sufficiently close to one. We find that average value of x_q for typical cuts of the RHIC experiments is very high.

Obviously it is more likely that two rather than one quark in a nucleon carry together x close to one. This suggests that in the discussed RHIC kinematics production the "double-scattering" contribution with two separate hard interactions in a single pp collision could become important. Hence though the discussed contribution is a "higher-twist", it is enhanced both by the probability of the relevant two quark configurations and the increase of the gluon density at small x which enters in the double-scattering in the second power.

One can derive the expression for the double-scattering mechanism based on the analysis of the corresponding Feynman diagrams and express it through the new rank two generalized parton densities (GPDs) in the nucleons, see Ref. 28 and refs. therein. The cross section can be written in the form

$$\frac{d^4\sigma}{dp_{T,1}d\eta_1 dp_{T,2}d\eta_2} = \frac{1}{\pi R_{int}^2} \sum_{abcda'b'c'd'} \int dx_a dx_b dz_c dx_{a'} dx_{b'} dz_{c'}$$

$$f_{aa'}^{H_1}(x_a, x_{a'}) f_b^{H_2}(x_b) f_{b'}^{H_2}(x_{b'}) D_c^{h_1}(z_c)\, D_{c'}^{h_1}(z_{c'})\, \frac{d^2\hat{\sigma}^{ab\to cd}}{dp_{T,1}d\eta_1} \frac{d^2\hat{\sigma}^{a'b'\to c'd'}}{dp_{T,2}d\eta_2}\,. \quad (24)$$

Here $f_{aa'}^{H_1}(x_a, x_{a'})$ is the double parton distribution. If the partons are not correlated, it is equal to the product of the single parton distributions. For simplicity we neglected here correlations in the target as in our case $x's$ for gluons are small. The dimensional factor πR_{int}^2 characterizes the transverse spread of the parton distributions in two nucleons as well as possible transverse parton–parton correlations and can be expressed through the integral of the product of two rank two GPDs (in many experimental papers and some theoretical papers this quantity is denoted as σ_{eff} though it has little to do with the interaction cross section of the colliding hadrons). In the approximation when partons are not correlated in the transverse plane one can

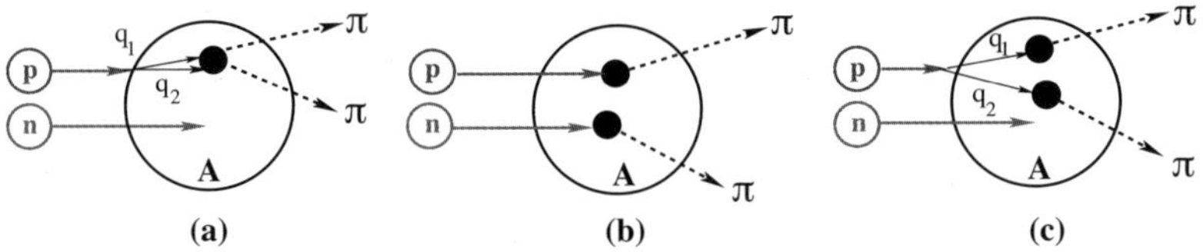

Fig. 3. Three double parton mechanisms of dipion production.

express πR_{int}^2 through the single parton GPDs, leading to $\pi R_{int}^2 \approx 34mb$.[29] This is larger than $\pi R_{int}^2 \approx 15$ mb observed at the Tevatron, indicating presence of the correlations in the nucleon. In our numerical calculations we will use this experimental value. We find that for the RHIC kinematics the only trivial correlation due to the fixed number of the valence quarks is important while the correlation between x_a and $x_{a'}$ remains a small correction if we follow the quark counting rules to estimate the $x_{a'}$ dependence off $f_{a,a'}(x_a, x_{a'})$ for fixed x_a.We find that that the LT and double parton mechanisms are comparable for the kinematics of the RHIC experiments. This provides a natural explanation of the experimental observation[25,26] that the number of events in the pedestal which does not depend on the azimuth angle between two pions, $\Delta\varphi$ is similar to — actually roughly a factor of two larger — than the number of events in the peak around $\Delta\varphi \sim \pi$.

Hence we conclude that the current experiments at RHIC have found a signal of double-parton interactions and that future experiments at RHIC will be able to obtain a unique information about double quark distributions in nucleons. It will be crucial for such studies to perform analyses for smaller bins in η and preferably switch to the analysis in bins of Feynman x.

Let us now consider the case of d-Au scattering studied at RHIC. There are three distinctive double-parton mechanisms depicted in Fig. 3. The first two are the same as in the pA scattering–scattering of two partons of the nucleon off two partons belonging to different nucleons (mechanism a), and off two partons belonging to the same nucleon of the target (mechanism b).[30] The third mechanism, which is not present for pA scattering is scattering of one parton of proton and one parton of the neutron off two partons of the nucleus. Let us consider the ratio of the double-parton and leading twist contributions for dA and pp collisions

$$r_{dA} = r_a + r_b + r_c = \frac{\sigma_{DP}(dA)}{\sigma_{LT}(dA)} / \frac{\sigma_{DP}(pp)}{\sigma_{LT}(pp)}. \tag{25}$$

The contribution to r_A of the mechanisms (a), (c) is given by:[30]

$$r_c = T(b)\sigma_{eff}; r_a = 1, \tag{26}$$

where $T(b)$ is the standard nuclear profile function ($\int d^2bT(b) = A$). Here we neglected nuclear gluon shadowing effect which is a small correction for the double-parton mechanism (cf. Ref. 19) but which is important for the LT mechanism where x_g maybe as low as 10^{-3} due to the leading twist shadowing. For the central d-Au collisions $T_A \approx 2.2 fm^{-2}$ and so $r_b/r_c \sim 1/3$ and $r_b + r_c \sim 4.4$. The contribution (b) can be calculated in a model independent way since no parton correlations enter in this case. $r_c \approx r_b$ for moderate rapidities where correlations between partons are not important. It reaches $r_c \sim 2r_b/3$ for the kinematics where only valence quarks contribute but $x_q + x_{q'} \ll 1$. For the very forward region $r_c \ll r_b$ since the kinematic constrain $x_q + x_{q'} \leq 1$ is not present in this case. As a result r_{dAu} for small b becomes of the order ten — r_A changes from ~ 9 to ~ 12 for $\pi R_{int}^2 = 15 \div 20$ mb.

Since the single inclusive pion spectrum for $\eta_2 \sim 2 \div 3$ is suppressed by a factor of the order $R_A(b) = 1/3 \div 1/4$ we find for the ratio of the pedestals in dAu and pp:

$$R_{pedestal} = r_A R_A(b) \sim 2.5 \div 4, \tag{27}$$

which should be compared with the experimental value of $R_{pedestal} \sim 3$. Hence we naturally explain the magnitude of the enhancement of the pedestal in the central $d - Au$ collision.

If most of the pedestal in the kinematics studied at RHIC is due to the double-parton mechanism, the uncertainties in the estimate of the rates due to this mechanism and uncertainties in the strength of the suppression of the single inclusive forward pion spectrum at $b \sim 0$ would make it very difficult to subtract this contribution with a precision necessary to find out whether all pedestal is due to double-parton mechanism or there is a room for a small contribution of the $2 \rightarrow 1$ broadening mechanism as it was assumed in Ref. 31.

Within the post selection picture the suppression of the away peak originating from the LT contribution is due to two effects: (i) the gluon shadowing for $x \sim 10^{-3}$ and $b \leq 3\,fm$ and $Q^2 \sim$ few GeV^2 reduces the cross section by a factor of about two (cf. Fig. 1), (ii) stronger effect of effective fractional energy losses due to larger x of the quark in the LT mechanism than in the double parton mechanism, leading to a suppression factor of the order two.[27] Combined these effects result in a suppression of the order of four as compared to the single pion trigger, and overall suppression of the order of ten. This is pretty close to the maximal possible suppression which could be estimated as the probability of the "punch through" mechanism

— contribution from the process where a quark scatters off one nucleon but does not encounter any extra nucleons at its impact parameter. Probability of such collisions at $b \sim 0$ for interaction with Au nucleus is of the order $5 \div 10\%$.[32]

The preliminary data[25] are described well as a sum of the away peak of the strength $\leq 1/4$ of the strength observed in pp scattering, and the pedestal enhanced by a factor of three relative to the pp case, corresponding to reduction of the away peak relative to pedestal of the order of ten.

8. Some implications for pp collisions at LHC

It has been understood a long ago that there are two distinctive classes of the hadron-hadron interactions: peripheral and central. The difference between the dynamics in two regimes increases with energy and at the LHC energies it becomes fact of life which should be taken into account in the analysis of any phenomena at the LHC.

Peripheral collisions are dominated by soft QCD–Pomeron physics where amplitude has the form: $\propto s^{\alpha_{I\!P}(t)-1}\exp(B_0 t/2)$. Within this physics essential impact parameters are increasing with energy: $b^2 \approx 2\alpha'_{I\!P}\ln(s/s_0)+2B_0$. At energies of LHC essential impact parameters (b^2) become larger by a factor ≈ 2 as compared to the fixed target energies. At achievable energies peripheral collisions give dominant contribution to the total inelastic pp cross. Central collisions are dominated by interplay of the Pomeron cuts and hard QCD.

Experimentally such a separation on peripheral and central pp interactions is manifested in the first LHC data on the inclusive spectrum $Ed\sigma/d^2p_T d\eta$. This spectrum is enhanced at small p_t and has a narrow peak at $p_t \leq 200$ MeV/c which cannot be explained only by decays of vector and heavier mesons.

Production of new particles is concentrated at much smaller impact parameters than average impact parameters in the inelastic collisions.[29] Thus background for new particles production should be strongly different from the minimal bias events.

At small impact parameters absorption reaches 100%. With increase of energy the regime of strong absorption is reached for interaction of partons with quite large p_t and hence unitarity should be relevant for jet production up to quite large p_t. Hence it looks plausible that small b collisions are dominated by high p_t processes.

9. Conclusions

The LT twist nuclear shadowing can be estimated with a small uncertainty using correspondence with hard diffraction leading to the expectation of a large gluon shadowing at $x \leq 10^{-3}$. This influences significantly estimates of the proximity to the black disk regime, so that for small b they are reached for proton and nuclei at comparable energies. For $b \sim 0$ black disk regime is effective for the RHIC energies leading to post selection effect (effective fractional energy losses) for the propagation of the leading partons. This is the only effect which currently allows to explain the suppression of the $2 \rightarrow 2$ mechanism for production of leading pions which dominates in the pp scattering including survival of forward – central correlations and suppression of forward – forward correlations. The processes of the production of two forward pions in pp, dAu scattering due to double -parton mechanism explain the pedestal observed at RHIC and make it difficult to search for the signal of $2 \rightarrow 1$ processes. Magnitude of the away peak is consistent with the expected magnitude of the suppression of the $2 \rightarrow 2$ mechanism due to LT gluon shadowing and fractional energy losses due to the post-selection mechanism. Also these experiments for the first time provide an opportunity to observe correlations of leading quarks in protons. At the LHC proximity to the BDR for the central pp collisions should lead to large post-selection effects including energy flow to the central rapidities, change of the p_t distributions, etc.

Acknowledgments

The research was supported by DOE grant No. DE-FG02-93ER40771 and the BSF grant 200811.

References

1. V. N. Gribov, *Sov. Phys. JETP* **29**, 483 (1969) [*Zh. Eksp. Teor. Fiz.* **56**, 892 (1969)].
2. V. A. Abramovsky, V. N. Gribov and O. V. Kancheli, *Yad. Fiz.* **18**, 595 (1973) [*Sov. J. Nucl. Phys.* **18**, 308 (1974)]
3. L. Frankfurt and M. Strikman, *Eur. Phys. J.* **A5**, 293 (1999); L. Frankfurt, V. Guzey and M. Strikman, *Phys. Rev.* **D71** 054001 (2005); for a detailed discussion see L. Frankfurt, V. Guzey and M. Strikman, to be submitted to *Phys. Rep.*
4. J. C. Collins, *Phys. Rev.* **D57**, 3051 (1998) [Erratum-ibid. **D61**, 019902 (2000)]
5. V. Guzey and M. Strikman, *Phys. Lett.* **B687**, 167 (2010).
6. M. Strikman, R. Vogt and S. N. White, *Phys. Rev. Lett.* **96**, 082001 (2006).

7. A. J. Baltz *et al.*, *Phys. Rept.* **458**, 1 (2008).
8. J. Rojo, G. Altarelli, R. D. Ball and S. Forte, arXiv:0907.0443 [hep-ph].
9. L. Frankfurt and M. Strikman, *Surveys High Energ. Phys.* **14**, 9 (1999).
10. L.D. Landau, E.M. Lifshitz, *Series: Course of Theoretical Physics, Vol. 3, Non-relativistic theory,* 1977.
11. L. Frankfurt, A. Radyushkin and M. Strikman, *Phys. Rev.* **D55**, 98 (2010).
12. T. Rogers, V. Guzey, M. Strikman and X. Zu, *Phys. Rev.* **D69**, 074011 (2004).
13. L.Frankfurt, T.Rogers and M.Strikman, in preparation.
14. B. Blok, L. Frankfurt and M. Strikman, *Phys. Rev.* **D80**, 114009 (2009).
15. V. N. Gribov, *Sov. Phys. JETP* **30**, 709 (1970) [*Zh. Eksp. Teor. Fiz.* **57**, 1306 (1969)].
16. L. Frankfurt, V. Guzey, M. McDermott and M. Strikman, *Phys. Rev. Lett.* **87**, 192301 (2001).
17. L. Frankfurt and M. Strikman, *Phys. Lett.* **B645**, 412 (2007).
18. F. Aversa, P. Chiappetta, M. Greco and J. P. Guillet, *Nucl. Phys.* **B327**, 105 (1989); B. Jager, A. Schafer, M. Stratmann and W. Vogelsang, *Phys. Rev.* **D67**, 054005 (2003); D. de Florian, *Phys. Rev.* **D67**, 054004 (2003).
19. V. Guzey, M. Strikman, and W. Vogelsang, *Phys. Lett.* **B603**, 173 (2004).
20. J. Adams *et al.* [STAR Collaboration], *Phys. Rev. Lett.* **97**, 152302 (2006).
21. B. Meredith, *Nucl. Phys.* **A830**, 595 (2009).
22. G. Rakness, private communication
23. D. Kharzeev, E. Levin and L. McLerran, *Phys. Lett.* **B561**, 93 (2003); D. Kharzeev, Y. V. Kovchegov and K. Tuchin, *Phys. Rev.* **D68**, 094013 (2003).
24. A. Dumitru, A. Hayashigaki and J. Jalilian-Marian, *Nucl. Phys.* **A765**, 464 (2006).
25. E. Braidot for the STAR collaboration, arXiv:1005.2378.
26. B. Meredith, DIS 2010 Conference Proceedings, PoS(DIS 2010)081.
27. M. Strikman and W. Vogelsang, arXiv:1009.6123 [hep-ph].
28. B. Blok, Yu. Dokshitzer, L. Frankfurt and M. Strikman, arXiv:1009.2714 [hep-ph].
29. L. Frankfurt, M. Strikman and C. Weiss, *Phys. Rev.* **D69**, 114010 (2004).
30. M. Strikman and D. Treleani, *Phys. Rev. Lett.* **88**, 031801 (2002).
31. J. L. Albacete and C. Marquet, arXiv:1005.4065 [hep-ph].
32. M. Alvioli, H. J. Drescher and M. Strikman, *Phys. Lett.* **B680**, 225 (2009), and to be published.

Bonn, 1995

THE PHENOMENOLOGY OF THE RENORMALIZED POMERON

URI MAOR

Department of Particle Physics, School of Physics and Astronomy
Raymond and Beverly Sackler Faculty of Exact Science
Tel Aviv University, Tel Aviv, 69978, Israel

An updated formulation of the Pomeron, in which s and t-channel unitarity screenings are included, is reviewed. The consequent soft scattering features are explored and it is suggested that the soft and hard Pomerons can be effectively presented by a single Pomeron trajectory. The interplay between theory and and data analysis is discussed.

1. Prologue

I met Volodya in the Summer of 1970 at the Kiev Rochester HEP Conference. Even though he was already a legendary person and I was just a young physics professor, we became fast friends. Considering the constraints imposed by the Soviet regime, we spent long nights in Kiev parks discussing every thing — personal, scientific, political and less provocative issues. For me this was one of the forming experiences of my youth for which I shall be indebted to Volodya forever.

2. Introduction

Pomeron Physics has re-emerged as a critical ingredient in the interpretation of high energy data. The present vigorous studies of the Pomeron ($I\!P$) and its dynamics are based on sophisticated utilization of relatively old theoretical ideas and models in which Gribov and his school played a decisive roll. Present day $I\!P$ physics is based on:

- Reggeon field theory, Gribov(1968).
- GW decomposition of the proton wave function, Feinberg and Pomeranchuk(1953), Good and Walker(1964).
- The eikonal approximation, Glauber(1959); Gribov(1969).

- Triple Pomeron formalism, Mueller(1971).
- Multi Pomeron interactions, Gribov(1969); Kaidalov et al.(1986).
- BFKL Pomeron, BFKL(1976-8)

The classical parametrization of the Pomeron Regge trajectory is

$$\alpha_{I\!P}(t) = 1 + \Delta_{I\!P} + \alpha'_{I\!P} t. \tag{1}$$

It has a straightforward interpretation in which the energy dependence of a given $I\!P$ exchange scattering process is controlled by $\Delta_{I\!P}$ and the the shrinkage of the corresponding forward angular cone is controlled by $\alpha'_{I\!P}$. A very successful determination of these parameters has been suggested by Donnachie and Landshoff (DL) in which $\Delta_{I\!P} = 0.08$ and $\alpha'_{I\!P} = 0.25 GeV^{-2}$. This parametrization provided an excellent reproduction of σ_{tot}, σ_{el} and B_{el} in the ISR-Tevatron energy range. Note, though, that DL analysis ignores the diffractive channels. Indeed, a significant change of $I\!P$ physics was triggered by the realization that the very mild energy dependence of σ_{sd} in the ISR-Tevatron range is associated with the onset of unitarity corrections.

Updated $I\!P$ models have several components:

- An input non screened $I\!P$ exchange amplitude.
- s-channel unitarity is induced by eikonal screening damping associated with the re-scattered incoming hadronic projectiles.
- These re-scatterings go through elastic and "low mass" diffraction implied by the GW mechanism.
- Compatibility with t-channel unitarity is attained through multi $I\!P$ interactions leading to "high mass" diffraction and a renormalization of the Pomeron. Mueller's triple Pomeron dynamics is the lowest order of this contribution.

In the following I shall present a summary of Tel Aviv GLMM Pomeron model and its high energy predictions. These will be compared with 2 versions of Durham KMR models, their simplified LKMR model and Ostapchenko (OS) model. All models discussed are based on the same principles but utilize different modelings, parametrization and data analysis procedures.

3. Good-Walker Eikonal Models

Current eikonal models are multi channels, including both elastic and diffractive re-scatterings of the initial projectiles. This is a consequence of the GW mechanism in which the proton (anti-proton) wave function has elastic and diffractive components.

Consider a system of two states, a hadron Ψ_h and a diffractive state Ψ_D, which are orthonormal. The GW mechanism stems from the observation that these states do not diagonalize the 2x2 interaction matrix $\mathbf{T}$. Lets introduce two wave functions Ψ_1 and Ψ_2 which diagonalize $\mathbf{T}$,

$$A_{i,k}^{i',k'} =< \Psi_i\,\Psi_k|\mathbf{T}|\Psi_{i'}\,\Psi_{k'} >= A_{i,k}^S\,\delta_{i,i'}\,\delta_{k,k'}. \tag{2}$$

$A_{i,k}^S$ is the elastic scattering amplitude of Ψ_i and Ψ_k. In this representation the observed hadronic and diffractive states are written as

$$\Psi_h = \alpha\,\Psi_1 + \beta\,\Psi_2 \quad \Psi_D = -\beta\,\Psi_1 + \alpha\,\Psi_2\,, \tag{3}$$

where $\alpha^2 + \beta^2 = 1$. The corresponding s-channel unitarity equations are

$$Im\,A_{i,k}^S\,(s,b) = |A_{i,k}^S\,(s,b)\,|^2 + G_{i,k}^{in}(s,b). \tag{4}$$

$G_{i,k}^{in}$ is the summed probability for all non GW inelastic processes induced by an initial (i,k) state. A general solution of Eq. (4) can be written as

$$A_{i,k}^S(s,b) = i\left(1 - \exp\left(-\frac{\Omega_{i,k}^S(s,b)}{2}\right)\right), \tag{5}$$

$$G_{i,k}^{in}(s,b) = 1 - \exp\left(-\Omega_{i,k}^S(s,b)\right). \tag{6}$$

In general the b space opacities,

$$\Omega_{i,k}^S\,(s,b) = \nu_{i,k}^S\,(s)\,\Gamma_{i,k}^S\,(s,b,...)\,, \tag{7}$$

are arbitrary. In the eikonal approximation, $\Omega_{i,k}^S$ are assumed to be real and are determined by the Born (non screened) input. From Eq. (6) we deduce that the probability that the initial projectiles (i,k) reach the final diffractive interaction unchanged, regardless of the initial re-scatterings, is

$$P_{i,k}^S(s,b) = \exp\left(-\Omega_{i,k}^S(s,b)\right). \tag{8}$$

In general, we have to consider four possible (i,k) elastic re-scattering options. For initial p-p (or $\bar{p}$-p) the two off diagonal amplitudes are equal, $A_{1,2}^S = A_{2,1}^S$. The corresponding elastic, SD and DD amplitudes are

$$a_{el}(s,b) = i\{\alpha^4 A_{1,1}^S + 2\alpha^2\beta^2 A_{1,2}^S + \beta^4 A_{2,2}^S\}, \tag{9}$$

$$a_{sd}(s,b) = i\alpha\beta\{-\alpha^2 A_{1,1}^S + (\alpha^2 - \beta^2)A_{1,2}^S + \beta^2 A_{2,2}^S\}, \tag{10}$$

$$a_{dd}(s,b) = i\alpha^2\beta^2\{A_{1,1}^S - 2A_{1,2}^S + A_{2,2}^S\}. \tag{11}$$

The GW mechanism was originally conceived so as to describe a system of a nucleon plus its diffractive N^* isobars. Obviously, this simplistic approach is not suitable for high energy diffraction where M^2_{diff} is bounded by 0.05s. This is an arbitrary bound commonly used, implying a continua of the diffractive Fock states. Two procedures were devised to comply with the above. GLMM lump together all GW diffractive states to an effective $|D>$ state with M^2_{diff} distribution bounded by 0.05s. The non GW $I\!P$ enhanced high mass diffraction is added to the GW diffractive mass to form the over all diffractive mass distribution and integrated cross section. In this approach the GW contribution is very significant and the mass distribution is smooth. KMR chose to confine GW diffraction to low $M^2_{diff} < 20GeV^2$, to which they add the high mass $I\!P$ enhanced contribution. In this approximation the bulk of the diffractive mass distribution is non GW, and its smoothness at $20GeV^2$ is not secured.

GLMM, KMR, LKMR and OS are multi channels eikonal models in which the initial re-scatterings of the incoming projectiles are obtained from a sum over the GW eigen states. To this end we need a parametrization of $\Omega^S_{i,k}$. In GLMM $\nu^S_{i,k}(s) = g_i g_k(\frac{s}{s_0})^{\Delta_{I\!P}}$. $\Gamma^S_{i,k}$ is given as a b-transform of a two t-poles expression ($t = -q^2$) Setting $\alpha'_{I\!P}$=0, these profiles are energy independent

$$\frac{1}{(1\,+\,q^2/m_i^2)^2} \times \frac{1}{(1\,+\,q^2/m_k^2)^2} \quad\Longrightarrow\quad \Gamma^S\left(b; m_i, m_k; \alpha'_{I\!P} = 0\right). \qquad (12)$$

GLMM introduce a small energy dependence

$$m_i^2 \;\Longrightarrow\; m_i^2(s) \;\equiv\; \frac{m_i^2}{1\,+\,4m_i^2\alpha'_{I\!P}\ln(s/s_0)}. \qquad (13)$$

This parametrization is compatible with the requirements of analyticity/crossing symmetry at large b, pQCD at large q^2 and Regge at small q^2. KMR, LKMR and OS use different parametrization for $\Gamma^S_{i,k}$, which are numerically compatible with GLMM. All these models well reproduce $d\sigma_{el}/dt$ in the forward $t \leq 0.5GeV^2$ angular cone.

4. Multi Pomeron Interactions

Consider a single diffraction channel $p + p \to p + M_{sd}$. Mueller's triple Pomeron mechanism, derived from 3 body unitarity, leads to high SD mass which is non GW. Neglecting the Regge contribution in the high energy limit, it depends on $g_{I\!Pp}$ and $G_{3I\!P}$. This approximation is valid for $s >> M^2_{sd} >> m^2_p$.

CDF analysis of SD suggests a relatively large value of $G_{3I\!P}$. Consequently, we need to consider a very large family of higher order multi Pomeron interactions which are not included in the GW mechanism. This dynamical feature is compatible with t-channel unitarity. It initiates additional screening to the s-channel unitarity screening, reducing the calculated soft cross sections and gap survival probabilities further more. **This feature becomes significant above the Tevatron energy.**

GLMM, KMR and OS treatment of multi $I\!P$ interactions stems from Gribov and Kaidalov et al. classical papers. The complexity of the relevant $I\!P$ diagrams requires summing algorithms which are model dependent. Indeed, the quoted models utilize different diagram summing procedures. Recall that LKMR confined themselves to Mueller's low order $3I\!P$ formalism. Consider the Pomeron Green function shown in Fig. 1. In low order the en-

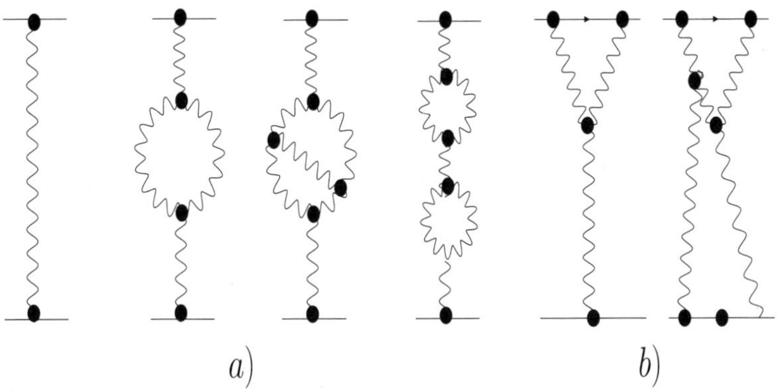

Fig. 1. Typical low order terms of the Pomeron Green's function. a) Enhanced diagrams. b) Semi-enhanced diagrams.

hanced diagrams re-normalize the $I\!P$ propagator while the semi enhanced diagrams re-normalize the $I\!P$ vertexes. This clean identification is lost at higher orders where the 2 sets of diagrams interfere.

The predicted energy dependence of a given $I\!P$ exchange process can be expressed in terms of $\Delta_{I\!P}^{eff}$ which is particular to the studied channel. The monotonic decrease of $\Delta_{I\!P}^{eff}$ with energy raises doubts if it may become negative. $\Delta_{I\!P}^{eff}(GLMM)$, obtained from σ_{tot}, is shown in the Table below. It is positive up to W = 100TeV, which is our validity bound.

Whereas σ_{tot} and σ_{el} are well reproduced by the non screened DL model in the ISR-Tevatron range, the energy dependence of SD and DD is much

W	$1.8 \to 7.0$	$7.0 \to 14.0$	$14.0 \to 30.0$
Δ_{eff}	0.059	0.049	0.043

W	$30.0 \to 60.0$	$60.0 \to 100.0$
Δ_{eff}	0.040	0.037

milder. This is a signature of the onsetting of s-channel unitarity corrections, commonly formulated in the eikonal approximation. These cross sections are further reduced by the t-channel multi Pomeron interactions some of which are shown in Fig. 2.

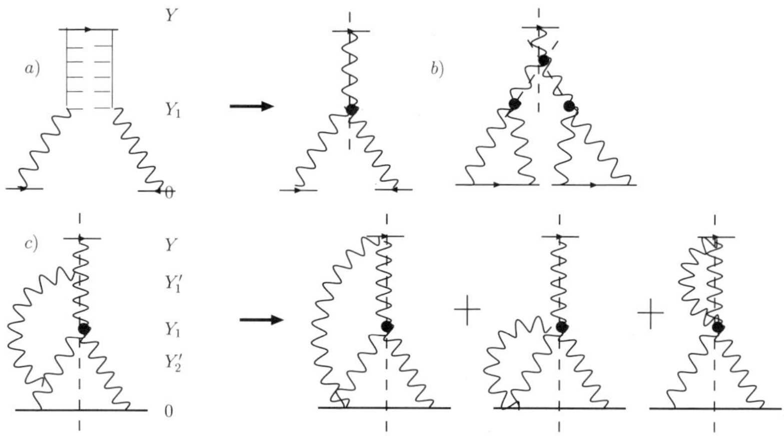

Fig. 2. Examples of $I\!P$ diagrams leading to diffraction. $Y - Y_1 = ln(M^2/s_0)$.

A key observation of GLMM is that the exceedingly small fitted value of $\alpha'_{I\!P}$ implies that the "soft" $I\!P$ is hard enough to be treated perturbatively. Following Gribov we identify the correlation between $\alpha'_{I\!P}$ and $< p_t >$, the mean transverse momentum of the partons (actually, color dipoles) associated with the Pomeron. $< p_t > \propto 1/\sqrt{\alpha'_{I\!P}}$, from which we deduce that the QCD running coupling constant $\alpha_S << 1$. Accordingly, we proceed from Gribov's parton model interpretation to pQCD. In this respect GLMM is different from KMR and OS which are partonic models.

Technically, GLMM have adopted the MPSI procedure in which the transition $nI\!P \to mI\!P$ is reduced to a sequence of triple $I\!P$ vertexes (Fan diagrams). GLMM approach is completely different from the approach of KMR and OS who, following Kaidalov et al., assume a point coupling for

$n\mathbb{P} \to m\mathbb{P}$ $(n+m>2)$,

$$g_m^n = \frac{1}{2} g_N \, nm \, \lambda^{n+m-2} = \frac{1}{2} nm \, G_{3\mathbb{P}} \, \lambda^{n+m-3}. \quad (14)$$

λ is a free parameter and $G_{3\mathbb{P}} = \lambda g_N$. Note that, KMR normalization is some what different from Kaidalov et al.

The updated $\mathbb{P}$ is different from the classical $\mathbb{P}$, in as much as its signature is an exceedingly small $\alpha'_{\mathbb{P}}$. As such, the effect of $\alpha'_{\mathbb{P}}$ on the forward angular cone is negligible. This deficiency is compensated by a large $\Delta_{\mathbb{P}}$ which induces strong screenings in both s and t-channels.

The consequent main features of the GLMM Pomeron are:

- GLMM have a single $\mathbb{P}$: $\alpha_{\mathbb{P}}(t) = 1.335 + 0.01t$.
- The fitted $\mathbb{P}$ parameters are remarkably close to the BFKL $\mathbb{P}$ after NLL corrections are summed.
- The $\mathbb{P}$ parameters are in accord with Zeus and H1 fits to HERA DIS data.
- The fitted $\alpha'_{\mathbb{P}} \to 0$ is a necessary condition to connect the npQCD soft interactions with hard pQCD.
- Recall that, $\alpha'_{\mathbb{P}}(BFKL)$ relates to the saturation scale $\alpha'_{\mathbb{P}} \propto 1/Q_s^2 \to 0$, as $s \to \infty$.

Regardless of their differences, GLMM and KMR $\mathbb{P}$ models are reasonably compatible. The starting point of both investigations is the observation that a GW eikonal model reproduces the elastic data well, but its reproduction of the diffractive sector is deficient. Both groups claim to achieve an improved reproduction of their data base once the contributions of enhanced Pomeron diagrams are included. They disagree, though, on their estimates of the LHC diffractive cross sections, gap survival probabilities and the approach of $a_{el}(s,b)$ toward the black disc bound. These disagreements can be traced to GLMM and KMR different diagram summation, to be discussed here, and data analysis, to be discussed in the next Section.

Having different diagram summations the resulting GLMM screening is stronger than KMR. This supposition is supported by a comparison of GLMM and KMR gap survival probabilities. Denote the gap survival factor initiated by s-channel eikonalization S^2_{2ch}, and the one initiated by t-channel multi $\mathbb{P}$ interactions S^2_{enh}. $S^2 = S^2_{2ch} \cdot S^2_{enh}$. The Table below compares the gap survival factors of a Higgs produced in a central exclusive diffractive production. GLMM and KMR S^2_{2ch} outputs are compatible. However, $S^2_{enh}(GLMM)$ is smaller and reduces faster than $S^2_{enh}(KMR)$.

This is compatible with GLMM screening being stronger and increasing faster than KMR.

	$1.8\,TeV$ GLMM	$1.8\,TeV$ KMR07	$14\,TeV$ GLMM	$14\,TeV$ KMR07	$14\,TeV$ KMR08	$100\,TeV$ GLMM	$100\,TeV$ KMR07
$S^2_{2ch}(\%)$	5.3	2.7-4.8	3.9	1.2-3.2	4.5	3.2	0.9-2.5
$S^2_{enh}(\%)$	28.5	100	6.3	100	33.3	3.3	100
$S^2(\%)$	1.51	2.7-4.8	0.24	1.2-3.2	1.5	0.11	0.9-2.5

5. The Interplay Between Theory and Data Analysis

Given a phenomenological model, its data analysis depends on the construction of a suitable data base to which we couple a reliable parameter adjustment procedure. KMR data base is more confined than GLMM. Consequently, the data analysis procedures applied by GLMM and KMR are different. GLMM adjusted their free parameters by fitting their complete data base in a single step. KMR chose to tune their free parameters through a 2 step procedure. In the first step they determine the GW free parameters by adjusting $d\sigma_{el}/dt$. These parameters are frozen in the second step in which the rest of the free parameters are determined by tuning $d\sigma_{sd}/dtd(M^2_{sd}/s)$. I am not clear at which stage $\Delta_{I\!P}$ was determined.

The final values KMR obtain for $\Delta_{I\!P}$ and $\alpha'_{I\!P}$ are compatible with GLMM. However, the final values of g_1 and g_2 are very different, in as much as $g_1 = g_2$ is assumed in KMR and $g_1 << g_2$ is fitted in GLMM. I presume that KMR choices reflect the deficiencies of their data base. The different outputs obtained in a single and two steps methods, when applied to the GLMM data base, are demonstrated in the Table below.

	$\Delta_{I\!P}$	β	$\alpha'_{I\!P}$ GeV^{-2}	g_1 GeV^{-1}	g_2 GeV^{-1}	m_1 GeV	m_2 GeV	χ^2 $/dof$
GW	0.120	0.46	0.012	1.27	3.33	0.913	0.98	0.87
GW+$I\!P$	0.335	0.34	0.010	5.82	239.6	1.54	3.06	1.00

A fit with a GW model (no $I\!P$-enh) provides excellent reproduction of the elastic sector while the the reproduction of the diffractive sector is very poor. The repeated fit with a GW+$I\!P$-enh model results with a very good χ^2. Checking the values of the fitted parameters, we note that the exceedingly small value of $\alpha'_{I\!P}$ is persistently obtained in both GLMM model versions, as well as in KMR and LKMR. This common feature is a consequence of the similar b-profiles $\Lambda^S_{i,k}$ which control the $\alpha'_{I\!P}$ output, decoupled from the proposed dynamics. The outputs of $\Delta_{I\!P}$ and g_2 change drastically once $I\!P$-enh is included. We note that $\Delta_{I\!P}$ in GLMM GW fit is compatible with LKMR. The common element in these models is that $\Delta_{I\!P}$ is obtained from a GW analysis of just the elastic sector. GLMM and KMR high energy Tevatron, LHC and Cosmic Rays predicted cross sections are summarized in the Table below. The elastic and total cross section outputs of the two models are compatible and, above the Tevatron, significantly lower than those obtained in models with no multi-Pomeron contributions. This is a consequence of $\Delta_{I\!P}$ renormalization due to the multi $I\!P$ contributions. KMR σ_{sd} and σ_{dd} predictions are systematically larger than GLMM, reflecting the KMR weaker screening.

	W = 1.8 TeV			**W = 14.0 TeV**		
	GLMM	**KMR07**	**KMR10**	**GLMM**	**KMR07**	**KMR10**
σ_{tot}	**73.3**	**74.0**	**73.9**	**92.1**	**88.0**	**86.3**
σ_{el}	**16.3**	**16.3**	**15.1**	**20.9**	**20.1**	**18.1**
σ_{sd}	**9.8**	**10.9**	**12.7**	**11.8**	**13.3**	**16.1**
σ_{dd}	**5.4**	**7.2**	**8.7**	**6.1**	**13.4**	**12.9**

	W = 100.0 TeV		
	GLMM	**KMR07**	**KMR10**
σ_{tot}	**108.0**	**98.0**	**94.3**
σ_{el}	**24.0**	**22.9**	**20.0**
σ_{sd}	**14.4**	**15.7**	**17.7**
σ_{dd}	**6.3**	**17.3**	**21.1**

The basic GW amplitudes of the GLM models are $A^S_{1,1}$, $A^S_{1,2}$ and $A^S_{2,2}$. These are the building blocks with which we construct a_{el}, a_{sd} and a_{dd} (Eq. (9)–Eq. (11)). The $A^S_{i,k}$ amplitudes are bounded by the s-channel unitarity black disc bound of unity. $a_{el}(s,b)$ reaches this bound at a given (s,b) when, and only when, $A^S_{1,1}(s,b) = A^S_{1,2}(s,b) = A^S_{2,2}(s,b) = 1$, independent of the value of β. Consequently, when $a_{el}(s,b) = 1$, $a_{sd}(s,b) = a_{dd}(s,b) = 0$.

Checking the GLMM fitted parameters, we observe that g_1 and g_2, are comparable. Indeed, the approach of $a_{el}(s, b = 0)$ to unity in GLMM GW version, is compatible with the results obtained by KMR, LKMR and OS. This picture changes radically in GLMM GW+$I\!P$-enh version, in which the fitted $g_2 >> g_1$ are coupled to the successful reproduction of the diffractive data base. Similar results were also obtained in a previous GW type GLM model, where we were able to reproduce the diffractive data only after adjusting $g_2 >> g_1$. As a consequence of the above, the three basic $A^S_{i,k}$ amplitudes reach the black disc bound at different energies. The net result is a very slow approach of $a_{el}(s, b = 0)$ toward the black disc bound in the GLMM model, reaching the bound well above the LHC energy.

6. Concluding Remarks

I conclude with a few comments:

1) The main conclusion of this summary is that the $I\!P$ is renormalized by Pomeron-enhanced dynamics. Consequently, the exceedingly high energy behaviour of a_{el}, a_{sd} and a_{dd} and the gap survival probabilities are determined by s and t unitarity considerations.
2) In my opinion, the GLMM and KMR compatible predictions of lower than expected total and elastic cross sections at the LHC and AUGER are of critical value. If verified by LHC data, they will provide a decisive support for the importance of multi-Pomeron interactions at high enough energies.
3) GLMM and KMR have rather different predictions of the gap survival probabilities at LHC. This disagreement originates from the difference between the 2 models diagram summations. It implies that GLMM diffractive cross sections are systematically lower than KMR. Hopefully, some relevant data on this issue would become available soon.
4) If multi $I\!P$ interaction play an important roll, a simple continuation from the Tevatron to LHC is not reliable. At this stage quite a bit of prudence is recommended to model and Monte Carlo builders.
5) The novelty of the GLMM approach is that it correlates the smallness of the fitted $\alpha'_{I\!P}$ with the hardness of the presumed "soft Pomeron". This allows one to treat the "soft Pomeron" perturbatively. consequently GLMM is a single $I\!P$ model, in which the traditional "soft" and BFKL "hard" Pomeron have the same effective trajectory.
6) As I demonstrated, data analysis may determine the output dynamics. I emphasise the importance of constructing a suitable data base and reliable data analysis procedures so as to test theoretical models and determine their free parameters.
7) There are 2 possible GLMM deficiencies:
   (i) It sums only the enhanced diagrams.
   (ii) The MPSI procedure does not explicitly depend on b.

These problems are currently addressed.

WEAK COUPLING OF POMERONS AND THE TRIPLE-REGGE ANALYSIS

E.G.S. LUNA[a], V.A. KHOZE[b,c], A.D. MARTIN[b] and M.G. RYSKIN[b,c]

[a] *Instituto de Física e Matemática, Universidade Federal de Pelotas, Pelotas, Caixa Postal 354, CEP 96010-900, Brazil*

[b] *Institute for Particle Physics Phenomenology, University of Durham, Durham, DH1 3LE, United Kingdom*

[c] *Petersburg Nuclear Physics Institute, Gatchina, St. Petersburg, 188300, Russia*

We study the case when the triple-Pomeron vertex is assumed to have a vectorial form, that is, the amplitude of high-mass diffractive dissociation vanishes as $V \propto \vec{q}_t \cdot \vec{e}$ as $q_t \to 0$. We find that the available data in the triple-Reggeon region may be well described in such a 'weak' coupling scenario, providing that absorptive effects are taken into account. We compare this weak (vector) coupling scenario with the strong and weak (scalar) coupling scenarios. Corresponding predictions are presented for an LHC energy of 14 TeV.

Keywords: Triple-Pomeron vertex; screening corrections.

1. Introduction

The energy behaviour of the scattering amplitude may be consistently described by two different scenarios for the asymptotic regime (for recent reviews see Refs. 1, 2). One is called the *weak coupling* of the Pomerons. In this case, at very high energy, $\sqrt{s}$, the cross sections tend to the universal constant value

$$\sigma_{\rm tot} \to \text{constant} \qquad \text{as} \qquad s \to \infty. \tag{1}$$

In order not to violate unitarity, the triple-Pomeron coupling must vanish with vanishing transverse momentum, q_t, transferred through the Pomeron[3,4]

$$g_{3P} \propto q_t^2 \qquad \text{as} \qquad q_t \to 0. \tag{2}$$

Another possibility is called the *strong coupling* scenario.[5] Here, at a very high energies, the cross sections grows as

$$\sigma_{\rm tot} \propto (\ln s)^{\eta} \qquad \text{with} \qquad 0 < \eta \leq 2, \tag{3}$$

and the bare vertex

$$g_{3P}|_{q_t \to 0} \quad \to \quad \text{constant.} \tag{4}$$

The present data are usually described within the Froissart-like limit of the second scenario (with $\eta = 2$). However to reach asymptotics we need **very** high energy – the energy at which the slope of the elastic amplitude, $B = B_0 + \alpha'_P \ln(s)$ is dominated by the second term, that is when $\alpha'_P \ln(\mathrm{s}) \gg B_0$. This is far beyond the energies available at present. Another possibility, to distinguish between the *weak* and *strong* approaches, is to study the q_t dependence of the bare triple-Pomeron vertex.[6,7] Thus, it is important to extract the *bare* vertex before its behaviour is affected by absorptive corrections.

In Ref. 8 we analysed the data in the triple-Pomeron region accounting for absorptive effects in the framework of a two-channel eikonal, see also Ref. 9. That is, we performed a triple-Regge analysis of the available $d^2\sigma/dtd\xi$ data for $pp \to pX$ and $\bar{p}p \to \bar{p}X$ (where $\xi = M^2/s$ and M is the mass of system X), allowing for screening (absorptive) effects. To be precise, we fitted the CERN-ISR,[10] FNAL fixed-target[11] and Tevatron[12] data for $pp \to pX$ and $\bar{p}p \to \bar{p}X$. The differential distributions for the FNAL fixed target and Tevatron experiments can be found in Ref.[13] Both the 'strong' and 'weak' coupling scenarios were considered. We found that the data favoured the 'strong' coupling scenario.[8]

However, there is another possibility which should be studied. In the 'weak' coupling scenario considered in Ref. 8 a 'scalar' form of the vanishing of the coupling as $q_t \to 0$, (2), was assumed. As was pointed out by V.N. Gribov,[14] it is natural to have a vector form of the triple-Pomeron vertex which vanishes as $q_t \to 0$. Indeed in Feynman diagrams for the $pp \to p + X$ amplitude we never deal with $\sqrt{q^2}$, but rather with vector $\vec{q}$ multiplied by some vector $\vec{e}$ that characterises the final state X. Thus we may have a weak coupling in which the vertex of $p \to X$ dissociation has the vectorial form

$$V \;\propto\; \vec{q}_t \cdot \vec{e} \qquad (\text{that is } \sqrt{g_{3P}} \propto \vec{q}_t). \tag{5}$$

An example is photon-exchange. In this case the vector form of the $p \to X$ vertex comes from current conservation (gauge invariance), and is clearly seen in the Weizsäcker-Williams approach,[15,16] where the polarisation vector of a Coulomb-like photon, $\vec{\epsilon}$, is replaced by $\vec{\epsilon} = \vec{q}_t/x$. Here, the (Coulomb) photon plays the role of Pomeron exchange and 'x' is the momentum fraction transferred through the photon (or Pomeron).

The 'polarisation' structure, (5), will change the predicted cross sections, since it leads to different screening corrections. However, can this 'vectorial weak' scenario describe the data in the triple-Pomeron domain?

2. Screening Corrections in the Triple-Regge Formalism

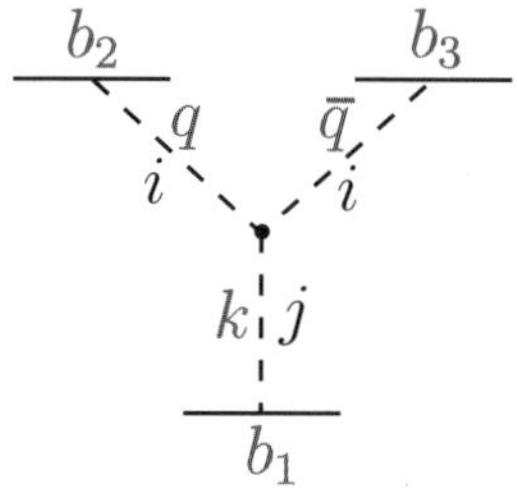

Fig. 1. A schematic diagram showing the notation of the impact parameters arising in the calculation of the screening corrections to the iij triple-Regge diagram. The conjugate momenta to b_1, b_2, b_3 are $k_t, q_t, \bar{q}_t$. If $k_t = 0$, then $\bar{q}_t = q_t$.

If, for the moment, we neglect the screening correction, then the iij triple-Regge diagram of Fig. 1 gives the contribution

$$\frac{M^2 d\sigma}{dt dM^2} = \beta_j(0)\beta_i^2(t) g_{iij}(t) \left(\frac{s}{M^2}\right)^{2\alpha_i(t)-2} \left(\frac{M^2}{s_0}\right)^{\alpha_j(0)-1}, \qquad (6)$$

where β_i is the coupling of Reggeon i to the proton, $\alpha_i(t)$ is the trajectory of Reggeon i, and M is the mass of the system X produced by proton dissociation (in Fig. 1 this system is described by Reggeon j). Let us first recall the 'strong' coupling case, (4), that we studied in Ref. 8. For this we used a simple exponential parametrisation of the triple-Regge vertices

$$g_{iij}(t) = g_{iij}(0) \exp(b'_{iij}(q^2 + \bar{q}^2 - k_t^2)), \qquad (7)$$

where the momenta are defined in Fig. 1, and where $q^2 = t_{\min} - q_t^2/x_L$.

Screening effects were then included by working in impact parameter, b, space and using suppression factors of the form $\exp(-\Omega(b))$. Recall that in the eikonal approach the elastic amplitude has the form

$$T_{\rm el} = 1 - e^{-\Omega/2}, \qquad (8)$$

where the opacity

$$\Omega = \sum_{i=P,R} \Omega_i \quad \text{with} \quad \Omega_P = \beta_P^2(0) \left(\frac{s}{s_0}\right)^{\alpha_i(0)-1} \frac{\exp(-b^2/4B_P)}{B_P}, \tag{9}$$

where $B_P = r_P + \alpha'_P \ln(s/s_0)$ is the t-slope of the Pomeron exchange amplitude. A similar expression holds for the Reggeon opacity Ω_R.

To determine the q_t or t dependence we took the Fourier transforms with respect to the impact parameters specified in Fig. 1. We then obtained*

$$\frac{M^2 d\sigma}{dtdM^2} = A \int \frac{d^2 b_2}{2\pi} e^{i\vec{q}_t\cdot\vec{b}_2} F_i(b_2) \int \frac{d^2 b_3}{2\pi} e^{i\vec{q}_t\cdot\vec{b}_3} F_i(b_3) \int \frac{d^2 b_1}{2\pi} F_j(b_1), \tag{10}$$

where

$$F_i(b_2) = \frac{1}{2\pi\beta_i(q_t=0)} \int d^2 q_t \beta_i(q_t) \left(\frac{s}{M^2}\right)^{-\alpha'_i q_t^2} e^{b'_{iij} q^2} e^{i\vec{q}_t\cdot\vec{b}_2}, \tag{11}$$

$$F_j(b_1) = \frac{1}{2\pi\beta_j(k_t=0)} \int d^2 k_t \beta_j(k_t) \left(\frac{M^2}{s_0}\right)^{-\alpha'_j k_t^2} e^{-b'_{iij} k_t^2}, \tag{12}$$

and where the q_t-independent factors are collected in A

$$A = \beta_j(0)\beta_i^2(0) g_{iij}(0) \left(\frac{s}{M^2}\right)^{2\alpha_i(t_{\min})-2} \left(\frac{M^2}{s_0}\right)^{\alpha_j(0)-1}. \tag{13}$$

These equations are relevant for the strong triple-Pomeron scenario, see (4). In Ref. 8 we also studied the 'scalar' weak triple-Pomeron coupling ansatz, (2). To do this we included a factor q_t in the integrand of the expression (11) for $F_i(b_2)$ when $i, j = P$, and also $\bar{q}_t$ $(= q_t)$ in the analogous formula for $F_i(b_3)$.

To obtain the screening corrections, for a single-channel eikonal, we then included in the integrands on the right-hand side of (10) the factors

$$\exp(-\Omega(\vec{b}_2 - \vec{b}_1)/2) \exp(-\Omega(\vec{b}_3 - \vec{b}_1)/2) \equiv S(\vec{b}_2 - \vec{b}_1)\, S(\vec{b}_3 - \vec{b}_1). \tag{14}$$

That is, we computed

$$\left.\frac{M^2 d\sigma}{dtdM^2}\right|_{iij} = A \int \frac{d^2 b_1}{2\pi} F_j(b_1) |I(b_1)|^2, \tag{15}$$

where I is given by

$$I(b_1) \equiv \int \frac{d^2 b_2}{2\pi} e^{i\vec{q}_t\cdot\vec{b}_2} F_i(b_2) S_i(\vec{b}_2 - \vec{b}_1). \tag{16}$$

*Note that $e^{i\vec{k}_t\cdot\vec{b}_1} = 1$ as $k_t = 0$.

Here q_t is the transverse momentum of the outgoing proton, which is now the sum of the transverse momentum of the Pomeron coupling to the triple-Pomeron vertex and that of the screening Pomeron.

These were the 'strong' and 'weak' scenarios studied in detail in Ref. 8. As mentioned above, we found that the data favoured the 'strong' coupling scenario. Our concern here is the possibility that we may have a 'weak' triple-Pomeron coupling with a vectorial structure. That is the coupling in Fig. 1 is of the form

$$g_{3P}(t) \;=\; g_{3P}(0)\,\exp(b'_{iij}(q^2+\bar{q}^2-k_t^2))\cdot\delta_{\mu,\nu}q_{t\mu}\bar{q}_{t\nu}. \qquad (17)$$

The final product means that an additional factor $\vec{q}_t$ will occur in the integrand of (11). The presence of a vectorial $\vec{q}_t$ means that after the angular integration in (11), we now obtain the Bessel function $J_1(q_t b)$, and not $J_0(q_t b)$ as was found in Ref. 8. Also, in the impact parameter, b, representation the amplitude corresponding to (11) takes a vector form

$$\vec{F}_i(\vec{b}) \;=\; \vec{b}f(b). \qquad (18)$$

Due to the factor $J_1(q_t b)$, the amplitude vanishes at $b = 0$, where the screening effect, $\exp(-\Omega)$, is at its maximum. Thus, we now anticipate a weaker screening for the triple-Pomeron term. Moreover, as the amplitude now has vector form (analogous to that occurring in the $\pi\pi P$ contribution discussed in Section 4.3 of Ref. 8) we have to consider the components I_x and I_y separately. That is I of (16) is given by

$$|I|^2 = |I_x|^2 + |I_y|^2, \qquad (19)$$

where it is convenient to direct x along $\vec{q}_t$.

The generalisation of the formalism to a two-channel eikonal is straightforward. It has been presented in Section 4.2 of Ref. 8.

3. Results

As can be seen from Figs. 2–4, the presently available data in the triple-Regge region are well described assuming that the triple-Pomeron vertex has a vectorial 'weak' form. Fig. 2 also shows the results of two previous fits, taken from Ref. 8, which were obtained with the assumption that the triple-Pomeron vertex had, first, a 'strong' coupling and, then, a scalar 'weak' coupling. All three fits used the same data sets.† Table 1 compares

†Recall that these three scenarios differ in how the triple-Pomeron coupling vanishes as $q_t \to 0$: (a) weak vector coupling – $\sqrt{g_{3P}} \propto \vec{q}_t$, (b) weak scalar coupling – $\sqrt{g_{3P}} \propto q_t$ and (c) strong coupling – $\sqrt{g_{3P}} \propto$ constant.

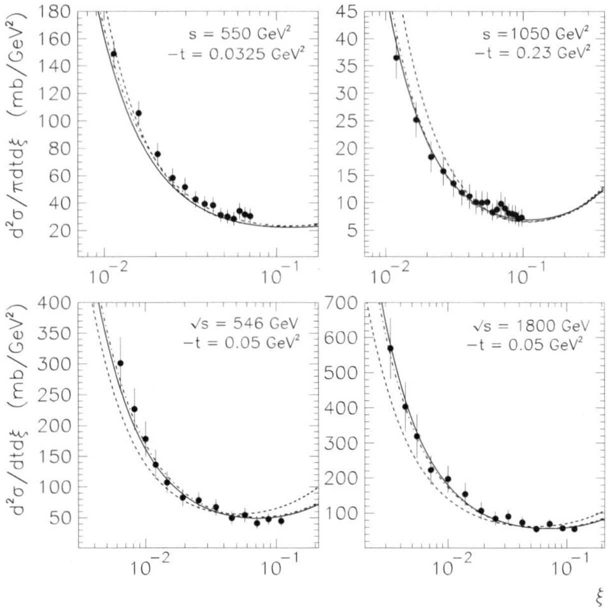

Fig. 2. The description of a sample of the $d^2\sigma/dtd\xi$ cross section data that are fitted using the weak-vector (continuous curves), weak-scalar (dashed curves) and strong (dot-dashed) triple-Pomeron coupling ansatzes. ($\xi \simeq M^2/s$). Here, the curves corresponding to the (strong, weak) coupling fits of the FNAL data have been normalised (down, up) by 15% at $\sqrt{s} = 546$ GeV and by 10% at $\sqrt{s} = 1800$ GeV, to allow for the normalisations found for these data in the respective fits.

the values of parameters obtained for the vectorial weak scenario with those obtained in Ref. 8 for the strong and scalar weak scenarios. Note that for the case of vectorial 'weak' coupling we get practically the same χ^2 as in the more popular 'strong' coupling scenario. The secondary Reggeon contributions, that is the RRP, RRR and PPR terms, coincide, within the error bars, in all three scenarios.

Unlike that for the weak-scalar, the weak-vector triple-Pomeron contribution has no dip at very low $|t| \sim 0.02$ GeV^2 (see Figs. 3 and 5). The *vector* contribution vanishes at $q_t = 0$ since we have no direction.

More rapid decrease of the PPP contribution at large $\xi > 0.1$, seen in Fig. 4, is due to the longitudinal part of the momentum transfer

$$-t = q_t^2 + q_{\|}^2 = (q_t^2 + \xi^2 m_p^2)/(1 - \xi). \tag{20}$$

In the vectorial 'weak' scenario the triple-Pomeron contribution vanishes as $q_t \to 0$, while for fixed t, the value of q_t decreases as ξ increases.

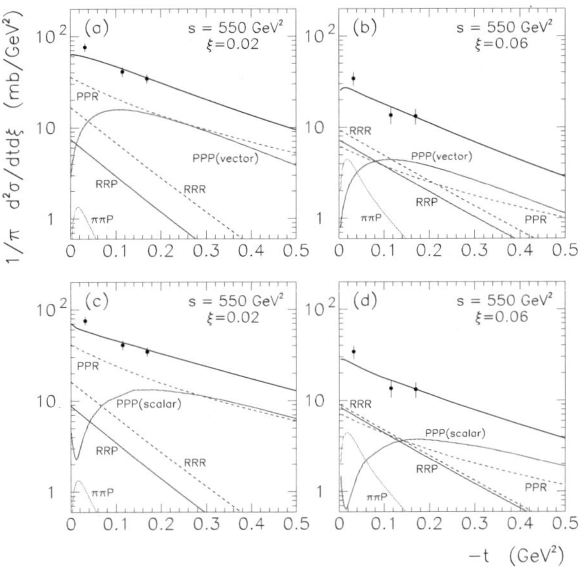

Fig. 3. Comparison of the t-dependence of $d^2\sigma/dtd\xi$ at $\xi = 0.02,\ 0.06$ and $s = 550\ \mathrm{GeV}^2$ obtained using the weak-vector triple-Pomeron coupling, with that corresponding to the weak-scalar triple-Pomeron coupling used in Ref. 8, together with the data available at these kinematic values.

Table 1. The values of the *"bare"* triple-Regge couplings $g_{iij}(0)$ of (13), and slopes b'_{iij} of (11, 12), obtained in the three optimum fits to the $d^2\sigma/dtd\xi$ data. GeV units are used; so, for example, the couplings g_{3P} have units of GeV^{-1}. The parameters for the strong and weak-scalar fits are taken from Ref. 8. Recall that all the slopes b'_{iij} are set to zero, except for those of the PPP and PPR vertices in the weak coupling fit.

	strong	weak-scalar	weak-vector
g^S_{3P}	0.44 ± 0.05	–	–
g^W_{3P}	–	3.0 ± 1.2	3.1 ± 0.5
b'^W_{PPP}	–	1.15 ± 0.3	0.9 ± 0.4
g_{PPR}	0.75 ± 0.10	0.76 ± 0.15	0.66 ± 0.11
b'^W_{PPR}	–	1.4 ± 1.7	0.8 ± 0.8
g_{RRP}	1.1 ± 0.3	1.3 ± 0.5	1.1 ± 0.4
g_{RRR}	2.6 ± 1.0	2.9 ± 1.4	3.0 ± 1.1
χ^2/DoF	0.83	1.40	0.86

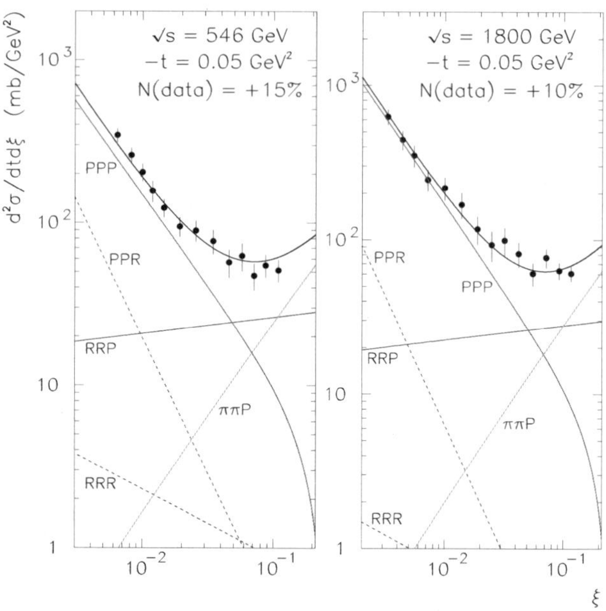

Fig. 4. The description of the $d^2\sigma/dtd\xi$, measured in the collider experiments at FNAL,[11–13] obtained in the weak-vector triple-Pomeron coupling fit. The individual triple-Regge contributions are also shown.

Measurements of $d^2\sigma/dtd\xi$ at the LHC, especially at small $\xi \sim 0.01$, should be able to distinguish between the three scenarios, see Fig. 5.

4. Discussion

Recall that in perturbative QCD, the leading-order BFKL triple-Pomeron vertex does not vanish, but takes a non-zero constant value at $q_t = 0$. This result corresponds to interactions at very small distances. However, at larger distances, relevant for $q_t \to 0$, the absorptive effects caused by enhanced multi-Pomeron diagrams could modify the perturbative QCD result leading to a vanishing triple-Pomeron coupling as $q_t \to 0$. This vanishing behaviour also looks natural if we bear in mind confinement, which does not allow colour-induced interactions at large distances. From this point of view it looks encouraging that the present triple-Regge data may be well described within the vectorial 'weak' approach, after accounting for absorptive corrections.

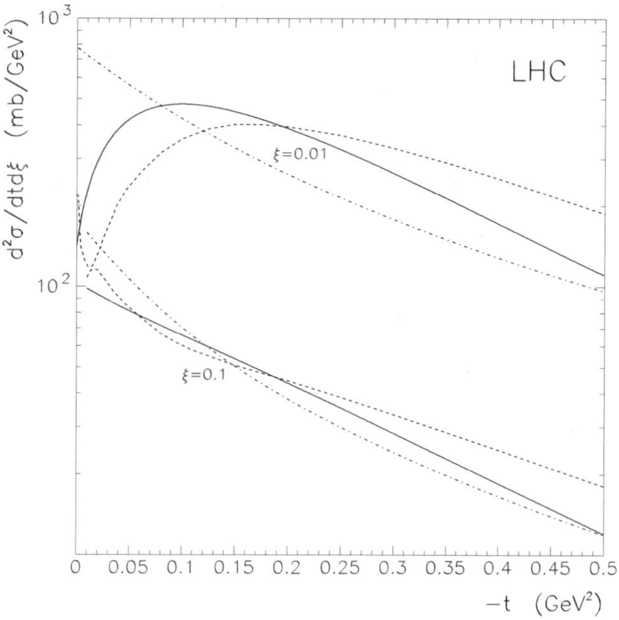

Fig. 5. The predictions for the t-dependence of the $d^2\sigma/dtd\xi$ at $\xi = 0.01$, 0.1 and $\sqrt{s} = 14$ TeV obtained using the weak-vector (continuous curves), weak-scalar (dashed) and strong (dot-dashed) triple-Pomeron coupling scenarios.

The only minor problem is the inelastic J/ψ diffractive photoproduction observed at HERA.[17] Due to the small cross section of the J/ψ-proton interaction, the absorptive effects in this case are very small. The ratio, r, of the cross section with proton dissociation, $\gamma p \to J/\psi + Y$, integrated over the mass region $M_Y < 30$ GeV, to that of 'elastic' photoproduction, $\gamma p \to J/\psi + p$, was measured as a function of momentum transfer t. At the smallest value of $-t = 0.2$ GeV^2 measured by ZEUS,[17] this ratio $r = 0.4 \pm 0.1$.[17–21] On the other hand, using the parameters of our 'weak-vector' fit from Table 1, we predict a smaller value

$$r = r_{PPP} + r_{PPR} \simeq 0.16 + 0.08 \ = 0.24.$$

Here r_{PPP} and r_{PPR} denote the contributions of the PPP and PPR terms respectively. This should be compared to the predictions $r \simeq 0.16 + 0.12 = 0.28$ and $r \simeq 0.14 + 0.07 = 0.21$ obtained using the strong and 'weak-scalar' fits of Ref. 8. Indeed, in Ref. 8 we concluded that the weak-scalar triple-Pomeron coupling was disfavoured both by the fit to the triple-Regge data and by the J/ψ data. On the other hand, the triple-Regge data can be reasonably accommodated by the weak-vector coupling, although this triple-Pomeron coupling is still a bit disfavoured, in comparison with the strong coupling regime, by the J/ψ data. Unfortunately, (a) we have no data at smaller $|t|$ and (b) the cross section of inelastic $\gamma p \to J/\psi + Y$ diffractive photoproduction has never been published in a journal. Thus, these data cannot be considered as a strong argument against the 'weak-vector' coupling scenario.

It would be very interesting to measure the t dependence of high-mass diffractive dissociation, $d^2\sigma/dtd\xi$, at the LHC in order to choose between the possible vectorial 'weak' and 'strong' coupling asymptotic regimes.

References

1. M. G. Ryskin, A. D. Martin and V. A. Khoze, in *Proc. of Gribov Memorial Workshop* (World Scientific, Singapoure, 2005), pp. 115, arXiv:hep-ph/0506272.
2. M. G. Ryskin, A. D. Martin, V. A. Khoze and A. G. Shuvaev, *J. Phys.* **G36**, 093001 (2009), arXiv:0907.1374 [hep-ph].
3. V. N. Gribov and A. A. Migdal, *Sov. J. Nucl. Phys.* **8**, 583 (1969).
4. V. N. Gribov, *Sov. J. Nucl. Phys.* **17**, 313 (1973).
5. V. N. Gribov and A. A. Migdal, *Sov. Phys. JETP* **28**, 784 (1969).
6. Y. I. Azimov, V. A. Khoze, E. M. Levin and M. G. Ryskin, *Nucl. Phys.* **B89**, 508 (1975).
7. Y. I. Azimov, V. A. Khoze, E. M. Levin and M. G. Ryskin, *Sov. J. Nucl. Phys.* **23**, 449 (1976) [*Yad. Fiz.* **23**, 853 (1976)].
8. E. G. S. Luna, V. A. Khoze, A. D. Martin and M. G. Ryskin, *Eur. Phys. J.* **C59**, 1 (2009).
9. A. B. Kaidalov and M. G. Poghosyan, arXiv:0909.5156 [hep-ph].
10. J. C. M. Armitage *et al.*, *Nucl. Phys.* **B194**, 365 (1982).
11. R. L. Cool *et al.*, *Phys. Rev. Lett.* **47**, 701 (1981).
12. F. Abe *et al.*, *Phys. Rev.* **D50**, 5535 (1994).
13. K. Goulianos and J. Montanha, *Phys. Rev.* **D59**, 114017 (1999).
14. V. N. Gribov, *Sov. J. Nucl. Phys.* **17**, 603 (1973).
15. C. F. von Weizsäcker, *Z. Phys.* **88**, 612 (1934).
16. E. J. Williams, *Phys. Rev.* **45**, 729 (1934).
17. ZEUS collaboration: Abstract 549, in *Int. Europhysics Conf. on HEP*, (Aachen, Germany, July 2003).
18. ZEUS collaboration: S. Chekanov *et al.*, *Nucl. Phys.* **B695**, 3 (2004).
19. H1 collaboration: C. Adloff *et al.*, *Eur. Phys. J.* **C10**, 373 (1999).
20. ZEUS collaboration: S. Chekanov *et al.*, *Eur. Phys. J.* **C24**, 345 (2002).
21. H1 collaboration: A. Aktas *et. al.*, *Eur. Phys. J.* **C46**, 585 (2006).

Como, 1994.

Erice, 1996.

DIFFRACTION, GOOD–WALKER, AND THE BFKL POMERON

GÖSTA GUSTAFSON

Dept. of Theoretical Physics, Lund University,
Lund, Sweden
gosta.gustafson@thep.lu.se

The proton has a substructure in terms of parton cascades, which in a high energy collision fills the whole rapidity range between projectile and target. In the Good–Walker formalism the fluctuations in the cascades will give rise to diffractive excitation. Within the Lund Dipole Cascade model, it is in this way possible to reproduce diffractive excitation in *pp* collisions and DIS. In central *pp* collisions the fluctuations, and thus diffractive excitation, is suppressed by saturation. This leads to factorization breaking when comparing *pp* scattering and DIS. The results shows a multi-regge behaviour, with a bare pomeron pole with $\alpha(0) = 1.21$, $\alpha' = 0.2\,\text{GeV}^{-2}$, and an almost constant triple-pomeron coupling $g_{3P} \approx 0.3\,\text{GeV}^{-1}$.

Keywords: Diffraction; saturation; dipole model; Good–Walker; BFKL.

1. Introduction

It is often assumed that high energy collisions are driven by partonic subcollisions. This is *e.g.* the case in the model implemented in the frequently used MC event generator PYTHIA.[1] In high energy *pp* scattering the cross section for minijet production becomes very large, and unitarity implies that multiple interactions, saturation, and diffraction become important. Parton rescattering is represented by a convolution in transverse momentum space, which corresponds to a simple multiplication in transverse coordinate space. Effects of saturation and multiple interactions are therefore most easily described in impact parameter space.

The proton has an internal substructure, which may be excited in a diffractive scattering process, and diffractive excitation represents large fractions of the cross section in *pp* collisions or DIS. In the Good–Walker formalism[2] diffractive excitation is described by the fluctuations in the scattering amplitude. In most analyses of *pp* collisions this mechanism is used only for low mass excitation, while high mass excitation is described by a triple-Regge formula,[3,4] where regge trajectories and couplings are fitted to experimental data (for recent analyses see *e.g.* Refs. 5–7).

The proton substructure is represented by a parton cascade, which at high energies is described by BFKL evolution. The fluctuations in this evolution are known to be very large.[8] An analysis of these fluctuations, within the Lund Dipole Cascade model, is able to reproduce the experimental cross sections for diffractive excitation in pp collisions or DIS. This implies that the effective pomeron couplings in the multi-pomeron formalism can be estimated without any new free parameters. The result corresponds to a bare pomeron pole with intercept $\alpha(0) = 1.21$ and slope $\alpha' = 0.2\,\text{GeV}^{-2}$, and a constant triple-pomeron coupling equal to 0.31 GeV^{-2}. The results presented here are obtained in collaboration with Christoffer Flensburg and Leif Lönnblad.

2. The Eikonal Approximation and the Good–Walker Formalism

As mentioned in the introduction, diffraction, saturation, and multiple interactions are most easily described in impact parameter space. If the interaction is driven by absorption into inelastic states i, with weights $2f_i$, the optical theorem gives an elastic amplitude given by

$$T = 1 - e^{-F}, \quad \text{with } F = \sum f_i. \tag{1}$$

For a structureless projectile we then find

$$\begin{cases} d\sigma_{tot}/d^2b = \langle 2T \rangle, \\ \sigma_{el}/d^2b \ = \langle T \rangle^2, \\ \sigma_{inel}/d^2b = \langle 1 - e^{-\sum 2f_i} \rangle = \sigma_{tot} - \sigma_{el}. \end{cases} \tag{2}$$

If the projectile has an internal structure, the mass eigenstates Ψ_k can differ from the eigenstates of diffraction Φ_n, which have eigenvalues T_n. With the notation $\Psi_k = \sum_n c_{kn}\Phi_n$ (with $\Psi_{in} = \Psi_1$) the elastic amplitude is given by $\langle \Psi_1 | T | \Psi_1 \rangle = \sum c_{1n}^2 T_n = \langle T \rangle$, while the amplitude for diffractive transition to mass eigenstate Ψ_k is given by $\langle \Psi_k | T | \Psi_1 \rangle = \sum_n c_{kn} T_n c_{1n}$. The corresponding cross sections become

$$d\sigma_{el}/d^2b = (\sum c_{1n}^2 T_n)^2 = \langle T \rangle^2 \tag{3}$$

$$d\sigma_{diff}/d^2b = \sum_k \langle \Psi_1 | T | \Psi_k \rangle \langle \Psi_k | T | \Psi_1 \rangle = \langle T^2 \rangle. \tag{4}$$

The diffractive cross section here includes elastic scattering. Subtracting this gives the cross section for diffractive excitation, which is thus determined by the fluctuations in the scattering process:

$$d\sigma_{diff\,ex}/d^2b = d\sigma_{diff} - d\sigma_{el} = \langle T^2 \rangle - \langle T \rangle^2. \tag{5}$$

3. Proton Substructure

A proton cascade is illustrated in Fig. 1. If one of the partons interacts with a target via gluon exchange, (parts of) the cascade can come on shell, and we get an inelastic event. According to the optical theorem this gives a contribution to the elastic scattering, and to diffractive excitation.

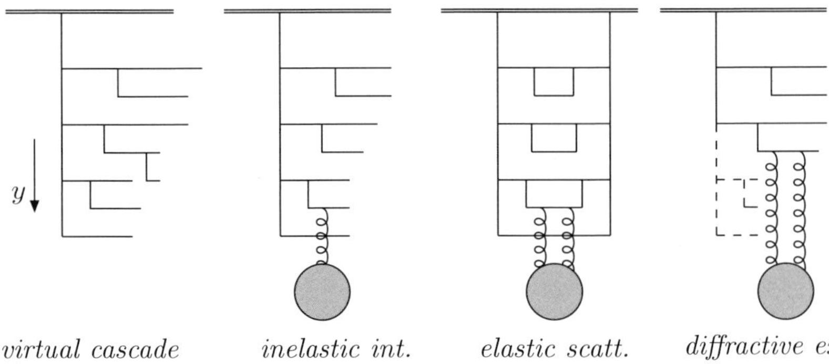

Fig. 1. A parton cascade in a proton can interact with a target via gluon exchange.

A high energy *pp* collision is sketched in Fig. 2. The cascades from the projectile and the target evolve from opposite sides, and interact via gluon exchange. This implies a colour connection between the projectile and the target, and thus gives an inelastic event. Multiple subcollisions lead to saturation, as a consequence of unitarity.

A diagram for diffractive excitation is shown in Fig. 3. Here the projectile is evolved a distance y_1 and the target y_2, in the frame in which the process is calculated. The BFKL evolution gives large fluctuations in the evolution of the projectile. If we first take the average over the target states, we get the amplitude for elastic scattering of the target. Squaring it gives the cross section, when the target is scattered elastically. If we after this take the average over the projectile states, we obtain the diffractive scattering of the projectile, including the elastic scattering. Thus the expression

$$\langle \langle T \rangle^2_{targ} \rangle_{proj} - \langle T \rangle^2_{targ,proj} \tag{6}$$

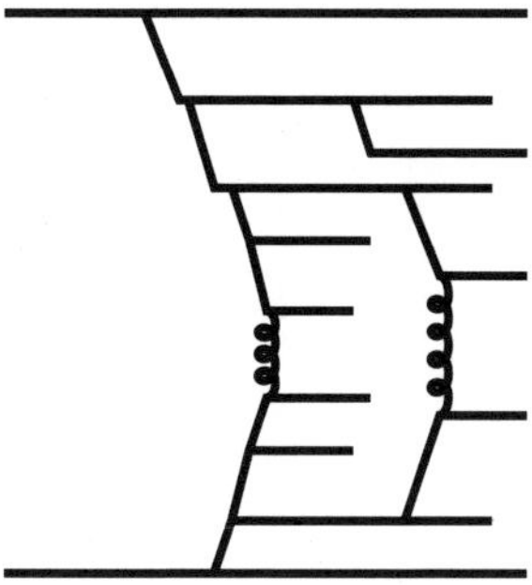

Fig. 2. A high energy inelastic pp collision.

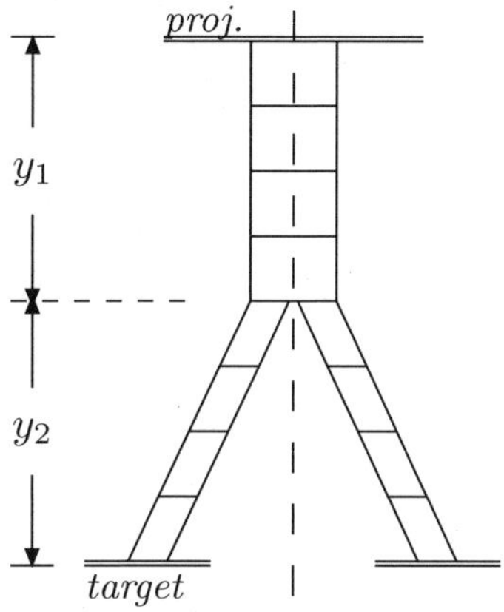

Fig. 3. A triple-pomeron diagram.

gives the cross section for single diffractive excitation of the projectile, with the excited mass limited to $M_X^2 < \exp(y_1)$. Varying y_1 gives then $d\sigma/dM_X^2$.

4. Dipole Cascade Models

Mueller's dipole cascade model[9–11] is a formulation of LL BFKL evolution in transverse coordinate space. Gluon radiation from the colour charge in a parent quark or gluon is screened by the accompanying anticharge in the colour dipole. This suppresses emissions at large transverse separation, which corresponds to the suppression of small $k_\perp$ in BFKL. For a dipole with charges in transverse points $\mathbf{x}$ and $\mathbf{y}$, the probability per unit rapidity (Y) for emission of a gluon at transverse position $\mathbf{z}$ is given by

$$\frac{d\mathcal{P}}{dY} = \frac{\bar{\alpha}}{2\pi} d^2\mathbf{z} \frac{(\mathbf{x}-\mathbf{y})^2}{(\mathbf{x}-\mathbf{z})^2(\mathbf{z}-\mathbf{y})^2}, \quad \text{with } \bar{\alpha} = \frac{3\alpha_s}{\pi}. \tag{7}$$

The dipole is split into two dipoles, which (in the large N_c limit) emit new gluons independently. The result is a cascade, where the number of dipoles grows exponentially with Y.

When two cascades collide, a pair of dipoles with coordinates $(\mathbf{x}_i, \mathbf{y}_i)$ and $(\mathbf{x}_j, \mathbf{y}_j)$ can interact via gluon exchange with the probability $2f_{ij}$, where

$$f_{ij} = f(\mathbf{x}_i, \mathbf{y}_i|\mathbf{x}_j, \mathbf{y}_j) = \frac{\alpha_s^2}{8}\left[\log\left(\frac{(\mathbf{x}_i-\mathbf{y}_j)^2(\mathbf{y}_i-\mathbf{x}_j)^2}{(\mathbf{x}_i-\mathbf{x}_j)^2(\mathbf{y}_i-\mathbf{y}_j)^2}\right)\right]^2. \tag{8}$$

Summing over all dipoles in the cascades then reproduces the LL BFKL result.

The *Lund dipole cascade model*[12–14] is a generalization of Mueller's model, which includes:

– NLL BFKL effects
– Nonlinear effects in the evolution
– Confinement effects

For an incoming virtual photon splitting into a $q\bar{q}$ pair, the initial state wavefunction is determined by perturbative QCD. For an incoming proton we make an ansatz in form of an equilateral triangle of dipoles. After evolution the result is rather insensitive to the exact form of the initial state. The model is also implemented in a MC program DIPSY. The model reproduces successfully the total and elastic cross sections for pp scattering and DIS, as shown in Figs. 4 and 5.

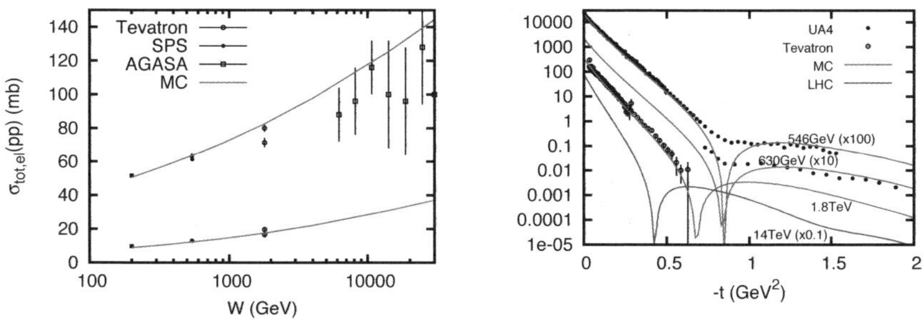

Fig. 4. Total and elastic cross sections in pp collisions in the dipole cascade model.

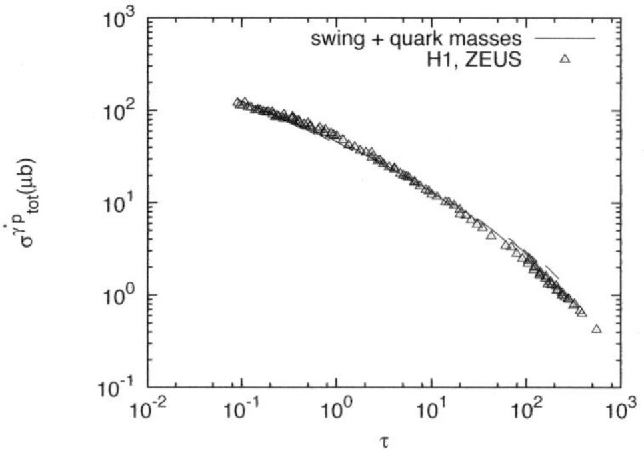

Fig. 5. Total $\gamma^* p$ cross section presented as a function of the Golec-Biernat–Wüsthoff scaling parameter $\tau = (Q^2/Q_0^2)(x/x_0)^\lambda$, with $Q_0 = 1\text{GeV}$, $x_0 = 3 \cdot 10^{-4}$, and $\lambda = 0.29$.

5. Fluctuations and Diffractive Excitation

The fluctuations in the evolution are large, and the model can also describe diffractive excitation within the Good–Walker formalism, without new parameters beyond those adjusted to the total and elastic cross sections.[15,16] This is similar in spirit to the early analysis by Miettinen and Pumplin.[17] In DIS saturation effects are not very important, while in *pp* collisions saturation effects strongly suppress the fluctuations, and thus the cross section for diffractive excitation.

$\gamma^* p$ **collisions.** As an example Fig. 6 shows the distribution in the non-saturated amplitude, F, for $\gamma^* p$ collisions at $Q^2 = 14$ GeV^2 and $W = 220$ GeV and different impact parameters b. The photon is here represented by a dipole with size $r = 1/Q$. The distribution can be approximately described by a power $\frac{dP}{dF} \approx A\,F^{-p}$ (with a cutoff for small F-values), which is illustrated by the straight lines in the figure. The relative width of this distribution is rather large, and the approximation gives the ratio $d\sigma_{diff.ex.}/d\sigma_{tot} \approx 1 - 1/2^{2-p}$. The power p is independent of the impact parameter, and therefore this result is also valid for the integrated cross sections. The resulting diffractive cross section is shown in Fig. 7, compared with data from ZEUS.[18] We see that the diffractive fraction of the cross section is fairly independent of W, but decreases with increasing Q^2.

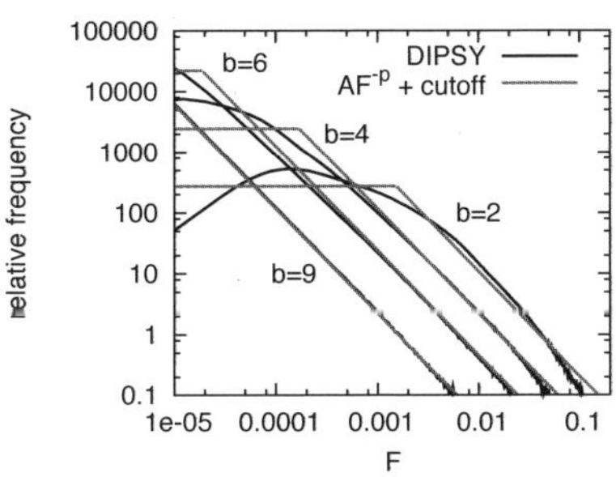

Fig. 6. Distribution in the one-pomeron amplitude F in DIS for $Q^2 = 14$ GeV^2 and $W = 220$ GeV. b is measured in GeV^{-2}.

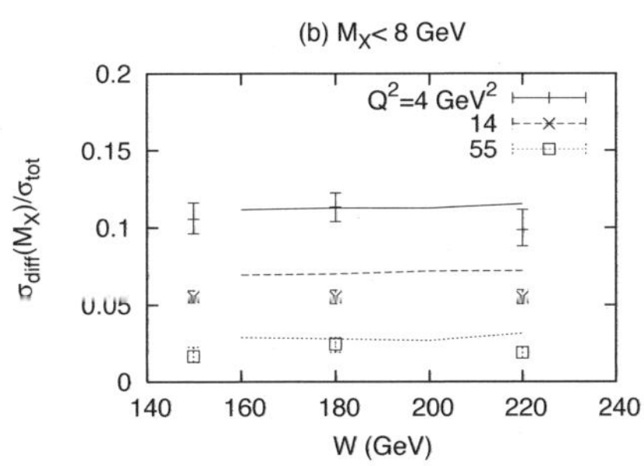

Fig. 7. The ratio between the diffractive excitation and the total cross sections in DIS, for $M_X < 8$ GeV. Data from ZEUS.[18]

pp **collisions.** In *pp* scattering the Born amplitude is large, and therefore unitarity effects are important. Fig. 8 shows both the Born amplitude and the unitarized amplitude at 2 TeV for different *b*-values. We see that the width of the Born amplitude is large, and without unitarization the fraction of diffractive excitation would be correspondingly large. (The smooth lines are fits of the form $AF^p e^{-aF}$.)

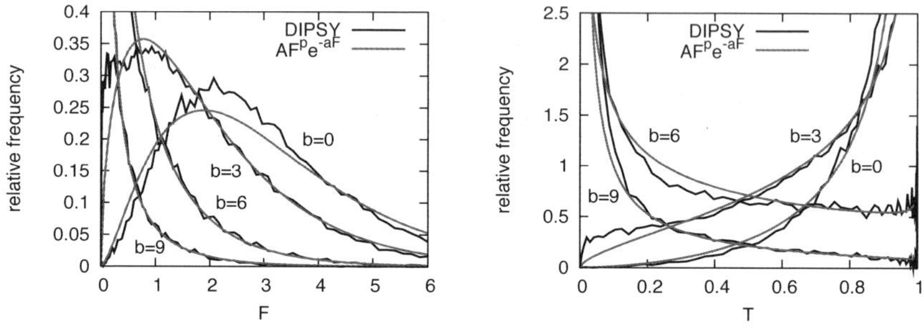

Fig. 8. Distribution in the one-pomeron amplitude F (left), and the unitarized amplitude T (right) in pp collisions at 2 TeV. b is in units of GeV^{-1}.

However, the unitarized amplitude is limited by 1, and the width, and therefore the diffractive excitation, is very much reduced. This is in particular the case for central collisions, where the amplitude approaches the black disc limit $T = 1$. This result corresponds to the effect of enhanced diagrams in the conventional triple-regge approach. The impact parameter profile is shown in Fig. 9. For central collisions the absorption is large, and diffractive excitation small. The cross section for diffractive excitation is largest in a ring with radius $b \sim 1\,\text{fm} \approx 5\,\text{GeV}^{-1}$, which grows slowly with energy. Another consequence of the large saturation in pp collisions, is that factorization is not satisfied when comparing diffractive excitation in DIS and pp scattering.

6. Comparison with Multi-Regge Analyses

It is also interesting to compare the results from the Good–Walker analysis with the multi-regge formalism. To this end we study the contribution from the *bare pomeron*, meaning the one-pomeron amplitude without contributions from saturation, enhanced diagrams or gap survival form factors.

When s, M_X^2, and s/M_X^2 are all large, pomeron exchange should dominate. If the pomeron is a simple pole we expect the following expressions

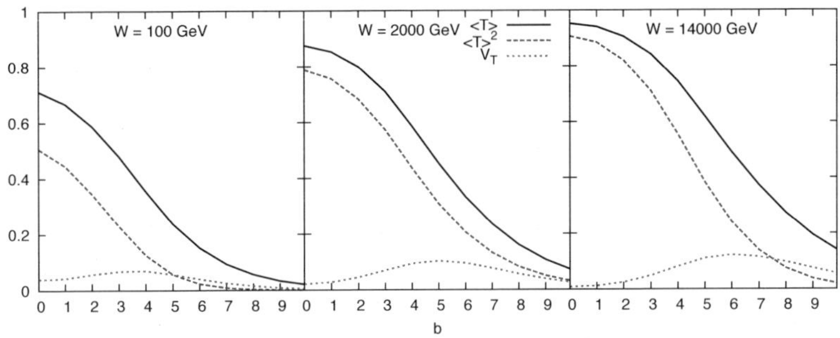

Fig. 9. Impact parameter distributions for $\langle T\rangle = (d\sigma_{\rm tot}/d^2b)/2$, $\langle T\rangle^2 = d\sigma_{\rm el}/d^2b$, and $V_T = d\sigma_{\rm diffex}/d^2b$ in pp collisions at $W = 2$ TeV. b is in units of GeV^{-1}.

for the pp total and diffractive cross sections:

$$\begin{aligned}
\sigma_{\rm tot} &= \beta^2(0)s^{\alpha(0)-1} = \beta^2(0)s^{\epsilon},\\
\frac{d\sigma_{\rm el}}{dt} &= \frac{1}{16\pi}\beta^4(t)s^{2(\alpha(t)-1)},\\
M_{\rm X}^2\frac{d\sigma_{\rm SD}}{dtd(M_{\rm X}^2)} &= \frac{1}{16\pi}\beta^2(t)\beta(0)g_{3\rm P}(t)\left(\frac{s}{M_{\rm X}^2}\right)^{2(\alpha(t)-1)}\left(M_{\rm X}^2\right)^{\epsilon}.
\end{aligned} \tag{9}$$

Here $\alpha(t) = 1 + \epsilon + \alpha' t$ is the pomeron trajectory, and $\beta(t)$ and $g_{3P}(t)$ are the proton-pomeron and triple-pomeron couplings respectively. (We have here omitted the scale s_0 in the powers $(s/s_0)^\alpha$ or $(M_{\rm X}^2/s_0)^\alpha$. This scale is in the following assumed to be 1 GeV^2.)

The results of the MC for the total, elastic, and single diffractive cross sections are shown by the crosses in Fig. 10. The elastic and diffractive cross sections are here integrated over t and $M_{\rm X}^2$. The single diffractive cross section is calculated in the total cms, which corresponds to an integration over masses in the range $M_{\rm X}^2 < \sqrt{s}\cdot 1$ GeV, and it corresponds to excitation of one side only. We see that the result indeed has the power-like increase with energy, which is characteristic for a Regge pole. We also note that in the one-pomeron approximation the elastic cross section is larger than the total for $\sqrt{s} > 15$ GeV.

The lines in Fig. 10 are obtained by integrating the elastic and single diffractive cross section in Eq. (9) over t, with the parameter values

$$\begin{aligned}
\alpha(0) &= 1 + \epsilon = 1.21, \quad \alpha' = 0.2\,\text{GeV}^{-2},\\
\beta^2(0) &= 12.6\,\text{mb}, \quad \beta(t) = \beta(0)\exp\left(\frac{2.5\,t}{1 - 1.8\,t}\right),\\
g_{3\rm P}(t) &= \text{const.} = 0.3\,\text{GeV}^{-1}.
\end{aligned} \tag{10}$$

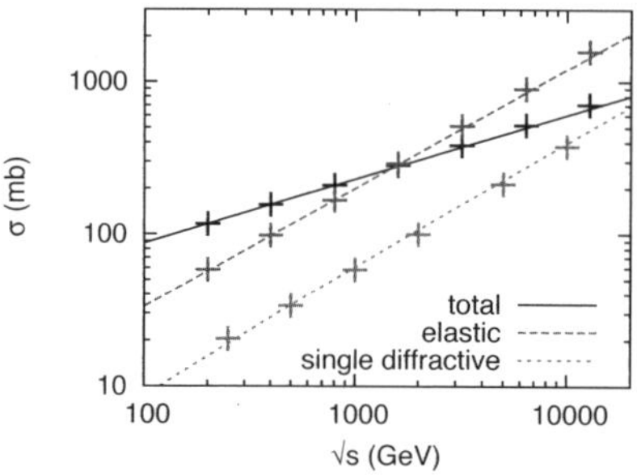

Fig. 10. The total, elastic and single diffractive cross sections in the one-pomeron approximation. The crosses are model calculations and the lines are from a tuned triple-regge parametrization.

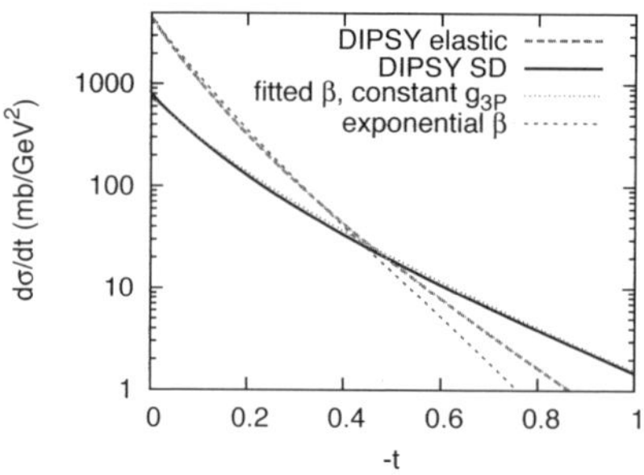

Fig. 11. t-dependence for elastic and single diffractive excitation without saturation effects, at 1800 GeV. The fit for σ_{el} obtained with an exponential proton-pomeron coupling β is shown by a dashed line.

The corresponding differential cross sections for elastic and single diffractive scattering, from the model and from the triple-regge fit, are shown in Fig. 11. A rather good fit is also obtained by an exponential proton-pomeron coupling $\beta(t) = \beta(0)\exp(b_{0,\mathrm{el}}\, t/4)$, with $b_{0,\mathrm{el}} = 8\,\mathrm{GeV}^{-2}$. The result for the elastic cross section is shown by a dashed line, and we see that it gives a good fit for $|t| < 0.4\mathrm{GeV}^2$, but undershoots the model result for larger t-values.

These results can also be compared with multi-regge analyses, where *e.g.* Ryskin *et al.*[5] obtain $\alpha(0) = 1.3$, $\alpha' \leq 0.05\,\mathrm{GeV}^{-2}$, Kaidalov *et al.*[6] find $\alpha(0) = 1.12$, $\alpha' = 0.22\,\mathrm{GeV}^{-2}$, while Gotsman *et al.*[7] find $\alpha(0) = 1.335$, $\alpha' = 0.01\,\mathrm{GeV}^{-2}$. Thus our values are somewhere in between.

We note here in particular that the Good–Walker results are reproduced by a single pomeron pole, *i.e.* not by a cut as expected in LL BFKL, or a series of poles as obtained with a running coupling.[19] Also the triple-regge coupling $g_{3\mathrm{P}}$ is approximately constant, while in LL BFKL it is proportional to $\sim 1/\sqrt{|t|}$.[10,20]

7. Conclusions

Diffractive excitation represents a large fraction of the cross section in pp collisions or DIS. In the Good–Walker formalism diffractive excitation is determined by the fluctuations in the scattering amplitude. The proton has a substructure consisting of partons, which fill the whole rapidity range between the projectile and the target. At high energies the dynamics of the parton cascades is determined by BFKL evolution, which is known to have very large fluctuations.

The Lund Dipole Cascade model, implemented in the MC DIPSY, is an implementation of BFKL evolution in transverse coordinate space. It is a generalization of Mueller's dipole cascade, which also includes NLL corrections, saturation within the evolution, and confinement effects. Besides the total and elastic cross sections, it also reproduces diffractive excitation in pp collisions and DIS within the Good–Walker formalism.

Saturation effects are relatively small in DIS, but are very strong in pp collisions. When the interaction approaches the black disc limit in central collisions, the fluctuations are reduced and diffractive excitation becomes suppressed. Therefore diffractive excitation is largest in a ring in impact parameter space, with a slowly increasing radius at very high energies. This effect also implies a breaking of factorization between pp collisions and DIS.

These results based on the Good–Walker formalism also have the expected form from triple-regge analysis. The results correspond to a bare pomeron pole with $\alpha(0) = 1.21$, $\alpha' = 0.2\,\mathrm{GeV}^{-2}$, and an almost constant triple-pomeron coupling $g_{3P} \approx 0.3\,\mathrm{GeV}^{-1}$. This contrasts to the result in LL BFKL, where the pomeron singularity is a cut, and the triple-pomeron coupling is proportional to $1/\sqrt{-t}$.

References

1. T. Sjöstrand and M. van Zijl, *Phys. Rev.* **D36**, 2019 (1987).
2. M.L. Good and W.D. Walker, *Phys. Rev.* **120**, 1857 (1960).
3. A.H. Mueller, *Phys. Rev.* **D2**, 2963 (1970).
4. C.E. DeTar *et al.*, *Phys. Rev. Lett.* **26**, 675 (1971).
5. M.G. Ryskin, A.D. Martin and V.A. Khoze, *Eur. Phys. J.* **C60**, 249 (2009), arXiv:0812.2407.
6. A.B. Kaidalov and M.G. Poghosyan, arXiv:0909.5156.
7. E. Gotsman, E. Levin, U.Maor and J.S.Miller *Eur. Phys. J.* **C57**, 689 (2008), arXiv:0805.2799.
8. A.H. Mueller and G.P. Salam, *Nucl. Phys.* **B475**, 293 (1996), arXiv:hep-ph/9605302.
9. A.H. Mueller, *Nucl. Phys.* **B415**, 373 (1994).
10. A.H. Mueller and B. Patel, *Nucl. Phys.* **B425**, 471 (1994), arXiv:hep-ph/9403256.
11. A.H. Mueller, *Nucl. Phys.* **B437**, 107 (1995), arXiv:hep-ph/9408245.
12. E. Avsar, G. Gustafson and L. Lönnblad, *JHEP* **07**, 062 (2005), arXiv:hep-ph/0503181.
13. E. Avsar, G. Gustafson and L. Lönnblad, *JHEP* **01**, 012 (2007), arXiv:hep-ph/0610157.

14. C. Flensburg, G. Gustafson and L. Lönnblad, *Eur. Phys. J.* **C60**, 233 (2009), arXiv:0807.0325.
15. E. Avsar, G. Gustafson and L. Lönnblad, *JHEP* **12**, 012 (2007) arXiv:0709.1368.
16. C. Flensburg and G. Gustafson, arXive:1004.5502 to be published in *JHEP*.
17. H.I. Miettinen and J. Pumplin, *Phys. Rev.* **D18**, 1696 (1978).
18. S. Chekanov *et al.*, *Nucl. Phys.* **B713**, 3 (2005), arXiv:hep-ex/0501060.
19. L.N. Lipatov, *Sov. Phys. JETP* **63**, 904 (1986).
20. J. Bartels, M.G. Ryskin and G.P. Vacca, *Eur. Phys. J.* **C27**, 101 (2003), arXiv:hep-ph/0207173.

PHOTON POLARIZATION ASYMMETRY IN MESONS AND MUON RADIATIVE DECAYS AND THE AXIAL ANOMALY*

LUCA TRENTADUE

Dipartimento di Fisica, Universitá degli Studi di Parma
and
INFN, Gruppo Collegato di Parma, Sezione di Milano Bicocca,
43100 Parma, Italy
luca.trentadue@cern.ch

We have computed meson and muon polarized radiative decays. The undergoing dynamics giving rise to lepton and photon polarizations is examined and analyzed in the soft and hard region of momenta. The particular configurations made by right-handed leptons with accompanying photons are investigated and interpreted as a manifestation of the axial anomaly. The photon polarization asymmetry is evaluated. Finiteness of polarized amplitudes against infrared and collinear singularities is shown to take place with mechanisms distinguishing between right handed and left handed final leptons.

1. Introduction

Radiative decays of light mesons and leptons have been widely studied both experimentally and theoretically. They represent an excellent source of information on the experimental side as well as a benchmark for theoretical speculations. Extensive comparisons have been carried on in the past between experiments and theoretical predictions for meson radiative decays.[1] A while ago, radiative polarized leptonic decays of mesons[2,3] and muons[4,5] have also been considered. Recently special attention has been given to the role played by the final lepton mass m_l in the threshold region of the decay and to the $m_l \to 0$ limit concerning the helicity amplitudes for mesons[3] and leptons.[4,5] The $O(\alpha)$ radiative corrections generate an helicity flip of the final lepton even in the zero mass limit[6] provided the lepton mass is kept from the beginning into account. Following the interpretation due to Dolgov

*Based on a work done in collaboration with Emidio Gabrielli.

and Zakharov[7] of the axial anomaly the final states with opposite helicity can be interpreted,[3,8,9] as a manifestation of the axial anomaly giving rise to a peculiar mass-singularity cancellation for the right-handed polarized final lepton amplitudes.

We consider in this work[10] the case of polarized radiative decays of the pion and kaon meson and of the muon more extensively by taking into account polarizations of final lepton and photon degrees of freedom. Contrary to the previous case,[2] in meson decays we consider the polarization states of both lepton and photon final states.

This approach, containing a complete description of the final momenta and helicities, may give further and more detailed information on the final state with respect to the inclusively polarized and unpolarized cases. Furthermore, the agreement with the more inclusive results previously obtained in the literature can be easily recovered by summing over the emitted final states polarizations. It is worth noticing that this approach allows to describe more closely the interplay between several peculiar features of the dynamics involved. As, for instance, to pinpoint the role played by angular momentum conservation and its connection with hard and soft photon momenta, and to consider the role played by the parity conservation in weak decays. All these aspects related to angular momentum dynamics may be effectively described in terms of the photon polarization asymmetry.

Here we emphasize that the knowledge of the helicity amplitudes of the final leptons and photons, in addition to an explicit test of the angular momentum conservation, shows the relative rates of the partial helicity amplitudes. Indeed, in the total rate, different helicity amplitudes, depending on the range of momenta, enter with varying weights. Therefore, this behavior gives the opportunity to isolate peculiar polarized configurations in order to maximize or minimize them according to favorable intervals of momenta. As far as phenomenological applications are concerned, this may be, as will be discussed later, an effective way to compare theory and experiment on a new basis. The case of the photon polarization asymmetry, proposed in this work, allows, in this respect, a new approach to inspect interaction dynamics via a finite and universal quantity which is also directly associated to parity violation. Moreover, the photon polarization asymmetry is very sensitive, in radiative meson decays, to the hadronic structure, allowing for a more precise determination of the electromagnetic form factors with respect to the one obtained so far.

Among the results achieved in this paper we mainly mention that: we explicitly calculate amplitudes and final distributions in terms of lepton and

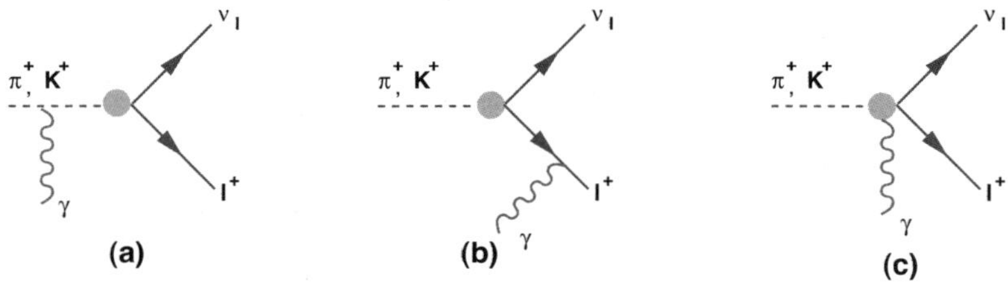

Fig. 1. Feynman diagrams for $(\pi^+, K^+) \to \nu_l l^+ \gamma$ decay, where $l = e, \mu$.

photon momenta at fixed final lepton and photon helicities. Double differential expressions in terms of lepton and photon momenta are also provided together with the partial helicity amplitudes for the meson and muon cases respectively. Moreover, we analyze how the cancellation pattern of mass singularities works on polarized processes. In the inclusive quantities this is a sensible test of the consistency of the results. Once inclusive distributions are obtained by integrating over final momenta, we observe the cancellation of all mass singularities both infrared and collinear. In particular, a peculiar pattern of mass singularity cancellation is shown to take place, which differs for the left-handed helicity final lepton states with respect to the right-handed ones. The same behavior can be observed for the meson as well as for the muon case.

2. The polarized radiative meson decay

We start with the calculation of the polarized amplitude for the process

$$M^+(p) \to \nu_l(p_\nu)\ l^+(p_l, \lambda_l)\ \gamma(k, \lambda_\gamma)\,, \tag{1}$$

where $M^+ = \pi^+$ (K^+) and $l = e$ (μ) stand for pion (kaon) and electron (muon) respectively, with $\nu_{l-e,\mu}$ the corresponding neutrinos. The four momenta p, p_ν, p_l correspond to meson M, neutrino, and charged lepton, while $\lambda_l, \lambda_\gamma$ indicate the charged lepton and photon helicity, respectively. The neutrino is assumed massless and therefore is a pure left-handed polarized state. The Feynman diagrams at tree-level for this process are shown in Fig. 1, where the green bubble just indicates the Fermi interaction.

The first two diagrams Figs. 1a–b correspond to the so-called inner bremsstrahlung (IB) diagrams, where the photon is emitted from external lines and the meson behaves as a point-like scalar particle. The third diagram Fig. 1c is the so-called structure-dependent (SD) diagram, where the photon is emitted from an intermediate hadronic state and the matrix element will depend on the vectorial (V) and axial (A) meson form factors.

The total amplitude for this process can be split in two gauge invariant contributions

$$\mathrm{M}^{(\lambda_l,\lambda_\gamma)} = \mathrm{M}_{IB}^{(\lambda_l,\lambda_\gamma)} + \mathrm{M}_{SD}^{(\lambda_l,\lambda_\gamma)}\,, \tag{2}$$

where M_{IB} and M_{SD} correspond to the IB and SD part of the amplitude. The IB amplitude is given by[11,12]

$$\begin{aligned}\mathrm{M}_{IB}^{(\lambda_l,\lambda_\gamma)} &= \frac{ieG_F}{\sqrt{2}} m_l f_M V_{uq}\ \epsilon_\mu^\star(k,\lambda_\gamma) \\ &\cdot \left[\bar{u}(p_\nu)\left(\frac{p^\mu}{(p\cdot k)} - \frac{\not{k}\gamma^\mu + 2p_l^\mu}{2\,(p_l\cdot k)}\right)(1+\gamma_5)\,v(p_l,\lambda_l)\right]\,, \end{aligned} \tag{3}$$

where $\not{k} = \gamma^\alpha k_\alpha$, $\epsilon_\mu(k,\lambda_\gamma)$ stands for the photon polarization vector of momentum k and helicity λ_γ, while $\bar{u}(p_\nu)$ and $v(p_l,\lambda_l)$ are the bispinors of final neutrino and charged lepton respectively. Explanations of other symbols appearing above are in order. The G_F is the Fermi constant, m_l is the charged lepton mass, f_M is the meson decay constant, where $f_\pi \simeq 131$ MeV and $f_K \simeq 161$ MeV, and V_{uq} is the Cabibbo-Kobayashi-Maskawa matrix element corresponding to $u \to q = d$ and $u \to q = s$ quark transitions for pion and kaon decays respectively.

The SD part of the amplitude contains vectorial (V) and axial (A) form factors, that clearly depend on the kind of initial meson, but not on the lepton final states. Indeed, they are connected to the matrix elements of the electromagnetic hadron current V^μ_{em} and the axial and vectorial weak currents A^μ and V^μ respectively, as

$$(\mathrm{V},\mathrm{A})^{\mu\nu}(p,k) \equiv \int d^4x e^{ikx} \langle 0|\ TV^\mu_{\mathrm{em}}(x)(V(0),A(0))^\nu\ |M^+(p)\rangle\,. \tag{4}$$

Using Lorentz covariance and electromagnetic gauge invariance, it follows that:

$$\begin{aligned}\mathrm{V}^{\mu\nu}(p,k) &= i\frac{V}{m_M}\epsilon^{\mu\nu\alpha\beta}k_\alpha p_\beta \\ \mathrm{A}^{\mu\nu}(p,k) &= (p\cdot k)\frac{A}{m_M}\left(\eta^{\mu\nu} - \frac{p^\mu k^\nu}{(p\cdot k)}\right) - f_M\left(\eta^{\mu\nu} + \frac{p^\mu\,(p^\nu - k^\nu)}{(p\cdot k)}\right)\,, \end{aligned} \tag{5}$$

where $\eta^{\mu\nu} = \mathrm{diag}(1,-1,-1,-1)$ is the Minkowski metric, and $\epsilon^{\mu\nu\alpha\beta}$ is the totally antisymmetric tensor †. Finally, the SD part of the amplitude is

† In our notation, the $\epsilon^{\mu\nu\alpha\beta}$ is defined as $\epsilon^{0123} = 1$ and $\epsilon_{0123} = -1$, when generic four-vectors v_μ are $v_\mu = (v_0, \vec{v})$ and $v^\mu = (p_0, -\vec{v})$.

given by[11,12]

$$\mathrm{M}_{SD}^{(\lambda_l,\lambda_\gamma)} = -\frac{ieG_F}{\sqrt{2}} V_{uq}\ \epsilon^\star_\mu(k,\lambda_\gamma)\left[\bar{u}(p_\nu)\gamma_\nu\left(1-\gamma_5\right)v(p_l,\lambda_l)\right]$$
$$\times\left\{(p\cdot k)\frac{A}{m_M}\left(-\eta^{\mu\nu}+\frac{p^\mu k^\nu}{(p\cdot k)}\right)+i\epsilon^{\mu\nu\alpha\beta}\frac{V}{m_M}k_\alpha p_\beta\right\} \tag{6}$$

where m_M stands for the meson mass, while V and A are the meson vectorial and axial form factors respectively. Notice that both the terms M_{IB} and M_{SD} are separately gauge invariant, as can be easily checked by making the substitution $\epsilon^\star_\mu(k,\lambda_\gamma)\to\epsilon^\star_\mu(k,\lambda_\gamma)+k_\mu$ in Eqs. (3) and (6).

Now we provide the corresponding expressions for the polarized amplitude in the center of mass (c.m.) frame of the fermion pair (neutrino and charged lepton), namely $\vec{p_l}+\vec{p_\nu}=0$. We choose a frame where the 3-momenta of neutrino and photon have the following components in polar coordinates

$$\vec{p_\nu}=E_\nu\ \left(\sin\theta\cos\varphi,\sin\theta\sin\varphi,\cos\theta\right),\quad \vec{p_l}=-\vec{p_\nu},\quad \vec{k}=E_\gamma\ (0,0,1)\,, \tag{7}$$

where E_ν and E_γ are the neutrino and photon energies respectively and θ,φ are the usual polar angles. For the photon polarization vectors we choose helicity eigenstates ($\epsilon(k,\lambda)$), which in this frame are given by

$$\epsilon_\mu(k,\lambda_\gamma)=\frac{1}{\sqrt{2}}\left(0,1,i\lambda_\gamma,0\right) \tag{8}$$

whose helicity eigenvalues correspond to $\lambda_\gamma=-1$ left-handed (L) and $\lambda_\gamma=1$ right-handed (R) circular polarizations. Photon polarization vectors satisfy the transversality condition $k^\mu\epsilon_\mu(k,\lambda_\gamma)=0$. Regarding the polarization vectors of fermions, it is convenient to use the solution of the Dirac equation for the particle (u) and antiparticle (v) bispinors in the momentum space.[13] In the standard basis[‡] we have:

$$u(p,\lambda)=\begin{pmatrix}\sqrt{E+m}\ \omega_\lambda(\vec{n})\\ \sqrt{E-m}\,(\vec{\sigma}\cdot\vec{n})\,\omega_\lambda(\vec{n})\end{pmatrix}\quad v(p,-\lambda)=\begin{pmatrix}\sqrt{E-m}\ (\vec{\sigma}\cdot\vec{n})\,\omega_\lambda(\vec{n})\\ \sqrt{E+m}\ \omega_\lambda(\vec{n})\end{pmatrix}$$

where the 2-component spinors $\omega_\lambda(\vec{n})$ (with helicity $\lambda=\pm1$) are the eigenstates of the helicity operator $(\vec{\sigma}\cdot\vec{n})\,\omega_\lambda(\vec{n})=\lambda\,\omega_\lambda(\vec{n})$, and σ_i are the Pauli matrices. Here, $\vec{n}\equiv\vec{p}/|\vec{p}|$, where $\vec{p}$ is the 3-momentum and $E=\sqrt{|\vec{p}|^2+m^2}$

[‡]In the standard basis representation, $\gamma_0=\mathrm{Diag}(\mathbf{1},-\mathbf{1})$, and $\vec{\gamma}=\begin{pmatrix}0&\vec{\sigma}\\-\vec{\sigma}&0\end{pmatrix}$, and $\gamma_5=\begin{pmatrix}0&\mathbf{1}\\\mathbf{1}&0\end{pmatrix}$, where $\mathbf{1}=\mathrm{Diag}(1,1)$ and $\vec{\sigma}$ are as usual the Pauli matrices.

is the corresponding energy. If $\vec{p} = |\vec{p}|\,(\sin\theta\cos\varphi, \sin\theta\sin\varphi, \cos\theta)$, then in polar coordinates, $\omega_\lambda(\vec{n})$ can be expressed as

$$\omega_{+1}(\vec{n}) = \begin{pmatrix} e^{-i\frac{\varphi}{2}}\cos\frac{\theta}{2} \\ e^{i\frac{\varphi}{2}}\sin\frac{\theta}{2} \end{pmatrix}, \qquad \omega_{-1}(\vec{n}) = \begin{pmatrix} -e^{-i\frac{\varphi}{2}}\sin\frac{\theta}{2} \\ e^{i\frac{\varphi}{2}}\cos\frac{\theta}{2} \end{pmatrix}. \tag{9}$$

At this point it is convenient to introduce the following Lorentz invariant quantities

$$x \equiv \frac{2p\cdot k}{m_M^2}, \qquad y \equiv \frac{2p\cdot p_l}{m_M^2}, \qquad z \equiv \frac{2p_l\cdot k}{m_M^2} = y - 1 + x - r_l \tag{10}$$

where in the meson rest frame, x and y are just proportional to the photon and charged lepton energies respectively and $r_l = m_l^2/m_M^2$. Finally, after a straightforward algebra, the IB and SD contributions to the polarized amplitude in the fermion pair c.m. frame are given by

$$\begin{aligned} \mathrm{M}_{IB}^{(\lambda_l,\lambda_\gamma)} &= eG_F m_l f_M V_{uq}\,\frac{2}{z}\Big\{\delta_{\lambda_l,-1}\Big(\delta_{\lambda_\gamma,-1}\,\hat{E}_\gamma + \hat{E}_\nu\Big)R_+\sin\theta \\ &\qquad + \delta_{\lambda_l,+1}\,\delta_{\lambda_\gamma,-1}\,\hat{E}_\gamma R_-\,(1-\cos\theta)\Big\}\,e^{i\lambda_\gamma\varphi} \\ \mathrm{M}_{SD^\pm}^{(\lambda_l,\lambda_\gamma)} &= eG_F m_M^2 V_{uq}\,\frac{(V\pm A)}{2}\,\delta_{\lambda_\gamma,\pm1}\,x\Big\{\mp\delta_{\lambda_l,-1}\,R_-\sin\theta \\ &\qquad \pm\delta_{\lambda_l,+1}R_+\,(\cos\theta\pm1)\Big\}\,e^{i\lambda_\gamma\varphi}, \end{aligned} \tag{11}$$

where the structure dependent part is given by $\mathrm{M}_{SD}^{(\lambda_l,\lambda_\gamma)} = \mathrm{M}_{SD^+}^{(\lambda_l,\lambda_\gamma)} + \mathrm{M}_{SD^-}^{(\lambda_l,\lambda_\gamma)}$ and the symbol $R_\pm \equiv \sqrt{\hat{E}_\nu}\Big(\sqrt{\hat{E}_l + \sqrt{r_l}} \pm \sqrt{\hat{E}_l - \sqrt{r_l}}\Big)$, with $\hat{E}_i \equiv E_i/m_M$ and E_l is the energy of the final charged lepton. In this frame, the energies normalized to the meson mass are given by

$$\begin{aligned} \hat{E}_\gamma &= \frac{x}{2\sqrt{1-x}}, \qquad \hat{E}_\nu = \frac{1-x-r_l}{2\sqrt{1-x}}, \qquad \hat{E}_l = \frac{1-x+r_l}{2\sqrt{1-x}}, \\ \cos\theta &= \frac{(x-2)(1-x+r_l) + 2y(1-x)}{x(1-r_l-x)} \end{aligned} \tag{12}$$

and

$$R_+ = \sqrt{1-r_l-x}, \qquad R_- = \sqrt{r_l\frac{1-r_l-x}{1-x}}. \tag{13}$$

Notice that, as expected from general arguments, the azimuthal angle φ factorizes in the overall phase of the amplitude. At this point it is important to stress that the SD terms in the amplitude, proportional to $V+A$ and $V-A$, correspond to pure right-handed and left-handed photon polarizations respectively, while the IB one is a mixture of both. In particular, the terms proportional to pure left-handed photon polarizations in the M_{IB}, come only from the tensorial structure in Eq. (3), namely from terms proportional to $[\bar{u}_\nu\ \sigma_{\mu\nu}(1+\gamma_5)\ v_l]$, while scalar contributions of type $[\bar{u}_\nu\ (1+\gamma_5)\ v_l]$ do not select any specific photon polarization.

By using Eqs. (11) and (12), it is now straightforward to evaluate the square modulus of the amplitude. Below we will provide its expression summed over the charged lepton polarizations, as a function of the photon helicities. After integrating over the phase space, we obtain for the photon polarized decay rate Γ^{λ_γ}, the following result:

$$\frac{d^2\Gamma^{(\lambda_\gamma)}}{dx\ d\lambda} = \frac{m_{\rm M}}{256\pi^3} \sum_{\lambda_l=\pm 1} |{\rm M}^{(\lambda_l,\lambda_\gamma)}|^2 = \rho^{(\lambda_\gamma)}(x,\lambda)\,. \tag{14}$$

Here $m_{\rm M}$ stands for the generic meson mass $m_{{\rm M}=\pi,K}$, and $\lambda \equiv z/x$. The Dalitz plot densities $\rho^{\lambda_\gamma}(x,\lambda)$ for the polarized decay are Lorentz invariant functions can be obtained.[10]

The differential branching ratio (BR) is finally obtained

$$\frac{d^2{\rm BR}}{dx\ d\lambda} = {\rm BR}(M \to l\nu_l)\,\frac{1}{\Gamma_0}\frac{d^2\Gamma}{dx\ d\lambda}\,, \tag{15}$$

where $\Gamma_0 = \Gamma(M \to l\nu_l)$ is Born contribution to the total width of non radiative decay $M \to l\nu_l$, in particular

$$\Gamma_0(M^+(p) \to \nu_l + l^+) = \frac{G_F^2\, f_M^2\, m_M}{8\pi}|V_{uq}]^2 r_l\,(1-r_l)^2 \tag{16}$$

where ${\rm BR}(M \to l\nu_l)$ is the total branching ratio of the corresponding non radiative decay. Finally, the total branching ratio BR is obtained by integrating Eq. (15) in the full kinematical range as follows

$${\rm BR} = \int dx \int d\lambda\ \frac{d^2{\rm BR}}{dx\ d\lambda} \tag{17}$$

$$0 \le x \le 1 - r_l, \qquad \frac{r_l}{1-x} \le \lambda \le 1\,. \tag{18}$$

In Fig. 2 the possible helicities of the initial and final states are shown. A detailed list of all the amplitudes for the pion, muon and K meson decays can be obtained[10] with the corresponding resulting right-handed and left-handed amplitudes. A series of curves has been also derived according to the different contributions from right-handed and left-handed amplitudes.[10] Let us shortly discuss the novel mechanism of mass singularities cancellation as it emerges from the results obtained for the polarized amplitudes.

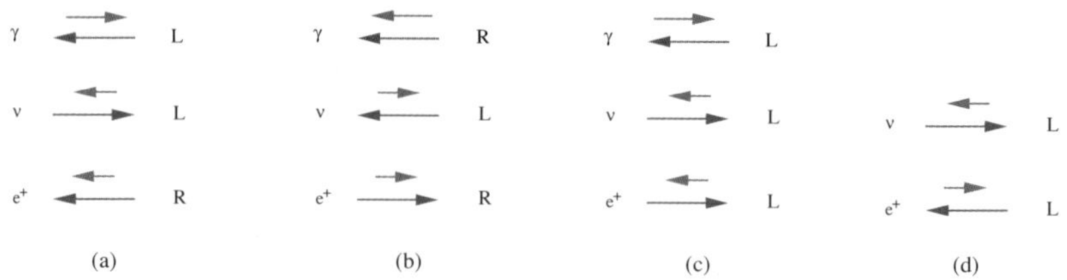

Fig. 2. Allowed helicity (upper arrows) configurations of γ, ν and e^+ for $\pi^+ \to e^+\nu_e\gamma$ decay in π^+ rest frame, figures (a), (b), (c), when all momenta (lower arrows) are aligned on the same axis. Direction of photon momentum is fixed by convention. Figure (d) corresponds to the non radiative decay $\pi^+ \to e^+\nu_e$. Analogous spin configurations hold for the corresponding K^+ decays as well.

3. Cancellation of mass singularities

Let us discuss the mechanism of mass singularities cancellation and the way it takes place in meson and muon polarized radiative decays. As will be seen a new peculiar cancellation pattern shows up in the particular case of the polarized amplitudes differently from the well known cancellation taking place in the inclusive unpolarized amplitudes.

In a theory with massless particles a crucial test of the consistency of the computation is represented by the absence of mass singularities in any obtained physical quantity. Mass singularities are of two types: infrared and collinear. Infrared divergences originate from massless particles with a vanishing momentum in the small energy soft limit. Physical states as, for example, a single charged particle, are degenerate with states made by the same particle accompanied by soft photons. This corresponds to the impossibility of distinguishing a charged particle from the one accompanied by given number of soft photons due to the finite resolution of any experimental apparatus. An infrared divergence appears in QED when the energy E_γ of the photon goes to zero as a factor of the form:

$$I = \int_0^1 \frac{d\epsilon}{\epsilon} \tag{19}$$

where $\epsilon = \frac{E_\gamma}{E}$ is the fraction of the energy of the photon with respect to the total available energy E for the process. The Bloch Nordsiek theorem[14] assures the cancellation of infrared divergences in any inclusive cross section. Collinear divergences, instead, come from massless particles having a vanishing value of the relative emission angle. In QED, specifically, when one or more photons, in the limit of zero fermion mass, are in a collinear configuration i.e. with emission angle $\theta \simeq 0$. Physical states containing a massless charged particle are degenerate with states containing the same particle and a number of collinear photons. Any experimental apparatus, having a finite angular resolution, cannot distinguish between them. The angular separation of two massless particles with momenta p and k is such that they move parallel to each other with a combined invariant mass for $\theta \to 0$:

$$q^2 = (p+k)^2 = 2\,p_0\,E_\gamma(1-\cos\theta) \to\ 0 \tag{20}$$

even though neither p nor E_γ are soft. Here θ is the emission angle of a photon with respect to the fermion. The inclusive procedure of integrating over the photon emission angles by keeping the fermion mass finite does not give rise to any collinear singularity. The divergence appears in the limit $\theta \to 0$ as the presence of a logarithm of the form $\log(\frac{E}{m}) \simeq \log(\theta)$.

For collinear singularities, as well as for infrared ones, the case for inclusive unpolarized processes is well known and it is governed by the Kinoshita-Lee-Nauenberg (KLN) theorem.[15] For the collinear singularities the KLN theorem guarantees that collinear divergences cancel out if one performs a sum of the amplitude over all the sets of degenerate states order by order in the perturbative expansion. For the amplitude of a single photon emission a combination of collinear and infrared singularities gives, for instance, contributions of the type:

$$R = \frac{\alpha}{\pi}\int_0^1 \frac{d\epsilon}{\epsilon}\int_0^1 \frac{d\theta}{1-\cos\theta}\,. \tag{21}$$

In order to discuss the above aspects on the cancellation of lepton mass singularity in the polarized pion decay, we first recall the mechanism taking place in the unpolarized case.[16,17] In general, the decay rate is made free from mass singularities in the ordinary way: the cancellation of divergences occurs in the total inclusive decay rate at order $O(\alpha)$, namely in the pion case

$$\Gamma^{(\mathrm{incl})} = \Gamma(\pi \to \nu e) + \Gamma(\pi \to \nu e\gamma)\,, \tag{22}$$

when the full $O(\alpha)$ order contributions are included, i.e. those relative to real and virtual photon emission.[16,17] However, for a pointlike (structureless) pion, due to the chirality flip of final charged lepton, the pion decay amplitude is always proportional to m_l and vanishes in the $m_l \to 0$ limit. In other words, in the limit $m_l \to 0$ the decay rate is made finite from mass singularities in a trivial way. For example, as we will see later on, a term proportional to $\mathrm{Log}(m_l)$ will remain in the inclusive width due to the mass renormalization of the charged lepton in the virtual contributions to $\Gamma(\pi \to \nu e)$. However, since it will be multiplied by m_l^2, it will give no troubles since $\Gamma_0 \to 0$ tends to zero at the same time. However, as pointed out by Kinoshita,[16] the leading $\log(m_l)$ terms in the IB contribution to $\Gamma(\pi \to \nu e \gamma)$ cancel out exactly when one adds the virtual contributions. In other words, the mass singularities in the $\log(m_l)$ terms should cancel independently from the fact that the effective coupling in the pion decay is proportional to the charged lepton mass or not. The cancellation mechanism of these Log terms shows a non trivial aspect of the KLN theorem in the pion decay. For this reason, in the following discussion we will consider the following ratios $\Gamma(\pi \to \nu e \gamma)/\Gamma_0$ and $\Gamma(\pi \to \nu e)/\Gamma_0$ which survives the limit $m_l \to 0$.

Let us now consider the case of radiative polarized decays within the soft and collinear region for the radiated photon. We will investigate in this section the mechanism which will assure the finiteness of the lepton distribution against the appearance of infrared and collinear singularities on the above ratios of widths. Let us start by the inclusive distributions in terms of the final lepton energy y. The Inner Bremsstrahlung contribution in the $r_l \to 0$ limit is composed by the four expressions corresponding to the various polarization states of the final photon and lepton respectively[10]

$$\lim_{r_l \to 0} \frac{1}{\Gamma_0} \frac{d\Gamma_{IB}^{(\mathrm{L,L})}}{dy} = \frac{\alpha}{2\pi} \frac{1}{y-1} \Big(1 + y - \hat{L}_1 - \hat{L}_2\Big)$$

$$\lim_{r_l \to 0} \frac{1}{\Gamma_0} \frac{d\Gamma_{IB}^{(\mathrm{L,R})}}{dy} = \frac{\alpha}{2\pi} \Big(1 - y\Big)$$

$$\lim_{r_l \to 0} \frac{1}{\Gamma_0} \frac{d\Gamma_{IB}^{(\mathrm{R,L})}}{dy} = \frac{\alpha}{2\pi} \frac{1}{y-1} \Big(y\,(y+1) - \hat{L}_1\, y^2 + \hat{L}_2\,(1 - 2\,y)\Big)$$

$$\lim_{r_l \to 0} \frac{1}{\Gamma_0} \frac{d\Gamma_{IB}^{(\mathrm{R,R})}}{dy} = 0 \qquad (23)$$

The last, RR polarized term is identically zero. The remaining three are related to the left-handed (first and third) and right-handed (second) lepton respectively.

The logarithms L_1, L_2 do correspond to collinear contributions. By integrating the double-inclusive distribution[10] one gets in the expression for the IB case that depend on the logarithms L_1 and L_2:[10]

$$L_1 = \log\left(\frac{y + \mathrm{A}_l - 2\,r_l}{y - \mathrm{A}_l - 2\,r_l}\right), \qquad L_2 = \log\left(\frac{y + \mathrm{A}_l - 2}{y - \mathrm{A}_l - 2}\right)$$

These terms do give rise to two kinds of "collinear" logarithms:

$$L_1 = \log \frac{E_l}{m_l}, \qquad L_2 = \log \frac{E_l + \sqrt{E_l^2 + m_l^2} - m_M}{E - \sqrt{E_l^2 + m_l^2} - m_M}.$$

The first logarithm represents the case of the photon being parallel to the lepton, the second collinear logarithm for $m_l \to 0$ and $m_M \to 0$ corresponds to the case where the photon is parallel to the decaying meson.[3] Clearly, it is only L_1 which is affected by the true collinear divergence in the limit $m_l \to 0$.

With respect to the unpolarized inclusive case some differences are worth to be noticed here:

- Different polarization amplitudes do represent independent observables in the decay. Therefore if we consider the two cases of a right-handed and left-handed lepton they have to be also separately finite.
- At zero order in the pion decay the angular momentum conservation imposes to the lepton to be left-handed. By radiating a photon a total zero angular momentum is assigned to the final state even if a right-handed lepton emits a left-handed polarized photon. This contribution is represented by the second term in Eq. (23).
- For $y \to 1$ only the second term in Eq. (23), corresponding to the LR polarization, contribution is finite, i.e. it is zero:

$$\lim_{r_l \to 0 \;\, y \to 1} \frac{1}{\Gamma_0} \frac{d\Gamma_{IB}^{(\mathrm{L,R})}}{dy} = \frac{\alpha}{2\pi} \Big(1 - y\Big) = 0\,.$$

This fact shows that the LR term, corresponding to the anomalous term,[18] it is finite by itself without the need of any cancellation mechanism in the infrared $y \to 1$ and collinear limit $r_l \to 0$. A detailed discussion of the undergoing dynamics has been given.[3]

Analogous conclusions, regarding the finiteness of the right-handed lepton contribution in the $r_l \to 0\,, y \to 1$ limits to the structure dependent terms $|SD|^2$ and the $IB \times SD$, hold there as well.[10]

Let us now consider the contributions of the type LL and RL giving rise to a left handed lepton. Manifestly the first and third term of Eq. (23) are divergent in the $r_l \to 0\ ,\ y \to 1$ limits. The expression obtained by adding first and third contributions in Eq. (23) is:

$$\frac{1}{\Gamma_0}\left[\frac{d\Gamma_{IB}^{(\mathrm{L,L})}}{dy}+\frac{d\Gamma_{IB}^{(\mathrm{R,L})}}{dy}\right]=\frac{\alpha}{2\pi}\frac{1}{y-1}\Big[\Big(1+y-\hat{L}_1-\hat{L}_2\Big) \qquad (24)$$
$$+\Big(y\,(y+1)-\hat{L}_1\,y^2+\hat{L}_2\,(1-2\,y)\Big)\Big]$$

which is divergent both in the collinear and in the infrared limit. The coefficients of the collinear logarithms remain different from zero as $r_l \to 0$ and $y \to 1$, leaving to a divergent expression. As for the unpolarized case for the left-handed lepton contributions one needs to consider the additional virtual contributions in order to cancel infrared singularities.[16]

The case of the left-handed lepton includes also the diagram of the virtual photon i.e. the one with a photon line connecting meson and charged lepton. This diagram does not add any angular momentum to the zeroth order term since a virtual particle does not add angular momentum to the final state. The amplitude containing the virtual photon gives rise, therefore, to a lepton neutrino final state having the same helicities as the ones of the tree level amplitude. The combination of real and virtual contributions should, in the left-handed lepton channel, add among each other to give a finite result. This mechanism is the same taking place for the cancellation of singularities for the inclusive, unpolarized amplitudes as we will discuss in more details below.

The total width for the unpolarized IB contribution to $\Gamma(\pi \to \nu e \gamma)$ is given by[16]

$$\frac{\Gamma_{IB}(x_0)}{\Gamma_0}=\frac{\alpha}{\pi}\Big\{b(r_l)\left(\log\frac{x_0}{2}-\log(1-r_l)-\frac{1}{4}\log(r_l)+\frac{3}{4}\right)$$
$$-\frac{r_l\,(10-7r_l)}{4\,(1-r_l)^2}\log(r_l)+\frac{2\,(1+r_l)}{1-r_l}L(1-r_l)+\frac{15-21r_l}{7(1-r_l)}\Big\}, \qquad (25)$$

where x_0 is the minimum photon energy which regularizes the infrared divergence in the photon mass, the function $b(x)=\frac{1+x}{(1-x)}\log(x)+2$, and $L(x)=\int_0^x \log(1-t)dt/t$. For a generalization of the result in Eq. (25) to the inclusion of the leading logarithmic terms to all orders in perturbation theory.[19–21]

For the virtual 1-loop contribution one has to consider the radiative corrections to the operator $gm_l^0 Q$, where $Q=\bar{\psi}_l(1-\gamma_5)\psi_\nu\varphi_\pi$, with φ_π is the pion field and m_l^0 is the 'bare' mass of the charged lepton. These corrections split in two separate contributions: $\Gamma^{(1)}$ given by the correction to the operator Q and $\Gamma^{(2)}$ arising when one try to express the bare mass

m_l^0 in terms of the renormalized lepton mass m_l, namely $m_l^0 = m_l - \delta m_l$ with $\delta m_l = \frac{3\alpha}{2\pi} m_l \left(\log(\Lambda/m_l) + \frac{1}{4}\right)$.[17] For $\Gamma^{(1)}$, one has[16]

$$\frac{\Gamma^{(1)}}{\Gamma_0} = \frac{\alpha}{\pi} \left\{ -b(r_l) \left(\log(\frac{x_0}{2}) - \frac{1}{4} \log(r_l) + \frac{3}{4} \right) + \frac{r_l}{2(1-r_l)} \log(r) + \frac{1}{2} \right\}$$
$$+ \frac{3\alpha}{2\pi} \log(\frac{\Lambda}{m_\pi}) \tag{26}$$

Notice that the last term, containing the ultraviolet cut-off Λ needed to regularize the UV divergency, can in principle be absorbed in a re-definition of f_π at order α.[17]

As can be seen by comparing the results in Eqs. (25) and (26), the $\log r_l$ terms surviving the limit $r_l \to 0$ cancel out in the sum of virtual and real emission contributions as a consequence of the KNL theorem. Finally, for the total contribution to the unpolarized inclusive decay rate at order α, including the contribution of $\Gamma^{(2)}$, one gets:[16,17]

$$\frac{\Gamma^{(\mathrm{incl})}}{\Gamma_0} = 1 + \frac{\alpha}{\pi} \left\{ \frac{3}{2} \log(r_l) + \frac{13}{8} - \frac{\pi^2}{3} \right\}, \tag{27}$$

where we retained only the leading terms in $m_e \to 0$ limit. As previously mentioned, the appearance of the $\log r_l$ term in (27) is due to the renormalization of the charged lepton mass which does not follow the same pattern of collinear mass singularities discussed above.[16] For simplicity, we omitted in (27) the term containing a $\log(\Lambda/m_\pi)$, since it can be absorbed into a re-definition of f_π at order α inside Γ_0.

As stated above, in the right-handed case, on the contrary, the mass singularities cancellation occurs with a different mechanism. Infrared and collinear limits in the ratio $\Gamma_{IB}^{(LR)}/\Gamma_0$ give separately a finite result. In particular, the coefficient of the collinear logarithms for the right handed lepton case is the lepton mass, instead of the usual correction factor coming from the soft and the virtual photon contributions.

The particular cancellation mechanism occurring in the right-handed radiative decay is originated by the combined constraints of the angular momentum conservation in the pion vertex and the one of the helicity flip in the photon-lepton vertex.[3]

Let us now consider the case of the muon decay. As for the meson case also in the muon decay lepton distribution we see that the LR-photon-lepton polarized distribution is free from collinear and infrared singularities and goes to zero in the infrared limit $y \to 1$. The remaining RL and LL distributions, apart from the identically zero RR term, do give a finite contributions in the $y \to 1$ limit, provided that the same x_0 cut-off is also set free to go to the soft kinematical limit i.e. $x_0 \to 0$. In the muon case the pattern of singularities cancellation repeats itself as for the meson case.

Acknowledgments and a personal memory of Vladimir Gribov

I wish to thank Julia and the organizers for inviting me to this Workshop. It gives me the opportunity to recall an episode which, I believe, goes well with the theme of my presentation and gives a further example of Volodya Gribov's spirit and personality as a man and as a physicist. In July 1984 I was invited by George Pócsik at the Institute for Theoretical Physics of the Eötvös University in Budapest. Pócsik had organized a meeting devoted mainly to QCD and to topics related to the CERN experiments at the $Sp\bar{p}S$ collider. Volodya was there and, among others, Bo Anderson was also there presenting his Lund Monte Carlo model. I still remember the continuos questioning of Volodya during Bo's talk. It was a new and quite stimulating experience to me. It showed to me a really new perspective about how talks and seminars could be lively and inspiring occasions to do physics. As I realized years later this was the Gribov's style and one of the most distinct peculiarities of the Gribov's School. The first time I spoke with him was the very last day of the meeting, during the farewell party. I dared to talk with Gribov, and, in some way, the conversation continued until when, later, we had to leave. The university meeting room was going to be closed and we were still talking. He asked me when I was supposed to leave. At my reply I had to leave the day after he gave me, looking to him with astonishment, a rendezvous, a couple of hours later, in a café in order to continue our conversation. Later, in the evening, Volodya explained to me, among other arguments we were talking about, the role played by the Anomaly within his picture of the confinement. As Arkady Vainshtein has also recalled in his talk at this workshop.[24] Volodya, with his pioneering article of 1981[22] and, later, with his 1987 paper,[23] did interpret the anomaly as a collective motion of the vacuum, a view that only later, in 1985, was presented in the anomaly book by S. Treiman, R. Jackiw, B. Zumino and E. Witten.[25] I've seen Volodya several times later and heard him and talked with him in several occasions. His enthusiasm, his scientific integrity and honesty, his endless quest for truth and his simple open attitude toward anyone interested in physics are still with us.

References

1. D.A. Bryman, P. Depommier and C. Leroy, *Phys. Rep.* **88**, 151(1982).
2. J. Choi, U. Won Lee, H.S. Song and J. H. Kim, *Phys. Rev.* **D39**, 2652 (1989).
3. L. Trentadue and M. Verbeni, *Phys. Lett.* **B478**, 137 (2000); *Nucl. Phys.* **B583**, 307 (2000), and references therein;
4. M. Fischer, S. Groote, J.G. Koerner and M.C. Mauser, *Phys. Rev.* **D67**, 113008 (2003), and references therein.
5. V.S. Schulz and L.M. Sehgal, *Phys. Lett.* **B594**, 153 (2004).
6. T.D. Lee and M. Nauenberg, *Phys. Rev.* **133**, 1549 (1964).
7. A.D. Dolgov and V.I. Zakharov, *Nucl. Phys.* **B27**, 525 (1971).
8. B. Falk and L.M. Sehgal, *Phys. Lett.* **B325**, 509 (1994); L.M. Sehgal, *Phys. Lett.* **B569**, 25 (2003).
9. A.V. Smilga, *Comments Nucl. Part. Phys.* **20**, 69 (1991).
10. L.Trentadue, E. Gabrielli, *Nucl. Phys.* **B792**, 48 (2008).
11. E. Gabrielli, *Phys. Lett.* **B301**, 409 (1993).
12. J. Bijnens, G. Colangelo, G. Ecker and J. Gasser, *Semileptonic Kaon Decays*, Published in 2nd DAPHNE Physics Handbook, 315 (1995), hep-ph/9411311.
13. L.D. Landau, E.M. Lifshitz, L.P. Pitaevskii, *Relativistic Quantum Theory*, Pergamon Press, Oxford 1971.
14. F. Bloch, A. Nordsieck, *Phys. Rev.* **52**, 54 (1937).
15. T. Kinoshita, *J. Math. Phys.* **3**, 650 (1962); T.D. Lee and M. Nauenberg, *Phys. Rev.* **133**, 1549 (1964).
16. S.M. Berman, *Phys. Rev. Lett.* **1**, 468 (1958); T. Kinoshita, *Phys. Rev. Lett.* **2**, 477 (1959).
17. W.J. Marciano and A. Sirlin, *Phys. Rev. Lett.* **71**, 3629 (1993).
18. A.D. Dolgov and V.I. Zakharov, *Nucl. Phys.* **B27**, 525 (1971); B. Falk and L.M. Sehgal, *Phys. Lett.* **B325**, 509 (1994); L. Trentadue and M. Verbeni, *Phys. Lett.* **B478**, 137 (2000) and *Nucl. Phys.* **B583**, 307 (2000).
19. E. Kuraev, *JETP Lett.* **65**, 127 (1997).
20. E.A. Kuraev, V. Fadin, *Sov. J. Nucl. Phys.* **41**, 466 (1985).
21. O. Nicrosini, L. Trentadue, *Phys. Lett.* **B196**, 551 (1987).
22. V. Gribov, *Anomalies, as a manifestation of the high momentum collective motion in the vacuum*, KFKI-1981–66 preprint 1981
23. V. Gribov, *Anomalies and a possible solution of problems of zero charge*, *Phys. Lett.* **B194**, 1 (1987).
24. A. Vainstein, Talk at this workshop.
25. S. Treiman, R. Jackiw, B. Zumino, E. Witten, *Current algebra and anomalies*, World Scientific Pub Co Inc., 520p. (1985)

A POSSIBLE ORIGIN OF THE JAFFE-WITTEN MASS GAP IN QCD

V. GOGOKHIA

HAS CRIP RMKI, Department of Theoretical Physics,
Budapest 114, P.O.B. 49, H-1525, Hungary
gogohia@rmki.kfki.hu

We explain how the Jaffe-Witten mass gap appears in QCD. It is responsible for the large-scale structure of its ground state.

Quantum Chromodynamics (QCD) is widely accepted as a realistic quantum field gauge theory of the strong interactions not only at the fundamental (microscopic) quark-gluon level but at the hadronic (macroscopic) level as well. It is a $SU(3)$ color gauge invariant theory. One of the important challenges of QCD is that its Lagrangian does not contain a mass scale parameter which could have a physical meaning even after the corresponding renormalization program is performed. Thus the only place where it may appear explicitly is the Schwinger-Dyson equation of motion for the full gluon propagator. Precisely this problem has been addressed and solved in our papers.[1–3]

The general scale parameter, having the dimensions of mass squared, is dynamically generated in the QCD gluon sector. It is introduced through the difference between the regularized full gluon self-energy and its value at some finite point. It violates transversality of the full gluon self-energy. The Slavnov-Taylor identity for the full gluon propagator, when it is given by the corresponding equation of motion, is also violated by it. So in order to maintain both transversality and the identity it should be disregarded from the very beginning, i.e., put formally zero everywhere. However, we have shown how to preserve the Slavnov-Taylor identity at non-zero mass squared parameter. This allows one to establish the structure of the full gluon propagator when it is explicitly present. Its contribution does not survive in the perturbation theory regime, when the gluon momentum goes to infinity. At the same time, its contribution dominates the structure of the full gluon propagator when the gluon momentum goes to zero. We have also proposed a method how to restore transversality of the relevant gluon propagator in a gauge invariant way, while keeping the mass squared parameter "alive". In this case, the two independent general types of formal solutions for the full gluon propagator as a function of the regularized mass gap have been found. The nonlinear iteration solution at which the gluons remain massless is explicitly present. It is nothing else but the Laurent expansion in inverse powers of the gluon momentum squared and multiplied by the corresponding powers of the regularized mass gap. The existence of the solution with an effective gluon mass is also demonstrated.

We have shown how precisely the renormalization program for the regularized mass gap should be performed. For this we have used the so-called Weierstrass-Sokhatsky-Casorati theorem which describes the behavior of meromorphic functions near their singularities. We have also shown how precisely severe infrared singularities (which inevitably appear in the above-mentioned general nonlinear iteration solution) should be correctly treated. For this we have used the theory of distributions into which the dimensional regularization method is to be correctly implemented. All this allows to analytically formulate the exact and gauge-invariant criteria of quark and gluon confinement (apparently, for the first time). It is explained how QCD exhibits the two phase transitions: in the strong and weak coupling limits, leading to intrinsically non-perturbative (INP) QCD and PT QCD, respectively. It has been shown explicitly how the regularized mass gap provides the existence of asymptotic freedom in PT QCD due to the renormalization program in the weak coupling limit. After the renormalization program is completed the gluon propagator applicable for the calculation of physical observables, processes, etc., in low-energy QCD (or, equivalently, INP QCD) from first principles is $D_{\mu\nu}(q) = iT_{\mu\nu}(q)(\Delta_R^2/q^4)$, where Δ_R^2 is the physical mass gap.

In comparison with the Jaffe and Witten theorem[4]

Yang-Mills Existence And Mass Gap: Prove that for any compact simple gauge group G, quantum Yang-Mills theory on $\mathbf{R^4}$ exists and has a mass gap $\Delta > 0$, the basic result obtained in our papers[1–3] can be formulated as follows:

Mass Gap Existence And Gluon Confinement: If quantum Yang-Mills theory with compact simple gauge group $G = SU(3)$ exists on $\mathbf{R^4}$, then undergoing the phase transition in the strong coupling regime it becomes INP QCD, which has a physical mass gap and confines gluons.

Our general conclusion is that the one of the main problems in quantum field gauge theories, in particular QCD, namely how a mass gap appears, is solved.

References

1. V. Gogokhia, *The color gauge invariance and a possible origin of a mass in QCD*, *Int. J. Theor. Phys.* **48**, 3061 (2009), arXiv:0806.0247.
2. V. Gogokhia, *Nonlinear iteration solution for the full gluon propagator as a function of the mass gap*, *Int. J. Theor. Phys.* **48**, 3470 (2009), arXiv:0904.2266.
3. V. Gogokhia, *Renormalization of the mass gap*, *Int. J. Theor. Phys.* **48**, 3449 (2009), arXiv:0907.0082.
4. A. Jaffe, E. Witten, *Yang-Mills Existence and Mass Gap*, *http://www.claymath.org/prize-problems/*, *http://www.arthurjaffe.com.*

CHIRALITY VIOLATING CONDENSATES IN QCD AND THEIR CONNECTION WITH ZERO MODE SOLUTIONS OF QUARK DIRAC EQUATIONS

B. L. IOFFE

A.I. Alikhanov Institute of Theoretical and Experimental Physics,
B. Cheremushkinskaya 25, 117218, Moscow, Russia

It is demonstrated, that chirality violating condensates in massless QCD arise entirely from zero mode solutions of Dirac equations in arbitrary gluon fields. A model is suggested, where the zero mode solutions are the ones for quarks, moving in the instanton field. Based on this model the quark condensate magnetic susceptibilities of dimensions $3(\chi)$ and 5 (κ and ξ) were calculated. The good coincidence of the values χ, κ and ξ, obtained in this approach, with those found from the hadronic spectrum is a serious argument in favour of the idea that instantons are the only source of chirality violating condensates in QCD. The temperature dependence of the quark condensate is discussed. It is shown that the phase transition, corresponding to the T-dependence of the quark condensate $\alpha(T)$ as an order parameter, is of the type of crossover.

It is well known that because of the small values of light quark masses the perturbative QCD possesses the property of chiral symmetry in cases, when the heavy quark contributions can be neglected. However, in the real hadronic world the chiral symmetry is badly violated. This statement evidently follows from the existence of the large proton mass and from the absence of the negative parity baryon, the partner of a proton. It is also well known that large values of light quark condensates indicate the violation of chiral symmetry in QCD. These two facts are deeply interconnected: the values of the proton mass can be expressed through the value of quark condensate.[1]

In this talk I discuss the appearance of quark condensate in nonperturbative QCD. The connection of the quark condensate value with zero mode solutions of quark Dirac equation is established.[2] In this way the nature of proton mass and, as a consequence, the nature of all the visible mass in the Universe is clarified.[3]

Consider the QCD action in Euclidean space-time

$$S=\frac{1}{4}\int d^4xG^2_{\mu\nu}-\int d^4x\sum_f\Big[\psi^+_f(i\gamma_\mu\nabla_\mu+im_f)\psi_f\Big] \tag{1}$$

where $G^n_{\mu\nu}$ is the gluon field tensor, the sum goes over quark flavours.

$$\nabla_\mu=\partial_\mu+ig\frac{\lambda^n}{2}A^n_\mu \tag{2}$$

and A^n_μ is the gluon field. Pay attention, that in the Euclidean formulation of QCD $\overline{\psi}$ is replaced by ψ^+. (The review of Euclidean formulation of QCD and instantons is given in Ref. 4, see especially Ref. 5.) The Dirac equation for massless quark in Euclidean space-time has the form:

$$-i\gamma_\mu\nabla_\mu\psi_n(x)=\lambda_n\psi_n(x) \tag{3}$$

where $\psi_n(x)$ and λ_n are the eigenfunctions and eigenvalues of the Dirac operator $-\boldsymbol{\nabla}=-i\gamma_\mu\nabla_\mu$. Expand the quark field operators into the left and right ones

$$\psi=\frac{1}{2}(1+\gamma_5)\psi_L+\frac{1}{2}(1-\gamma_5)\psi_R$$

$$\psi^+=\psi^+_L\frac{1}{2}(1+\gamma_5)+\psi^+_R\frac{1}{2}(1-\gamma_5), \tag{4}$$

where

$$\gamma_5\psi_L=\psi_L,\quad \gamma_5\psi_R=-\psi_R \tag{5}$$

Then for nonzero λ_n the Lagrangian and the action reduces to the sum of two terms

$$L=-\int\Big[\ \psi^+_L\boldsymbol{\nabla}\psi_R+\psi^+_R\boldsymbol{\nabla}\psi_L\ \Big]d^4x \tag{6}$$

completely symmetric under the interchange $L\longleftrightarrow R$. Therefore the solutions of the equations for left and right quark fields are also the same – the states, constructed from left and right quarks are completely symmetrical. This conclusion was obtained for a fixed gluon field. It is evident, that the averaging over the gluon fields does not change it. A quite different situation arises in the case $\lambda_0=0$. The contribution of this term to the Lagrangian:

$$\Delta L=\int d^4x[\ \psi^+_L+\psi^+_R\ ]\boldsymbol{\nabla}\psi_0 \tag{7}$$

is equal to zero and no conclusion can be made about the symmetry of states build from left and right quark fields. One of the consequences from what is said above is that all chirality violating vacuum condensates in QCD arise from zero mode solutions of the Dirac equations (3).

These general arguments are supported by the well-known facts:
1. The general representation of the trace of quark propagator $S(x)$ is expressed through the spectral function $\rho(\lambda)$ as a function of eigenvalues λ (Källen-Lehmann representation):

$$TrS(x^2) = \frac{1}{\pi} \int d\lambda \rho(\lambda) \Delta(x^2, \lambda) \tag{8}$$

At $x^2 = 0$ $\Delta(x^2, \lambda)$ reduces to $\delta(\lambda)$ and we have (in Minkowski space-time):

$$\rho(0) = -\pi \langle 0 \mid \overline{\psi}(0)\psi(0) \mid 0 \rangle. \tag{9}$$

(The Banks-Casher relation[6]).
2. The zero-mode solution of (3) for a massless quark in the instanton field is the right wave function – $\psi_R(x) = (1 - \gamma_5)\psi(x)$ and in the field of anti-instanton is the left one – $\psi_L(x) = (1 + \gamma_5)\psi(x)$.[7,8]

Based on the statements, presented above, let us formulate the model for the calculation of chirality violating vacuum condensates in QCD. Suppose, that the vacuum expectation value (v.e.v.) of the chirality violating operator $O_{c.v.}$ is proportional to the matrix element $\psi_0^+ O_{c.v.} \psi_0$, where ψ_0 is the zero-mode solution of Eq. (3) in Euclidean space-time:

$$\langle 0 \mid \overline{\psi} O_{c.v.} \psi \mid 0 \rangle \sim \psi_0^+ O_{c.v.} \psi_0. \tag{10}$$

ψ_0 depends on x, on the position of the center of the solution x_c, as well as on its size ρ: $\psi_0 = \psi_0(x - x_c, \rho)$. Equation (10) must be integrated over x_c, what is equivalent to integration over $x - x_c$. (In what follows the notation x will be used for $x - x_c$.) We assume, that ρ =const and find its value from the comparison with the known v.e.v.s. Finally, we introduce in (10) the coefficient of proportionality n. So, our assumption has the form:

$$\langle 0 \mid \overline{\psi}(0) O_{c.v.} \psi(0) \mid 0 \rangle = -n \int d^4x \psi_0^+(x, \rho) O_{c.v.} \psi_0(x, \rho) \tag{11}$$

Our model is similar to the dilute instanton gas model,[9] where x_c is the position of the instanton center. Unlike the latter, where the instanton density has dimension 4, n has dimension 3 and may be interpreted as the density of zero-modes centers in 3-dimension space. Note, that the left-hand side of (11) is written in the Minkowski space-time, while the right-hand side in the Euclidean one. (The minus sign is put in order to have n positive.) In perturbative calculation in case, when the number of flavours is more than 1, $N_f > 1$, the contribution of instantons to the action is suppressed by a factor, proportional to the product of $N_f - 1$ light quark masses. Therefore, the assumption (11) means that such a suppression is absent in non-perturbative calculation. For the x and ρ-dependence of $\psi_0(x,\rho)$ we take the form of the zero-mode solution in the field of instanton in the $SU(2)$ colour group:

$$\psi_0(x,\rho) = \frac{1}{2}(1-\gamma_5)\frac{1}{\pi}\frac{\rho}{(x^2+\rho^2)^{3/2}}\chi_0, \tag{12}$$

where χ_0 is the spin-colour isospin ($\mid \mathbf{T} \mid = 1/2$) wave function, corresponding to the total spin $\mathbf{I} + \mathbf{T} = \mathbf{J}$ equal to zero, $J = 0$. $\psi_0(x,\rho)$ is normalized to 1:

$$\int d^4x\psi^+(x,\rho)\psi(x,\rho) = 1 \tag{13}$$

Consider first the quark condensate $\langle 0 \mid \bar{q}q \mid 0\rangle$, the most important chirality violating v.e.v., determining the values of baryon masses.[1–11] (Here $q = u, d$ are the fields of u, d-quarks). In this case $O_{c.v.} = 1$ and, in accord with (11), (13) we have

$$n = -\langle 0 \mid \bar{q}q \mid 0\rangle = (1.65 \pm 0.15) \times 10^{-2} \ \ \mathrm{GeV}^3 \ \ (\mathrm{at} \ \ 1 \ \ \mathrm{GeV}) \ [10] \tag{14}$$

(The integration over the $SU(2)$ subgroup in the $SU(3)$ colour group as well as the anti-instanton contribution are included in the definition of n.) The anomalous dimension of quark condensate is equal to 4/9. According to (14) n has the same anomalous dimension. The size ρ of the zero-mode wave function can be found by calculations in the framework of our model of the v.e.v.

$$-g\langle 0 \mid \overline{\psi}\sigma_{\mu\nu}\frac{\lambda^n}{2}G^n_{\mu\nu}\psi \mid 0\rangle \equiv m_0^2\langle 0 \mid \bar{q}q \mid 0\rangle. \tag{15}$$

The parameter m_0^2 is equal to[13]: $m_0^2 = 0.8$ GeV^2 at 1 GeV. The m_0^2 anomalous dimension is equal to $-14/27$. Working in the $SU(2)$ colour group, substitute λ^n by $\tau^a (a = 1, 2, 3)$ and take for $G^a_{\mu\nu}$ the instanton field

$$G^a_{\mu\nu}(x,\rho) = \frac{4}{g}\eta_{a\mu\nu}\frac{\rho^2}{(x^2+\rho^2)^2}, \tag{16}$$

where the parameter $\eta_{a\mu\nu}$ was defined by 't Hooft[14] (see also Ref. 5). The substitution of (12) and (16) into (11) gives after some simple algebra

$$\frac{1}{2}n\frac{1}{\rho^2} = m_0^2 n. \tag{17}$$

Therefore,

$$\rho = \frac{1}{\sqrt{2}m_0} = 0.79 \ \ \mathrm{GeV}^{-1} = 0.256\mathrm{fm}(\mathrm{at}\ 1\ \mathrm{GeV}). \tag{18}$$

This value of ρ is close to those used in the delute instanton gas or instanton liquid models.

We are now in a position to calculate less well known quantities – the magnetic susceptibilities of the quark condensate, induced by an external constant electromagnetic field.

The dimension 3 quark condensate magnetic susceptibility is defined by:[15]

$$\langle 0 \mid \bar{q}\sigma_{\mu\nu} q \mid 0\rangle_F = e_q \chi \langle 0 \mid \bar{q}q \mid 0\rangle F_{\mu\nu}, \quad q = u, d, \tag{19}$$

where quarks are considered as moving in an external constant weak electromagnetic field $F_{\mu\nu}$ and e_q is the charge of the quark q in units of the proton charge (the proton charge e is included in the definition of $F_{\mu\nu}$). The left-hand side of (19) violates chirality, so it is convenient to separate explicitly the factor $\langle 0 \mid \bar{q}q \mid 0\rangle$ in the right-hand side. It was demonstrated in Ref. 15 that $\langle 0 \mid \bar{q}\sigma_{\mu\nu} q \mid 0\rangle_F$ is proportional to the charge e_q of the quark q. A universal constant χ is called the quark condensate magnetic susceptibility.

Let us determine the value of χ in our approach. For this goal it is necessary to consider Eq. (3) in the presence of the external constant electromagnetic field $F_{\mu\nu}$ and to find the first order in $F_{\mu\nu}$ correction to the zero mode solution (12). This can be easily done by representing ψ as

$$\psi(x,\rho) = \psi_0(x,\rho) + \psi_1(x,\rho), \tag{20}$$

where ψ_0 is given by (12) and ψ_1 represents the proportional to $F_{\mu\nu}$ correction. Substitute (20) in Eq. (3) added by the term of interaction with electromagnetic field, neglect ψ_1 in this term and solve the remaining equation for $\psi_1(x,\rho)$). The result is:

$$\psi_1(x,\rho) = \frac{1}{16} e_q \eta_{a\mu\nu} \sigma_a F_{\mu\nu} x^2 \left(1 + \frac{1}{2}\frac{x^2}{\rho^2}\right) \psi_0(x,\rho), \tag{21}$$

where σ_a are Pauli matrices. The matrix element $\psi^+ \sigma_{\mu\nu} \psi$ appears to be

$$\psi^+ \sigma_{\mu\nu} \psi = -\frac{1}{2} e_q F_{\mu\nu} \psi_0^+ x^2 \left(1 + \frac{1}{2}\frac{x^2}{\rho^2}\right) \psi_0. \tag{22}$$

(The properties of $\eta_{a\mu\nu}$ symbols[5,14] were exploited.) The v.e.v. (19) in the Minkowski space-time is given by:

$$\langle 0 \mid \overline{\psi}\sigma_{\mu\nu}\psi \mid 0\rangle_F = e_q F_{\mu\nu} n \frac{1}{\pi^2} \int d^4x x^2 \left(1 + \frac{1}{2}\frac{x^2}{\rho^2}\right) \frac{\rho^2}{(x^2+\rho^2)^3} \tag{23}$$

(The normalization condition (13) for $\psi_0(x,\rho)$ was used.) It is convenient to express n through quark condensate by (14), use the notation $x^2 = r^2$, where r is the radius-vector in 4-dimensional space. Then according to (19) we have:

$$\chi = -\rho^2 \int_0^{R^2} dr^2 r^4 \left(1 + \frac{1}{2}\frac{r^2}{\rho^2}\right) \frac{1}{(r^2+\rho^2)^3} \tag{24}$$

The integral (24) is quadratically divergent at large r. So, the cut-off R is introduced. Its value can be estimated in following way. The volume occupied by one zero-mode in 3-dimensional space is approximately equal to $1/n$ (the volume of the Wigner-Seitz cell). So, for cut-off radius square R^2 in four-dimensions we put

$$R^2 = \frac{4}{3}\left(\frac{3}{4\pi n}\right)^{2/3} = 7.92 \text{ GeV}^{-2} \tag{25}$$

where the factor 4/3 corresponds to transition from 3 to 4 dimensions. The calculation of the integral (24) at the values of parameters ρ (18) and R^2 (25) gives

$$\chi = -3.52 \text{ GeV}^{-2} \tag{26}$$

The quark condensate magnetic susceptibility was previously calculated by the QCD sum rule method[16–19] and expressed through the masses and coupling constants of mesonic resonances. The recent results are:

$$\begin{aligned} \chi(\text{ 1 GeV}) &= -3.15 \pm 0.3 \text{ GeV}^{-2} \text{ [18]} \\ \chi(\text{ 1 GeV}) &= -2.85 \pm 0.5 \text{ GeV}^{-2} \text{ [19]} \end{aligned} \tag{27}$$

(The earlier results, obtained by the same method, were: $\chi(\text{ 0.5 GeV}) = -\,5.7 \text{ GeV}^{-2}$, see Ref. 16, and $\chi(\text{ 1 GeV}) = -\,4.4 \pm 0.4 \text{ GeV}^{-2}$, see Ref. 17.) The anomalous dimension of χ is equal to -16/27. It was accounted in Refs. 16–19, but not in the presented above calculation. (In some of these papers, the α_s-corrections and continuum contribution, were also accounted.) One can believe, that the value (26) refer to 1 GeV, because the value of quark condensate (14) refer to this scale and also because the scale 1 GeV is a typical scale, where, on the one hand, the zero-modes and quark condensates are quite important (see, e.g. Ref. 12) and, on the other, the instanton gas model is valid.[9] Since the integral is quadratically divergent it is hard to estimate the accuracy of (26). I guess, that it is not worse, than 30-50%. In the limit of this error the result (26) is in an agreement with those found in phenomenological approaches.

Let us turn now to quark condensate magnetic susceptibilities of dimension 5, κ and ξ defined in Ref. 15

$$g\langle 0 \mid \bar{q}\frac{1}{2}\lambda^n G^n_{\mu\nu}\bar{q} \mid 0\rangle_F = e_q \kappa F_{\mu\nu}\langle 0 \mid \bar{q}q \mid 0\rangle, \tag{28}$$

$$-ig\varepsilon_{\mu\nu\rho\tau}\langle 0 \mid \bar{q}\gamma_5 \frac{1}{2}\lambda^n G^n_{\rho\tau} q \mid 0\rangle_F = e_q \xi F_{\mu\nu}\langle 0 \mid \bar{q}q \mid 0\rangle \tag{29}$$

Perform first the calculation of κ. In this case the expression of $\psi_1(x,\rho)$ (21) must be multiplied by the additional factor: $\frac{1}{2}\tau^b G^b_{\mu\nu}$ where $G^b_{\mu\nu}$ is given by (16) and the indices μ, ν in (21) are changed to λ, σ. In the further calculation it will be taken into account, that χ_0 in (12) corresponds to total spin-colour isospin $J = 0$ and consequently

$$\boldsymbol{\sigma\tau}\chi_0 = -3\chi_0, \quad \sigma^a \tau^b \chi_0 = -\delta^{ab}\chi_0. \tag{30}$$

In the relation

$$\eta_{b\mu\nu}\eta_{b\lambda\sigma} = \delta_{\mu\lambda}\delta_{\nu\sigma} - \delta_{\mu\sigma}\delta_{\nu\lambda} + \varepsilon_{\mu\nu\lambda\sigma} \tag{31}$$

the last term drops out after summation of zero-modes from the instanton and anti-instanton configuration. The final result for κ is:

$$\begin{aligned}\kappa &= -\int\limits_0^z u^2 du \frac{1}{(u+1)^4}\left(1 + \frac{1}{2}u\right) \\ &= -\frac{1}{2}\left[\ln(z+1) - \frac{13}{6} + \frac{1}{z+1} + \frac{1}{2}\frac{1}{(z+1)^2} - \frac{1}{3}\frac{1}{(z+1)^3}\right],\end{aligned} \tag{32}$$

where $z = R^2/\rho^2 = 12.7$. Numerically, we have:

$$\kappa = -0.26 \tag{33}$$

The calculation of ξ is very similar to that of κ and the result is

$$\xi = 2\kappa = -0.52 \tag{34}$$

The values of κ and ξ only logarithmically depend on the cut-off. But unfortunately the logarithm in (32) is not very large and its main part is compensated by the term $-13/6$, appearing in (32). So, the accuracy of (33), (34) can be estimated as about 30%. The phenomenological determination of 5-dimensional quark condensate magnetic susceptibilities was performed by Kogan and Wyler[20] along the same lines, as it was done in Refs. 16, 17. No anomalous dimensions were accounted. The results of Ref. 20 are:

$$\kappa = -0.34 \pm 0.1, \quad \xi = -0.74 \pm 0.2 \tag{35}$$

As can be seen, they are in a good agreement with (33), (34). The 5-dimensional quark condensate magnetic susceptibilities play a remarkable role in determination of Λ-hyperon magnetic moment.[21]

I conclude this part of my talk. It was argued, that chiral symmetry violation in QCD arises due to zero-mode solution of Dirac equation for massless quark in arbitrary gluon field. The model is proposed similar to the delute instanton gas model, in which the zero-mode solution is the same as in the field of instanton. The parameters of the model: the density of zero-modes and their size are determined from the values of quark and quark-gluon condensates. In the framework of this model the values of quark condensates magnetic susceptibilities of dimensions 3 and 5 were calculated in agreement with ones found by QCD sum rules method using the properties of hadronic spectrum. The agreement of these two approaches gives the strong argument in favour that the instantons are the only source of chirality violating condensates in QCD.

Turn now to the temperature dependence of the quark condensate, considered as an order parameter. It is expected that quark condensate vanishes at high temperatures and the chiral symmetry is restored. Two possibilities are discussed: the second order phase transition and the crossover. In our approach the normalization condition (13) takes place at any temperature.[22] So, the temperature dependence of quark condensate reduces to

$$\alpha(T) \equiv \langle 0 \mid \bar{q}q \mid \rangle_T = n(T) \tag{36}$$

One may expect that $n(T)$ vanishes only in the case, when the quantum number carried by the instanton is vanishing. This quantum number is the topological charge, which is temperature independent. So, in our approach $n(T)$ never vanishes and the phase transition is of the type of crossover. The order parameter temperature dependence of the second order phase transition near the critical point T_c is smeared by fluctuations[23] (The same statement refers, surely, to the crossover.) The fluctuations are determined by long wave oscillations of the fields. In QCD they are given by small frequency gluonic field $G^n_{\mu\nu}$, which is a constant. We have:

$$\frac{\Delta\alpha(T_c)}{\alpha(T_c)} \sim \langle 0 \mid G^2_{\mu\nu} \mid 0 \rangle T_c^2 \rho^6, \tag{37}$$

where $\Delta\alpha(T_c)$ is of the order of magnitude of the variation of $\alpha(T)$ near the crossover critical point. It is evident, that the left hand side should vanish, if T_c would be zero. I guess, that the factor corresponding to this circumstance is T_c^2. The factor ρ^6 is added for dimensional reasons.

All the discussion of the $\alpha(T)$ temperature dependence is based on the implicit assumption, that this object has a physical meaning at finite T. It is unclear, however, how it can be defined. I did not succeed to formulate the gedanken-experiments, in which this object could be measured at finite T. Therefore, may be, such an object has no physical sense at $T \neq 0$.

Acknowledgement

This work is supported by RFBR grant 09-02-00732 and in part by CRDF Cooperative Program grant RUP2-2961-MO-09. I acknowledge the support of the European Community-Research Infrastructure Integrating Activity "Study of Strongly Interacting Matter" under the Seventh Framework Program of EU.

References

1. B.L. Ioffe, *Nucl. Phys.* **B188**, 317 (1981), (E) *ibid.* **B192**, 591 (1982).
2. B.L. Ioffe, *Phys. Lett.* **B678**, 512 (2009).
3. B.L. Ioffe, *Usp. Fiz. Nauk* **176**, 1103 (2008), *Sov. Phys. Usp.* **49**, 1077 (2006).
4. *Instantons in Gauge Theories*, ed. by M. Shifman, World Scientific, 1994.
5. A.I. Vainshtein, V.I. Zakharov, V.A. Novikov and M.A. Shifman, *Usp. Fiz. Nauk* **136**, 553 (1982) (*Sov. Phys. Usp.* **25**, 195 (1982)), Reproduced in [1], p.468.
6. T. Banks and A. Casher, *Nucl. Phys.* **B169**, 103 (1980).
7. L.S. Brown, R. Carlitz and C. Lee, *Phys. Rev.* **D16**, 417 (1977).
8. R. Jackiw and C. Rebbi, *Phys. Rev.* **D16**, 1052 (1977).
9. T. Schäfer and E.V. Shuryak, *Rev. Mod. Phys.* **70**, 323 (1998).
10. B.L. Ioffe, *Z. Phys.* **C18**, 67 (1983).
11. B.L. Ioffe, *Phys. Atom. Nucl.* **72**, 1214 (2009), arXiv:0810.4234.
12. B.L. Ioffe, *Prog. Part. Nucl. Phys.* **56**, 232 (2006).
13. V.M. Belyaev and B.L. Ioffe, *Sov. Phys. JETP* **56**, 493 (1982).
14. G. 't Hooft, *Phys. Rev.* **D16**, 3432 (1976).
15. B.L. Ioffe and A.V. Smilga, *Nucl. Phys.* **B232**, 109 (1984).
16. V.M. Belyaev and I.I. Kogan, *Yad. Fiz.* **40**, 1035 (1984).
17. I.I. Balitsky, A.V. Kolesnichenko and A.V. Yung, *Sov. J. Nucl. Phys.* **41**, 138 (1985).
18. P. Ball, V.M. Braun and N. Kivel, *Nucl. Phys.* **B649**, 263 (2003).
19. J. Rohrwild, *JHEP* **0709**, 073 (2007).
20. I.I. Kogan and D. Wyler, *Phys. Lett.* **B274**, 100 (1992).
21. J. Pasupathy, J.P. Singh, S.L. Wilson and C.B. Chiu, *Phys. Rev.* **D36**, 1442 (1987).
22. D.J. Gross, R.D. Pisarski and L.G. Yaffe, *Rev. Mod. Phys.* **53**, 43 (1981).
23. L.D. Landau and E.M. Lifshitz, *Statistical Physics*, Part I, Pergamon Press, 1980.

Trieste, 2010

NONPERTURBATIVE QUARK SEA ASYMMETRIES

H. DAHIYA* and N. SHARMA

Department of Physics, Dr. B.R. Ambedkar National Institute of Technology, Jalandhar, 144011, India
**harleen_dahiya@yahoo.com*

The effects of nonperturbatively generated "quark sea" have been investigated to determine the flavor structure of the octet baryons. The chiral constituent quark model (χCQM), which is known to provide a satisfactory explanation of the proton spin and related issues in the nonperturbative regime, is able to explain the qualitative generation of the requisite amount of quark sea. The importance of quark sea has been studied at different values of the Bjorken scaling variable x by including it phenomenologically in the sea quark distribution functions. The results for the quark sea asymmetries like $\bar{d}(x) - \bar{u}(x)$, $\bar{d}(x)/\bar{u}(x)$ and Gottfried integral for the octet baryons strengthen the significance of quark sea at lower values of x.

Keywords: Chiral symmetry breaking; nonperturbative regime of QCD; Octet baryons.

1. Introduction

After the first direct evidence for the point-like constituents in the nucleon,[1] identified as the valence quarks with spin-1/2 in the naive constituent quark model (NQM),[2–4] a lot of experiments have been conducted to probe the structure of the proton in the deep inelastic scattering (DIS) experiments. Surprisingly, the DIS results in the early 80's[5] indicated that the valence quarks of the proton carry only about 30% of its spin and is referred to as the "proton spin crisis" in the NQM. These results provided the first evidence for the proton being composed of three valence quarks surrounded by an indistinct sea of quark-antiquark pairs (henceforth referred to as the "quark sea"). In the present day, the study of the composition of hadrons can be said to be primarily the study of the quark sea and gluons and is considered as one of the active areas in hadronic physics.

The conventional expectation that the quark sea perhaps can be obtained through the perturbative production of the quark-antiquark pairs by gluons produces nearly equal numbers of $\bar{u}$ and $\bar{d}$. Until early 90's a symmetric sea w.r.t. $\bar{u}$ and $\bar{d}$ was assumed, however, the famous New Muon Collaboration in 1991[6] established the quark sea asymmetry of the unpolarized quarks in the case of nucleon by measuring $\bar{d} - \bar{u}$ giving first clear evidence for the nonperturbative origin of the quark sea. This was later confirmed by the Drell-Yan experiments[7] which measured a large quark sea asymmetry ratio $\bar{d}/\bar{u}$ reminding us that the study of the quark sea is intrinsically a nonperturbative phenomena and it is still a big challenge to perform these calculations from the first principles of QCD.

One approach to account for the observed quark sea asymmetry is the pion cloud mechanism[8] where the quark sea is believed to originate from process such as virtual pion production. It is suggested that in the deep inelastic lepton-nucleon scattering, the lepton probe also scatters off the pion cloud surrounding the target proton. The $\pi^+(\bar{d}u)$ cloud, dominant in the process $p \to \pi^+ n$, leads to an excess of $\bar{d}$ sea. However, this effect should be significantly reduced by the emissions such as $p \to \Delta^{++} + \pi^-$ with $\pi^-(\bar{u}d)$ cloud. Therefore, the pion cloud idea is not able to explain the significant $\bar{d} > \bar{u}$ asymmetry. This approach can be improved upon by adopting a mechanism which operates in the *interior* of the hadron.

The chiral constituent quark model (χCQM)[9] can yield an adequate description of the quark sea generation through the chiral fluctuations. The basic idea is based on the possibility that chiral symmetry breaking takes place at a distance scale much smaller than the confinement scale. In this region, the effective degrees of freedom are the valence quarks and the internal Goldstone bosons (GBs) which are coupled to the valence quarks[10–12] allowing a simple and intuitive method to investigate the principle features of the hadron structure. In the case of spin dependent quantities, the χCQM is not only successful in giving a satisfactory explanation of "proton spin crisis"[10,13] but is also able to account for the baryon magnetic moments[14] and hyperon $\beta-$decay parameters.[11,15,16] However, in the case of quark distribution functions, the latest developments by the NuSea (E866)[17] and HERMES[18] to determine the variation of the sea-antiquark ratio $\bar{d}(x)/\bar{u}(x)$ and the difference $\bar{d}(x) - \bar{u}(x)$ with Bjorken scaling variable x have renewed considerable interest in the quark sea asymmetries.

Recently, there has been substantial theoretical progress to take into account the effects of quark sea in determining the flavor structure of the baryons and the question of sea asymmetry has been investigated by several authors using various phenomenological models. Calculations have been carried out in the meson cloud models,[19] chiral quark-soliton model,[20] effective chiral quark model,[21] statistical models,[22] bag model,[23] model for parton densities,[24] radiative parton model[25] etc. However, the inclusion of $x-$dependence has not yet been successfully included in the quark distribution functions. Therefore, pending further experiments, it would be interesting to examine the flavor structure of the octet baryons at low energy, thereby giving vital clues to the nonperturbative effects of QCD. The study of $x-$dependence in the quark distribution functions becomes particularly interesting for the χCQM where the effects of quark sea and valence quarks can separately be calculated.

The purpose of the present communication is to determine the sea quark distribution functions and their asymmetries in the octet baryons by phenomenologically incorporating $x-$dependence in the χCQM. The extent of contributions coming from the different sea quarks for the octet baryons can also be compared. To understand the relation of the Bjorken scaling variable and quark sea, it would be significant to study its implications in the region $x < 0.3$ which is a relatively clean region to test the quark sea structure as well as to estimate their structure functions and related quantities.[26]

2. Chiral Constituent Quark Model

The key to understand the "proton spin crisis", in the χCQM formalism, is the fluctuation process[11] $q^{\pm} \rightarrow \text{GB} + q^{'\mp} \rightarrow (q\bar{q}^{'}) + q^{'\mp}$, where $q\bar{q}^{'} + q^{'}$ constitute the "quark sea".[11–13,15] The effective Lagrangian describing interaction between quarks and a nonet of GBs, can be expressed as $\mathcal{L} = g_8 \bar{\mathbf{q}} \left(\Phi + \zeta \frac{\eta'}{\sqrt{3}} I \right) \mathbf{q} = g_8 \bar{\mathbf{q}} \left(\Phi' \right) \mathbf{q}$, where $\zeta = g_1/g_8$, g_1 and g_8 are the coupling constants for the singlet and octet GBs, respectively, I is the 3×3 identity matrix. In terms of the SU(3) and axial U(1) symmetry breaking parameters, introduced by considering $M_s > M_{u,d}$, $M_{K,\eta} > M_\pi$ and $M_{\eta'} > M_{K,\eta}$,[11,12,15] the GB field can be expressed as

$$\Phi' = \begin{pmatrix} \frac{\pi^0}{\sqrt{2}} + \beta \frac{\eta}{\sqrt{6}} + \zeta \frac{\eta^{'}}{\sqrt{3}} & \pi^+ & \alpha K^+ \\ \pi^- & -\frac{\pi^0}{\sqrt{2}} + \beta \frac{\eta}{\sqrt{6}} + \zeta \frac{\eta^{'}}{\sqrt{3}} & \alpha K^0 \\ \alpha K^- & \alpha \bar{K}^0 & -\beta \frac{2\eta}{\sqrt{6}} + \zeta \frac{\eta^{'}}{\sqrt{3}} \end{pmatrix}. \quad (1)$$

The parameter $a(= |g_8|^2)$ denotes the probability of chiral fluctuation $u(d) \rightarrow d(u) + \pi^{+(-)}$, whereas $\alpha^2 a$, $\beta^2 a$ and $\zeta^2 a$ respectively denote the probabilities of fluctuations $u(d) \rightarrow s + K^{-(0)}$, $u(d,s) \rightarrow u(d,s) + \eta$, and $u(d,s) \rightarrow u(d,s) + \eta^{'}$.

For the sake of simplification, the GB field can also be expressed in terms of the quark contents of the GBs and is expressed as

$$\begin{pmatrix} \phi_{uu}u\bar{u}+\phi_{ud}d\bar{d}+\phi_{us}s\bar{s} & \varphi_{ud}u\bar{d} & \varphi_{us}u\bar{s} \\ \varphi_{du}d\bar{u} & \phi_{du}u\bar{u}+\phi_{dd}d\bar{d}+\phi_{ds}s\bar{s} & \varphi_{ds}d\bar{s} \\ \varphi_{su}s\bar{u} & \phi_{sd}s\bar{d} & \phi_{su}u\bar{u}+\phi_{sd}d\bar{d}+\phi_{ss}s\bar{s} \end{pmatrix}, \tag{2}$$

where

$$\phi_{uu}=\phi_{dd}=\frac{1}{2}+\frac{\beta}{6}+\frac{\zeta}{3}, \phi_{ss}=\frac{2\beta}{3}+\frac{\zeta}{3}, \phi_{us}=\phi_{ds}=\phi_{su}=\phi_{sd}=-\frac{\beta}{3}+\frac{\zeta}{3},$$
$$\phi_{du}=\phi_{ud}=-\frac{1}{2}+\frac{\beta}{6}+\frac{\zeta}{3}, \varphi_{ud}=\varphi_{du}=1, \varphi_{us}=\varphi_{ds}=\varphi_{su}=\varphi_{sd}=\alpha. \tag{3}$$

The quark sea content of the baryon can be calculated in χCQM by substituting for every constituent quark $q \to \sum P_q q + |\psi(q)|^2$, where $\sum P_q$ is the transition probability of the emission of a GB from any of the q quark and $|\psi(q)|^2$ is the transition probability of the q quark. The flavor structure for the baryon of the type $B(xxy)$ is expressed as $2P_x x + P_y y + 2|\psi(x)|^2 + |\psi(y)|^2$ and for the type $B(xyz)$ it is expressed as $P_x x + P_y y + P_z z + |\psi(x)|^2 + |\psi(y)|^2 + |\psi(z)|^2$, where $x, y, z = u, d, s$.

3. Quark Distribution Functions

There are no simple or straightforward rules which could allow incorporation of $x-$dependence in χCQM. To this end, instead of using an *ab initio* approach, we have phenomenologically incorporated the $x-$dependence getting clues from Eichten *et al.*,[10] Isgur *et al.*[3] and Le Yaouanc *et al.*[4] The $x-$dependent sea quark distribution functions can be now expressed as $\bar{u}^B(x) = \bar{u}^B(1-x)^{10}$, $\bar{d}^B(x) = d^B(1-x)^7$, $\bar{s}^B(x) = s^B(1-x)^8$ which together with the valence quark distribution functions give the flavor structure of the baryon as

$$q^B(x) = q^B_{\rm val}(x) + \bar{q}^B(x)\,, \tag{4}$$

where $q = u, d, s$. Using the sea quark distribution functions as explained above, the quark sea asymmetries $\bar{u}(x) - \bar{d}(x)$ and $\bar{d}(x)/\bar{u}(x)$ can also be calculated at different x values. We have already discussed the inclusion of x-dependence in detail and compared our results with the experimental data for the case of nucleon in Ref.[27] In the present communication however, we have extended our calculations to the case of other octet baryons for which experimental data is not yet available.

The $x-$dependence of the structure functions F_1 and F_2 can be calculated from

$$F_2^B(x) = x \sum_{u,d,s} e_q^2[q^B(x) + \bar{q}^B(x)], \quad F_1^B(x) = \frac{1}{2x} F_2^B(x), \tag{5}$$

where e_q is the charge of the quark q ($e_u = \frac{2}{3}$ and $e_d = e_s = -\frac{1}{3}$). In terms of the quark distribution functions, the structure function F_2 for any baryon can be expressed as

$$F_2^B(x) = \frac{4}{9}x(u^B(x) + \bar{u}^B(x)) + \frac{1}{9}x(d^B(x) + \bar{d}^B(x) + s^B(x) + \bar{s}^B(x)). \tag{6}$$

Several important quantities can be obtained from the structure functions of different isospin multiplets. For example, for the case of proton and neutron we have

$$\frac{F_2^p(x) - F_2^n(x)}{x} = \frac{4}{9}\left(u^p_{\rm val}(x) - u^n_{\rm val}(x) + 2\bar{u}^p(x) - 2\bar{u}^n(x)\right) + \frac{1}{9}\left(d^p_{\rm val}(x)\right.$$

$$\left. + s^p_{\rm val}(x) - d^n_{\rm val}(x) - s^n_{\rm val}(x) + 2\bar{d}^p(x) + 2\bar{s}^p(x) - 2\bar{d}^n(x) - 2\bar{s}^n(x)\right). \tag{7}$$

The Gottfried integral I_G^{pn} [28] can be expressed in terms of the sea quarks as follows

$$I_G^{pn} = \int_0^1 \frac{F_2^p(x) - F_2^n(x)}{x} dx = \frac{1}{3} + \frac{2}{3}\int_0^1 \left[\bar{u}^p(x) - \bar{d}^p(x)\right] dx, \tag{8}$$

where we have used the following normalization conditions

$$\int_0^1 u^p_{\rm val}(x)dx = 2, \quad \int_0^1 d^p_{\rm val}(x)dx = 1, \int_0^1 s^p_{\rm val}(x)dx = 0,$$

$$\int_0^1 u^n_{\rm val}(x)dx = 1, \int_0^1 d^n_{\rm val}(x)dx = 2, \int_0^1 s^n_{\rm val}(x)dx = 0,$$

$$\int_0^1 \bar{d}^n(\bar{u}^n)(x)dx = \int_0^1 \bar{u}^p(\bar{d}^p)(x)dx, \quad \int_0^1 \bar{s}^n(x)dx = \int_0^1 \bar{s}^p(x)dx. \tag{9}$$

Similarly, for the case of other octet baryons the following normalization conditions

$$\int_0^1 u^{\Sigma^+}_{\rm val}(x)dx = 2, \quad \int_0^1 d^{\Sigma^+}_{\rm val}(x)dx = 0, \quad \int_0^1 s^{\Sigma^+}_{\rm val}dx = 1,$$

$$\int_0^1 u^{\Sigma^0}_{\rm val}(x)dx = 1, \quad \int_0^1 d^{\Sigma^0}_{\rm val}(x)dx = 1, \quad \int_0^1 s^{\Sigma^0}_{\rm val}(x)dx = 1,$$

$$\int_0^1 u^{\Xi^0}_{\rm val}(x)dx = 1, \quad \int_0^1 d^{\Xi^0}_{\rm val}(x)dx = 0, \quad \int_0^1 s^{\Xi^0}_{\rm val}(x)dx = 2, \tag{10}$$

lead to the other Gottfried integrals in terms of the sea quarks

$$
\begin{aligned}
I_G^{\Sigma^+\Sigma^0} &= \frac{1}{3} + \frac{2}{9}\int_0^1 \left[4\bar{u}^{\Sigma^+}(x) + \bar{d}^{\Sigma^+}(x) - 4\bar{u}^{\Sigma^0}(x) - \bar{d}^{\Sigma^0}(x)\right] dx\,,\\
I_G^{\Sigma^0\Sigma^-} &= \frac{1}{3} + \frac{2}{9}\int_0^1 \left[4\bar{u}^{\Sigma^0}(x) + \bar{d}^{\Sigma^0}(x) - 4\bar{d}^{\Sigma^+}(x) - \bar{u}^{\Sigma^+}(x)\right] dx\,,\\
I_G^{\Xi^0\Xi^-} &= \frac{1}{3} + \frac{2}{3}\int_0^1 \left[\bar{u}^{\Xi^0}(x) - \bar{d}^{\Xi^0}(x)\right] dx\,.
\end{aligned}
\tag{11}
$$

It is clear from Eqs. (8) and (11), the flavor symmetric sea leads to the Gottfried sum rule $I_G = \frac{1}{3}$ with $\bar{u}^B{=}\bar{d}^B$.

4. Results and Discussion

After having detailed the contribution of the quark sea and the various asymmetries in the octet baryons of different quark structure, we now discuss the variation of these quantities with the Bjorken variable x. For the numerical calculation of the sea quark distribution functions of the octet baryons, we have used the same set of input parameters as detailed in our earlier calculations.[13,16,27,29] One can easily find out that

$$\bar{d}^p(x) > \bar{u}^p(x) > \bar{s}^p(x),$$

$$\bar{d}^{\Sigma^+}(x) > \bar{u}^{\Sigma^+}(x) \approx \bar{s}^{\Sigma^+}(x),$$

$$\bar{d}^{\Sigma^0}(x) > \bar{u}^{\Sigma^0}(x) > \bar{s}^{\Sigma^0}(x),$$

$$\bar{d}^{\Xi^0}(x) > \bar{u}^{\Xi^0}(x) > \bar{s}^{\Xi^0}(x),$$

showing a clear quark sea asymmetry as observed in the DIS experiments.[6,17,18] These distributions clearly indicate that our results pertaining to the quark sea asymmetry seem to be well in line with the expected results. As already mentioned in the introduction, the sea quarks do not contribute at higher values of x, therefore in this region, the contributions should be completely dominated by the valence quarks. The difference between the various sea distributions is observed to be maximum at $x \approx 0.1$. As the value of x increases, the difference between the sea contributions decreases in all the cases which is in line with the observations of other models.[20,22–24]

The general aspects of the variation of the magnitudes of the sea quark distribution functions $\bar{u}(x)$, $\bar{d}(x)$ and $\bar{s}(x)$ for the octet baryons are able to explain some of the well known experimentally measurable quantities, for example, $\bar{d}^B(x) - \bar{u}^B(x)$, $\bar{d}^B(x)/\bar{u}^B(x)$ and the Gottfried integral. These quantities not only provide important constraint on a model that attempts to describe the origin of the quark sea but also provide a direct determination of the presence of significant amount of quark sea in the low x region. When x is small, $\bar{d}^B(x) - \bar{u}^B(x)$ asymmetries are large implying the dominance of sea quarks in the low x region. In fact, the sea quarks dominate only in the region where x is smaller than 0.3. At the values $x > 0.3$, $\bar{d} - \bar{u}$ tends to 0 implying that there are no sea quarks in this region. The contribution of the quark sea in the case of Σ^0 is particularly interesting because of its flavor structure which has equal numbers of u, d and s quarks in its valence structure. Unlike the other octet baryons, where the $\bar{d}(x) - \bar{u}(x)$ asymmetry decreases continuously with the x values, the asymmetry in this case first increases and then for values of $x > 0.1$ it decreases. However, it is interesting to observe the the asymmetry peak in this case which matches with our other predictions where the contribution of the quark sea is maximum at $x \approx 0.1$

A measurement of the Gottfried integral[6,17] for the case of nucleon has shown a clear violation of Gottfried sum rule from $\frac{1}{3}$ which can find its explanation in a global quark sea asymmetry $\int_0^1 (\bar{d}(x) - \bar{u}(x))dx$. Similarly, for the case of Σ^+, Σ^0, and Ξ^0, the Gottfried sum rules should read $I_G^{\Sigma^+\Sigma^0} = \frac{1}{3}$, $I_G^{\Sigma^0\Sigma^-} = \frac{1}{3}$ and $I_G^{\Xi^0\Xi^-} = \frac{1}{3}$ if the quark sea was symmetric. However, due to the $\bar{d}(x) - \bar{u}(x)$ asymmetry in the case of octet baryons, a lower value of the Gottfried integrals is obtained. In the case of nucleon the results are in good agreement with the experimental data[17] as already presented in.[27] The quality of numerical agreement in the other cases can be assessed only after the data gets refined. Further, this phenomenological analysis strongly suggests an important role for the quark sea at low value of x. New experiments aimed at measuring the flavor content of the other octet baryons are needed for profound understanding of the nonperturbative properties of QCD.

5. Summary and Conclusions

To summarize, in order to investigate the effects of "quark sea", we have calculated the sea quark distribution functions for the octet baryons in the chiral constituent quark model (χCQM). The Bjorken scaling variable x has been incorporated phenomenologically to enlarge the scope of model and to understand the range of x where quark sea effects are important. Implications of the quark sea have also been studied to estimate the quark sea asymmetries like $\bar{d}(x) - \bar{u}(x)$, $\bar{d}(x)/\bar{u}(x)$ and Gottfried integral. The results justify our conclusion regarding the importance of quark sea at small values of x.

In conclusion, the results obtained for the quark distribution functions reinforce our conclusion that χCQM is able to generate qualitatively as well as quantitatively the requisite amount of quark sea. This can perhaps be substantiated by a measurement of the quark distribution functions of the other octet baryons.

Acknowledgments

H.D. would like to thank the organizers, Gribov80 Workshop held at ICTP and Department of Science and Technology, Government of India, for financial support.

References

1. E.D. Bloom *et al.*, *Phys. Rev. Lett.* **23**, 930 (1969); M. Breidenbach *et al.*, *Phys. Rev. Lett.* **23**, 935 (1969).
2. A. De Rujula, H. Georgi and S.L. Glashow, *Phys. Rev.* **D12**, 147 (1975).
3. N. Isgur, G. Karl and R. Koniuk, *Phys. Rev. Lett.* **41**, 1269 (1978); N. Isgur and G. Karl, *Phys. Rev.* **D21**, 3175 (1980); P. Geiger and N. Isgur, *Phys. Rev.* **D55**, 299 (1997); N. Isgur, *Phys. Rev.* **D59**, 034013 (1999).
4. A. Le Yaouanc, L. Oliver, O. Pene and J.C. Raynal, *Phys. Rev.* **D12**, 2137 (1975); **15**, 844 (1977).
5. J. Ashman *et al.*, (EMC Collaboration), *Phys. Lett.* **B206**, 364 (1988); *Nucl. Phys.* **B328**, 1 (1989); B. Adeva *et al.* (SMC Collaboration), *Phys. Lett.* **B302**, 533 (1993); P. Adams *et al.*, *Phys. Rev.* **D56**, 5330 (1997); P.L. Anthony *et al.* (E142 Collaboration), *Phys. Rev. Lett.* **71**, 959 (1993); K. Abe *et al.* (E143 Collaboration), *Phys. Rev. Lett.* **75**, 391 (1995).
6. P. Amaudruz *et al.* (New Muon Collaboration), *Phys. Rev. Lett.* **66**, 2712 (1991); M. Arneodo *et al.*, *Phys. Rev.* **D50**, R1 (1994).
7. A. Baldit *et al.* (NA51 Collaboration), *Phys. Lett.* **B253**, 252 (1994).
8. J.D. Sullivan, *Phys. Rev.* **D5**, 1732 (1972); A.W. Thomas, *Phys. Lett.* **B126**, 97 (1983); J. Magnin and H.R. Christiansen, *Phys. Rev.* **D61**, 054006 (2000).
9. S. Weinberg, *Physica* **A96**, 327 (1979); A. Manohar and H. Georgi, *Nucl. Phys.* **B234**, 189 (1984).
10. E.J. Eichten, I. Hinchliffe and C. Quigg, *Phys. Rev.* **D45**, 2269 (1992).
11. T.P. Cheng and L.F. Li, *Phys. Rev. Lett.* **74**, 2872 (1995); T.P. Cheng and L.F. Li, *Phys. Rev.* **D57**, 344 (1998).
12. X. Song, J.S. McCarthy and H.J. Weber, *Phys. Rev.* **D55**, 2624 (1997); X. Song, *Phys. Rev.* **D57**, 4114 (1998).
13. H. Dahiya and M. Gupta, *Phys. Rev.* **D64**, 014013 (2001); **67**, 074001 (2003); *Int. J. Mod. Phys.* **A19**, 19 5027 (2004); H. Dahiya, M. Gupta and J.M.S. Rana, *Int. J. Mod. Phys.* **A21**, 21 4255 (2006); H. Dahiya and M. Gupta, *Phys. Rev.* **D78**, 014001 (2008).

14. H. Dahiya and M. Gupta, *Phys. Rev.* **D66**, 051501(R) (2002); **D67**, 114015 (2003).
15. J. Linde, T. Ohlsson and H. Snellman, *Phys. Rev.* **D57**, 452 (1998); **D57**, 5916 (1998).
16. N. Sharma, H. Dahiya, P.K. Chatley and M. Gupta, *Phys. Rev.* **D79**, 077503 (2009); N. Sharma, H. Dahiya and P.K. Chatley, *Eur. Phys. J.* **A44**, 125 (2010); N. Sharma, H. Dahiya, P.K. Chatley and M. Gupta, *Phys. Rev.* **D81**, 073001 (2010).
17. E.A. Hawker *et al.* (E866/NuSea Collaboration), *Phys. Rev. Lett.* **80**, 3715 (1998); J.C. Peng *et al.*, *Phys. Rev.* **D58**, 092004 (1998); R.S. Towell *et al.*, *Phys. Rev.* **D64**, 052002 (2001).
18. K. Ackerstaff *et al.* (HERMES Collaboration), *Phys. Rev. Lett.* **81**, 5519 (1998).
19. M. Alberg, E.M. Henley and G.A. Miller,*Phys. Lett.* **B471**, 396 (2000); S. Kumano and M. Miyama, *Phys. Rev.* **D65**, 034012 (2002); F.G. Cao and A.I. Signal, *Phys. Rev.* **D68**, 074002 (2003); F. Huang, R.G. Xu and B.Q. Ma, *Phys. Lett.* **B602**, 67 (2004); B. Pasquini and S. Boffi, *Nucl. Phys.* **A782**, 86 (2007).
20. M. Wakamatsu, *Phys. Rev.* **D44**, R2631 (1991); M. Wakamatsu, *Phys. Rev.* **D46**, 3762 (1992); H. Weigel, *Phys. Rev.* **D55**, 6910 (1997); M. Wakamatsu and T. Kubota, *Phys. Rev.* **D57**, 5755 (1998); M. Wakamatsu, *Phys. Rev.* **D67**, 034005 (2003).
21. Y. Ding, R.G. Xu and B.Q. Ma, *Phys. Rev.* **D71**, 094014 (2005); L.J. Shao, Y.J. Zhang and B.Q. Ma, *Phys. Lett.* **B686**, 136 (2010).
22. L.A. Trevisan, C. Mirez, T. Frederico and L. Tomio, *Eur. Phys. J.* **C56**, 221 (2008); Y.H. Zhang, L.J. Shao and B.Q. Ma, *Phys. Lett.* **B671**, 30 (2009); Y.H. Zhang, L.J. Shao and B.Q. Ma, *Nucl. Phys.* **A828**, 390 (2009).
23. A.I. Signal and A.W. Thomas, *Phys. Rev.* **D40**, 2832 (1989).
24. J. Alwall and G. Ingelman, *Phys. Rev.* **D71**, 094015 (2005).
25. M. Glück, E. Reya and A. Vogt, *Z. Phys.* **C67**, 433 (1995); M. Glück, E. Reya, M.Stratmann and W. Vogelsang, *Phys. Rev.* **D53**, 4775 (1996); D. de Florian, C.A. Garcia Canal and R. Sassot, *Nucl. Phys.* **B470**, 195 (1996).
26. H. Abramowicz *et al.* (CDHS Collaboration), *Z. Phys.* **C17**, 283 (1983); G. Costa *et al.*, *Nucl. Phys.* **B297**, 244 (1988).
27. H. Dahiya and M. Gupta, *Eur. Phys. J.* **C52**, 571 (2007).
28. K. Gottfried, *Phys. Rev. Lett.* **18**, 1174 (1967).
29. C. Amsler *et al.*, *Phys. Lett.* **B667**, 1 (2008).

STRONGLY INTERACTING SYSTEMS

CP SYMMETRY AND PHASE TRANSITIONS

M. AGUADO

Max-Planck-Institut für Quantenoptik
Hans-Kopfermann-Str. 1, D-85748 Garching, Germany

M. ASOREY

Departamento. de Física Teórica. Facultad de Ciencias
Universidad de Zaragoza, E-50009 Zaragoza, Spain

The analytic behaviour of θ-vacuum energy at $\theta = 0$ and $\theta = \pi$ in QCD is related to the existence of phase transitions which might involve CP symmetry breaking. The appearance of cusp singularities is a signal of CP symmetry breaking by non-perturbative effects. The only cusp singularities that could arise in the vacuum energy density are due to the presence of Lee-Yang zeros but these singularities are always $\wedge$ cusp singularities and never $\vee$ cusps, which in the case $\theta = 0$ is incompatible with the Vafa-Witten diamagnetic inequality. The argument is very similar to that used in the derivation of Bank-Casher formula. In this case the topological charge condensate is proportional to the density of Lee-Yang zeros at the CP invariant points. The fact that this density is always positive provides a key missing link in the Vafa-Witten proof of parity symmetry conservation in vector-like gauge theories like QCD. However, this property does not exclude the existence of a first phase transition at $\theta = \pi$ with a $\wedge$ cusp singularity, or a second order phase transition at $\theta = 0$, which might be very relevant for interpretation of the anomalous behavior of the topological susceptibility in the $\mathbb{C}\mathrm{P}^1$ sigma model.

Keywords: Phase transitions; CP symmetry; Vafa-Witten theorem; chiral fermions.

1. Introduction

In the Standard Model CP symmetry is broken by the electroweak interactions. The theory provides a consistent mechanism for this phenomenon due to the existence of three quark generations. However, the simplest way of breaking CP symmetry appears in strong interactions where the fundamental QCD theory has room for a CP violating θ-term. Nevertheless, such a strong violation of CP has been almost excluded by the absence of a significant electric dipole momentum for neutrons. The search of a physical argument for the cancellation of θ-term is known as strong CP problem. One proposal for a possible mechanism of cancellation of the θ-term perturbation is due to Peccei-Quinn and is based of an extra field, which dynamically will force the effective θ-parameter to vanish. However, the Peccei-Quinn mechanism[1] involves a new type of particle, the axion, which at present has not been observed. In some sense one problem has been translated to another problem.

However, even if one assumes that there is no explicit breaking of CP symmetry by θ terms there still exists the possibility of a non-perturbative spontaneous symmetry breaking. That this phenomenon does not occur in QCD is once again a surprising feature. In this paper we clarify why there is no first order phase transition at $\theta = 0$ and CP symmetry is preserved in QCD by using the basic argument of Vafa-Witten[2] and some analytic properties of the partition function in the complex θ plane.[3] In particular, we analyze the structure of Lee-Yang zeros,[4,5] which appear to play a role similar to the eigenvalues of Dirac operator in the Banks-Casher approach[6] to chiral symmetry breaking for massless fermions.

2. CP symmetry and θ–term

Quantum Chromodynamics, as any gauge theory with Dirac fermions, is CP invariant in the absence of θ–term interactions. However, CP symmetry is broken by such a perturbation, with only two exceptions. In the quantum case the theory is not only formally CP symmetric for $\theta = 0$ but also for $\theta = \pi$. This is a consequence of the periodicity in θ of the Euclidean partition function $\mathcal{Z}(\theta) = \mathcal{Z}(\theta + 2\pi)$. Indeed, under CP transformation $\mathcal{Z}(\theta)$ is mapped into $\mathcal{Z}(-\theta)$, which in the case $\theta = \pi$ is identical to $\mathcal{Z}(\theta)$ because $-\pi$ and π are congruent by the periodicity property of $\mathcal{Z}(\theta)$. The CP invariance of $\mathcal{Z}(\theta)$ at $\theta = 0$ and $\theta = \pi$ has as extra consequence that these two values of θ are extremal points of the partition function.

In a strict sense, the functional integral $\mathcal{Z}(\theta)$ is UV divergent but it can be regularized in such a way that its positivity properties are preserved.[7] On the other hand, in the infinite volume limit, the regularized partition function $\mathcal{Z}_\theta(g)$ vanishes for any value of θ. Therefore, we will consider throughout the paper a compact space-time with large but finite space-time volume $VT \ll \infty$. The vacuum energy density $\mathcal{E}(\theta)$ can be extracted from the asymptotic expansion

$$\mathcal{Z}(\theta) \sim \mathrm{e}^{-VT\mathcal{E}(\theta)}$$

of the partition function in the large Euclidean time $T \gg V \gg 1$ limit of the space-time volume VT.

The vacuum energy density presents quartic divergences in the UV limit, but within a fixed UV gauge invariant regularization scheme $\mathcal{E}(\theta)$ has the same symmetries as the partition function, i.e. periodicity $\mathcal{E}(\theta) = \mathcal{E}(\theta + 2\pi n)$, reflection symmetry $\mathcal{E}(\theta) = \mathcal{E}(-\theta)$ and Bragg symmetry $\mathcal{E}(\pi+\theta) = \mathcal{E}(\pi-\theta)$. Thus, the two CP symmetric values $\theta = 0$ and $\theta = \pi$ are also extremals of the vacuum energy density $\mathcal{E}$.

The existence of a first order phase transition at $\theta = 0$ or $\theta = \pi$ will imply the spontaneous breaking of CP symmetry. Such a phase transition may be signaled by the lack of regularity of the vacuum energy on θ.

Indeed if

$$\left.\frac{d\mathcal{E}(\theta)}{d\theta}\right|_{\theta=0^+,\pi^+} = -\left.\frac{d\mathcal{E}(\theta)}{d\theta}\right|_{\theta=0^-,\pi^-} = -\frac{i}{16\pi^2}\langle\, F\widetilde{F}(x)\,\rangle_{0^+,\pi^+} \neq 0, \qquad (1)$$

the vacuum energy density $\mathcal{E}$ has a first order discontinuity at the corresponding CP invariant point. In this case the CP odd order parameter is the topological charge density,

$$\langle\, Q(x)\,\rangle_{0^\pm,\pi^\pm} = -i\langle\, F\widetilde{F}(x)\,\rangle_{0^\pm,\pi^\pm} \neq 0.$$

Vafa-Witten argued that parity and CP symmetries cannot be broken for any gauge theory with Dirac fermions in 3+1 space-time dimensions.[2] We will analyze the proof of this Vafa-Witten theorem[3] and show that the possible pathologies associated with first order transitions[8–11] do not appear in QCD at $\theta = 0$.

In a similar way, a discontinuity in the second order derivative of the vacuum energy will signal the existence of a second order quantum phase transition

$$\left.\frac{d^2\mathcal{E}(\theta)}{d\theta^2}\right|_{\theta=0^+,\pi^+} \neq \left.\frac{d^2\mathcal{E}(\theta)}{d\theta^2}\right|_{\theta=0^-,\pi^-}. \qquad (2)$$

In this case the order parameter will be the topological susceptibility

$$\chi_{\tau} = \int dx\, \left[\langle F\widetilde{F}(0)\; F\widetilde{F}(x)\rangle_{0,\pi} - \langle F\widetilde{F}(0)\rangle_{0,\pi}\langle F\widetilde{F}(x)\rangle_{0,\pi}\right], \qquad (3)$$

which in the case of first order phase transitions is divergent because the topological charge density is non-vanishing $\langle Q\rangle \neq 0$ and there is a jump in the first derivative. But even with vanishing $\langle Q\rangle = 0$, χ_{τ} can diverge for a quantum phase transition of fractional order.

This should not happen for $\theta = 0$ in the large N limit, where the topological susceptibility is connected with physical quantities by the celebrated

Witten-Veneziano formula

$$\chi_{\tau} \sim f_{\pi}^2(m_{\eta'}^2 + m_{\eta}^2 - 2m_K^2)/6. \tag{4}$$

In general, the topological susceptibility is the average of the connected 2-point correlation function of the topological charge density $Q(x)$. However, for $\theta = 0$ the main contribution to the topological susceptibility χ_{τ}, even in absence of phase transitions, comes from the ultraviolet singular contact point of this correlation function $\langle Q(x)^2\rangle_0$.[12–14] This arises because, by definition, the topological susceptibility is a formally positive quantity, but the 2-point correlation function $\langle Q(0)\ Q(x)\rangle$ is always negative for any $x \neq 0$ due to the Osterwalder-Schrader reflection positivity property[15]

$$\langle Q(0)\ Q(x)\rangle = -\langle F\widetilde{F}(0)\, F\widetilde{F}(x)\rangle = \langle \Theta_T(F\widetilde{F})(0)\, F\widetilde{F}(x)\rangle \geq 0 \text{ for } x \neq 0, \tag{5}$$

where Θ_T is the Osterwalder-Schrader time-reflection operator.

The only way that these two opposite behaviours can be made compatible is by assuming that there is a leading contact term in the 2-point correlation function of the form[12,13] $\langle F\widetilde{F}(0)\ F\widetilde{F}(x)\rangle = a\delta(x) + K(x)$ with

$$\int dx\, K(x) \leq a. \tag{6}$$

On the other hand, if the cluster property holds, which occurs when the vacuum is unique, the integral (6) is divergent because

$$\lim_{x=\infty} \langle F\widetilde{F}(0)\ F\widetilde{F}(x)\rangle = \langle F\widetilde{F}\rangle^2 \leq 0, \tag{7}$$

which is possible even in the case of broken CP symmetry because in that case $\langle F\widetilde{F}\rangle = iQ \neq 0$ has a pure imaginary expectation value.

3. The Vafa-Witten theorem

Vafa and Witten introduced a very elegant argument[2] to explain why we do not observe spontaneous violation of parity symmetry, not only in QCD but in any gauge theory with Dirac fermions in 3 + 1 space-time dimensions. The argument also implies the absence of first order phase transitions associated with the breaking process of the symmetry. Vafa and Witten claimed that their argument should hold for any parity violating order parameter, but this might not be true for non-renormalizable or non-local order parameters.[9,10] One of the order parameters which does not suffer from these problems is the topological charge density $Q(x)$. In this case, as already pointed out in the previous section, because of the reflection and Bragg symmetries, if $\mathcal{E}(\theta)$ is a smooth function of θ it should have a critical behaviour at $\theta = 0$ and $\theta = \pi$, i.e.

$$\left.\frac{d\mathcal{E}(\theta)}{d\theta}\right|_{\theta=0,\pi} = 0, \tag{8}$$

which would imply the vanishing of the expectation value of the order parameter $\langle Q(x)\rangle$ and, thus, the absence of CP symmetry breaking. However, the smoothness assumption is just equivalent to the non-existence of first order phase transitions.[8,11] Thus, one extra argument is needed to complete the proof of the Vafa-Witten theorem.

It has been shown in Ref. 3 that this argument together with some general assumptions of the theory (unitarity and renormalizability) are enough to rigorously prove the Vafa-Witten result. This is one of the very few non-perturbative analytic results of QCD. The extra assumptions guarantee the proper existence of the vacuum energy density, escaping from other more pathological scenarios.[8,11] However, one needs to exclude the existence of a cusp in the energy density at $\theta = 0$. The only extra property one needs to prove is the smoothness of $\mathcal{E}(\theta)$ at $\theta = 0$, which would make the existence of such a cusp impossible. In other terms, $\theta = 0$ is always an extremal of the free energy density but only if $\mathcal{E}(\theta)$ is smooth we will have $\mathcal{E}'(0) = 0$. Only in that case we can make sure that the vacuum expectation value of the topological density vanishes and the Vafa-Witten theorem holds for this particular order parameter.

Although the two classically CP symmetric points $\theta = 0, \pi$ are extremals of the vacuum energy density, the behaviour of $\mathcal{E}(\theta)$ around these values of θ might be very different. The main difference is that, due to the Vafa-Witten diamagnetic inequality ($\mathcal{E}(0) \leq \mathcal{E}(\theta)$) the vacuum energy density has an absolute minimum value at $\theta = 0$, whereas $\theta = \pi$ might be a minimal point or not. The forthcoming argument only applies to minimal points, therefore in the cases where $\theta = \pi$ is not a minimal point the argument cannot applied. In fact there are cases where CP is spontaneously broken at $\theta = \pi$, but never at $\theta = 0$. That the pathology does not appear at $\theta = 0$ is remarkable effect encoded in the Vafa-Witten argument.

In order to prove smoothness, the role of Lee-Yang singularities becomes very relevant. Because of Bragg symmetry and unitarity, non-analyticities are harder to trace in the pure real θ sector. Non-analyticity in θ can be best inferred from the lack of an analytic continuation of the energy density into the complex θ plane. Indeed, in that case Osterwalder-Schrader positivity (i.e. unitarity) is not preserved and there is even no guarantee that the free energy is well defined, in which case it will not make sense to talk about smoothness. Essentially, there are two possible ways in which the θ theory might give rise to pathological non-analyticities in the complex sector. One is that the analytic continuation to complex values of θ is not defined at all, i.e. the partition function itself becomes divergent. We shall see that this does not occur in gauge theories. The other possibility is that the partition function exhibits a sequence of zeros converging to $\theta = 0$ in the infinite volume limit. This is the scenario advocated by Lee and Yang,[4,5] which signals in many cases the existence of a phase transition like in the Ising model in the presence of an external (imaginary) magnetic field.

The QCD partition function splits into a sum of functional integrals over the different topological sectors

$$\mathcal{Z}_\theta(g) = \sum_{q=-\infty}^{\infty} \mathrm{e}^{-q \operatorname{Im}\theta + iq \operatorname{Re}\theta} \int_{c_2(A)=q} \delta A \, \mathrm{e}^{-S_{\mathrm{YM}}(g)} \det(\not{D}_A + m), \tag{9}$$

where $S_{\mathrm{YM}}(g)$ denotes the standard Yang-Mills action with coupling g and q the topological charge of the corresponding gauge fields.

Positivity of fermionic determinants[2] implies the existence of a real effective action

$$S_{\mathrm{eff}}(g) = S_{\mathrm{YM}}(g) - \log \det(\not{D}_A + m).$$

On the other hand the Yang-Mills action $S_{\mathrm{YM}}(g)$ is bounded below by the BPS bound

$$S_{\mathrm{YM}}(g) \geq \frac{8\pi^2 |q|}{g^2}$$

in each q-topological sector. We shall see that the total effective action $S_{\mathrm{eff}}(g)$ is also bounded below in the sector of topological charge q by a similar linear bound involving the absolute value $|q|$ of the topological charge. In the regularized theory, the coefficient of this linear bound increases with the UV regulating scale. This growth of the effective action is enhanced in the chiral limit of light quark masses by the contribution of fermionic determinants. In general, from the positivity of fermionic determinants and the Yang-Mills Euclidean measure[7,16] and the BPS bounds we have the inequality

$$|\mathcal{Z}_\theta(g)| \leq \sum_{q=-\infty}^{\infty} \mathrm{e}^{|q \operatorname{Im}\theta|} \int_{c_2(A)=q} \delta A \, \mathrm{e}^{-S_{\mathrm{eff}}(g)} \leq \sum_{q=-\infty}^{\infty} \int_{c_2(A)=q} \delta A \, \mathrm{e}^{-S_{\mathrm{eff}}(\tilde{g})} = \mathcal{Z}_0(\tilde{g}), \tag{10}$$

where

$$\frac{1}{\tilde{g}^2} = \frac{1}{g^2} - \frac{1}{8\pi^2} |\operatorname{Im}\theta|.$$

This shift of the gauge coupling constant can be considered as a change of the renormalization scale.

Thus, the partition function of the original theory with a complex θ-term is bounded by the partition function of a similar theory with $\theta = 0$ but with a different coupling constant $\tilde{g}$. Since the theory at $\theta = 0$ is unitary and renormalizable, its partition function $\mathcal{Z}_0$ is finite and from (10) it follows that $|\mathcal{Z}_\theta| < \infty$ is also finite for small values of $\operatorname{Im}\theta$. If in addition the theory is asymptotically free, then the renormalization of g^2 in the UV fixed point can absorb any value of $\operatorname{Im}\theta$, which implies that the radius of convergence of the sum (9) in the complex θ plane, in fact, becomes infinity. To make the argument more precise one should consider the regularized theory. The analyticity of $\mathcal{Z}_\theta$ also appears naturally in lattice regularizations, even with non-topological regularizations of the θ–term. However, to prove the absence of poles the topological charge bound (10) is essential. The bare coupling constant g has to be fine tuned according to the renormalization group to yield the appropriate continuum limit. But because of asymptotic freedom this coupling goes to zero as the UV regulator is removed and the shift from g to $\tilde{g}$ induced by $\operatorname{Im}\theta$ simply implies a change in the effective scale of the continuum theory. However, the proof of the Vafa-Witten theorem only requires a finite radius of convergence of $\mathcal{Z}_\theta$ around $\theta = 0$. This feature does not require asymptotic freedom, just renormalizability.

We remark that one cannot use the same kind of arguments for other order parameters. For instance, the partition function associated to higher odd powers of the topological density might diverge for any complex value of the perturbation parameter, as they are non-renormalizable. A similar remark applies to the perturbation by a chemical potential term in theories at finite density. In that case the missing property is the lack of a BPS bound for the perturbation.

4. Lee-Yang zeros and phase transitions

The only remaining possible source of non-analyticity is the presence of Lee-Yang singularities, i.e. zeros in the partition function $\mathcal{Z}_\theta$ which could prevent the existence of a unique limit of the free energy $\log \mathcal{Z}_\theta$ or its derivatives at $\theta = 0$.

Because of Bragg symmetry and θ–periodicity $\mathcal{Z}_\theta$ is a function of $\sin^2(\theta/2)$ and its zeros come in 4-plets $\{\vartheta_n, -\vartheta_n, \vartheta_n^*, -\vartheta_n^*\}$ in the modular strip $\theta \in (-\pi, \pi] \times \mathbb{R}$ of the complex plane. They can be parametrized by their representatives in the positive quadrant ϑ_n , which cannot lie on the real or imaginary axes for any finite volume.

Only zeros converging to the origin in the thermodynamic limit can give rise to a cusp. Moreover, only an infinity of such zeros can produce the cusp.

The Weierstrass factorization theorem implies that the vacuum energy density

$$\mathcal{E}_V(\theta) = -\frac{1}{VT} \ln Z_V(\theta), \tag{11}$$

can be rewritten in the functional form[17,18]

$$\begin{aligned}\mathcal{E}_V(\theta) = & -\frac{2}{VT} \operatorname{Re} \sum_{n=1}^{\infty} \rho_n \left\{ \ln\left(1 - \frac{\sin^2(\theta/2)}{\sin^2(\vartheta_n/2)}\right) \right. \\ & \left. + h_n(\sin^2(\theta/2), VT) \right\} + g_V(\sin^2(\theta/2)),\end{aligned} \tag{12}$$

where ϑ_n denotes the order n Lee-Yang zero in the positive quadrant, ρ_n its degeneracy and h_n and g are, respectively, polynomial and analytic functions of $\sin^2(\theta/2)$. The polynomial functions h_n are necessary to compensate the divergent effect of zeros with slow approach to the infinite modulus for a given volume.

In the thermodynamic limit the degeneracy factor ρ_n of zeros converging to the origin defines a density of zeros $\rho(z)$ as a function of the continuous variable $z = \vartheta_n/VT$. Then, in the thermodynamic limit

$$\begin{aligned}\mathcal{E}(\theta) = & -2 \operatorname{Re} \int_0^{\pi} \int_0^{\infty} \mathrm{d}z_1\, \mathrm{d}z_2\, \rho(\sin^2(z/2)) \left\{ \ln\left(1 - \frac{\sin^2(\theta/2)}{\sin^2(z/2)}\right) \right. \\ & \left. + h(\sin^2(\theta/2),\, z) \right\} +\; g(\sin^2(\theta/2)),\end{aligned} \tag{13}$$

where $z = z_1 + i z_2$. In most of the cases $\rho(z)$ is concentrated along a singular Lee-Yang curve.

Notice that only if $\rho(0) \neq 0$ the contribution from the logarithm at $z = 0$ could give rise to a cusp at $\theta = 0$. The rest of the contributions from the integral have continuous derivatives for real θ at $\theta = 0$.

In fact, the logarithmic term in (13) corresponds to a Coulomb potential, in two dimensions, generated by the charge density $-2\rho(z)$. The cusp at $\theta = 0$ is only possible if there is a non-vanishing linear charge density at $\theta = 0$,[19,20] i.e. $\rho(z) \sim \rho_0\, \delta(z_1 - m z_2^{\beta})$ for $|z| \ll 1$ with $\rho_0 \neq 0$. A cusp corresponds to a gap in the potential derivative, which is proportional to the linear charge density $-2\rho_0$. In fact, in the thermodynamic limit we have a relation

$$Q = -i\langle F\widetilde{F}\rangle = \pi \rho_0 \tag{14}$$

between the density of instantons Q and the density of Lee-Yang zeros ρ_0 at $\theta = 0$. This formula is quite similar to the Banks-Casher formula for chiral fermions. In that case, by a similar argument one obtains a relation[6] between the chiral condensate and the density of Lee-Yang zeros of the partition function in the complex m plane in the chiral limit (m=0)

$$\langle \bar{\psi}\psi \rangle = \pi \rho_0. \tag{15}$$

In this case the analyticity of the partition function in the mass parameter m follows from the positivity and regularity properties of fermionic determinants of Dirac operators. The absence of poles is also a consequence of the particular nature of the partition function defined by fermionic determinants.[21,22] The only singularities in the vacuum energy density can arise from Lee-Yang zeros. However, there is a difference between the UV behaviors of the two cases. The expectation value of the vacuum energy density is quartic divergent whereas the chiral condensate is cubic divergent. But once an UV regularization is fixed the analysis of the phase structure governed by both condensates based on the Lee-Yang zeros densities is very similar.

In both cases the charge density ρ_0 is non-negative because there are no poles, the corresponding potential is always repulsive, which means that any cusp in the vacuum energy density is always a $\wedge$ cusp instead of a $\vee$ cusp. However, the main difference is that in the θ-vacuum case the vacuum energy density $\mathcal{E}(\theta)$ has an absolute minimum at $\theta = 0$, because of the Vafa-Witten diamagnetic inequality, which implies that a $\wedge$ cusp cannot exist at $\theta = 0$. The argument completes the proof of the Vafa-Witten theorem because no first order cusp is possible at $\theta = 0$. The theorem follows from two very simple facts, the Vafa-Witten inequality and the repulsive character of the negative charge density defined by $-2\rho(z)$, which follows from the fact that the partition function $Z_V(\theta)$ is an analytic function with zeros but no poles. The only way of having a a $\vee$ cusp at $\theta = 0$ is with an attractive charge density which can only appear from poles in $Z_V(\theta)$.

In fact, this is what happens in 1+1 massless scalar field theories with pseudo-periodic boundary conditions[23] where the partition function is non-analytic in the complex plane. This allows the vacuum energy density to present a cusp at $\theta = 0$, which corresponds to pure periodic boundary conditions:

$$\mathcal{E}(\theta) = \frac{1}{2V}\left(\frac{|\theta|}{2\pi} - \frac{\theta^2}{4\pi^2} - \frac{1}{6}\right); \quad \theta \in [-\pi, \pi]. \tag{16}$$

However, in QCD the analyticity of the partition function is based on

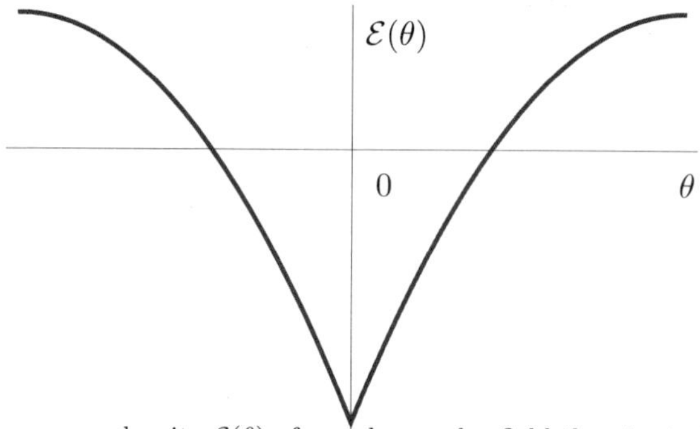

Fig. 1. Vacuum energy density $\mathcal{E}(\theta)$ of massless scalar field theories in 1+1 dimensions for pseudo-periodic boundary conditions with $\theta \in [-\pi, \pi]$.

the existence of BPS bounds, which do not exist in the 1+1 dimensional theory. This makes the existence of a cusp in the vacuum energy density at $\theta = 0$ impossible. In summary, the theory cannot undergo a first order phase transition at $\theta = 0$ with parity symmetry breaking, which confirms the claim of the Vafa-Witten theorem. The result is in agreement with numerical simulations.[24]

In the case of chiral fermions there is no obstruction to the appearance of a $\wedge$ cusp and chiral symmetry can be spontaneously broken.

From the above arguments we cannot, however, exclude the existence of first order phase transitions for other values of $\theta \neq 0$.[25,26] In particular, we cannot discard CP symmetry breaking for $\theta = \pi$, where $\mathcal{E}(\theta)$ can have a local maximum.

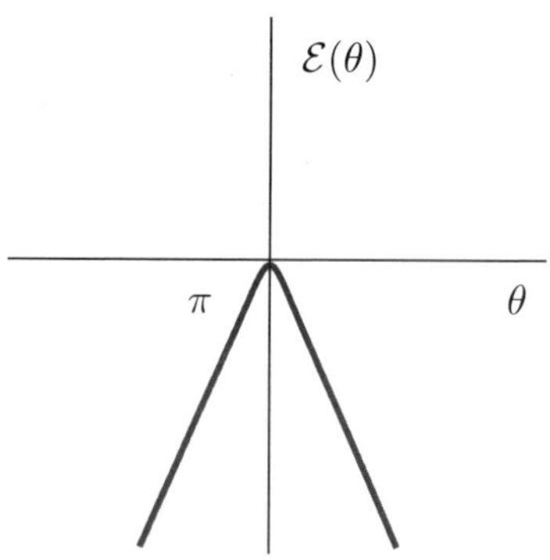

Fig. 2. Possible behaviour of the vacuum energy density in a theory undergoing a first order phase transition at $\theta = \pi$ with parity symmetry breaking

5. CP symmetry in $\mathbb{CP}^N$ sigma models

We remark that many other theories share the same properties used in the above arguments and therefore the Vafa-Witten theorem can also be extended for all of them, e.g. random matrix theories.[22] One particularly interesting case is that of $\mathbb{CP}^N$ non-linear sigma models (see Ref. 27 for a review). Indeed, from a similar analysis one can conclude that there is no spontaneous CP symmetry breaking in $\mathbb{CP}^N$ models at $\theta = 0$.[3,28]

This result can be analytically checked for the $\mathbb{CP}^1$ model, where the exact solution is known for $\theta = 0$.[29,30] This solution describes the interaction of a massive scalar particle in the adjoint representation of $SU(2)$, and CP symmetry is preserved confirming the result of the Vafa-Witten theorem. Numerical simulations indicate that the same behaviour holds for larger values of N.[27,31,32]

Another case where an exact solution is also known is the $\mathbb{CP}^N$ model in the large N limit.[33] In this case the system also describes a weakly interacting and parity preserving massive scalar particle in the adjoint representation of $SU(N)$, with no first order phase transition at $\theta = 0$

The behaviour at the other CP invariant point $\theta = \pi$ is rather different. In the large N limit the system undergoes a first order phase transition with spontaneous CP symmetry breaking. The existence of a cusp points out the existence of a non-vanishing density of Lee-Yang zeros at $\theta = \pi$. In this case the existence of the cusp (see Fig. 2) implies that the vacuum energy-density reaches a local maximum at $\theta = \pi$. On the other hand the role of instantons is enhanced for low values of N. This suggests the existence of a critical value N_c of N which splits the models into two families.[34] The models with $N < N_c$ preserve parity symmetry also at $\theta = \pi$, whereas the symmetry is broken for all models with $N > N_c$. The critical value N_c could be as low as $N_c = 1$ or 2.[35–37]

The existence of first order phase transitions for values of $\theta \neq 0, \pi$ which arise in some simulations[25] may be due to large volume lattice artifacts.[38] They can be neither pointed out nor excluded by the previous arguments, because in any case they cannot be related to the behaviour of the system under CP transformations.

The previous arguments cannot exclude the existence of higher-order phase transitions which do not require spontaneous parity symmetry breaking. Indeed, if $\rho_0 = 0$ but $\rho(z) \sim \rho_1\, z_2^\alpha \delta(z_1 - \gamma z_2^\beta)$ with $0 < \alpha \leq 1$ for $|z| \ll 1$, the vacuum energy density presents a second order singularity and the topological susceptibility diverges at $\theta = 0$.[39] This behaviour is compatible with the Vafa-Witten inequality and might explain the behaviour of the topological susceptibility of the $\mathbb{CP}^1$ model in numerical simulations.[39–41] A remaining interesting question is whether or not similar second order phase transitions can occur at $\theta = \pi$ for gauge theories and $\mathbb{CP}^{\rm N}$ sigma models with $N < N_c$, which in the case of QCD could be associated with a quark deconfining phase transition.

Acknowledgments

M. Asorey thanks Julia Nyíri and the organizers of the meeting in honor of Volodya Gribov for their hospitality. This work of M. Asorey was partially supported by the Spanish CICYT grant FPA2009-09638 and DGIID-DGA (grant 2009-E24/2).

References

1. R.D. Peccei and H.R. Quinn, *Phys. Rev. Lett.* **38**, 1440 (1977).
2. C. Vafa and E. Witten, *Phys. Rev. Lett.* **53**, 535 (1984).
3. M. Aguado and M. Asorey, *Phys. Rev.* **D80**, 127702 (2009).
4. C.N. Yang and T.D. Lee, *Phys. Rev.* **87**, 404 (1952).
5. T.D. Lee and C. N. Yang, *Phys. Rev.* **87**, 410 (1952).
6. T. Banks and A. Casher, *Nucl. Phys.* **B169**, 103 (1980).
7. M. Asorey and F. Falceto, *Nucl. Phys.* **B327**, 427 (1989).
8. V. Azcoiti and A. Galante, *Phys. Rev. Lett.* **83**, 1518 (1999).
9. T.D. Cohen, *Phys. Rev.* **D64**, 047704 (2001).
10. M.B. Einhorn and J. Wudka, *Phys. Rev.* **D67**, 045004 (2003).
11. X. Ji, *Phys. Lett.* **B554**, 33 (2003).
12. E. Seiler, *Lect. Notes Phys.* **159**, 1 (1982).
13. K. Osterwalder and E. Seiler, *Ann. Phys. (N.Y.)* **110**, 440 (1978).
14. M. Aguado and E. Seiler, *Phys. Rev.* **D72**, 094502 (2005).
15. K. Osterwalder and R. Schrader, *Commun. Math. Phys.*, **31**, 83 (1973); *Commun. Math. Phys.*, **42**, 281 (1975).
16. M. Asorey and P.K. Mitter, *Commun. Math. Phys.* **80**, 43 (1981).
17. E.T. Whittaker and G.N. Watson, *A course of modern analysis*, Cambridge University Press (1986).
18. M. Aguado and M. Asorey, In preparation
19. R.A. Blythe and M.R. Evans, *Phys. Rev. Lett.* **89**, 080601 (2002).
20. R.A. Blythe and M.R. Evans, *Physica* **A 313**, 110 (2002); *Braz. J. Phys.* **33**, 464 (2003).
21. J. Vink, *Nucl. Phys.* **B323**, 399 (1989).
22. M.A. Halasz, A.D. Jackson and J.J.M. Verbaarschot, *Phys. Lett.* **B395**, 293 (1997); *Phys. Rev.* **D56**, 5140 (1997).
23. M. Asorey, D. García-Alvarez and J.M. Muñoz-Castañeda, *J. Phys.* **A39**, 6127 (2006).
24. B. Alles, M. D'Elia and A. Di Giacomo, *Nucl. Phys.* **B494**, 281 (1997); *Phys. Rev.* **D 71**, 034503 (2005).

25. S. Olejnik and G. Schierholz, *Nucl. Phys. Proc. Suppl.* **34**, 709 (1994).
26. V. Azcoiti, A. Galante and V. Laliena, *Prog. Theor. Phys.* **109**, 843 (2003).
27. E. Vicari and H. Panagopoulos, *Phys. Rep.* **470**, 93 (2009).
28. M. Aguado and M. Asorey, arXiv:1009.2629 [hep-th] (2010).
29. A. Zamolodchikov and A.B. Zamolodchikov, *Ann. Phys. (N.Y.)*, **120**, 253 (1979).
30. A. Polyakov and P.B. Wiegmann, *Phys. Lett.* **B131**, 121 (1983).
31. M. Campostrini, P. Rossi and E. Vicari, *Phys. Rev.* **D46**, 2647 (1992).
32. B. Alles, L. Cosmai, M. D'Elia and A. Papa, *Phys. Rev.* **D62**, 094507 (2000).
33. A. D'Adda, P. Di Vecchia and M. Lüscher, *Nucl. Phys.* **B152**, 125 (1979).
34. R. Burkhalter, M. Imachi, Y. Shinno and H. Yoneyama, *Prog. Theor. Phys.* **106**, 613 (2001).
35. B.B. Beard, M. Pepe, S. Riederer and U.-J. Wiese, *Phys. Rev. Lett.* **94**, 010603 (2005).
36. Y. Lian and H. B. Thacker, *Phys. Rev.* **D75**, 065031 (2007).
37. M. Imachi, S. Kanou and H. Yoneyama, *Prog. Theor. Phys.* **102**, 653 (1999).
38. J.C. Plefka and S. Samuel, *Phys. Rev.* **D56**, 44 (1997).
39. M. Lüscher, *Nucl. Phys.* **B200**, 61 (1982).
40. M. Blatter, R. Burkhalter, P. Hasenfratz and F. Niedermayer, *Phys. Rev.* **D53**, 923 (1996).
41. S. Ahmad, J.T. Lenaghan and H.B. Thacker, *Phys. Rev.* **D72**, 114511 (2005).

TESTING CPT INVARIANCE WITH ANTIPROTONS

DEZSŐ HORVÁTH

KFKI Research Institute for Particle and Nuclear Physics, H–1525 Budapest, Hungary and Institute of Nuclear Research (ATOMKI), Debrecen, Hungary
horvath@rmki.kfki.hu

The structure of matter is related to symmetries at every level of study. CPT symmetry is one of the most important laws of field theory: it states the invariance of measurable physical properties when one simultaneously changes the signs of the charges and of the spatial and time coordinates of free elementary particles. Although in general opinion CPT symmetry is not violated in Nature, there are theoretical attempts to develop CPT-violating models. The Antiproton Decelerator at CERN has been built to test CPT invariance. The ASACUSA experiment compares the properties of particles and antiparticles by studying the antiprotonic helium atom via laser spectroscopy and measuring the mass, charge and magnetic moment of the antiproton as compared to those of the proton.

1. Introduction: Symmetries in particle physics

Symmetries in particle physics are even more important than in chemistry or solid state physics. Just like in any theory of matter, the inner structure of the composite particles are described by symmetries, but in particle physics everything is deduced from the symmetries (or invariance properties) of the physical phenomena or from their violation: the conservation laws, the interactions and even the masses of the elementary particles (see Ref. 1 for all general references).

The conservation laws are related to symmetries: the Noether theorem states that a global symmetry leads to a conserving quantity. The conservation of momentum and energy are deduced from the translational invariance of space-time: the physical laws do not depend upon where we place the zero point of our coordinate system or time measurement; and the fact that we are free to rotate the coordinate axes at any angle is the origin of angular momentum conservation.

Spin is one of the most important properties of the particles: those having half-integer spins are the fermions whereas the integer-spin particles are bosons. The different symmetries of the fermions and bosons lead to dramatic differences in their behaviour, e.g. the numbers of fermions are conserved whereas the numbers of bosons are not. The basic building blocks of the visible matter of the Universe, the quarks and leptons are fermions and all known interactions are mediated by bosons.

All fermions have *antiparticles*, anti-fermions which have identical properties but with charges of the opposite sign. The different abundance of particles and antiparticles in our Universe is one of the mysteries of astrophysics: apparently there is no antimatter in the Universe in significant quantities, see, e.g., Ref. 2. If there were antimatter galaxies they would radiate antiparticles and we would see zones of strong radiation at their borders with matter galaxies, but the astronomers do not see such a phenomenon anywhere.

An extremely interesting property of *antiparticles* is that they can be treated mathematically as if they were *particles* of the same mass and of oppositely signed charge of the same absolute value *going backward in space and time*. This is the consequence of one of the most important symmetries of Nature: CPT invariance.[3] CPT reflection means the following simultaneous operations:

- charge conjugation (i.e. changing particles into antiparticles), $C\psi(r,t) = \overline{\psi}(r,t)$;
- parity change (i.e. mirror reflection), $P\psi(r,t) = \psi(-r,t)$, and
- time reversal, $T\psi(r,t) = \psi(r,-t)$.

The principle of CPT invariance states that this transformation does not change the measurable physical properties of the system. This means that, e.g., the annihilation of a positron with an electron can be described as if an electron came to the point of collision, irradiated two or three photons and then went out backwards in space-time. Although the weak interaction, unlike the other interactions, violates both P and CP conservation, CPT invariance is assumed to be absolute. In the author's opinion, the strongest argument for the validity of CPT invariance is the fact that using Feynman diagrams gives theoretical results in quantum electrodynamics which agree with experiment with an incredible precision.

2. Testing CPT invariance

All such laws have to be and are checked experimentally. However, CPT invariance is so deeply embedded in field theory that many theorists claim it is impossible to test within the framework of present-day physics. Indeed, in order to develop CPT-violating models one has to reject such fundamental axioms as Lorentz invariance or the locality of interactions (i.e. causality) or unitarity.[4–6]

About 20 years ago an American colleague was giving a seminar in Budapest on the D0 experiment at Fermilab's Tevatron. His first slide entitled "Motivation" quoted among others "Testing CPT invariance". Vladimir Gribov, whom I believe to have been the greatest scientist who ever worked or even spent more than a few hours within the walls of our institute, jumped up and announced "CPT invariance cannot be tested!". A lively discussion started in the audience about this question while the nominal speaker stood there forgotten. As we just started an experiment at CERN to measure the mass of the antiproton, I asked Gribov, what happens in the very unlikely event, if we find a difference between the masses of the proton and the antiproton. His reply was that he would rather believe a tiny non-conservation of the electric charge than CPT-violation. This must have left such a deep impression in our theoreticians, that ten years later, when organizing a little workshop on CPT tests, I could not persuade any of our theorist colleagues to present an introductory talk, so at the end there was none.

As far as we know, the Standard Model is valid up to the Planck scale, $\sim 10^{19}$ GeV. Above this energy scale we expect to have new physical laws which may allow for Lorentz and CPT violation as well.[4] Quantum gravity[5,6] could cause fluctuations leading to Lorentz violation, or to loss of information in black holes, i.e. unitarity violation. Also, a quantitative expression of Lorentz and CPT invariance needs a Lorentz and CPT violating theory.[4] On the other hand, testing CPT invariance at low energy should be able to limit possible high energy violation. This makes experimental CPT tests physically valuable in spite of the fact that most of us do not expect its violation at any degree.

On the other hand, symmetry violation is not unimaginable in particle physics. The weak interaction violates parity conservation maximally and CP-invariance a little. We have to introduce the Higgs field in the Standard Model to break local $U(1) \otimes SU(2)$ gauge invariance in order to obtain masses for the quarks and leptons and also to create a scalar boson. And the problems of the Standard Model seem to get solved when supersymmetry is introduced, which is broken at laboratory energies.

CPT invariance is so far fully supported by the available experimental evidence and it is absolutely fundamental in field theory. Nevertheless, there are many experiments trying to test it. This principle requires, e.g., that particles and antiparticles have the same mass and have additive quantum numbers (like charge) of the same absolute value but opposite sign. Thus a straightforward *CPT* test is measuring the mass and charge of particles and antiparticles. The most precise such measurement is that of the relative mass difference of the neutral K meson and its antiparticle: it is less than 10^{-18}.[3] Another good pair of candidates should be the proton and the antiproton as the heaviest stable particles.

In spite of its huge success in describing all experimental data the Standard Model of elementary particles has several theoretical difficulties. Gravity does not fit into its system of gauge interactions. We do not know why there are exactly three fermion families and what interaction causes the mixing of neutrino types. The prevalence of matter against antimatter in the Universe and the left-right asymmetry of weak interactions are also unexplained and the Universe seems to be dominated by the mysterious dark matter and the even more mysterious dark energy. There are extensions of the Standard Model trying to explain these effects, but so far we have no experimental evidence supporting any of them in spite of great efforts of the particle physics community.

Phenomenologically, the oscillations observed for solar and atmospheric neutrinos can be incorporated using two neutrino mass differences and a mixing matrix with three angles and a phase.[3] The *Liquid Scintillator Neutrino Detector (LSND)* experiment at Los Alamos found a strange excess of $\overline{\nu}_e$ in a $\overline{\nu}_\mu$ beam[8] which cannot be explained within the Standard Model with three neutrinos, but can be interpreted in terms of Lorentz violation.[9] Recently, the Kostelecký group developed a Lorentz-violating extension of the Standard Model called *Tandem model*[10] which could explain all of the observed neutrino oscillations using massive neutrinos in a Hamiltonian with Lorentz-violating terms. Another recent attempt[11] deduced the neutrino mass differences from unitarity-violating quantum-gravity induced decoherence. The MiniBooNE experiment, designed to test the LSND anomaly, in 2007 announced[12] that their study excludes the LSND result assuming the same mechanisms for neutrino and antineutrino oscillation, but in 2009 practically retracted this statement.[13]

Moreover, Lorentz violation may involve a dependence of the result of an experiment on its spatial direction. The BaBar collaboration made a systematic sidereal study of $B^0 - \overline{B}^0$ oscillations and excluded any Lorentz violation[7] with a significance of 2.8σ.

3. Antimatter creation

The first antimatter atoms were created at the Low Energy Antiproton Ring (LEAR) at CERN in 1996,[14] the last year of LEAR's operation. The fast antiprotons of the ring were crossed by a xenon atomic beam. In the electromagnetic field of the xenon nuclei the antiproton created electron-positron pairs and with some low probability the positron got attached to the antiproton, producing antihydrogen atoms. The neutral atoms left the ring at a straight section, where the positron was separated in a foil and the antiproton and the positron were separately identified. 12 such events were detected with an estimated background of 3 events.

CERN has constructed its *Antiproton Decelerator (AD)* facility[15] in 1999 in order to test the CPT invariance by comparing the properties of proton and antiproton and those of hydrogen and antihydrogen. The AD was constructed mainly using outside funds and started to operate at the end of 1999. By the end of 2000 it was brought to specifications. 25 GeV/c protons from the Proton Synchrotron are shot in an iridium target where they produce particle–antiparticle pairs. Antiprotons are collected at 3.5 GeV/c momentum and slowed down in the AD ring in three steps to 100 MeV/c while keeping them in orbit using stochastic and electron cooling.

The aim of the present work is to briefly summarize some of the results and plans of the ASACUSA[16] experiment; the interested reader can find a comprehensive review.[17] The other two AD-experiments testing the CPT invariance are ALPHA[18] and ATRAP,[19] they are preparing spectroscopic studies of antihydrogen, the bound state of an antiproton with a positron.[20]

4. Antiprotonic Helium Atoms

An exotic atom is formed when a fast negative particle — muon, pion, kaon or antiproton — penetrates matter: it first slows down in atomic collisions (mostly via ionization), then gets captured in an atomic orbit *replacing* the last knocked-out electron. The capture cross section is related to the overlap between the wave functions of the particle and the atomic electron so the heavy particle will initially populate atomic states with radii close to that of the electrons. Thus an antiproton captured, e.g., in a helium atom will initially populate the $\overline{p}He^+$ states with principal quantum numbers $n_0 = \sqrt{M/m} \approx 38$ where $M \approx 0.8m_p$ and $m \approx m_e$ are the reduced

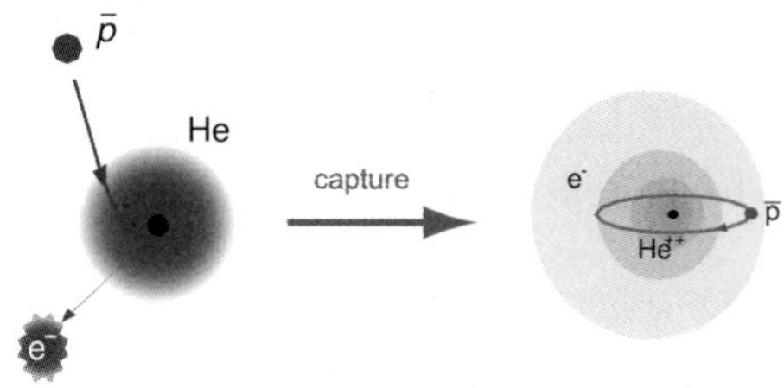

Fig. 1. Formation of the metastable antiprotonic helium atom, a three-body bound state.

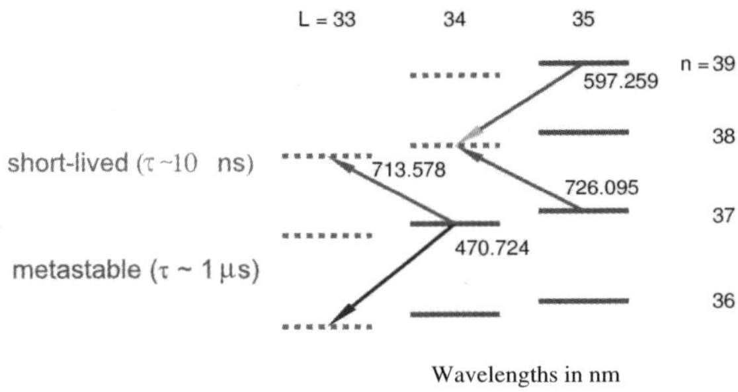

Fig. 2. Principle of laser spectroscopy: forcing fast annihilation by inducing transitions between long-lived and short-lived antiprotonic states.

masses of the $\overline{\mathrm{p}}\mathrm{He}^{++}$ and $\mathrm{e}^{-}\mathrm{He}^{++}$ systems.[21] A high n, naturally, involves orbital quantum numbers in the region $0 \leq \ell \leq n-1$; and although experiments found deviations of the initial populations from a purely statistical $2\ell+1$ distribution, the states with higher ℓ will be populated with higher probability.

The freshly formed, highly excited exotic atom has two basic ways to step down. Between high-n states, where the energy spacing is low, the Auger mechanism dominates whereas lower lying levels will preferably decay via radiation. Approaching the ground state a strongly interacting hadron like the antiproton gets absorbed by the nucleus from higher nS levels and it hardly reaches ground state in heavier atoms. Both in condensed media and in gases at higher pressures (about standard conditions) slowing down, atomic capture, de-excitation and nuclear absorption proceeds quite fast: theoretical calculations and experimental measurements agree upon total lifetimes below or around 1 ps (10^{-12} s).

The only exception is helium: while 97% of the antiprotons stopped in a dense helium target annihilate with the usual short lifetimes, 3% live as long as several microseconds, sufficiently long to do laser spectroscopy.

They form a $\overline{\rm p}{\rm He}^+$ 3–body system (Fig. 1) where the antiprotonorbit is protected against collisions by the electron, and the antiprotonic states of the same n but different ℓ lose the energy degeneracy and so cannot undergo Stark transitions. The model and its experimental proof are described in reviews.[17,22–24]

The principle of the spectroscopy method is simple. We stop a bunch of antiprotons in helium, wait until the promptly annihilating states disappear and then stimulate the transition from a long–living state to a short–lived one with a tunable laser system. At the resonant frequency corresponding to the transition energy the laser shot will be followed by immediate annihilation (Fig. 2).

4.1. *The Mass and Charge of the Antiproton*

The TRAP group measured the charge/mass ratio of the proton and the antiproton at LEAR.[25] They kept a single antiproton and a single H^- ion in the same Penning–trap simultaneously at different orbits and measured their cyclotron frequencies. After having made corrections for the H^- — p deviation they limited the relative difference to $9 \cdot 10^{-11}$. Achieving this precision took 10 years' work.

The aim of the ASACUSA experiment is to supply additional data for facilitating the separation of charge and mass information. This is done via studying antiprotonic transition energies in the $\overline{\rm p}{\rm He}^+$ system as those are proportional to $m(\overline{\rm p}) \cdot q(\overline{\rm p})^2$:

$$E_n \approx -\frac{m_{\rm red} c^2 (Z\alpha)^2}{2n^2} \, .$$

The precision of determining the transition frequencies is limited among others by the laser bandwidth, the density shift of the lines and the Doppler effect. Our first measurement was performed at LEAR for different helium densities and we extrapolated to zero[26] in order to make the comparison with the theoretical calculations for isolated atoms.[27,28] The way to obtain a limit on the antiprotonic mass and charge is illustrated in Fig. 3: the intersection of the regions allowed by the two measurements constitutes the limit. The result was $[m({\rm p}) - m(\overline{\rm p})]/m({\rm p}) < 5 \cdot 10^{-7}$ and similarly $[q({\rm p}) - q(\overline{\rm p})]/q({\rm p}) < 5 \cdot 10^{-7}$.

In 2000, the first year of the AD we lowered the deteriorating effects of the laser bandwidth and gained almost an order of magnitude in precision,[29] but the collisional effects were still significant. In 2002 we have installed a

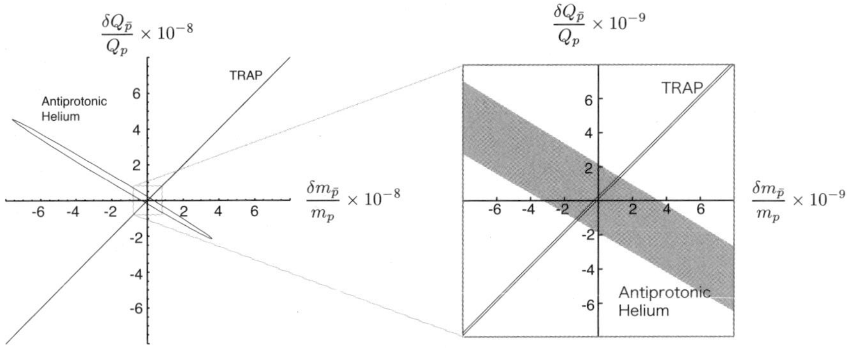

Fig. 3. Limits on the difference of mass and charge between proton and antiproton. The charge/mass (Q/M) ratio was measured by the TRAP group[25] whereas $M \cdot Q^2$ by ASACUSA.[26,29–31] With the improvement of the experimental technique the allowed region was step-by-step reduced: the present limit is 2 ppb (2×10^{-9}).[31]

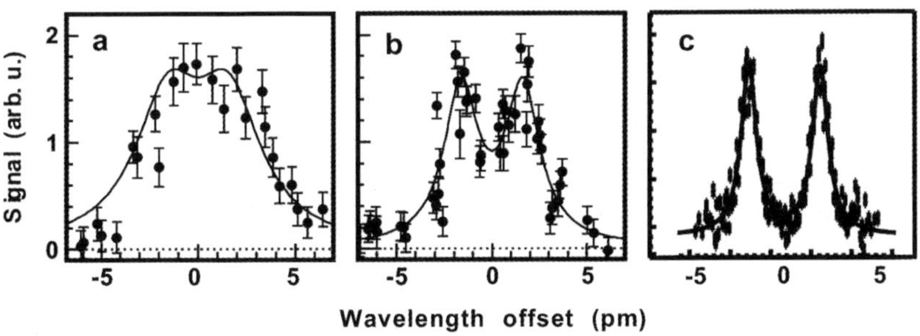

Fig. 4. The development of the energy resolution of $\bar{\text{p}}$He spectroscopy. The $(n, \ell) = (37, 35) \rightarrow (38, 34)$ split transition at 726.1 nm as measured in (a) 2000; (b) 2002 and (c) 2004.

radio-frequency quadrupole post-decelerator (RFQD) which decelerated the AD beam from 5.6 MeV to 100 keV with a $\sim 30\%$ efficiency. The RFQD made it possible to substantially reduce the density effect by using a low-pressure (< 1 mbar), cryogenic ($T = 6 - 10$ K) target (Fig. 4). Our latest published limit of possible CPT-violation on the antiproton charge and mass is 2 ppb ($< 2 \times 10^{-9}$).[31]

Further improvement is expected from two-photon spectroscopy which helps to reduce the Doppler broadening. In a counterpropagating geometry two laser beams of the same frequency can completely eliminate the first-order Doppler effect. However, this is not feasible to use in our

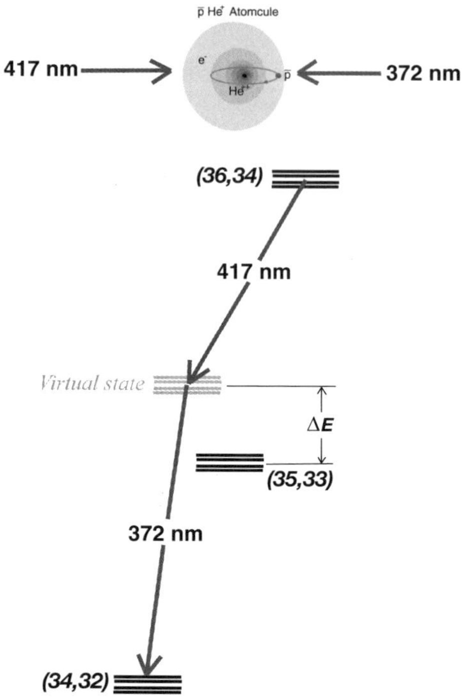

Fig. 5. Portion of the energy level diagram of $\overline{p}{}^4\mathrm{He}^+$, indicating the principle of sub-Doppler two-photon spectroscopy. The ΔE energy difference should be as small as possible, but larger than the involved laser frequency uncertainties.

case as it should involve unattainably high laser powers.[17] In the practical case shown in Fig. 5 the frequencies of the two laser beams are adjusted so that their sum, $\nu_1 + \nu_2$ is tuned to the two-photon transition $(n, \ell) = (36, 34) \to (34, 32)$ involving an angular momentum change of $\Delta\ell = 2$, and the virtual intermediate state involved in the two-photon transition is tuned close (within a few GHz) to a real state $(n, \ell) = (35, 33)$. The nearness of the virtual state to a real one should lead to an enhancement of the two-photon transition probability by several orders of magnitude compared to the case $\nu_1 = \nu_2$ and laser energy densities of ~ 1 mJ/cm^2 would be sufficient to efficiently drive the two-photon transition. The observed width $\Delta\nu_{2\gamma}$ of the resonance line will decrease by a factor 20 compared to the thermal Doppler width $\Delta\nu_D$ according to the equation,

$$\Delta\nu_{2\gamma} = \left| \frac{\nu_1 - \nu_2}{\nu_1 + \nu_2} \right| \Delta\nu_D. \tag{1}$$

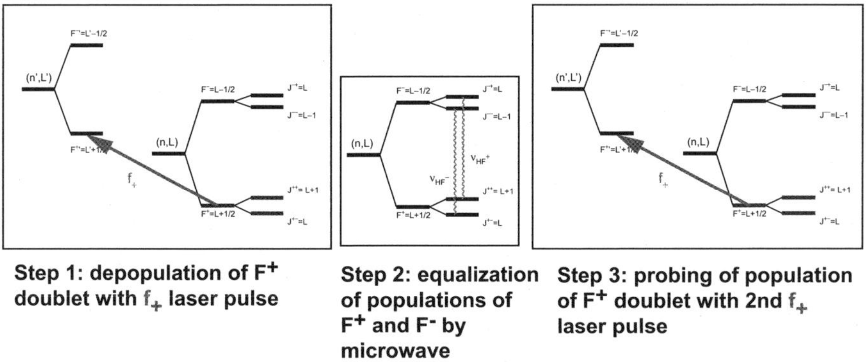

Fig. 6. The scheme of a split transition line in the $\overline{\text{p}}$-^{4}He$^+$ atom as studied using the laser–microwave–laser resonance method in order to measure the magnetic moment of the antiproton[32] in the atomic bound state.

4.2. *Level Splitting and Magnetic Moment*

As seen in Fig. 4 the $(n = 37, \ell = 35) \rightarrow (38, 34)$ line is split due to interaction between the antiproton magnetic moment and the electron spin. This resonance was used by the ASACUSA Collaboration to measure the magnetic moment of the antiproton: of course, as that is in a highly excited state, the measured momentum is mostly orbital.

The level scheme is presented in Fig. 6. Both levels involved in the transition are split, scanning the laser frequency will show the difference between the transition frequencies f_- and f_+. We have measured the splitting $\nu_{\rm HF}$ directly by emptying one of the split levels with a suitably tuned laser pulse, irradiating the system with a variable microwave pulse and again with a laser pulse of the same frequency as before. When the microwave was correctly tuned to the resonance, the second resonance had a reduced amplitude. The result is presented in Fig. 6.[32,36] It agrees within 6×10^{-5} with the recent theoretical calculations performed with the properties of the proton assumed for the antiproton. We also checked the possible effect of collisions on the result by studying its density dependence and found it was negligible.

5. Conclusion

Our latest published limit of CPT violation is 2×10^{-9} (2 ppb). However, I must point out that these experiments can only confirm the validity of CPT, nobody would believe us if we reported discovering a violation as there are theoretical calculations are involved in the analysis. On the other hand, if we assume the validity of CPT, our data can be used to increase the precision of the proton mass: the official CODATA uncertainty for those measurements decreased from 0.46 ppb to 0.43 ppb as a result of including our data.

Acknowledgements

The present work was supported by Monbukagakusho (Grant No. 15002005) and by the Hungarian National Research Foundation (Contracts OTKA K72172, NK67974, NK81447 and TeT JP-21/2006). The author is indebted to Dr Júlia Nyiri for illuminating suggestions and to MD's Dr Lidia Harmati and Dr Márta Kiss-Leizer for keeping him in good health.

References

1. F. Halzen and A.D. Martin, *Quarks and Leptons: An Introductory Course in Modern Particle Physics*, Wiley, New York, 1984.
2. A.G. Cohen, A. De Rujula and S.L. Glashow, *Astrophys. J.* **495**, 539 (1998).
3. C. Amsler *et al.* (Particle Data Group), *Phys. Lett.* **B667**, 1 (2008), http://pdg.lbl.gov.
4. V. A. Kostelecký, *Phys. Rev.* **D69**, 105009 (2004).
5. N. E. Mavromatos, *Lecture Notes in Physics*, **669**, 245 (2005).
6. F. R. Klinkhamer and Ch. Rupp, *Phys. Rev.* **D70**, 045020 (2004).
7. B. Aubert *et al.* (The BaBar Collaboration), arXiv:0711.2713v1 (2007).
8. A. Aguilar *et al.* (LSND Collaboration), *Phys. Rev.* **D64**, 112007 (2001).
9. L. B. Auerbach *et al.* (LSND Collaboration), *Phys. Rev.* **D72**, 076004 (2005).
10. T. Katori, A. Kostelecky and R. Tayloe, *Phys. Rev.* **D74**, 105009 (2006).
11. G. Barenboim, N. E. Mavromatos, S. Sarkar and A. Waldron-Lauda, *Nucl. Phys.* **B758**, 90 (2006).
12. A. A. Aguilar-Arevalo *et al.* (The MiniBooNE Collaboration), *Phys. Rev. Lett.* **98**, 231801 (2007).
13. A. A. Aguilar-Arevalo *et al.* (The MiniBooNE Collaboration), *Phys. Rev. Lett.* **103**, 111801 (2009).
14. G. Baur *et al. Phys. Lett.* **B368**, 251 (1996).
15. The Antiproton Decelerator (AD) at CERN, `http://psdoc.web.cern.ch/PSdoc/acc/ad/index.html`
16. ASACUSA (*Atomic Spectroscopy And Collisions Using Slow Antiprotons*) Collaboration,[a] `http://www.cern.ch/ASACUSA`.
17. R.S. Hayano, M. Hori, D. Horváth and E. Widmann, *Reports on Progress in Physics* **70**, 1995 (2007).

[a]Asakusa is one of the oldest districts of Tokyo; the name was proposed by our non-Japanese collaborators to honour the dominant Japanese contribution to the experiment.

18. ALPHA (*Antihydrogen Laser PHysics Apparatus*) Collaboration, `http://alpha.web.cern.ch/alpha`.
19. ATRAP (*Antimatter TRAP*) Collaboration, `http://hussle.harvard.edu/ atrap/`.
20. M. Charlton, J. Eades, D. Horváth, R. J. Hughes and C. Zimmermann, *Physics Reports* **241**, 65 (1994).
21. K. Tőkési, B. Juhász and J. Burgdörfer, *J. Phys.* **B38**, 401 (2005).
22. D. Horváth, *Radiochimica Acta* **77**, 75 (1997).
23. J. Eades and J. F. Hartmann, *Rev. Mod. Phys.* **71**, 373 (1999).
24. T. Yamazaki, N. Morita, R. S. Hayano, E. Widmann and J. Eades, *Physics Reports* **366**, 183 (2002).
25. G. Gabrielse, A. Khabbaz, D.S. Hall, C. Heimann, H. Kalinowsky and W. Jhe, *Phys. Rev. Lett.* **82**, 3198 (1999).
26. H. A. Torii, R. S. Hayano, M. Hori, T. Ishikawa, N. Morita, M. Kumakura, I. Sugai, T. Yamazaki, B. Ketzer, F. J. Hartmann, T. von Egidy, R. Pohl, C. Maierl, D. Horváth, J. Eades and E. Widmann, *Phys. Rev.* **A 59**, 223 (1999).
27. V. I. Korobov, *Phys. Rev.* **A 54**, R1749 (1996).
28. V.I. Korobov and D.D. Bakalov, *Phys. Rev. Lett.* **79**, 3379 (1997).
29. M. Hori, J. Eades, R. S. Hayano, T. Ishikawa, J. Sakaguchi, E. Widmann, H. Yamaguchi, H. A. Torii, B. Juhász, D. Horváth, T. Yamazaki, *Phys. Rev. Lett.*, **87**, 093401 (2001).
30. M. Hori, J. Eades, R. S. Hayano, T. Ishikawa, W. Pirkl, E. Widmann, H. Yamaguchi, H. A. Torii, B. Juhász, D. Horváth, T. Yamazaki, *Phys. Rev. Lett.* **91**, 123401 (2003).
31. M. Hori, A. Dax, J. Eades, K. Gomikawa, R. S. Hayano, N. Ono, W. Pirkl, E. Widmann, H. A. Torii, B. Juhász, D. Barna, D. Horváth, *Phys. Rev. Lett.* **96**, 243401 (2006).
32. E. Widmann, R.S. Hayano, T. Ishikawa, J. Sakaguchi, H. Yamaguchi, J. Eades, M. Hori, H.A. Torii, B. Juhász, D. Horváth, T. Yamazaki, *Phys. Rev. Lett.*, **89**, 243402 (2002).
33. K. Y. Franzen, N. Kuroda, H. A Torii, M. Hori, Z. Wang, H. Higaki, S. Yoneda, B. Juhász, D. Horváth, A. Mohri, K. Komaki, Y. Yamazaki, *Rev. Sci. Instrum.* **74**, 3305 (2003)
34. N. Kuroda, H. A. Torii, K. Yoshiki Franzen, Z. Wang, S. Yoneda, M. Inoue, M. Hori, B. Juhász, D. Horváth, H. Higaki, A. Mohri, J. Eades, K. Komaki, Y. Yamazaki, *Phys. Rev. Lett.* **94**, 023401 (2005).
35. E. Widmann, R. S. Hayano, M. Hori and T. Yamazaki, *Nucl. Instrum. Meth.* **B214**, 31 (2004).
36. T. Pask, D. Barna, A. Dax, R.S. Hayano, M. Hori, D. Horváth, S. Friedreich, B. Juhász, O. Massiczek, N. Ono, A. Sótér, E. Widmann, *Phys. Lett.* **B678**, 55 (2009).
37. R. Bluhm, V. A. Kostelecky and N. Russell, *Phys. Rev. Lett.* **82**, 2254 (1999).
38. CODATA *Internationally recommended values of the fundamental physical constants* 2006. http://physics.nist.gov/cuu/Constants/index.html.

NOVEL QCD PHENOMENOLOGY

STANLEY J. BRODSKY

SLAC National Accelerator Laboratory
Stanford University, Stanford, California 94309
and CP[3]-Origins, Southern Denmark University
Odense, Denmark
sjbth@slac.stanford.edu

I review a number of topics where conventional wisdom in hadron physics has been challenged. For example, hadrons can be produced at large transverse momentum directly within a hard QCD subprocess, rather than from jet fragmentation. Such "direct" higher-twist processes can explain the deviations from perturbative QCD predictions in measurements of inclusive hadron cross sections at fixed $x_T = 2p_T/\sqrt{s}$, as well as the "baryon anomaly, the anomalously large proton-to-pion ratio seen in high centrality heavy ion collisions. Initial-state and final-state interactions of the struck quark, soft-gluon rescattering associated with its Wilson line lead to Bjorken-scaling single-spin asymmetries, diffractive deep inelastic scattering, the breakdown of the Lam-Tung relation in Drell-Yan reactions, as well as nuclear shadowing and anti-shadowing. The Gribov-Glauber theory predicts that antishadowing of nuclear structure functions is not universal, but instead depends on the flavor quantum numbers of each quark and antiquark, thus explaining the anomalous nuclear dependence measured in deep-inelastic neutrino scattering. Since shadowing and antishadowing arise from the physics of leading-twist diffractive deep inelastic scattering, one cannot attribute such phenomena to the structure of the nucleus itself. It is thus important to distinguish "static" structure functions, the probability distributions computed from the square of the target light-front wavefunctions, versus "dynamical" structure functions which include the effects of the final-state rescattering of the struck quark. The importance of the $J = 0$ photon-quark QCD contact interaction in deeply virtual Compton scattering is also emphasized. The scheme-independent BLM method for setting the renormalization scale is discussed. The elimination of the renormalization scale ambiguity would greatly improve the precision of QCD predictions and increase the sensitivity of searches for new physics at the LHC. Other novel features of QCD are discussed, including the consequences of confinement for quark and gluon condensates.

1. Introduction

Volodya Gribov, whose work we are honoring at this workshop, was never satisfied with conventional wisdom. In this contribution I will review a number of topics where new, and in some cases surprising, perspectives for QCD physics have emerged.

(1) It is natural to assume that the nuclear modifications to the structure functions measured in deep inelastic lepton-nucleus and neutrino-nucleus interactions are identical; in fact, the Gribov-Glauber theory predicts that the antishadowing of nuclear structure functions is not universal, but depends on the quantum numbers of each struck quark and antiquark.[1] This observation can explain the recent analysis of Schienbein et al.[2] which shows that the NuTeV measurements of nuclear structure functions obtain from neutrino charged current reactions differ significantly from the distributions measured in deep inelastic electron and muon scattering.

(2) It is conventional to assume that high transverse momentum hadrons in inclusive high energy hadronic collisions, such as $pp \to HX$, only arise from jet fragmentation. In fact, a significant fraction of high $p_\perp^H$ events can emerge directly from the hard subprocess itself.[3,4] This phenomena can explain[5] the "baryon anomaly" observed at RHIC– the ratio of baryons to mesons at high $p_\perp^H$, as well as the power-law fall-off $1/p_\perp^n$ at fixed $x_\perp = 2p_\perp/\sqrt{s}$, increases with centrality,[6] opposite to the usual expectation that protons should suffer more energy loss in the nuclear medium than mesons.

(3) The effects of final-state interactions of the scattered quark in deep inelastic scattering have been traditionally assumed to be power-law suppressed. In fact, the final-state gluonic interactions of the scattered quark lead to a T-odd non-zero spin correlation of the plane of the lepton-quark scattering plane with the polarization of the target proton.[7] This leading-twist"Sivers effect" is nonuniversal since QCD predicts an opposite-sign correlation[8,9] in Drell-Yan reactions due to the initial-state interactions of the annihilating antiquark. The final-state interactions of the struck quark with the spectators[10] also lead to diffractive events in deep inelastic scattering (DDIS) at leading twist, such as $\ell p \to \ell' p' X$, where the proton remains intact and isolated in rapidity; in fact, approximately 10% of the deep inelastic lepton-proton scattering events observed at HERA are diffractive.[11,12] The presence of a rapidity gap between the target and diffractive system requires that

the target remnant emerges in a color-singlet state; this is made possible in any gauge by the soft rescattering incorporated in the Wilson line or by augmented light-front wavefunctions.

(4) It is usually assumed – following the intuition of the parton model – that the structure functions measured in deep inelastic scattering can be computed in the Bjorken-scaling leading-twist limit from the absolute square of the light-front wavefunctions, summed over all Fock states. In fact, dynamical effects, such as the Sivers spin correlation and diffractive deep inelastic lepton scattering due to final-state gluon interactions, contribute to the experimentally observed DIS cross sections. Diffractive events also lead to the interference of two-step and one-step processes in nuclei which in turn, via the Gribov-Glauber theory, lead to the shadowing and the antishadowing of the deep inelastic nuclear structure functions;[1] such phenomena are not included in the light-front wavefunctions of the nuclear eigenstate. This leads to an important distinction between "dynamical" vs. "static" (wavefunction-specific) structure functions.[13]

(5) As noted by Collins and Qiu,[14] the traditional factorization formalism of perturbative QCD fails in detail for many hard inclusive reactions because of initial- and final-state interactions. For example, if both the quark and antiquark in the Drell-Yan subprocess $q\bar{q} \to \mu^+\mu^-$ interact with the spectators of the other hadron, then one predicts a $\cos 2\phi \sin^2 \theta$ planar correlation in unpolarized Drell-Yan reactions.[15] This "double Boer-Mulders effect" can account for the large $\cos 2\phi$ correlation and the corresponding violation[15,16] of the Lam Tung relation for Drell-Yan processes observed by the NA10 collaboration. An important signal for factorization breakdown at the LHC will be the observation of a $\cos 2\phi$ planar correlation in dijet production.

(6) It is conventional to assume that the charm and bottom quarks in the proton structure functions only arise from gluon splitting $g \to Q\overline{Q}$. In fact, the proton light-front wavefunction contains *ab initio* intrinsic heavy quark Fock state components such as $|uudc\bar{c}>$.[17–20] The intrinsic heavy quarks carry most of the proton's momentum since this minimizes the off-shellness of the state. The heavy quark pair $Q\overline{Q}$ in the intrinsic Fock state is primarily a color-octet, and the ratio of intrinsic charm to intrinsic bottom scales scales as $m_c^2/m_b^2 \simeq 1/10$, as can easily be seen from the operator product expansion in non-Abelian QCD. Intrinsic charm and bottom explain the origin of high x_F open-charm and open-bottom hadron production, as well as the single and double J/ψ

hadroproduction cross sections observed at high x_F. The factorization-breaking nuclear $A^\alpha(x_F)$ dependence of hadronic J/ψ production cross sections is also explained. Kopeliovich, Schmidt, Soffer, Goldhaber, and I[21] have proposed a novel mechanism for Inclusive and diffractive Higgs production $pp \to pHp$ in which the Higgs boson carries a significant fraction of the projectile proton momentum. The production mechanism is based on the subprocess $(Q\overline{Q})g \to H$ where the $Q\overline{Q}$ in the $|uudQ\overline{Q}>$ intrinsic heavy quark Fock state of the colliding proton has approximately 80% of the projectile protons momentum.

(7) It is often stated that the renormalization scale of the QCD running coupling $\alpha_s(\mu_R^2)$ cannot be fixed, and thus it has to be chosen in an *ad hoc* fashion. In fact, as in QED, the scale can be fixed unambiguously by shifting μ_R so that all terms associated with the QCD β function vanish. In general, each set of skeleton diagrams has its respective scale. The result is independent of the choice of the initial renormalization scale μ_{R0}, thus satisfying Callan-Symanzik invariance. Unlike heuristic scale-setting procedures, the BLM method[22] gives results which are independent of the choice of renormalization scheme, as required by the transitivity property of the renormalization group. The divergent renormalon terms of order $\alpha_s^n \beta^n n!$ are transferred to the physics of the running coupling. Furthermore, one retains sensitivity to "conformal' effects which arise in higher orders, physical effects which are not associated with QCD renormalization. The BLM method also provides scale-fixed, scheme-independent high precision connections between observables, such as the "Generalized Crewther Relation",[23] as well as other "Commensurate Scale Relations".[24,25] Clearly the elimination of the renormalization scale ambiguity would greatly improve the precision of QCD predictions and increase the sensitivity of searches for new physics at the LHC.

(8) It is usually assumed that the QCD coupling $\alpha_s(Q^2)$ diverges at $Q^2 = 0$; i.e.,"infrared slavery". In fact, determinations from lattice gauge theory, Bethe-Salpeter methods, effective charge measurements, gluon mass phenomena, and AdS/QCD all lead (in their respective scheme) to a finite value of the QCD coupling in the infrared.[26] Because of color confinement, the quark and gluon propagators vanish at long wavelength: $k < \Lambda_{QCD}$, and consequently the quantum loop corrections underlying the QCD β-function – decouple in the infrared, and the coupling freezes to a finite value at $Q^2 \to 0$,[27] This observation underlies the use of conformal methods in AdS/QCD.

(9) It is conventionally assumed that the vacuum of QCD contains quark $< 0|q\bar{q}|0 >$ and gluon $< 0|G^{\mu\nu}G_{\mu\nu}|0 >$ vacuum condensates, although the resulting vacuum energy density leads to a 10^{45} order-of-magnitude discrepancy with the measured cosmological constant.[28] However, a new perspective has emerged from Bethe-Salpeter and light-front analyses where the QCD condensates are identified as "in-hadron" condensates, rather than vacuum entities, but consistent with the Gell Mann-Oakes- Renner relation.[29] The "in-hadron" condensates become realized as higher Fock states of the hadron when the theory is quantized at fixed light-front time $\tau = x^0 + x^3/c$.

(10) In nuclear physics nuclei are composites of nucleons. However, QCD provides a new perspective:[30,31] six quarks in the fundamental 3_C representation of $SU(3)$ color can combine into five different color-singlet combinations, only one of which corresponds to a proton and neutron. The deuteron wavefunction is a proton-neutron bound state at large distances, but as the quark separation becomes smaller, QCD evolution due to gluon exchange introduces four other "hidden color" states into the deuteron wavefunction.[32] The normalization of the deuteron form factor observed at large Q^2,[33] as well as the presence of two mass scales in the scaling behavior of the reduced deuteron form factor,[30] suggest sizable hidden-color Fock state contributions in the deuteron wavefunction.[34] The hidden-color states of the deuteron can be materialized at the hadron level as $\Delta^{++}(uuu)\Delta^{-}(ddd)$ and other novel quantum fluctuations of the deuteron. These dual hadronic components become important as one probes the deuteron at short distances, such as in exclusive reactions at large momentum transfer. For example, the ratio $d\sigma/dt(\gamma d \to \Delta^{++}\Delta^{-})/d\sigma/dt(\gamma d \to np)$ is predicted to increase to a fixed ratio $2:5$ with increasing transverse momentum p_T. Similarly, the Coulomb dissociation of the deuteron into various exclusive channels $ed \to e' + pn, pp\pi^-, \Delta\Delta, \cdots$ will have a changing composition as the final-state hadrons are probed at high transverse momentum, reflecting the onset of hidden-color degrees of freedom.

(11) It is usually assumed that the imaginary part of the deeply virtual Compton scattering amplitude is determined at leading twist by generalized parton distributions, but that the real part has an undetermined "D-term" subtraction. In fact, the real part is determined by the local two-photon interactions of the quark current in the QCD light-front Hamiltonian.[35,36] This contact interaction leads to a real energy-independent contribution to the DVCS amplitude which is in-

dependent of the photon virtuality at fixed t. The interference of the timelike DVCS amplitude with the Bethe-Heitler amplitude leads to a charge asymmetry in $\gamma p \to \ell^+ \ell^- p$.[36–38] Such measurements can verify that quarks carry the fundamental electromagnetic current within hadrons.

(12) A long-sought goal in hadron physics is to find a simple analytic first approximation to QCD analogous to the Schrödinger-Coulomb equation of atomic physics. This problem is particularly challenging since the formalism must be relativistic, color-confining, and consistent with chiral symmetry. de Teramond and I have shown that the soft-wall AdS/QCD model, modified by a positive-sign dilaton metric, leads to a simple Schrödinger-like light-front wave equation and a remarkable one-parameter description of nonperturbative hadron dynamics.[39–41] The model predicts a zero-mass pion for zero-mass quarks and a Regge spectrum of linear trajectories with the same slope in the (leading) orbital angular momentum L of the hadrons and their radial quantum number N. Light-Front Holography maps the amplitudes which are functions of the fifth dimension variable z of anti-de Sitter space to a corresponding hadron theory quantized on the light front. The resulting Lorentz-invariant relativistic light-front wave equations are functions of an invariant impact variable ζ which measures the separation of the quark and gluonic constituents within the hadron at equal light-front time. The result is a semi-classical frame-independent first approximation to the spectra and light-front wavefunctions of meson and baryon light-quark bound states, which in turn predicts the behavior of the pion and nucleon form factors. The theory implements chiral symmetry in a novel way: the effects of chiral symmetry breaking increase as one goes toward large interquark separation, consistent with spectroscopic data, and the hadron eigenstates generally have components with different orbital angular momentum; e.g., the proton eigenstate in AdS/QCD with massless quarks has $L = 0$ and $L = 1$ light-front Fock components with equal probability. The AdS/QCD soft-wall model also predicts the form of the non-perturbative effective coupling $\alpha_s^{AdS}(Q)$ and its β-function, and the AdS/QCD light-front wavefunctions also lead to a method for computing the hadronization of quark and gluon jets at the amplitude level.[42]

I will review several of these topics in these proceedings.

2. Direct Production of High $p_\perp$ Hadrons

A fundamental test of leading-twist QCD predictions in high transverse momentum hadronic reactions is the measurement of the power-law fall-off of the inclusive cross section[43] $Ed\sigma/d^3p(AB \to CX) = F(\theta_{cm}, x_T)/p_T^{n_{eff}}$ at fixed $x_T = 2p_T/\sqrt{s}$ and fixed θ_{CM}, where $n_{eff} \sim 4+\delta$. Here $\delta = \mathcal{O}(1)$ is the correction to the conformal prediction arising from the QCD running coupling and the DGLAP evolution of the input distribution and fragmentation functions.[3,4,44] The usual expectation is that leading-twist subprocesses will dominate measurements of high p_T hadron production at RHIC and Tevatron energies. Indeed, the data for isolated photon production $pp \to \gamma_{\text{direct}} X$ as well as jet production agrees well with the leading-twist scaling prediction $n_{eff} \simeq 4.5$ as seen in Fig. 1.[3] However, as seen in Fig. 1, measurements of n_{eff} for $pp \to \pi X$ are not consistent with the leading twist predictions. Striking deviations from the leading-twist predictions were also observed at lower energy at the ISR and Fermilab fixed-target experiments.[43,45,46] The high values n_{eff} with x_T seen in the data indicate the presence of an array of higher-twist processes, including subprocesses where the hadron enters directly, rather than through jet fragmentation.[47] The predicted deviations for the experimental and NLO scaling exponent at RHIC and the LHC with PHENIX preliminary measurements are shown in Fig. 2.

It should be emphasized that the existence of dynamical higher-twist processes in which a hadron interacts directly within a hard subprocess is a prediction of QCD. For example, the subprocess $\gamma^* q \to \pi q$, where the pion is produced directly through the pion's $\bar{q}q \to \pi$ distribution amplitude $\phi_\pi(x, Q)$ underlies deeply virtual meson scattering $\gamma p \to \pi X$. The corresponding timelike subprocess $\pi q \to \gamma^* q$ dominates the Drell-Yan reaction $\pi p \to \ell^+ \ell^- X$ at high x_F,[48] thus accounting for the change in angular distribution from the canonical $1 + \cos^2\theta$ distribution, for transversely polarized virtual photons, to $\sin^2\theta$, corresponding to longitudinal photons; the virtual photon thus becomes longitudinally polarized at high x_F, reflecting the spin of the pion entering the direct QCD hard subprocess. Crossing predicts reactions where the final-state hadron appears directly in the subprocess such as $e^+e^- \to \pi X$ at $z = 1$. The nominal power-law fall-off at fixed x_T is set by the number of elementary fields entering the hard subprocess $n_{\text{eff}} = 2n_{\text{active}} - 4$. The power-law fall-off $(1 - x_T)^F$ at high x_T is set by the total number of spectators $F = 2n_{\text{spectators}} - 1$,[47] up to spin corrections.

The direct higher-twist subprocesses, where the trigger hadron is produced within the hard subprocess avoid the waste of same-side energy, thus allowing the target and projectile structure functions to be evaluated

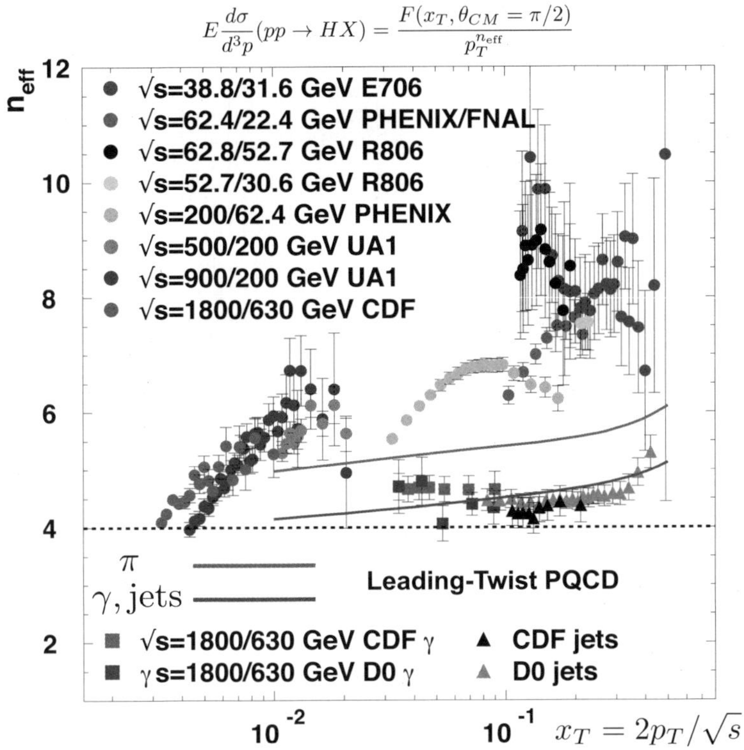

Fig. 1. Comparison of RHIC and fixed-target data for hadron, isolated photon, and jet production with the leading-twist pQCD predictions for the power-falloff of the semi-inclusive cross section $Ed\sigma/d^3p(pp \to HX) = F(x_T, \theta_{CM} = \pi/2)/p_T^{n_{\rm eff}}$ at fixed x_T. The data from R806, PHENIX, ISR/FNAL, E706 are for charged or neutral pion production, whereas the CDF, UA1 data at small x_T are for charged hadrons. The blue curve is the prediction of leading-twist QCD for isolated photon and jet production, including the scale-breaking effects of the running coupling and evolution of the proton structure functions. The red curve is the QCD prediction for pion production, which also includes the effect from the evolution of the fragmentation function. The dashed line at $n_{\rm eff} = 4$ is the prediction of the scale-invariant parton model. From Arleo, et al.[3]

at the minimum values of x_1 and x_2 where they are at their maximum. Examples of direct baryon and meson higher-twist subprocesses are: $ud \to \Lambda\bar{s}, u\bar{d} \to \pi^+ g, ug \to \pi^+ d, u\bar{s} \to K^+ g, ug \to K^+ s$. These direct subprocesses involve the distribution amplitude of the hadron which has dimension Λ_{QCD} for mesons and Λ^2_{QCD} for baryons; thus these higher-twist contributions to the inclusive cross section $Ed\sigma/d^3p$ at fixed x_T nominally scale as Λ^2_{QCD}/p_T^6 for mesons and Λ^4_{QCD}/p_T^8 for baryons.

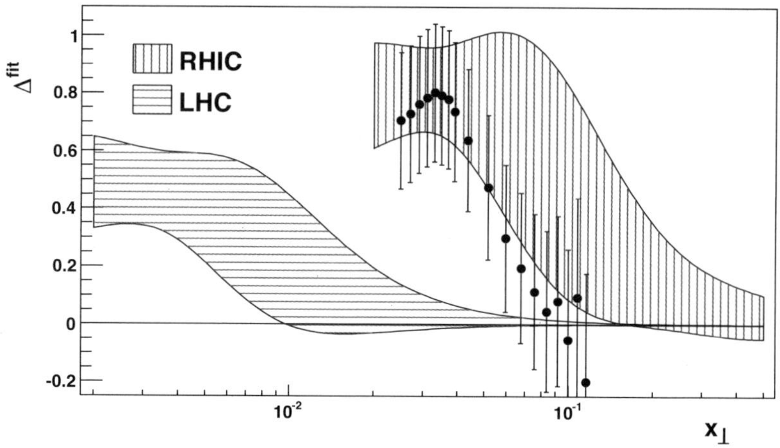

Fig. 2. Predicted difference between the experimental and NLO scaling exponent at RHIC ($\sqrt{s} = 200, 500$ GeV) and the LHC ($\sqrt{s} = 7$ TeV as compared to $\sqrt{s} = 1.8$ TeV), compared to PHENIX preliminary measurements. From Arleo, et al.[3]

The behavior of the single-particle inclusive cross section will be a key test of QCD at the LHC, since the leading-twist prediction for $n_{\rm eff} \sim 4 + \delta$ is independent of the detailed form of the structure and fragmentation functions.

The fixed x_T scaling of the proton production cross section $Ed\sigma/d^3p(pp \to ppX)$ is particularly anomalous, far from the $1/p_T^4$ to $1/p_T^5$ scaling predicted by pQCD.[44] See Fig. 1. Sickles and I have argued that the anomalous features of inclusive high p_T proton production is due to hard subprocesses[44] where the proton is created directly within the hard reaction, such as $uu \to p\bar{d}$, such as the mechanism illustrated in Fig. 3. The fragmentation of a gluon or quark jet to a proton requires that the underlying 2 to 2 subprocess occurs at a higher transverse momentum than the p_T of the observed proton because of the fast-falling $(1-z)^3$ quark-to-proton fragmentation function; in contrast, the direct subprocess is maximally energy efficient. Such "direct" reactions thus can explain the fast-falling power-law falloff observed at fixed x_T and fixed-θ_{cm} at the ISR, FermiLab and RHIC.[44]

Since the proton is initially produced as a small-size $b_\perp \sim 1/p_T$ color-singlet state, it is "color transparent",[50] and it can thus propagate through dense nuclear matter with minimal energy loss. In contrast, the pions which are produced from jet fragmentation have a normal inelastic cross section.

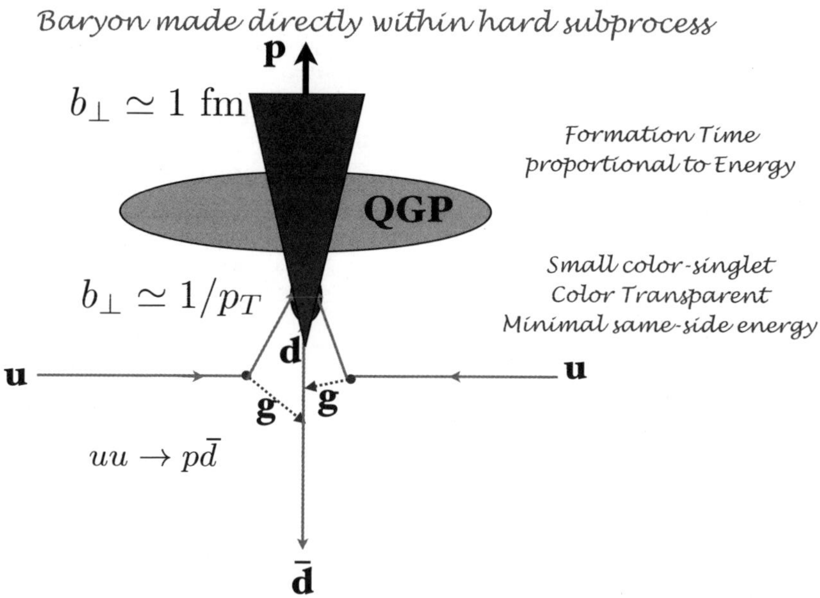

Fig. 3. Direct production of a proton in QCD. The proton is initially produced as a color-transparent small-size color singlet hadron.

This provides a plausible explanation[5] of the RHIC data,[6] which shows a dramatic rise of the p to π ratio with increasing p_T when one compares peripheral with central heavy ion collisions, as illustrated in Fig. 4. The color transparency of the proton produced in the direct process also can explain why the index n_{eff} rises with centrality, as seen in Fig. 5, – the higher-twist color-transparent subprocess dominates in the nuclear medium.[44] In addition, the fact that the proton tends to be produced alone in a direct subprocess explains why the yield of same-side hadrons along the proton trigger is diminished with increasing centrality. Thus the QCD color transparency of directly produced baryons can explain the baryon anomaly seen in heavy-ion collisions at RHIC: the color-transparent proton state is not absorbed, but a pion produced from fragmentation is diminished in the nuclear medium.[49] The increase of n_{eff} with centrality is consistent with the nuclear survival of direct higher-twist subprocesses for both protons and antiprotons, and to a lesser extent, for mesons.

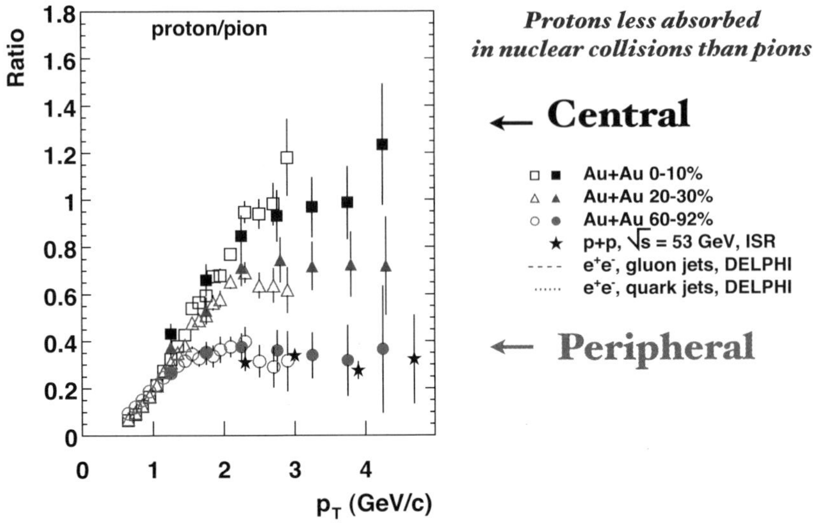

Fig. 4. The baryon anomaly observed by the PHENIX experiment at RHIC,[6] The anomalous rise of the proton to pion ratio with centrality at large p_T.

3. Leading-Twist Shadowing and Anti-Shadowing of Nuclear Structure Functions

The shadowing of the nuclear structure functions: $R_A(x, Q^2) < 1$ at small $x < 0.1$ can be readily understood in terms of the Gribov-Glauber theory. Consider a two-step process in the nuclear target rest frame. The incoming $q\overline{q}$ dipole first interacts diffractively $\gamma^* N_1 \to (q\overline{q})N_1$ on nucleon N_1 leaving it intact. This is the leading-twist diffractive deep inelastic scattering (DDIS) process which has been measured at HERA to constitute approximately 10% of the DIS cross section at high energies. The $q\overline{q}$ state then interacts inelastically on a downstream nucleon $N_2 : (q\overline{q})N_2 \to X$. The phase of the pomeron-dominated DDIS amplitude is close to imaginary, and the Glauber cut provides another phase i, so that the two-step process has opposite phase and destructively interferes with the one-step DIS process $\gamma * N_2 \to X$ where N_1 acts as an unscattered spectator. The one-step and-two-step amplitudes can coherently interfere as long as the momentum transfer to the nucleon N_1 is sufficiently small that it remains in the nuclear target; *i.e.*, the Ioffe length[51] $L_I = 2M\nu/Q^2$ is large compared to the inter-nucleon separation. In effect, the flux reaching the interior nucleons is diminished, thus reducing the number of effective nucleons and $R_A(x, Q^2) < 1$. The Bjorken-scaling diffractive contribution to DIS arises

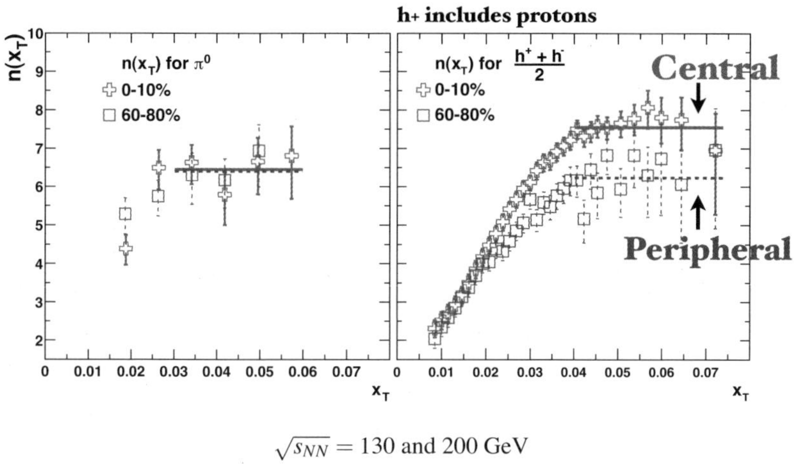

Fig. 5. The power-law scaling index n_{eff} at fixed x_T as a function of centrality versus peripheral collisions, using spectra at $\sqrt{s} = 130$ GeV and $\sqrt{s} = 200$ GeV.[49] The positive-charged hadron trigger is dominated by protons at high p_T for central collisions, consistent with the color transparency of direct higher-twist baryon production processes.

from the rescattering of the struck quark after it is struck (in the parton model frame $q^+ \leq 0$), an effect induced by the Wilson line connecting the currents. Thus one cannot attribute DDIS to the physics of the target nucleon computed in isolation.[10]

One of the novel features of QCD involving nuclei is the *anti-shadowing* of the nuclear structure functions as observed in deep inelastic lepton-nucleus scattering. Empirically, one finds $R_A(x, Q^2) \equiv \left(F_{2A}(x, Q^2)/(A/2)F_d(x, Q^2)\right) > 1$ in the domain $0.1 < x < 0.2$; *i.e.*, the measured nuclear structure function (referenced to the deuteron) is larger than the scattering on a set of A independent nucleons. Ivan Schmidt, Jian-Jun Yang, and I[1] have extended the analysis of nuclear shadowing to the shadowing and antishadowing of the electroweak structure functions. We note that there are leading-twist diffractive contributions $\gamma^* N_1 \to (q\bar{q})N_1$ arising from Reggeon exchanges in the t-channel.[52] For example, isospin–non-singlet $C = +$ Reggeons contribute to the difference of proton and neutron structure functions, giving the characteristic Kuti-Weisskopf $F_{2p} - F_{2n} \sim x^{1-\alpha_R(0)} \sim x^{0.5}$ behavior at small x. The x dependence of the structure functions reflects the Regge behavior $\nu^{\alpha_R(0)}$ of the virtual Compton amplitude at fixed Q^2 and $t = 0$. The phase of the diffractive amplitude is determined by analyticity and crossing to be proportional to $-1+i$ for $\alpha_R = 0.5$, which together with the phase from the Glauber cut, leads to *constructive* interference of the diffractive and nondiffractive multi-step nuclear amplitudes. The nuclear structure function is predicted to be enhanced precisely in the domain $0.1 < x < 0.2$ where antishadowing is empirically observed. The strength of the Reggeon amplitudes is fixed by the fits to the nucleon structure functions, so there is little model dependence. Since quarks of different flavors will couple to different Reggeons; this leads to the remarkable prediction that nuclear antishadowing is not universal; it depends on the quantum numbers of the struck quark. This picture implies substantially different antishadowing for charged and neutral current reactions, thus affecting the extraction of the weak-mixing angle θ_W. The ratio of nuclear to nucleon structure functions $R_{A/N}(x,Q) = \frac{F_{2A}(x,Q)}{AF_{2N}(x,Q)}$ is thus process independent. We have also identified contributions to the nuclear multi-step reactions which arise from odderon exchange and hidden color degrees of freedom in the nuclear wavefunction.

Schienbein et al.[2] have recently given a comprehensive analysis of charged current deep inelastic neutrino-iron scattering, finding significant differences with the nuclear corrections for electron-iron scattering. See Fig. 6. The measured nuclear effect measured in the NuTeV deep inelastic scattering charged current experiment is distinctly different from the nuclear modification measured at SLAC and NMC in deep inelastic scattering electron and muon scattering. This implies that part of of the anomalous NuTeV result[53] for θ_W could be due to the non-universality of nuclear antishadowing for charged and neutral currents.

A new understanding of nuclear shadowing and antishadowing has emerged based on multi-step coherent reactions involving leading twist diffractive reactions.[1,52] The nuclear shadowing of structure functions is a consequence of the lepton-nucleus collision; it is not an intrinsic property of the nuclear wavefunction. The same analysis shows that antishadowing is *not universal*, but it depends in detail on the flavor of the quark or antiquark constituent.[1] Detailed measurements of the nuclear dependence of individual quark structure functions are thus needed to establish the distinctive phenomenology of shadowing and antishadowing and to make the NuTeV results definitive. A comparison of the nuclear modification in neutrino versus anti-neutrino interactions is clearly important. There are other ways in which this new view of antishadowing can be tested; for example, antishadowing can also depend on the target and beam polarization.

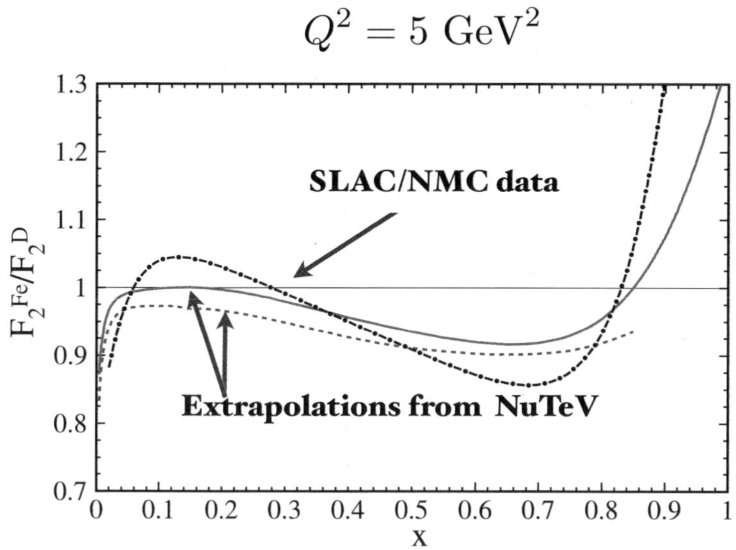

Fig. 6. Comparison of the Nuclear Modification of Charged vs. Neutral Current Deep Inelastic Structure Functions. From I. Schienbein et al.[2]

4. Dynamic versus Static Hadronic Structure Functions

The nontrivial effects from rescattering and diffraction highlight the need for a fundamental understanding the dynamics of hadrons in QCD at the amplitude level. This is essential for understanding phenomena such as the quantum mechanics of hadron formation, the remarkable effects of initial and final interactions, the origins of diffractive phenomena and single-spin asymmetries, and manifestations of higher-twist semi-exclusive hadron subprocesses. A central tool in these analyses is the light-front wavefunctions of hadrons, the frame-independent eigensolutions of the Heisenberg equation for QCD $H^{LF}|\Psi>= M^2|\Psi>$ quantized at fixed light-front. Given the light-front wavefunctions $\psi_{n/H}(x_i, \vec{k}_{\perp i}, \lambda_i)$, one can compute a large range of exclusive and inclusive hadron observables. For example, the valence, sea-quark and gluon distributions are defined from the squares of the LFWFS summed over all Fock states n. Form factors, exclusive weak transition amplitudes[54] such as $B \to \ell\nu\pi$, and the generalized parton distributions[55] measured in deeply virtual Compton scattering are (assuming the "handbag" approximation) overlaps of the initial and final LFWFS with $n = n'$ and $n = n' + 2$.

Static	Dynamic
• Square of Target LFWFs	Modified by Rescattering: ISI & FSI
• No Wilson Line	Contains Wilson Line, Phases
• Probability Distributions	No Probabilistic Interpretation
• Process-Independent	Process-Dependent - From Collision
• T-even Observables	T-Odd (Sivers, Boer-Mulders, etc.)
• No Shadowing, Anti-Shadowing	Shadowing, Anti-Shadowing, Saturation
• Sum Rules: Momentum and J^z	Sum Rules Not Proven
• DGLAP Evolution; mod. at large x	DGLAP Evolution
• No Diffractive DIS	Hard Pomeron and Odderon Diffractive DIS

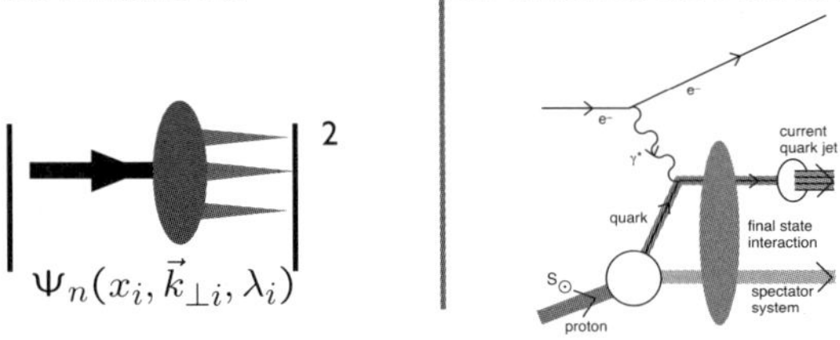

Fig. 7. Static versus dynamic structure functions.

It is thus important to distinguish "static" structure functions which are computed directly from the light-front wavefunctions of a target hadron from the nonuniversal "dynamic" empirical structure functions which take into account rescattering of the struck quark in deep inelastic lepton scattering. See Fig. 7. The real wavefunctions underlying static structure functions cannot describe diffractive deep inelastic scattering nor single-spin asymmetries, since such phenomena involve the complex phase structure of the $\gamma^* p$ amplitude. One can augment the light-front wavefunctions with a gauge link corresponding to an external field created by the virtual photon $q\bar{q}$ pair current,[56,57] but such a gauge link is process dependent,[8] so the resulting augmented wavefunctions are not universal.[10,56,58] The physics of rescattering and nuclear shadowing is not included in the nuclear light-front wavefunctions, and a probabilistic interpretation of the nuclear DIS cross section is precluded.

5. Novel Intrinsic Heavy Quark Phenomena

Intrinsic heavy quark distributions are a rigorous feature of QCD, arising from diagrams in which two or more gluons couple the valence quarks to the heavy quarks. The probability for Fock states of a light hadron to have an extra heavy quark pair decreases as $1/m_Q^2$ in non-Abelian gauge theory.[18,20] The relevant matrix element is the cube of the QCD field strength $G^3_{\mu\nu}$, in contrast to QED where the relevant operator is $F^4_{\mu\nu}$ and the probability of intrinsic heavy leptons in an atomic state is suppressed as $1/m_\ell^4$. The maximum probability occurs at $x_i = m^i_\perp / \sum_{j=1}^n m^j_\perp$ where $m_{\perp i} = \sqrt{k^2_{\perp i} + m^2_i}$.; *i.e.*, when the constituents have minimal invariant mass and equal rapidity. Thus the heaviest constituents have the highest momentum fractions and the highest x_i. Intrinsic charm thus predicts that the charm structure function has support at large x_{bj} in excess of DGLAP extrapolations;[17] this is in agreement with the EMC measurements.[19] Intrinsic charm can also explain the $J/\psi \to \rho\pi$ puzzle.[59] It also affects the extraction of suppressed CKM matrix elements in B decays.[60] The dissociation of the intrinsic charm $|uudc\bar{c}>$ Fock state of the proton can produce a leading heavy quarkonium state at high $x_F = x_c + x_{\bar{c}}$ in $pN \to J/\psi X$ since the c and $\bar{c}$ can readily coalesce into the charmonium state. Since the constituents of a given intrinsic heavy-quark Fock state tend to have the same rapidity, coalescence of multiple partons from the projectile Fock state into charmed hadrons and mesons is also favored. For example, one can produce a leading Λ_c at high x_F and low p_T from the coalescence of the udc constituents of the projectile $|uudc\bar{c}>$ Fock state.

The operator product analysis of the IC matrix element shows that the IC Fock state has a dominant color-octet structure: $|(uud)_{8C}(c\bar{c})_{8C}>$. The color octet $c\bar{c}$ converts to a color singlet by gluon exchange on the front surface of a nuclear target and then coalesces to a J/ψ which interacts weakly through the nuclear volume.[21] Thus the rate for the IC component has $A^{2/3}$ dependence corresponding to the area of the front surface. This is illustrated in fig 8. This forward contribution is in addition to the A^1 contribution derived from the usual perturbative QCD fusion contribution at small x_F. Because of these two components, the cross section violates perturbative QCD factorization for hard inclusive reactions.[61] This is consistent with the two-component cross section for charmonium production observed by the NA3 collaboration at CERN[62] and more recent experiments.[63] The diffractive dissociation of the intrinsic charm Fock state leads to leading charm hadron production and fast charmonium production in agreement with measurements.[64] The hadroproduction cross sections for

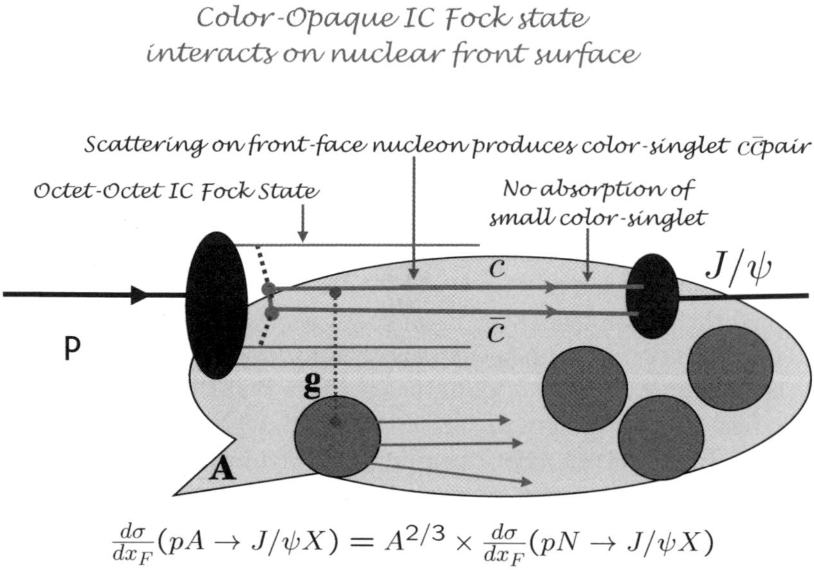

Fig. 8. Color-Octet intrinsic charm mechanism for the nuclear dependence of J/ψ production.

double-charm Ξ^+_{cc} baryons at SELEX[65] and the production of J/ψ pairs at NA3 are consistent with the diffractive dissociation and coalescence of double IC Fock states.[66] These observations provide compelling evidence for the diffractive dissociation of complex off-shell Fock states of the projectile and contradict the traditional view that sea quarks and gluons are always produced perturbatively via DGLAP evolution or gluon splitting. It is also conceivable that the observations[67] of Λ_b at high x_F at the ISR in high energy pp collisions could be due to the dissociation and coalescence of the "intrinsic bottom" $|uudb\bar{b}>$ Fock states of the proton.

As emphasized by Lai, Tung, and Pumplin,[68] there are strong indications that the structure functions used to model charm and bottom quarks in the proton at large x_{bj} have been underestimated, since they ignore intrinsic heavy quark fluctuations of hadron wavefunctions. Furthermore, the neglect of the intrinsic-heavy quark component in the proton structure function will lead to an incorrect assessment of the gluon distribution at large x if it is assumed that sea quarks always arise from gluon splitting. It is thus critical for new experiments (HERMES, HERA, COMPASS) to definitively establish the phenomenology of the charm structure function at large x_{bj}.

6. Vacuum Effects and Light-Front Quantization

The vacuum in quantum field theories is remarkably simple in light-cone quantization because of the restriction $k^+ \geq 0$. For example in QED, vacuum graphs such as $e^+e^-\gamma$ associated with the zero-point energy do not arise. In the Higgs theory, the usual Higgs vacuum expectation value is replaced with a $k^+ = 0$ zero mode;[69] however, the resulting phenomenology is identical to the standard analysis.

Hadronic condensates play an important role in quantum chromodynamics. It is widely held that quark and gluon vacuum condensates have a physical existence, independent of hadrons, measurable spacetime-independent configurations of QCD's elementary degrees-of-freedom in a hadron-less ground state. However, a non-zero spacetime-independent QCD vacuum condensate poses a critical dilemma for gravitational interactions because it would lead to a cosmological constant some 45 orders of magnitude larger than observation. As noted in Ref. 28, this conflict is avoided if strong interaction condensates are properties of the light-front wavefunctions of the hadrons, rather than the hadron-less ground state of QCD.

The usual assumption that non-zero vacuum condensates exist and possess a measurable reality has long been recognized as posing a conundrum for the light-front formulation of QCD. In the light-front formulation, the ground-state is a structureless Fock space vacuum, in which case it would seem to follow that dynamical chiral symmetry breaking (CSB) is impossible. In fact, as first argued by Casher and Susskind[70] dynamical CSB must be a property of hadron wavefunctions, not of the vacuum in the light-front framework. This thesis has also been explored in a series of recent articles.[27,28,42]

Conventionally, the quark and gluon condensates are considered to be properties of the QCD vacuum and hence to be constant throughout spacetime. A new perspective on the nature of QCD condensates $\langle \overline{q}q \rangle$ and $\langle G_{\mu\nu}G^{\mu\nu} \rangle$, particularly where they have spatial and temporal support, has recently been presented.[27–29,71,72] Their spatial support is restricted to the interior of hadrons, since these condensates arise due to the interactions of quarks and gluons which are confined within hadrons. For example, consider a meson consisting of a light quark q bound to a heavy antiquark, such as a B meson. One can analyze the propagation of the light q in the background field of the heavy $\overline{b}$ quark. Solving the Dyson-Schwinger equation for the light quark one obtains a nonzero dynamical mass and, via the connection mentioned above, hence a nonzero value of the condensate $\langle \overline{q}q \rangle$. But this is not a true vacuum expectation value; instead, it is the matrix element of the operator $\overline{q}q$ in the background field of the $\overline{b}$ quark. The change in the (dynamical) mass of the light quark in this bound state is somewhat reminiscent of the energy shift of an electron in the Lamb shift, in that both are consequences of the fermion being in a bound state rather than propagating freely. Similarly, it is important to use the equations of motion for confined quarks and gluon fields when analyzing current correlators in QCD, not free propagators, as has often been done in traditional analyses of operator products. Since the distance between the quark and antiquark cannot become arbitrarily large, one cannot create a quark condensate which has uniform extent throughout the universe. Thus in a fully self-consistent treatment of the bound state, this phenomenon occurs in the background field of the $\overline{b}$-quark, whose influence on light-quark propagation is primarily concentrated in the far infrared and whose presence ensures the manifestations of light-quark dressing are gauge invariant.

In the case of the pion one finds that the vacuum quark condensate that appears in the Gell Mann-Oakes Renner formula, is, in fact, a chiral-limit value of an 'in-pion' condensate.[29] This condensate is no more a property of the "vacuum" than the pion's chiral-limit leptonic decay constant. One can connect the Bethe-Salpeter formalism to the light-front formalism, by fixing the light-front time τ. This then leads to the Fock state expansion. In fact, dynamical CSB in the light-front formulation, expressed via 'in-hadron' condensates, can be shown to be connected with sea-quarks derived from higher Fock states. This solution is similar to that discussed in Ref. 70. Moreover, Ref. 73 establishes the equivalence of all three definitions of the vacuum quark condensate: a constant in the operator product expansion,[74,75] via the Banks-Casher formula,[76] and the trace of the chiral-limit dressed-quark propagator.

Acknowledgments

Invited talk, presented at the Gribov-80 Memorial Workshop on Quantum Chromodynamics and Beyond, held at the Abdus Salam International Centre for Theoretical Physics, Trieste, Italy. I am grateful to Julia Nyíri and Yuri Dokshitser for their invitation to this meeting, and I thank all of my collaborators whose work has been cited in this report. This research was supported by the Department of Energy, contract DE–AC02–76SF00515. SLAC-PUB 14265.

References

1. S. J. Brodsky, I. Schmidt and J. J. Yang, *Phys. Rev.* **D70**, 116003 (2004).
2. I. Schienbein, J. Y. Yu, C. Keppel, J. G. Morfin, F. I. Olness and J. F. Owens, arXiv:0806.0723 [hep-ph].
3. F. Arleo, S. J. Brodsky, D. S. Hwang and A. M. Sickles, *Phys. Rev. Lett.* **105**, 062002 (2010).
4. F. Arleo, S. J. Brodsky, D. S. Hwang and A. M. Sickles, arXiv:1006.4045 [hep-ph].
5. S. J. Brodsky and A. Sickles, *Phys. Lett.* **B668**, 111 (2008).
6. S. S. Adler *et al.* (PHENIX Collaboration), *Phys. Rev. Lett.* **91**, 172301 (2003).
7. S. J. Brodsky, D. S. Hwang and I. Schmidt, *Phys. Lett.* **B530**, 99 (2002).
8. J. C. Collins, *Phys. Lett.* **B536**, 43 (2002).
9. S. J. Brodsky, D. S. Hwang and I. Schmidt, *Nucl. Phys.* **B642**, 344 (2002).
10. S. J. Brodsky, P. Hoyer, N. Marchal, S. Peigne and F. Sannino, *Phys. Rev.* **D65**, 114025 (2002).
11. C. Adloff *et al.* (H1 Collaboration), *Z. Phys.* **C76**, 613 (1997).
12. J. Breitweg *et al.* (ZEUS Collaboration), *Eur. Phys. J.* **C6**, 43 (1999).
13. S. J. Brodsky, *Nucl. Phys.* **A827**, 327C (2009).
14. J. Collins and J. W. Qiu, *Phys. Rev.* **D75**, 114014 (2007).
15. D. Boer, S. J. Brodsky and D. S. Hwang, *Phys. Rev.* **D67**, 054003 (2003).
16. D. Boer, *Phys. Rev.* **D60**, 014012 (1999).
17. S. J. Brodsky, P. Hoyer, C. Peterson and N. Sakai, *Phys. Lett.* **B93**, 451 (1980).
18. S. J. Brodsky, J. C. Collins, S. D. Ellis, J. F. Gunion and A. H. Mueller, "Intrinsic Chevrolets At The Ssc." Proc. of 1984 Summer Study on the SSC, Snowmass, CO.
19. B. W. Harris, J. Smith and R. Vogt, *Nucl. Phys.* **B461**, 181 (1996).
20. M. Franz, M. V. Polyakov and K. Goeke, *Phys. Rev.* **D62**, 074024 (2000).
21. S. J. Brodsky, B. Kopeliovich, I. Schmidt and J. Soffer, *Phys. Rev.* **D73**, 113005 (2006).
22. S. J. Brodsky, G. P. Lepage and P. B. Mackenzie, *Phys. Rev.* **D28**, 228 (1983).
23. S. J. Brodsky, G. T. Gabadadze, A. L. Kataev and H. J. Lu, *Phys. Lett.* **B372**, 133 (1996).
24. S. J. Brodsky and H. J. Lu, *Phys. Rev.* **D51**, 3652 (1995).
25. S. J. Brodsky, E. Gardi, G. Grunberg and J. Rathsman, *Phys. Rev.* **D63**, 094017 (2001).
26. S. J. Brodsky, G. F. de Teramond and A. Deur, *Phys. Rev.* **D81**, 096010 (2010).
27. S. J. Brodsky and R. Shrock, *Phys. Lett.* **B666**, 95 (2008).
28. S. J. Brodsky and R. Shrock, arXiv:0905.1151 [hep-th].
29. S. J. Brodsky, C. D. Roberts, R. Shrock and P. C. Tandy, *Phys. Rev.* **C82**, 022201 (2010).
30. S. J. Brodsky and B. T. Chertok, *Phys. Rev.* **D14**, 3003 (1976).
31. V. A. Matveev and P. Sorba, *Lett. Nuovo Cim.* **20**, 435 (1977).
32. S. J. Brodsky, C. R. Ji and G. P. Lepage, *Phys. Rev. Lett.* **51**, 83 (1983).

33. R. G. Arnold *et al.*, *Phys. Rev. Lett.* **35**, 776 (1975).
34. G. R. Farrar, K. Huleihel and H. Y. Zhang, *Phys. Rev. Lett.* **74**, 650 (1995).
35. S. J. Brodsky, F. J. Llanes-Estrada and A. P. Szczepaniak, *Phys. Rev.* **D79**, 033012 (2009).
36. S. J. Brodsky, F. E. Close and J. F. Gunion, *Phys. Rev.* **D5**, 1384 (1972).
37. S. J. Brodsky, F. E. Close and J. F. Gunion, *Phys. Rev.* **D8**, 3678 (1973).
38. S. J. Brodsky, F. E. Close and J. F. Gunion, *Phys. Rev.* **D6**, 177 (1972).
39. G. F. de Teramond and S. J. Brodsky, *Phys. Rev. Lett.* **94**, 201601 (2005).
40. G. F. de Teramond and S. J. Brodsky, *Phys. Rev. Lett.* **102**, 081601 (2009).
41. S. J. Brodsky and G. F. de Teramond, arXiv:1009.4232 [hep-ph].
42. S. J. Brodsky, G. de Teramond and R. Shrock, *AIP Conf. Proc.* **1056**, 3 (2008).
43. D. W. Sivers, S. J. Brodsky and R. Blankenbecler, *Phys. Rept.* **23**, 1 (1976).
44. S. J. Brodsky and M. Rijssenbeek, arXiv:hep-ph/0511178.
45. J. W. Cronin, H. J. Frisch, M. J. Shochet, J. P. Boymond, P. A. Piroue and R. L. Sumner, *Phys. Rev. Lett.* **31**, 1426 (1973).
46. D. Antreasyan, J. W. Cronin, H. J. Frisch, M. J. Shochet, L. Kluberg, P. A. Piroue and R. L. Sumner, *Phys. Rev.* **D19**, 764 (1979).
47. R. Blankenbecler, S. J. Brodsky and J. F. Gunion, *Phys. Rev.* **D12**, 3469 (1975).
48. E. L. Berger and S. J. Brodsky, Phys. Rev. Lett. **42**, 940 (1979).
49. A. M. Sickles, *Nucl. Phys.* **A830**, 131C (2009).
50. S. J. Brodsky and A. H. Mueller, *Phys. Lett.* **B206**, 685 (1988).
51. B. L. Ioffe, *Phys. Lett.* **B30**, 123 (1969).
52. S. J. Brodsky and H. J. Lu, *Phys. Rev. Lett.* **64**, 1342 (1990).
53. G. P. Zeller *et al.* [NuTeV Collaboration], *Phys. Rev. Lett.* **88**, 091802 (2002); *Erratum-ibid.* **90**, 239902 (2003).
54. S. J. Brodsky and D. S. Hwang, *Nucl. Phys.* **B543**, 239 (1999).
55. S. J. Brodsky, M. Diehl and D. S. Hwang, *Nucl. Phys.* **B596**, 99 (2001).
56. A. V. Belitsky, X. Ji and F. Yuan, *Nucl. Phys.* **B656**, 165 (2003).
57. J. C. Collins and A. Metz, *Phys. Rev. Lett.* **93**, 252001 (2004).
58. J. C. Collins, *Acta Phys. Polon.* **B34**, 3103 (2003).
59. S. J. Brodsky and M. Karliner, *Phys. Rev. Lett.* **78**, 4682 (1997).
60. S. J. Brodsky and S. Gardner, *Phys. Rev.* **D65**, 054016 (2002).
61. P. Hoyer, M. Vanttinen and U. Sukhatme, *Phys. Lett.* **B246**, 217 (1990).
62. J. Badier *et al.* (NA3 Collaboration), *Phys. Lett.* **B104**, 335 (1981).
63. M. J. Leitch *et al.* (FNAL E866/NuSea Collaboration), *Phys. Rev. Lett.* **84**, 3256 (2000).
64. J. C. Anjos, J. Magnin and G. Herrera, *Phys. Lett.* **B523**, 29 (2001).
65. A. Ocherashvili *et al.* (SELEX Collaboration), *Phys. Lett.* **B628**, 18 (2005).
66. R. Vogt and S. J. Brodsky, *Phys. Lett.* **B349**, 569 (1995).
67. G. Bari *et al.*, *Nuovo Cim.* **A104**, 1787 (1991).
68. J. Pumplin, H. L. Lai and W. K. Tung, *Phys. Rev.* **D75**, 054029 (2007).
69. P. P. Srivastava and S. J. Brodsky, *Phys. Rev.* **D66**, 045019 (2002).
70. A. Casher and L. Susskind, *Phys. Rev.* **D9**, 436 (1974).
71. S. J. Brodsky and R. Shrock, arXiv:0803.2541 [hep-th].

72. S. J. Brodsky and R. Shrock, arXiv:0803.2554 [hep-th].
73. K. Langfeld, H. Markum, R. Pullirsch, C. D. Roberts and S. M. Schmidt, *Phys. Rev.* **C67**, 065206 (2003).
74. K. D. Lane, *Phys. Rev.* **D10**, 2605 (1974).
75. H. D. Politzer, *Nucl. Phys.* **B117**, 397 (1976).
76. T. Banks and A. Casher, *Nucl. Phys.* **B169**, 103 (1980).

VISCOSITY NEAR PHASE TRANSITIONS*

ANTONIO DOBADO, FELIPE J. LLANES-ESTRADA and
JUAN M. TORRES-RINCON

Departamento de Física Teórica I, Universidad Complutense de Madrid
28040 Madrid, Spain

Probably the most enticing observation in theoretical physics during the last decade was the discovery of the great amount of consequences obtained from the AdS/CFT conjecture put forward by Maldacena. In this work we review how this correspondence can be used to address hydrodynamic properties such as the viscosity of some strongly interacting systems. We also employ the Boltzmann equation for those systems closer to low-energy QCD, and argue that this kind of transport coefficients can be related to phase transitions, in particular the QGP/hadronic phase transition studied in heavy ion collisions.

1. Holography and the Maldacena conjecture

One of the most successful tools developed in the last years in order to explore the strongly interacting regime of gauge theories is the idea of holography, or more properly the AdS/CFT (Anti de Sitter/Conformal Field Theory) correspondence introduced by Maldacena and refined by Witten more than one decade ago.[1] The correspondence states the equivalence of two theories defined in spaces of different dimension. In its original version the two theories were Type-IIB string theory defined on $AdS_5 \times S_5$ and $N = 4$ supersymmetric $SU(N)$ Yang-Mills (SYM) theory defined in the four dimensional boundary of this space. Although this correspondence has not been proven so far, there is a vast body of evidence supporting it and thus it stands as a very strong conjecture that has also been generalized to many other cases.

*This work has been partially supported by the DGICYT (Spain) under grants FPA 2008-00592, FIS2008-01323 plus 227431, HadronPhysics2 (EU) and PR34-1856-BSCH, UCM-BSCH GR58/08, 910309, PR34/07-15875. J. M. T. is an FPU scholarship recipient. A. D. thanks Julia Nyiri for her kind invitation to participate in this celebration of Gribov's work.

Let us try to understand the rationale of the conjecture in the following. The low energy excitations of a Type IIB string theory in the presence of a stack of N 3D-branes can be seen in two very different ways.

- The first description of the system reduces to a $N = 4$ SYM theory in four dimensions (which is a conformal quantum field theory) plus free gravity in ten dimensions.
- The second way to look at the low energy excitations of the system is by thinking of an observer in the asymptotic, flat Minkowski region. Then the only relevant modes are those belonging to the ten dimensional graviton multiplet. Near the branes the geometry becomes effectively $AdS_5 \times S_5$. Deep in this "throat" region an arbitrarily high-energy closed string (graviton) may appear to have an arbitrarily low energy as seen from an observer at the Minkowski infinity. Therefore we can conclude that the second description of the system is given by interacting closed strings in the mentioned $AdS_5 \times S_5$ background.

This observation supports the idea that it is reasonable to conjecture that four dimensional $N = 4$ $SU(N)$ SYM theory and Type IIB string theory on $AdS_5 \times S_5$ are two different (dual) descriptions of the same physical system. It can also be checked that the symmetries of both theories are exactly the same which gives additional support to the conjecture.

In principle, the parameters needed to define the $N = 4$ $SU(N)$ SYM theory (CFT in the following) are the gauge coupling g and N. It is useful to define also the so called 't Hooft parameter $\lambda = g^2 N$ specially when dealing with the large N limit. On the other hand, to define the string theory one needs to specify the string length l_s, the string coupling g_{st} and the common size of the AdS_5 and the S_5 spaces R. Then it is possible to relate the two sets of parameters by using the renowned "Maldacena dictionary"

$$g^2 = 4\pi g_{st} \tag{1}$$

and

$$\lambda = g^2 N = \frac{R^4}{l_s^4}. \tag{2}$$

This last equation is quite interesting. If one considers the strong coupling limit of the gauge theory (large t'Hooft coupling) we have also $l_s \ll R$, or in other words, we are in the region where string theory can be approximated by Supergravity (SUGRA) or even plain Einstein theory. Therefore it is possible to make reliable computations in a strongly coupled theory by using its weakly coupled gravity dual. This highly non trivial result is one of the reasons for the enormous attention paid to the Maldacena conjecture.

The holographic correspondence can also be extended to finite temperature $T > 0$. In order to do that one needs to introduce a black hole in the AdS space. Then it is possible to find the well known Bekenstein-Hawking formulae relating

- First the horizon area to the entropy S and the horizon area A viz. $S = A/4G$, G being the Newton-Cavendish constant,
- and second, the temperature with the horizon radius r_0 and the whole space size,

$$T = r_0/\pi R^2. \tag{3}$$

In spite of the big theoretical appeal of the AdS/CFT ideas, it is fair to ask what is their real relevance for well established quantum field theories such as QCD. So far, no gravity dual has been found for this theory. However it is possible that the $N = 4$ SYM theory could still bring some light about the QCD plasma studied at heavy ion colliders like RHIC. As we will see in the next sections, this is in particular the case of some transport coefficients like the shear viscosity.

2. η/s from the AdS/CFT correspondence

One interesting (and little stressed) feature of hydrodynamics beyond its classical applications is that it can be used to describe the low energy (long wave) behavior of certain quantum field theories. The standard way to proceed is by expanding the energy-momentum tensor in terms of the number of space-time derivatives. At lowest order one gets the well known ideal fluid equations.[2] The introduction of dissipative processes requires going to the next order where transport or diffusion coefficients such as the shear viscosity η, the bulk viscosity ζ and the heat conductivity κ appear, together with other possible diffusion coefficients D_i related with different conserved quantities such as flavors. The linearized equations of motion yields the dispersion relations corresponding for example to the two transverse modes

$$\omega(k) = -i \frac{\eta}{\epsilon + P} k^2, \tag{4}$$

(where ϵ is the energy density and P the pressure) or to the longitudinal sound mode:

$$\omega(k) = c_s k - \frac{i}{2}\left(\frac{4}{3}\eta + \zeta\right)\frac{k^2}{\epsilon + P}, \tag{5}$$

where $c_s = \sqrt{dP/d\epsilon}$ is the speed of sound. By using the thermodynamic relation $\epsilon + P = Ts$ valid for vanishing chemical potential, with s being the entropy density, Eq. (4) shows that η/s characterizes the intrinsic ability of the system to relax towards equilibrium. Let us recall also that $\zeta = 0$ for scale-invariant theories.

The traditional way to compute transport coefficients is kinetic theory. The Boltzmann equation or its quantum version, the Uehling-Uhlenbenck equation, written in terms of the elastic cross sections can be solved by the Chapman-Enskog method to first order in the perturbation out of equilibrium[3] and from the different perturbations one can find the different transport coefficients. However this method applies typically only for weakly-interacting, dilute enough systems.

A more modern way for computing transport coefficients is by using the Kubo formulae. By analyzing linear response theory and coupling the system to gravity by using an appropriate lightly curved space-time background it is possible to find:

$$\eta = \lim_{\omega \to 0}\left(\frac{1}{2\omega}\int dt d\bar{x}\; e^{i\omega t}\langle [T_{xy}(t,\bar{x}), T_{xy}(0,\bar{0})]\rangle\right) . \tag{6}$$

One of the nice things about the Maldacena conjecture is that it is supposed to provide a complete description of the CFT even in the strong interaction regime. This makes possible to study the theory, not only at finite temperature, but also in the hydrodynamic regime. In particular the Kubo equation can be used to compute η/s in the context of the AdS/CFT correspondence. To do that one starts from a CFT with gravity dual. For example for $N = 4, SU(N)$ SYM one can consider the metric:

$$ds^2 = \frac{r^2}{R^2}\left[-\left(1 - \frac{r_0^4}{r^4}\right)dt^2 + dx^2 + dy^2 + dz^2\right] + \frac{R^2}{r^2(1 - r_0^4/r^4)}dr^2 . \tag{7}$$

The dual theory is a CFT at temperature T given by Eq. (3). Then it is possible to consider[4] a graviton polarized in the x-y direction propagating perpendiculary to the brane. The absorption cross section of the graviton by the brane measures, in the dual CFT, the imaginary part of the retarded Green's function of the operator coupled to the metric i.e. the energy-momentum tensor. Then it is possible to find:

$$\eta = \frac{\sigma(0)}{16\pi G}, \tag{8}$$

where $\sigma(0)$ is the graviton absorption cross section at zero energy. This cross section can be computed classically by using linerized Einstein equations and it turns out to equals the horizon area. Finally one arrives to the remarkable result:[5]

$$\frac{\eta}{s} = \frac{1}{4\pi}. \tag{9}$$

Curiously enough this result is quite independent on the particular form of the metric and it is the same for Dp, M2 and M5 branes the reason being the universality of the graviton absorption cross section.

The above result for η/s, together with the absence of an empirical counterexample in spite of the many fluids known, was the inspiration for Kovtun, Son and Starinets (KSS) to set the conjecture that, for a very wide class of systems, including those that can be described by a sensible (i.e. ultraviolet complete) quantum field theory, the above ratio has the lower bound $1/4\pi$. There are many theoretical arguments which favour this bound, both theoretical and experimental, including the plasma produced in heavy ion collisions which seems to be near the saturation of the bound. However some controversy has been raised recently about the universal applicability of the bound.[6]

If correct, one of the most obvious consequences of the KSS conjecture is the absence of perfect fluids in Nature. (In fact this could be welcome since perfect fluids are known to give rise to problems such as the d'Alembert paradox). More recently Bekenstein and collaborators have pointed out that the accretion of an ideal fluid onto a black hole could violate the Generalized Second Law of Thermodynamics,[7] suggesting a possible connection between this law and the KSS bound.

3. The RHIC case

During the last years the Relativistic Heavy Ion Collider (RHIC) has been producing a large number of Au+Au collisions (A= 197), at a center of mass energy per nucleon of $E = 200$ GeV with a luminosity $\mathcal{L} = 2 \times 10^{26}$ cm^{-2} s^{-1}. From its four experiments STAR, PHENIX, BRAHMS and PHOBOS it has been possible to obtain a lot of experimental information with important phenomenological consequences.

First of all thermochemical models describe well the different particle yields fitting $T = 177$ MeV and the baryon chemical potential $\mu_B = 29$ MeV for $E_{CM} = 200$ GeV per nucleon. From the observed transverse and rapidity distributions, the Bjorken model predicts an energy density at time $t_0 = 1$ fm of 4 GeV fm^{-3}, which strongly suggest that the produced matter may be well above the threshold for Quark Gluon Plasma (QGP) formation. In addition, hydrodynamical models with very low viscosity reproduces the measurements of radial and elliptic flow up to transverse momenta of 1.5 GeV. The collective flow is probably generated early in the collision before hadronization. The QGP seems to be more strongly interacting than expected on the basis of perturbative QCD and asymptotic freedom (hence the low viscosity) and some preliminary estimations of η/s based on elliptic flow[8,9] and transverse momentum correlations[10] seem to be compatible with a value close to 0.08 (the KSS bound). This would make hadronic matter in this regime the most perfect fluid known.

4. η/s and the phase transition

Recently Csernai, Kapusta and McLerran[11] made the observation that, in all systems whose $\frac{\eta}{s}(T)$ plot has been examined, the minimum of η/s and the liquid-gas phase transition seem to happen at the same temperature.

Since no general demonstration of this coincidence is known either, we set out to clarify it in a controlled model setup. We chose the large-N Linear Sigma Model (LσM)[12] that is a theory sufficiently ressembling the pion gas that our hadron-physics computer codes to solve Boltzmann's equations can be directly used. At the same time one knows when and how the phase transition occurs, unlike the case of chiral perturbation theory where one is bound to the low-energy phase.

The model Lagrangian is

$$\mathcal{L}[\Phi, \partial_\mu \Phi] = \frac{1}{2}\, \partial_\mu \Phi^T \partial^\mu \Phi + \overline{\mu}^2 (\Phi^T \Phi) - \lambda (\Phi^T \Phi)^2 \tag{10}$$

where $\overline{\mu}^2$ is positive (opposite in sign to a scalar field mass term) and $\lambda > 0$. With this choice of parameters the LσM presents spontaneous symmetry breaking from $SO(N+1)$ to $SO(N)$. The field Φ acquires a vacuum expectation value (VEV) where the field configuration of minimum energy verifies (at tree level)

$$\Phi^T \Phi = \frac{\overline{\mu}^2}{2\lambda} = f_\pi^2 = N F^2. \tag{11}$$

Denoting π^a $(a = 1, \ldots, N)$ to the N first components of Φ and σ the $N + 1$ component we can choose the VEV in the direction of the latter. Thus we have $\langle \pi^a \rangle = 0$ but $\langle \sigma \rangle = f_\pi$. The pions are the N massless Nambu-Goldstone bosons; on the other hand, the field σ acquires a mass equal to $m_\sigma^2 = 8\lambda N F^2$. Taking the limit $m_\sigma^2 \to \infty$ one can express σ in terms of the pions as $\sigma = \sqrt{f_\pi^2 - \pi^a \pi^a}$. This is the non-linear sigma model in which one can eliminate explicitly the σ degree of freedom.

Our first finding is that η/s presents a non-analyticity as a function of T at $T = T_c$ where a very different qualitative behaviour of η/s begins. This is due to the dependence of η/s on the thermal pion mass $m_\pi(T)$ and the σ condensate, both non-analytical at T_c. Below T_c, the thermal pion mass is identically zero (the Goldstone theorem protects the pion masslessness from radiative corrections). In this phase the symmetry is still broken and the condensate $\langle \sigma \rangle$ is non-zero. Above T_c the symmetry is restored and $\langle \sigma \rangle$ vanishes. The "classical" pion mass m_π is zero as well. However, quantum thermal corrections force a temperature-dependent thermal mass, $m_\pi(T)$,

$$\langle \sigma(T) \rangle = \begin{cases} f_\pi \left(1 - \dfrac{T^2}{T_c^2}\right)^{1/2} & T < T_c \\ 0 & T \geq T_c \end{cases} , \tag{12}$$

$$m_\pi^2(T) = \begin{cases} 0 & T < T_c \\ \dfrac{N}{3}\lambda_R (T^2 - T_c^2) & T \geq T_c \end{cases} . \tag{13}$$

This mass is continuous at T_c but non-analytic. The behaviour of the condensate $\langle \sigma \rangle$ itself is also continuous but with a discontinuous derivative at T_c. These two quantities influence both η and s and their non-analiticity is inherited by the KSS ratio.

By varying f_π or N, the minimum of η/s moves according to the N and f_π dependences of T_c, providing evidence for our claim that T_c and the minimum of η/s are related. In Fig. 1 we zoom in the η/s plot near the critical temperature. As can be seen, the minimum is not reached at the critical temperature, but right before T_c. This shows that the minimum of η/s is controlled by the rapid variation of the order parameter. To understand this result one needs to keep in mind the diffusive nature of the transport in a gas. With increasing temperature, the gas particles carry transverse momentum between different parts of the gas more efficiently, and thus increase the shear viscosity. However their interactions hamper transport. As $\langle \sigma \rangle \propto F$ decreases rapidly, the pion elastic cross section increases. Since

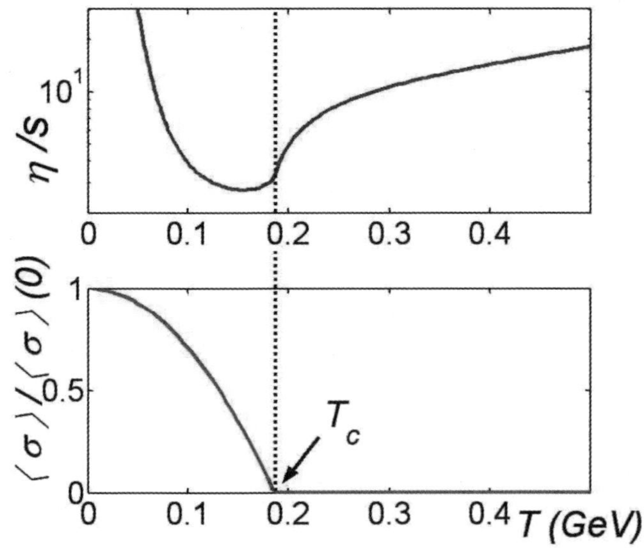

Fig. 1. The minimum of η/s occurs just before the critical temperature for the phase transition in the Linear Sigma Model. This is where the condensate varies rapidly approaching zero. The phase transition is marked by a non–analyticity of the ratio.

$\eta \propto 1/\sigma_{\pi\pi}$ in kinetic theory, the viscosity must drop. Eventually growth is regained as the temperature increases.

5. Outlook

One could argue that our mean field calculation of the vacuum structure could be improved by employing the 2PI formalism to compute the effective action. However we think the result of these, more elaborate computations, would be the same around the physical minimum of the effective potential. The phase transition is marked by a non-analiticity of η/s, as for many other thermodynamic quantities, but the minimum does not need to exactly coincide.

A similar result has been obtained for the bulk viscosity[13] in the $1+1$-dimensional Gross-Neveu model, where the maximum of the trace anomaly does not coincide with the maximum of the bulk viscosity, that has just a sharp change of derivative there.

We are looking forward to further computations of transport coefficients in various systems, and to what insight the AdS/CFT conjecture can bring us.

References

1. J. M. Maldacena, *Adv. Theor. Math. Phys.* **2**, 231 (1998) [*Int. J. Theor. Phys.* **38**, 1113 (1999)];
   E. Witten, *Adv. Theor. Math. Phys.* **2**, 253 (1998);
   E. Witten, *Adv. Theor. Math. Phys.* **2**, 505 (1998).
2. L.D. Landau and E. M. Lifshitz, *Fluid Mechanics* Oxford: Pergamon Press (1987).
3. A. Dobado and F. J. Llanes-Estrada, *Phys. Rev.* **D69**, 116004 (2004).
4. I. R. Klebanov, *Nucl. Phys.* **B496**, 231 (1997).
5. G. Policastro, D. T. Son and A. O. Starinets, *Phys. Rev. Lett.* **87**, 081601 (2001);
   P. Kovtun, D. T. Son and A. O. Starinets, *JHEP* **0310**, 064 (2003);
   P. Kovtun, D. T. Son and A. O. Starinets, *Phys. Rev. Lett.* **94**, 111601 (2005).
6. A. Dobado, F. J. Llanes-Estrada and J. M. Torres-Rincon, *AIP Conf. Proc.* **1031**, 221 (2008).
7. I. Fouxon, G. Betschart and J. D. Bekenstein, *Phys. Rev.* **D77**, 024016 (2008).
8. E. Shuryak, *Prog. Part. Nucl. Phys.* **53**, 273 (2004).
9. D. Teaney, *Phys. Rev.* **C68**, 034913 (2003).
10. S. Gavin and M. Abdel-Aziz, *Phys. Rev. Lett.* **97**, 162302 (2006).
11. L. P. Csernai, J. I. Kapusta and L. D. McLerran, *Phys. Rev. Lett.* **97**, 152303 (2006).
12. A. Dobado, F. J. Llanes-Estrada and J. M. Torres-Rincon, *Phys. Rev.* **D80**, 114015 (2009).
13. D. Fernandez-Fraile, arXiv:hep-ph/1009.2741 (2009).

SCREENING IN STRONGLY COUPLED PLASMAS: UNIVERSAL PROPERTIES FROM STRINGS IN CURVED SPACE

CARLO EWERZ and KONRAD SCHADE

Institut für Theoretische Physik, Universität Heidelberg
Philosophenweg 16, D-69120 Heidelberg, Germany
and
ExtreMe Matter Institute EMMI
GSI Helmholtzzentrum für Schwerionenforschung
Planckstraße 1, D-64291 Darmstadt, Germany

We use the gauge/gravity correspondence to study the screening of a heavy quark-antiquark pair in various strongly coupled plasmas. Besides $\mathcal{N} = 4$ super Yang-Mills theory and the corresponding AdS_5 space we also study theories obtained as deformations of AdS_5, among them in particular a class of deformations solving supergravity equations of motion. We consider the dependence of the screening distance on the velocity and the orientation of the pair in the plasma. The value of the screening distance in $\mathcal{N} = 4$ SYM is found to be a minimum in the class of theories under consideration for all kinematic parameters.

1. Introduction

The finding that the quark-gluon plasma created in heavy-ion collisions at RHIC (and soon to be created at the LHC) appears to be strongly coupled calls for a better theoretical understanding of strongly coupled gauge theories at finite temperature. Until recently, lattice QCD constituted the only viable approach to this problem, albeit only for static observables. With the discovery of the gauge/gravity (or AdS/CFT) correspondence[1–3] a new path has opened to attack the problem of strongly coupled gauge theories. It even allows one to study dynamical processes in the plasma. In our study we make use of the correspondence to study the screening of a heavy quark-antiquark pair in a strongly coupled plasma. In particular, we will consider the maximal distance for which the quark and antiquark form a bound state, also called the screening distance, and its dependence on the velocity and the orientation of the pair with respect to the plasma.

A good understanding of this observable might be helpful for diagnosing the properties of the quark-gluon plasma with the help of heavy probes like charmonium or bottomonium.

In its original form the gauge/gravity correspondence is a holographic duality between supergravity on a five-dimensional AdS_5 space and four-dimensional $\mathcal{N} = 4$ super Yang-Mills (SYM) theory with gauge group $SU(N_c)$ in the large-N_c limit. The most interesting property of the duality is that the weak-coupling (small curvature) limit on the gravity side corresponds to the strong-coupling limit on the gauge theory side. This makes it possible to solve hard problems in a gauge theory by doing simple calculations on the gravity side. Finite temperature can be accommodated by introducing a black hole in the bulk of the AdS space.

Although the AdS/CFT duality was a major leap from a theoretical point of view its use for the phenomenology of QCD is far from obvious. Clearly, $\mathcal{N} = 4$ SYM is very different from QCD: it is maximally supersymmetric, features only particles in the adjoint representation of the gauge group, does not have a running coupling (i. e. is conformal), and exhibits neither confinement nor chiral symmetry breaking. It hence requires some optimism to apply the duality to QCD. The situation appears somewhat more promising when one considers finite temperature. Here, QCD is no longer confining and above $2T_c$ even appears close to conformal. At the same time, finite temperature breaks the exact conformal invariance of $\mathcal{N} = 4$ SYM. In addition, it is feasible that some properties of a gauge theory plasma are to some extent independent of the microscopic degrees of freedom. This is supported by the apparent validity of the hydrodynamical description of various observables in the quark-gluon plasma. Still, it remains unknown whether a gravitational theory dual to QCD exists.

In an attempt to come closer to a potential dual of QCD one can introduce deformations of the AdS_5 space which break the conformal invariance of the dual theory. Obviously one can think of a vast variety of deformations of this kind, and it will be very difficult to find a particular deformation that reproduces all properties of QCD. Therefore it seems more interesting to ask for universal properties of large classes of such deformations. In particular, a given observable can be *robust* under deformations and change only very little. In this case one might hope that the value of an observable in $\mathcal{N} = 4$ SYM might already be a good approximation to the value in actual QCD. An observable can also be *universal* in the sense that it does not change at all under deformations, or changes consistently in one direction. A famous example for a universal observable is the ratio of viscosity to entropy density η/s which acquires the value $1/(4\pi)$ in all theories with gravity duals.[4] This value has even been conjectured to be a lower bound for all possible theories.[5]

In the following we will show that the screening distance of a heavy quark-antiquark pair which moves in a hot plasma is in fact a universal observable in a large class of theories. We will restrict ourselves to the main results, a more detailed account of our study will be published elsewhere.[6]

2. AdS/CFT at Finite Temperature and its Deformations

The gravity dual of $\mathcal{N} = 4$ SYM at finite temperature is an $AdS_5 \times S^5$ space with a Schwarzschild black hole. The S^5 factor will not be relevant for our considerations and will be suppressed from the beginning. The 5-dimensional AdS black hole metric is given by

$$ds^2 = G_{\mu\nu}dx^\mu dx^\nu = -f(r)dt^2 + \frac{r^2}{R^2}(dx_1^2 + dx_2^2 + dx_3^2) + \frac{1}{f(r)}dr^2 \tag{1}$$

with the curvature radius R of AdS_5, and with

$$f(r) = \frac{r^2}{R^2}\left(1 - \frac{r_0^4}{r^4}\right) . \tag{2}$$

For each fixed value of the fifth coordinate r the metric describes a Minkowski space in the remaining four coordinates. The coordinate r has an interpretation as an energy variable in the boundary theory. The holographically dual $\mathcal{N} = 4$ SYM 'lives' at $r = \infty$, that is at the boundary of AdS_5. The location r_0 of the black hole horizon is related to the temperature T of the boundary theory via $T = r_0/(\pi R^2)$. The latter coincides with the Hawking temperature of the black hole. A large curvature radius R and thus the applicability of classical gravity on the AdS side requires large 't Hooft coupling $\lambda = g_{\rm YM}^2 N_c$ on the gauge theory side.

A simple class of deformations of AdS_5 which break conformal invariance in the dual theory is the KTY model[7] in which the original metric is multiplied by an exponential factor depending on r such that

$$ds^2 = \exp\left(\frac{29}{20}c\frac{R^4}{r^2}\right)\left[-f(r)dt^2 + \frac{r^2}{R^2}(dx_1^2 + dx_2^2 + dx_3^2) + \frac{1}{f(r)}dr^2\right] . \tag{3}$$

The function f is as in (2), and also here the temperature of the dual field theory is given by $T = r_0/(\pi R^2)$. Several properties of QCD thermodynamics are well reproduced with $T_c \simeq 170\,\text{MeV}$ when the dimensionful parameter c is chosen as $c \simeq 0.127\,\text{GeV}^2$. In order to see how an observable depends on the deformation it is interesting to study how it changes with the actual conformality-breaking parameter c/T^2. (A 'realistic' range for that parameter is $0 \leq c/T^2 \leq 4$.) However, the KTY model suffers from the problem that it does not solve supergravity equations of motion. Thus, its thermodynamic consistency is questionable.

More recently, also the construction of thermodynamically consistent deformations has been explored.[8] It is found that the dilaton potential $V(\Phi)$ in the 5-dimensional gravitational action

$$S_5 = \frac{1}{16\pi G_5} \int d^5x \sqrt{-g} \left(R - \frac{1}{2}(\partial_\mu \Phi)^2 - V(\Phi) \right) \tag{4}$$

can be chosen such that one obtains a 2-parameter model[9] of the form

$$ds^2 = e^{2A(r)}(-h(r)dt^2 + d\vec{x}^2) + \frac{e^{2B(r)}}{h(r)} dr^2 \tag{5}$$

with two parameters c/T^2 and $\alpha = c/\phi$, where $\Phi = \sqrt{\frac{3}{2}}\phi \frac{R^2}{r^2}$. One can use a residual gauge freedom to identify $r = \Phi$. The temperature of the dual theory is given by

$$T = \frac{e^{A(\Phi_h) - B(\Phi_h)} |h'(\Phi_h)|}{4\pi}, \tag{6}$$

where Φ_h is the location of the horizon defined by the zero of $h(r)$. Defining

$$\begin{aligned} A(\Phi) &= \frac{1}{2} \ln \left(\sqrt{\frac{3}{2}} c \frac{R^2}{\alpha} \right) - \frac{1}{2} \ln \Phi - \frac{\alpha}{\sqrt{6}} \Phi \\ B(\Phi) &= \ln \left(\frac{R}{2} \right) + \frac{1 + 2\alpha^2}{2\alpha^2} \ln \left(1 + \alpha \sqrt{\frac{2}{3}} \Phi \right) - \ln \Phi - \frac{1}{\alpha\sqrt{6}} \Phi \end{aligned} \tag{7}$$

one can calculate h from supergravity equations of motion. This model becomes similar to the KTY model for the choice $\alpha = \alpha_{\text{KTY}} = 20/49$, when the exponential factor of (3) is reproduced in the first term of (5).

From the two-parameter model one can obtain a further model if one treats Φ not as the dilaton but as an additional scalar field, assuming a trivial dilaton instead. In that case the Einstein frame and the string frame for the calculation of a string on this background coincide. In our figures below we denote this construction as 'Einstein frame' model, while the model described above (with Φ being the dilaton) is denoted as 'string frame' model.

Clearly, the deformed AdS metrics described here define dual four-dimensional theories at strong coupling. However, the Lagrangians of these holographically dual theories are not known, and it is not even clear that they correspond to gauge theories. Nevertheless, they are perfectly fine if one is interested in the effects of conformality-breaking on various observables.

3. Screening Distance and Free Energy of a $Q\bar{Q}$ Pair

The free energy $E(L)$ of a heavy quark-antiquark pair separated by a distance L in a gauge theory plasma is obtained from the temporal Wegner-Wilson loop

$$W(\mathcal{C}) = \mathrm{Tr}\,\mathcal{P}\,\exp\left[i \oint_{\mathcal{C}} dx_\mu\, A^\mu(x)\right] \tag{8}$$

via its expectation value $\langle W(\mathcal{C})\rangle = \exp\left[-i\mathcal{T}\,E(L)\right]$, where $\mathcal{T}$ is the (large) temporal extension of the closed curve $\mathcal{C}$. On the gravity side, one has

$$\langle W(\mathcal{C})\rangle \propto \exp\left[-i(S - S_0)\right], \tag{9}$$

where S is the Nambu-Goto action for an open string hanging down into the bulk of the AdS-type space. Its ends are attached to the quark and antiquark at the boundary $r = \infty$. For a plasma moving in x_3-direction this situation is illustrated in Fig. 1. S_0 is twice the Nambu-Goto action for an open string hanging down from a single quark.

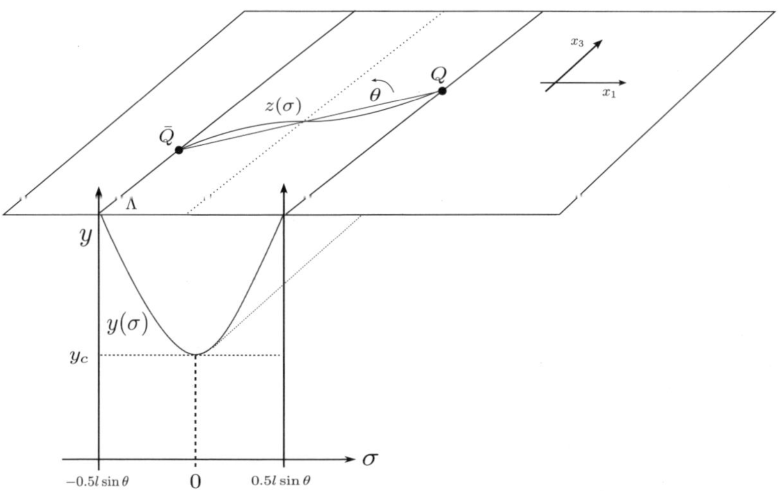

Fig. 1. String configuration for a $Q\bar{Q}$ pair in a moving plasma. The heavy quarks are located at $r = \Lambda$, and the limit $\Lambda \to \infty$ is implied.

The calculation of $E(L)$ and of related observables had been performed for $\mathcal{N} = 4$ SYM[10] and for the KTY model.[11] In our work we have considered the full dependence on the velocity and on the orientation of the pair in the plasma and have extended it to the two-parameter models given above. In the following we outline the calculation for the simple case of $\mathcal{N} = 4$ SYM. The calculation for the other models is done along the same lines but leads to formulae less suited for a short exposition.

The moving plasma is accommodated by boosting the metric with velocity $v = \tanh\eta$ in x_3-direction. The $Q\bar{Q}$ pair can be rotated w.r.t. the x_3-direction by an angle θ. We parametrize the string world sheet as indicated in Fig. 1 and extremize the resulting Nambu-Goto action

$$S = \frac{\mathcal{T}}{2\pi\alpha'} \int_{-\frac{L}{2}}^{\frac{L}{2}} d\sigma \sqrt{A\left(\frac{(\partial_\sigma r)^2}{f} + \frac{r^2}{R^2}\right)} \tag{10}$$

where

$$A = \frac{r^2}{R^2}\left[1 - \frac{r_0^4 \cosh^2\eta}{r^4}\right]. \tag{11}$$

The solutions can be parametrized by the conserved Hamiltonian

$$\mathcal{H} \equiv \mathcal{L} - y'\frac{\partial \mathcal{L}}{\partial y'} = \frac{y^4 - \cosh^2\eta}{\mathcal{L}} = q\,, \tag{12}$$

and one can solve for the coordinate function $r = r_0 y$ of the string,

$$y' = \frac{1}{q}\sqrt{(y^4-1)(y^4-y_c^4)} \quad \text{with} \quad y_c^4 \equiv \cosh^2\eta + q^2\,. \tag{13}$$

Using the boundary conditions one finally obtains the quark-antiquark distance as a function of q,

$$\frac{L\pi T}{2} = \int_0^{\frac{L\pi T}{2}} d\sigma = q\int_{y_c}^{\Lambda} dy \frac{1}{\sqrt{(y^4-1)(y^4-y_c^4)}}\,. \tag{14}$$

In AdS_5 and the deformations discussed above one finds that $L(q)$ has a maximum for all values of the rapidity η and for all orientation angles. For all L up to this $L_{\max}$ there are two solutions with different q. For $L > L_{\max}$, on the other hand, no string configuration connecting the quark and the antiquark exists. We call this maximally possible distance between the quark and the antiquark the screening distance. It depends on the rapidity η and the orientation angle θ with respect to the moving plasma, and on the parameters of the deformation of the AdS metric.

For the case of $\mathcal{N} = 4$ SYM the behavior of the string configurations up to $L_{\max}$ is shown in Fig. 2. For each distance L up to $L_{\max}$ there are two

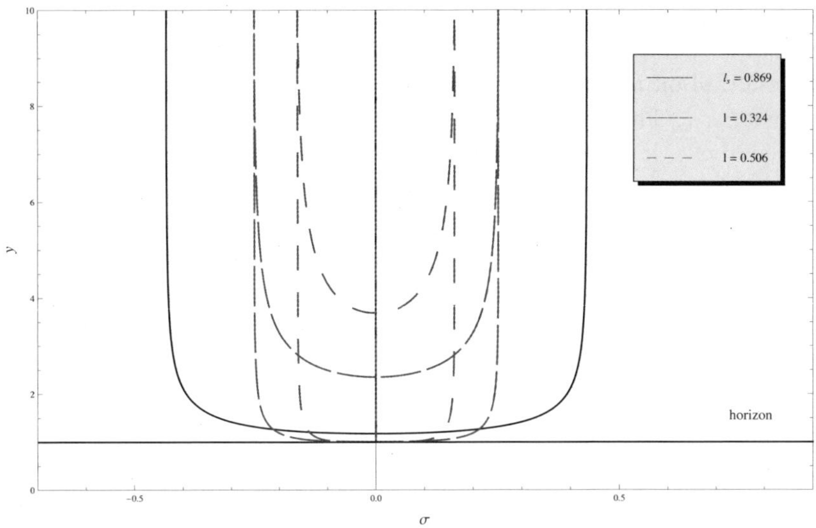

Fig. 2. String configurations for different quark-antiquark distances. The distance is given as the dimensionless $l = L\pi T$ and is parametrized by σ. The strings end on the quark and antiquark located at $y \to \infty$.

solutions. The configuration that stays higher up in the bulk has a smaller energy $E(L)$ than the configuration coming closer to the horizon. The latter can therefore be identified as the unstable solution.* With increasing distance L the two configurations move towards each other until for the screening distance $L_{\max}$ there is only one solution. Note that none of the solutions touches the horizon.

The computation of the free energy $E(L)$ of the quark-antiquark pair from (9) requires the knowledge of the action S_0 of a single string hanging down from a moving quark into the bulk of AdS_5. This can be obtained from an AdS/CFT calculation of the drag force acting on the quark.[12,13] This calculation had been extended to the KTY model.[14] We have computed the drag force also for the two-parameter models solving supergravity equations of motion. Knowing S_0 we obtain the free energy $E(L)$ also for the latter models and find its behavior to be qualitatively the same as in $\mathcal{N} = 4$ SYM and in the KTY model.

*This configuration might also be metastable. A more precise statement would require an understanding of the dynamical mechanism of the transition between the two solutions.

The screening distance should not be confused with the inverse Debye mass which is often called screening length. The latter describes the exponential fall-off of the free energy at distances *larger* than our screening distance $L_{\rm max}$. The Debye mass cannot be obtained from the calculations outlined above. It has been argued[15] that it can be related to the exchange of the lowest supergravity modes between two open strings hanging into the bulk at a separation $L > L_{\rm max}$.

We have computed the screening distance for all deformations of the AdS space presented in section 2. We find that in all cases the screening distance has a very weak dependence of up to about 10 % on the orientation angle of the $Q\bar{Q}$ pair with respect to the plasma wind. The dependence on the velocity is dominated by a factor $(\sqrt{\cosh\eta})^{-1}$. The screening distance approaches this dominant behavior at large velocities in all models.

Finally, we have studied how the screening distance changes when deformations of the AdS space are introduced. This is illustrated in Fig. 3 where we show the dimensionless quantity $\pi T L_{\rm max}\sqrt{\cosh\eta}$ for the case $\theta = 0$ ($Q\bar{Q}$ oriented parallel to the plasma wind). The last factor in this product compensates the dominant behavior of $L_{\rm max}$ with η just discussed so that the differences between the models become visible more clearly. In the KTY

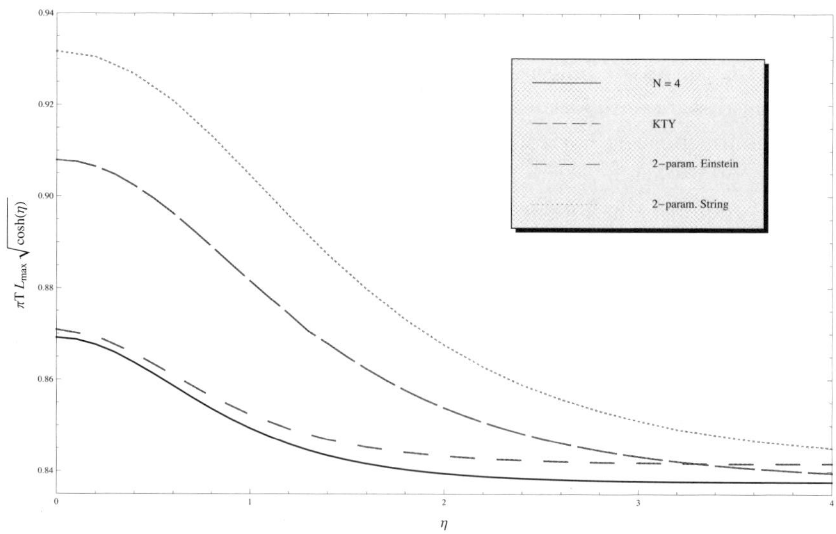

Fig. 3. Dimensionless screening distance $\pi T L_{\rm max}\sqrt{\cosh\eta}$ of the $Q\bar{Q}$ pair in the plasma as a function of rapidity η.

model we have chosen $c/T^2 = 1$ for this figure, while in the two-parameter deformations $c/T^2 = 1$ and $\alpha = \alpha_{\text{KTY}}$.

We first find that the screening distance is a robust observable. Under 'realistic' deformations of the AdS space, i. e. deformations with thermodynamic observables not drastically different from those of QCD, its value changes only by up to 30 %.

The most remarkable observation is that the screening distance has a universal behavior under the class of deformations studied here. For any given velocity and orientation angle of the $Q\bar{Q}$ pair with respect to the plasma and for all deformations considered here the value of L_{max} is larger than in $\mathcal{N} = 4$ SYM. In other words: the screening distance in $\mathcal{N} = 4$ SYM is minimal in the class of theories under consideration for all kinematic parameters. It suggests itself to speculate that this might also apply to (all?) other theories obtained holographically as deformations of AdS_5.

4. Screening of Heavy Baryons

In the framework of the gauge/gravity correspondence heavy baryons can be constructed out of N_c heavy quarks situated at the boundary of AdS_5. An open string is attached to each of the quarks and ends at a D5-brane that fills the 5 dimensions of S^5 and is located at a point r_e in the bulk of AdS_5. For such a baryon configuration a similar analysis of screening in a moving plasma can be performed.[16] We have extended this study, previously done for $\mathcal{N} = 4$ SYM, to the deformation models described in section 2. In analogy to the screening distance of a $Q\bar{Q}$ pair one can define the maximally possible radius of the baryon configuration as a screening distance of the baryon. Also here we find that the screening distance of the baryon in $\mathcal{N} = 4$ SYM is minimal in the class of theories under consideration.

5. Summary

We have calculated the screening distance of a heavy quark-antiquark pair moving in different strongly coupled plasmas which are obtained holographically as duals of deformations of AdS_5. Our study includes deformations solving supergravity equations of motion. We observe that the screening distance is a robust observable and changes only little when deformations are introduced. We find the screening distance in $\mathcal{N} = 4$ SYM to be a minimum among the theories under consideration for all velocities and orientation angles of the pair in the plasma. A similar behavior is found for heavy baryons moving in strongly coupled plasmas. We conjecture that the screening distance found in $\mathcal{N} = 4$ SYM constitutes a lower bound for an even wider range of theories. It would obviously be interesting to show this analytically although that appears to be a challenging problem.

Acknowledgments

C. E. would like to thank the organizers and especially Julia Nyíri for the kind invitation to give a talk at this pleasant meeting in honor of Volodya Gribov from whom he had learned so much.

K. S. acknowledges support by the International Max Planck Research School for Precision Tests of Fundamental Symmetries. This work was supported by the Alliance Program of the Helmholtz Association (HA216/EMMI).

References

1. J. M. Maldacena, *Adv. Theor. Math. Phys.* **2**, (1998) 231.
2. S. S. Gubser, I. R. Klebanov and A. M. Polyakov, *Phys. Lett.* **B428**, 105 (1998).
3. E. Witten, *Adv. Theor. Math. Phys.* **2**, 253 (1998).
4. G. Policastro, D. T. Son and A. O. Starinets, *Phys. Rev. Lett.* **87**, 081601 (2001).
5. P. Kovtun, D. T. Son and A. O. Starinets, *Phys. Rev. Lett.* **94**, 111601 (2005).
6. C. Ewerz and K. Schade, in preparation
7. K. Kajantie, T. Tahkokallio and J. T. Yee, *JHEP* **0701**, 019 (2007).
8. S. S. Gubser and A. Nellore, *Phys. Rev.* **D78**, 086007 (2008).
9. O. DeWolfe and C. Rosen, arXiv:0903.1458 [hep-th].
10. H. Liu, K. Rajagopal and U. A. Wiedemann, *Phys. Rev. Lett.* **98**, 182301 (2007).
11. H. Liu, K. Rajagopal and Y. Shi, *JHEP* **0808**, 048 (2008).
12. C. P. Herzog, A. Karch, P. Kovtun, C. Kozcaz and L. G. Yaffe, *JHEP* **0607**, 013 (2006).
13. S. S. Gubser, *Phys. Rev.* **D74**, 126005 (2006).
14. E. Nakano, S. Teraguchi and W. Y. Wen, *Phys. Rev.* **D75**, 085016 (2007).
15. D. Bak, A. Karch and L. G. Yaffe, *JHEP* **0708**, 049 (2007).
16. C. Athanasiou, H. Liu and K. Rajagopal, *JHEP* **0805**, 083 (2008).

SCREENING IN PLASMA WITH CHARGED BOSE CONDENSATE

A.D. DOLGOV

Department of Physics, University of Ferrara,
Ferrara, FE44100, Italy
ITEP, Moscow, 117218, Russia
dolgov@fe.infn.it

It is shown that the screening of impurities in plasma with Bose-Einstein condensate of electrically charged bosons is drastically different from the usual Debye screening. The polarization operator of photons in plasma becomes infrared singular and the screened potential drops down as a power of distance and even has an oscillating behavior, similar to the Friedel oscillations in plasma with degenerate fermions. The properties of the cosmological plasma with condensed W-bosons are also studied and it is shown that W-bosons condense in ferromagnetic state. The spontaneously generated magnetic fields in such ferromagnetic state may seed the large scale magnetic fields observed in the present-day universe.

Keywords: Plasma physics; screening; cosmological magnetic fields.

1. Introduction

The screening of impurities in plasma is usually described by the well known text-book Debye formula.[1,2] Instead of the long range Coulomb potential a test charge creates an exponentially decreasing Yukawa potential:

$$U(r) = \frac{Q}{4\pi r} \to \frac{Q\exp(-m_D r)}{4\pi r}, \tag{1}$$

because the time-time component of the photon propagator acquires the "mass" term. Instead of the vacuum k^2-term it becomes:

$$k^2 \to k^2 + \Pi_{00}(k) = k^2 + m_D^2, \tag{2}$$

where e.g. for relativistic fermions:[3]

$$m_D^2 = e^2\left(T^2/3 + \mu^2/\pi^2\right). \tag{3}$$

These classical results are true if the fermions in plasma are not strongly degenerate and if the charged bosons are not condensed. The modification of the Debye screening in the case of degenerate fermions was studied half a century ago,[4] but the impact of Bose-Einstein condensate (BEC) on the screening of impurities was considered only very recently[5,6] . Surprisingly in the presence of BEC the screened potential drastically changes and becomes even an oscillating function of distance.

The first part of this talk about the effects of BEC of a charged scalar field is based on the works done in collaboration with A. Lepidi and G. Piccinelli.[6] The second part about condensation of charged vector bosons is based on paper[7] of the same group.

Let us consider electrically neutral plasma with large electric charge density of bosons compensated by charged fermions. Bosons condense when their chemical potential reaches the maximum allowed value:

$$\mu_B = m_B\,. \tag{4}$$

It can be easily seen that the equilibrium distribution of bosons, if and only if their chemical potential is equal to their mass, takes the form:

$$f_B^{(eq)} = C\delta^{(3)}(\mathbf{q}) + \frac{1}{\exp\left[(E - m_B)/T\right] - 1}\,, \tag{5}$$

where the constant C is the amplitude of the condensate. One can check that $f_B^{(eq)}$ annihilates the collision integral for an arbitrary C. Thus it is indeed the equilibrium distribution. It is worth noting that the equilibrium distributions are always functions of two parameters: temperature, T, and chemical potential, μ, when $\mu < m$, and temperature T, and the amplitude of the condensate, C, when $\mu = m$.

We calculate the time-time component of the photon polarization operator in a simple straightforward way perturbatively solving operator equation of motion for the electromagnetic field (Maxwell equations) and averaging them over medium. One can reach the goal without applying to more refined real or imaginary time methods (for a review of these methods see e.g. Ref. 2).

2. QED in medium

The Lagrangian of interacting electromagnetic field and charged scalar and fermion fields with masses m_B and m_F respectively and with opposite electric charges $\pm e$ has the form:

$$\mathcal{L} = -\frac{1}{4}F_{\mu\nu}F^{\mu\nu} - m_B^2|\phi|^2 + |(\partial_\mu + i\,eA_\mu)\phi|^2 + \bar{\psi}(i\partial\!\!\!/ - eA\!\!\!/ - m_F)\psi. \tag{6}$$

The corresponding equations of motion for the electromagnetic and charged scalar and spinor fields are:

$$(i\partial\!\!\!/ - m)\psi(x) = eA\!\!\!/\psi(x)\,, \tag{7}$$

$$(\partial_\mu\partial^\mu + m^2)\phi(x) = \mathcal{J}_\phi(x)\,, \tag{8}$$

$$\partial_\nu F^{\mu\nu}(x) = \mathcal{J}^\mu(x)\,, \tag{9}$$

where the currents $\mathcal{J}$ can be written as:

$$\mathcal{J}_\phi(x) = -i\,e\Big[\partial_\mu A^\mu(x) + 2A_\mu(x)\partial^\mu\Big]\phi(x) + e^2A^\mu(x)A_\mu(x)\phi(x)\,, \tag{10}$$

$$\mathcal{J}^\mu(x) = -i\,e\Big[(\phi^\dagger(x)\partial^\mu\phi(x)) - (\partial^\mu\phi^\dagger(x))\phi(x)\Big] \tag{11}$$

$$+\,2e^2A^\mu(x)|\phi(x)|^2 - e\bar{\psi}(x)\gamma^\mu\psi(x). \tag{12}$$

Here $F_{\mu\nu} = \partial_\mu A_\nu - \partial_\nu A_\mu$ and $\mathcal{J}^\mu$ (11) is the total electromagnetic current of bosons and fermions.

Operator equations (7) and (8) can be formally solved as:

$$\phi(x) = \phi_0(x) + \int d^4y\, G_B(x-y)\mathcal{J}_\phi(y)\,,$$

$$\psi(x) = \psi_0(x) + \int d^4y\, G_F(x-y)eA\!\!\!/(y)\psi(y)\,, \tag{13}$$

where G_B and G_F are the Green's functions of bosons and ferminos respectively and the zeroth order fields satisfy the free equations of motion:

$$(\partial_\mu\partial^\mu + m_B^2)\phi_0(x) = 0, \qquad (i\partial\!\!\!/ - m_F)\psi_0(x) = 0 \tag{14}$$

and are quantized in the usual way.

As a result we obtain the Maxwell equations with the lowest order corrections to the electromagnetic current:

$$\begin{aligned}
\partial_\nu F^{\mu\nu}(x) &= -i\,e\Big[(\phi_0^\dagger(x)\partial^\mu\phi_0(x)) - (\partial^\mu\phi_0^\dagger(x))\phi_0(x)\Big] - e\bar{\psi}_0(x)\gamma^\mu\psi_0(x)\\
&- ie\,\phi_0^\dagger(x)\partial^\mu\Big[\int d^4y\, G_B(x-y)\mathcal{J}_{\phi_0}(y)\Big] - ie\Big[\int d^4y\, G_B(x-y)\mathcal{J}_{\phi_0}(y)\Big]^\dagger\partial^\mu\phi_0(x)\\
&+ ie\,\partial^\mu\phi_0^\dagger(x)\Big[\int d^4y\, G_B(x-y)\mathcal{J}_{\phi_0}(y)\Big] + ie\,\partial^\mu\Big[\int d^4y\, G_B(x-y)\mathcal{J}_{\phi_0}(y)\Big]^\dagger\phi_0(x)\\
&- e\bar{\psi}_0(x)\gamma^\mu\int d^4y\, G_F(x-y)eA\!\!\!/(y)\psi(y) - e\Big[\int d^4y\,\bar{\psi}_0(y)A\!\!\!/(y)\,G_F^*(x-y)\Big]\gamma^\mu\psi_0(x)\\
&+ 2e^2A^\mu(x)|\phi_0(x)|^2.
\end{aligned} \tag{15}$$

To derive the Maxwell equations with the account of the impact of medium on the photon propagator we have to average operators ϕ and ψ over the medium. The first term in Eq. (15), linear in e, is non-zero if the medium is either electrically charged or possesses an electric current. We assume here that this is not the case, i.e. the medium is electrically neutral and "current-less".

The products of creation-annihilation operators averaged over the medium have the standard form:

$$\langle a^\dagger(\mathbf{q})a(\mathbf{q}')\rangle = f_B(E_q)\delta^{(3)}(\mathbf{q}-\mathbf{q}'), \tag{16}$$
$$\langle a(\mathbf{q})a^\dagger(\mathbf{q}')\rangle = [1+f_B(E_p)]\delta^{(3)}(\mathbf{q}-\mathbf{q}'), \tag{17}$$
$$\langle c^\dagger(\mathbf{q})c(\mathbf{q}')\rangle = f_F(E_p)\delta^{(3)}(\mathbf{q}-\mathbf{q}'), \tag{18}$$
$$\langle c(\mathbf{q})c^\dagger(\mathbf{q}')\rangle = [1-f_F(E_p)]\delta^{(3)}(\mathbf{q}-\mathbf{q}'), \tag{19}$$

where $f_{F,B}(E_q)$ is the energy dependent fermion/boson distribution function, which may be arbitrary since we assumed only that the medium is homogeneous and isotropic. We also assumed, as it is usually done, that non-diagonal matrix elements of creation-annihilation operators vanish on the average due to decoherence. For the vacuum case $f_{F,B}(E)=0$ and we obtain the usual vacuum average values of $aa^\dagger$ and $a^\dagger a$, which from now on will be neglected because we are interested only in the matter effects. As a result we obtain linear but non-local equation for electromagnetic field $A_\mu(x)$, for which it is convenient to perform the Fourier transform:

$$A^\mu(k) = \int \frac{d^4x}{(2\pi)^3} \exp^{-ikx} A^\mu(x). \tag{20}$$

Finally we find that the field $A^\mu(k)$ satisfies the equation

$$\left[k^\rho k_\rho g^{\mu\nu} - k^\mu k^\nu + \Pi^{\mu\nu}(k)\right] A_\nu(k) = \mathcal{J}^\mu(k), \tag{21}$$

which is equivalent to the photon equation of motion (9) but in momentum space.

In this way the photon polarization tensor, which contains contributions from the charged bosons and fermions, $\Pi_{\mu\nu}(k) = \Pi^B_{\mu\nu}(k) + \Pi^F_{\mu\nu}(k)$, according to Eq. (15), can explicitly found in the lowest order in e^2:

$$\Pi^B_{\mu\nu}(k) = e^2\int \frac{d^3q}{(2\pi)^3E}\left[f_B(E)+\bar{f}_B(E)\right]\left[\frac{1}{2}\frac{(2q-k)_\mu(2q-k)_\nu}{(q-k)^2-m_B^2}\right.$$
$$\left. + \frac{1}{2}\frac{(2q+k)_\mu(2q+k)_\nu}{(q+k)^2-m_B^2} - g_{\mu\nu}\right], \tag{22}$$

$$\Pi^F_{\mu\nu}(k) = 2e^2 \int \frac{d^3q}{(2\pi)^3 E} \left[f_F(E) + \bar{f}_F(E)\right] \left[\frac{q_\nu (k+q)_\mu - q^\rho k_\rho g_{\mu\nu} + q_\mu (k+q)_\nu}{(k+q)^2 - m_F^2}\right.$$
$$\left. + \frac{q_\nu (q-k)_\mu + q^\rho k_\rho g_{\mu\nu} + q_\mu (q-k)_\nu}{(k-q)^2 - m_F^2}\right] . \tag{23}$$

The static properties of the medium are determined by the time-time component of the polarization tensor in the limit of $\omega = 0$, which can be easily calculated from the above expressions:

$$\Pi^B_{00}(k) = -\frac{e^2}{2\pi^2} \int_0^\infty \frac{dqq^2}{E} (f_B + \bar{f}_B) \left(1 + \frac{E^2}{kq} \ln \left|\frac{2q+k}{2q-k}\right|\right) , \tag{24}$$

$$\Pi^F_{00}(k) = -\frac{e^2}{\pi^2} \int_0^\infty \frac{dqq^2}{E} (f_F + \bar{f}_F) \left(1 + \frac{E^2}{kq} \ln \left|\frac{2q+k}{2q-k}\right|\right) . \tag{25}$$

Here and in what follows k and q are respectively the absolute values of the spatial component of the photon and the charged particle momenta. Expressions (24,25) coincide with the well known ones found by other methods. Our new results for screening come from an addition of the condensate term to f_B, Eq. (5).

After straightforward calculations we find that the time-time component of the charge boson contribution into the photon polarization tensor at zero frequency and small k (but high T) has the form:

$$\Pi_{00} = \left[k^2 + e^2(m_0^2 + m_1^3/k + m_2^4/k^2)\right] , \tag{26}$$

where

$$\begin{aligned} m_0^2 &= 2T^2/3 + C/[(2\pi)^3 m_B] \\ m_1^3 &= m_B^2 T/2 \\ m_2^4 &= 4C m_B/(2\pi)^3 . \end{aligned} \tag{27}$$

In fact the same dependence on k is true for any temperature but the coefficients m_j may be different.

3. Screened electric potential

To calculate the screened potential we have to make the Fourier transformation of the photon propagator $(k^2 - \Pi_{00})^{-1}$:

$$U(r) = q \int \frac{d^3k}{(2\pi)^3} \frac{\exp(i\mathbf{kr})}{k^2 - \Pi_{00}(k)} = \frac{q}{2\pi^2} \int_0^\infty \frac{dkk^2}{k^2 - \Pi_{00}(k)} \frac{\sin kr}{kr} . \tag{28}$$

If Π_{00} is an even function of k, as is usually the case, the integration over k can be extended to the interval from $-\infty$ to $+\infty$ and the integral can be taken as a sum over residues of the integrand. For example, if the term proportional to $1/k$ can be neglected (low temperature case), Π_{00} is evidently even and its poles can be easily found:

$$k_j^2 = -\frac{e^2 m_0^2}{2} \pm \sqrt{\left[\frac{e^4 m_0^4}{4} - e^2 m_2^4\right]} \approx \pm i e m_2^2. \tag{29}$$

The last approximate equality is formally true in the limit of small e.

Thus in the presence of the charged Bose condensate the "Debye" poles acquire the non-zero real parts:

$$k_j = \pm\sqrt{e} m_2 \exp(\pm i\pi/4) \equiv k_j' + i k_j'', \tag{30}$$

Non-zero k' leads to the oscillating behavior of the potential

$$U(r)_j \sim q\,\frac{\exp(-\sqrt{e/2} m_2 r)\cos(\sqrt{e/2} m_2 r)}{r}. \tag{31}$$

If the term proportional to $1/k$ is present in Π_{00}, the calculations of the potential are slightly more complicated. Now the integration path in the complex k plane cannot be extended to $-\infty$ but the integration should be done along the real axis from 0 to ∞, then along infinitely large quarter-circle, and along the imaginary axis from $-\infty$ to 0. The result would contain the usual contribution from the poles in complex k-plane and the integral over imaginary axis. The former gives the usual exponentially decreasing potential, while the latter gives a power law decrease:

$$U \sim q\, m_1^3/(e^2 m_2^8 r^6). \tag{32}$$

Notice that the potential is inversely proportional to the electric charge squared. This is because we consider asymptotical behaviour of the screened potential at large distance, r, when the parameter er formally tends to infinity.

There are some other singularities in the integrand of Eq. (28), which arise from the pinching of the contour of integration over q in Eq. (24) by the poles of the distribution function, f_B (5), and the branch point of the logarithm. The induced singularities in the complex k-plane give rise to the screening effects analogous to the Friedel oscillations.[4] If the first "pinch" dominates, the screened potential is:

$$U_1(r) = -\frac{32\pi Q}{e^2 m_B r^2}\,\frac{e^{-z}}{\ln^2(2\sqrt{2}z)}\,\sin z\,, \tag{33}$$

where $z = 2r\sqrt{2\pi T m_B}$. Notice that $U_1(r)$ is inversely proportional to e^2 and formally vanishes at $T \to 0$, but remains finite if $\sqrt{T m_B} r \neq 0$.

If all pinches are comparable, the screened potential drops down as a power of distance:

$$U(r) \approx -\frac{3Q}{2e^2 T^2 m_B^3 r^6 \ln^3(\sqrt{8 m_B T} r)} . \tag{34}$$

More details can be found in the second paper of Ref. 6.

4. Vector boson condensate

Now we discuss the condensation of charged vector bosons,[7] which might take place in the early universe if the cosmological lepton asymmetry is sufficiently high. Condensation of vector particles differs from that of the scalars due to an additional degree of freedom, their spin states. Depending upon the interactions between the spins, they can be either aligned or anti-aligned. These states are called respectively ferromagnetic and anti-ferromagnetic ones, see e.g. Ref. 9. We show that W-bosons of the minimal electroweak theory condense in the ferromagnetic state and spontaneous magnetization of the primeval plasma could generate strong primordial magnetic fields on macroscopically large scales.

Recently somewhat similar problem of condensation of deuterium nuclei in astrophysics has been studied in Ref. 8. The authors argue that the interaction between deuterium nuclei forces them into the lowest spin antiferromagnetic state.

In the minimal standard electroweak model the spin-spin interaction of W-bosons is determined by the interaction between their magnetic moments and their contact quartic coupling. The former can be found from the analogue of the Breit equation for vector particles which leads to the spin-spin potential of the form[7]:

$$U_{em}^{spin}(r) = \frac{\alpha \rho^2}{m_W^2} \left[\frac{(\mathbf{S}_1 \cdot \mathbf{S}_2)}{r^3} - 3\,\frac{(\mathbf{S}_1 \cdot \mathbf{r})\,(\mathbf{S}_2 \cdot \mathbf{r})}{r^5} - \frac{8\pi}{3}\,(\mathbf{S}_1 \cdot \mathbf{S}_2)\,\delta^{(3)}(\mathbf{r}) \right] \tag{35}$$

where $\alpha = e^2/4\pi \approx 1/137$ and ρ is the ratio of the real magnetic moment of W to its value in the electroweak theory. Since the plasma is supposed to be neutral, the Coulomb interaction between the condensed W is compensated by the charged leptons.

To find the energy shift of a pair of W-bosons due to this interaction we need to average potential (35) over the W-wave function. The wave function of the condensate is supposed to be angle independent S-wave state. Thus the energy shift induced by the spin-spin interaction is equal to:

$$\delta E = \int \frac{d^3 r}{V} \, U_{em}^{spin}(r) = -\frac{2\, e^2 \rho^2}{3\, V m_W^2} \left(\mathbf{S}_1 \cdot \mathbf{S}_2 \right), \tag{36}$$

where V is the normalization volume.

Since $S_{tot}^2 = (S_1 + S_2)^2 = 4 + 2S_1 S_2$, the average value of $S_1 S_2$ is equal to

$$S_1 S_2 = S_{tot}^2/2 - 2\,. \tag{37}$$

For $S_{tot} = 2$ this term is $S_1 S_2 = 1 > 0$, while for $S_{tot} = 0$ it is $S_1 S_2 = -2 < 0$. Thus, if the spin-spin interaction is dominated by the interactions between the magnetic moments of W bosons, the state with their maximum total spin is energetically more favorable and W-bosons should condense in the ferromagnetic state. This interaction could lead to the spontaneous magnetization in the early universe.

Another contribution to the spin-spin interactions of W comes from their quartic self-coupling:

$$L_{4W} = \frac{e^2}{2 \sin^2 \theta_W} \left[W_\mu^\dagger W^{\mu \dagger} W_\nu W^\nu - (W_\mu^\dagger W^\mu)^2 \right] = \frac{e^2 \left(\mathbf{W}^\dagger \times \mathbf{W} \right)^2}{2 \sin^2 \theta_W}. \tag{38}$$

It is assumed here that $\partial_\mu W^\mu = 0$ and thus only the spatial 3-vector $\mathbf{W}$ is non-vanishing, while $W_t = 0$. The Fourier transform of this term with proper (nonrelativistic) normalization leads to

$$U_{4W}^{(spin)} = \frac{e^2}{8 m_W^2 \sin^2 \theta_W} \left(\mathbf{S}_1 \mathbf{S}_2 \right) \delta^{(3)}(\mathbf{r}). \tag{39}$$

Thus the quartic self-coupling of W contributes only to the spin-spin interaction whose sign is antiferromagnetic.

In the minimal standard model the interaction between the magnetic moments of W dominates and the condensed W-bosons have ferromagnetic behavior. However, in some modification of the standard model antiferromagnetic behavior is possible and W would condense in the state with zero or microscopically small spin.

The exchange of Z^0-bosons may also contribute to spin-spin interactions of W. It can be shown that for non-relativistic Z this contribution vanishes.[7] However, if the momentum carried by the virtual Z is non-negligible in comparison with its mass, the contributions of Z and photon exchanges are similar.

In principle, the long range interactions between magnetic moments can be screened by the plasma effects. However, in contrast to electric interactions, which are Debye-like screened, magnetic interactions in pure electrodynamics (or in any other Abelian theory) are known to remain unscreened. On the other hand, in non-Abelian theories the screening may occur in higher orders of perturbation theory due to the violent infrared singularities, which make impossible perturbative calculations.[10] The screening may potentially change the relative strength of the electromagnetic spin-spin coupling, which is affected by screening effects, with respect to the local quartic, W^4-coupling which is not screened. However, in the broken phase the system is reduced to the usual electrodynamics, where screening is absent and W-bosons would condense in the ferromagnetic state. In the unbroken phase of the electroweak theory the answer is not yet known.

If W-bosons make a ferromagnetic state, the primeval plasma, where such bosons condensed (possibly due to a large cosmological lepton asymmetry), could be spontaneously magnetized, as it happens in the usual ferromagnet. The typical size of the magnetic domains is determined by the cosmological horizon at the moment of the condensate evaporation. The latter takes place when the neutrino chemical potential, which scales as temperature in the course of cosmological cooling down, becomes smaller than the W mass at this temperature.

The large scale magnetic field, produced by the ferromagnetism of W-bosons, might survive after the decay of the condensate due to the conservation of the magnetic flux in the primeval plasma because of its high electric conductivity. Such magnetic fields, which were unifom at macroscopically large scales at formation, may be the seeds for the observed galactic or intergalactic magnetic fields at astronomically large scales. Evidently the characteristic size of magnetic domains at production is much smaller than the galactic size, even with an account of the cosmological stretching. Nevertheless, magnetic fields homogeneous at astronomical scales may be created by chaotic reconneciton (Brownian motion) of the magnetic field lines at much smaller scales but by an expence of the field amplitude.

References

1. E.M. Lifshitz and L.P Pitaevskii, *Landau and Lifshitz, Course of Theoretical Physics, Volume 10 - Physical Kinetics*, Elsevier, 1981.
2. J.I. Kapusta and C. Gale, *Finite temperature field theory: Principles and Applications*, Cambridge Monographs on Mathematical Physics, 2006.
3. I.A. Akhiezer and C.V. Peletminsky, *ZhETF* **38**, 1829 (1960); *Sov. Phys. JETP*, **11**, 1316 (1960); E.S. Fradkin, *Proc. Lebedev Inst.* **29**, 7 (1965).
4. J. Friedel, *Phil. Mag.* **43**, 153 (1952); Nuovo Cim. **7**, 287 (1958), suppl. 2; J.S. Langer and S.H. Vosko, *J. Phys. Chem. Solids* **12**, 196 (1960);

A. Fetter and J. Walecka, *Quantum Theory of Many-Particle Systems*, McGraw-Hill, San Francisco, 1971.

5. G. Gabadadze and R.A. Rosen, *JCAP* **0810**, 030 (2008), arXiv:0806.3692 [astro-ph].
   G. Gabadadze and R.A. Rosen, *JCAP* **0902**, 016 (2009), arXiv:0811.4423 [hep-th].
   G. Gabadadze and R.A. Rosen, *JCAP* **1004**, 028 (2010), arXiv:0912.5270 [hep-ph].
6. A.D. Dolgov, A. Lepidi and G. Piccinelli, *JCAP* **0902**, 027 (2009), arXiv:0811.4406 [hep-th].
   A.D. Dolgov, A. Lepidi and G. Piccinelli, *Phys. Rev. D* **80**, 125009 (2009), arXiv:0905.4422 [hep-ph].
7. A.D. Dolgov, A. Lepidi and G. Piccinelli, *JCAP* **08**, 031 (2010), arXiv:1005.2702 [astro-ph.CO].
8. L. Berezhiani, G. Gabadadze and D. Pirtskhalava, *JHEP* **1004**, 122 (2010), arXiv:1003.0865 [hep-ph].
9. C. J. Pethick and H. Smith, *Bose-Einstein Condensation in Dilute Gases*, Cambridge University Press, 2002.
10. A.D. Linde, *Rep. Prog. Phys.* **42**, 389 (1979) ;
    A.D. Linde, *Phys. Lett.* **96B**, 293 (1980);
    D.J. Gross, R. Pisarski and L. Yaffe, *Rev. Mod. Phys.* **53**, 43 (1981).

BARYON AND DIBARYON MASSES IN QUASI-PARTICLE APPROACH

A. BHATTACHARYA, B. CHAKRABARTI, S. MANI and A. SAGARI

Department of Physics, Jadavpur University
Calcutta 700032, India
pampa@phys.jdvu.ac.in

In the framework of the quasi-particle picture of diquark, the possibility of describing the baryons as a cluster of the quark and diquark has been explored. Different diquark masses have been estimated and these are used to estimate baryon masses. The results are found to be in reasonable agreement with experimental data. The model has been extended to describe the dibaryon masses such as Λ_6, Σ_6^0, etc which are yet to be confirmed.

1. Introduction

Baryons in the conventional quark model are color singlets composed of three quarks. However Gell-Mann[1] himself suggested the idea that q-q bound states or diquarks can be formed within the hadron. The diquark picture allows us to look at a baryon as being composed of a diquark and a quark, thus reducing a 3-body (q-q-q) problem to a 2-body (q-qq) one. The simplification has resulted in a vast volume of work on various baryon properties using the concept of diquark. Recently, we[2] have successfully estimated mass, binding energy, compressibility and excitation energy for Roper resonance of the proton assuming it to be quark–diquark system. We have also studied the mass and decay properties of the Θ^+ pentaquark in the framework of the suggested model and reasonable result have been obtained[2] although it must be mentioned that the experimental observation of Θ^+ is still in under some shadow. Assuming proton as quark-diquak system Bialas et al.[3] have investigated small momentum transfer elastic p-p cross- section at high energies. Oka[4] has made a detailed study on the diquark correlation in the context of different models. Maris[5] has calculated charge radius for scalar *ud* diquark and compared it to corresponding pion radius. Diquark mass too has been a subject of study. Ram and Kriss[6] and de Castro et al.[7] have investigated diquark mass and radius and used them to reproduce hadron properties successfully further strengthening the diquark theory. QCD, as the underlying theory of strong interaction allows a rich baryon spectrum that includes hybrid baryons(qqqg) and multiquark baryons such as pentaquarks(qqqq$\bar{q}$) and dibaryons(qqqqqq), among others. Jaffe's[8] proposal of the H-dibaryon as a possible metastable state, on the basis of detailed calculation in the framework of MIT bag model, invited a lot of interest in studying the properties of the dibaryons. Experimental searches to find the dibaryons[9–11] have produced varied results and been accompanied by much theoretical speculation.[12–14]

Recently[15] we have suggested a model in which two quarks are assumed to be correlated to form a low energy configuration, forming a diquark, and behaving like a quasi-particle in analogy with the electron behaving as a quasi-particle in the crystal lattice.[16] We have estimated the diquark mass in the framework of the quasi-particle model. In the present work the baryon is described as a diquark-quark system considering diquark as quasi-particle. The mass of the baryon has then been calculated employing the mass of the constituent quarks and diquarks. The mass of the dibaryon has been expressed as the sum of the masses of the three constituent diquarks. The results have been found to compare favourably with experimental values and current estimates.

The paper begins with Section I: Introduction, following which the calculation and approach have been clearly outlined in Section II: Formalism. In Section III: Conclusions and Discussions, the results have been discussed and compared with other available estimates. A tabular representation of the obtained data has been provided.

2. Formalism

The quasi-particle concept is of much importance in condensed matter physics. A quasi-particle is a low-lying excitation and corresponds to a single particle whose motion is modified by the interactions within the system. An electron in a crystal is subjected to two types of forces, namely, the effect of the crystal field (∇V) and an external force (F) which accelerates the electron.[16] Under the influence of these two forces, an electron in a crystal behaves like a quasi-particle whose effective mass m^* reflects the inertia of electrons which are already in a crystal field such that:

$$m^* \frac{dV}{dt} = F \tag{1}$$

The bare electrons (with normal mass) are affected by the lattice force -∇V(where V is the periodic potential) and the external force F so that:

$$m\frac{dV}{dt} = F - \frac{dV}{dx} \tag{2}$$

Hence the ratio of the normal mass (m) to the effective mass (m^*) can be expressed as:

$$m/m^* = 1 - \frac{1}{F}\left[\frac{\delta V}{\delta x}\right] \tag{3}$$

An elementary particle in vacuum may be suggested to be in a situation exactly resembling that of an electron in a crystal.[16] We have proposed a similar type of picture for the diquark $[ud]_0$ as a quasi-particle inside a nucleon. We assume that the diquark is an independent body which is under the influence of two types of forces. One is due to the background meson cloud which is represented by potential V =-α/r, where $\alpha = \frac{2}{3}\alpha_s$, α_s being the strong coupling constant, and this potential resembles the crystal field on a crystal electron. On the other hand for the external force we have considered an average force F = -ar, where a is a suitable constant, which is of confinement type. It has been assumed that under the influence of these two types of interactions the diquark is behaving like a quasi-particle, a low lying excited state and its mass gets modified. The ratio of the constituent mass and the effective mass of the diquark can be expressed by using the same formalism as in equation (3) and is obtained as:

$$\frac{m_q + m_{q'}}{m_D} = 1 + \frac{\alpha}{2a < r^3 >} \tag{4}$$

Here $m_q + m_{q'}$ represents the normal constituent mass of the diquark and m_D is the effective mass of the diquark, V being the average value of the one gluon exchange type of potential. r is the radius parameter of the diquark. To calculate the effective mass of the diquark from the above expression we need the radius parameter r of the diquark. The input of the effective radius of the diquark system is taken to be similar to the corresponding meson radius (having same constituent quarks)as exact value of the radius parameter is unknown. These meson radii have all been calculated by us previously employing the Statistical Model for hadrons[17] fitting against the charge radius or the decay constant of the respective meson. For the ud diquark we substitute pion radius = 5.38 GeV^{-1}, for us diquark we take kaon radius= 4.77 GeV^{-1}.[18] The cs diquark is taken to have radius equal to that of (D_s) meson = 4.3 GeV^{-1} and that of the bs diquark is equal to that of the B_s meson = 3.4 GeV^{-1}.[19] The masses considered are $m_u = m_d = 0.35\ GeV$, $m_s = 0.54\ GeV$, $m_c = 1.24\ GeV$ and $m_b = 4.2\ GeV$.[20] α_S has been taken to be 0.59 for the light sector and 0.2 for the heavier quarks[21] and a$= 0.077\ GeV^3$.[22] Reyes et al.[22] have investigated the hadron spectra in context of nonrelativistic model using harmonic oscillator potential between the quarks. However a wide range of values of interaction parameters of the same potential have been given by Dmitrasinovic et al.[23] Substituting these values in Eq. (4) we estimate the diquark masses which are displayed in Table 1. The binding energies of these calculated diquark masses have also been displayed in Table 1. The estimated diquark masses have been compared with other's works.

Table 1. Diquark masses and binding energy of diquarks.

Diquark	B.E. of Diquark GeV	Mass (m_D) GeV	Others' Work GeV
ud/uu/dd	0.042	0.658	0.59-0.62[6]
us/ds	0.074	0.816	0.911[7]
sc	0.060	1.72	–
bu/bd	0.033	4.22	–
bs	0.037	4.37	–

In current description of the quasi-particle picture of diquark, the diquarks are considered as low- lying excitations and regarded as separate elementary entities, behaving like scalar bosons analogous to the quasi-particles.[24] They are independent of each other and do not interact with each other or the single quark. Therefore as in an ideal gas of quasi-particles, the energies of the diquark and quark within the hadron are assumed to be simply additive[24] and the effective mass characterises the dynamic properties of the quasi-particles. Hence the mass of corresponding baryons M_B can be obtained by summing the diquark mass (m_D) and quark mass (m_q).

$$M_B = m_D + m_q \tag{5}$$

The masses of the baryons thus calculated are displayed in Table 2 where they have been compared to corresponding experimental data[20] and other estimates. The mass of the dibaryons (M_{DB}) can be written as the sum of three diquark masses that make up the dibaryon.

$$M_{DB} = m_D + m_{D'} + m_{D''} \tag{6}$$

where m_D, $m_{D'}$ and $m_{D''}$ represent the three diquark masses which may be same or different. The masses of the dibaryons Λ_6,$\Sigma_6^0[ud, us, ds]$, $\Sigma_6^\pm$

$[us, us, ud]$, $\Xi_6[ud, ud, us]$, $N_6[us, us, us]$, H$[ud, us, ds]$ have been estimated and been displayed in Table 3 along with other estimates.

Table 2. Baryon masses calculated using Diquark masses.

Baryon	Mass(Calc) GeV	Mass(Expt)[11] GeV	Others' Work GeV
P	1.008	0.938	1.024[36]
N	1.008	0.939	1.018[7]
Λ_c^+	1.898	2.286	–
Ξ_c^+	2.056	2.468	–
Λ^0	1.198	1.115	1.167[7]
Ξ^0	1.356	1.314	1.321[7]

Table 3. Dibaryon masses using Diquark masses.

Dibaryon	Mass GeV	Others' Work GeV
M_{Λ_6,Σ_6^0}	2.291	2.270-2.310[12]
M_{Ξ_6}	2.132	2.129[31]
M_{N_6}	2.449	–
$M_{\Sigma_6^\pm}$	2.291	2.365[35]
M_H	2.291	$>$ 2.2037[12]

3. Conclusion

In the present quasi-particle picture of diquark resembling an electron in a crystal lattice has been used to estimate the masses of the baryons and dibaryons. This quasi-particle description of hadrons may have some relevance to the recently developed idea that diquarks may be the building blocks of hadrons,pentaquarks and exotics.[25] Experimental results are not available for comparison of diquark mass. However a comparison with calculations performed by others is found to be satisfactory. The diquark mass of ud has been calculated to lie within a range 0.5910 to 0.5958 GeV by Ram et al.[6] and our value is 0.658 GeV. de Castro et al.[7] have calculated mass of us/ds diquark to be 0.911 GeV while the value obtained by us are 0.816 GeV. It may be noted that Jaffe and Wilczek[27] have extensively studied the exotics using a diquark model.

In calculating dibaryon masses we have assumed them to be constituted of three diquarks. For example, we consider the particle Λ_6, Σ_6^0 mass to be the sum of $2m_{us} + m_{ud}$. For nucleon type dibaryon N_6 the quark content suggests a mass $3m_{us}$.[12] Our calculations of dibaryon masses predict the mass to range between 2.13 and 2.29 GeV. Goyal and Mishra,[29] Piekarz et al.,[30] Pigot et al.[31] have all suggested dibaryon masses that are well within this range or very close to it. Shu Qin Xie et al.[32] have suggested masses for a 27-plet of dibaryons, many of which also conform to our estimates. Using the MIT bag model and the color- magnetic interaction, the H dibaryon mass was predicted to be 2150 MeV, whereas the $\Lambda\Lambda$ threshold is 2231 MeV.[8] The lower limit of H- dibaryon mass can be concluded from the double $-\Lambda$ hypernuclei experiments. If the H-dibaryon mass is below the Λ masses in the nucleus, two Λ hyperons will form a H-dibaryon. The H dibaryon must be a loosely bound system, if it exists, and in the vicinity of the $\Lambda\Lambda$ threshold. The binding energy of such a particle must be less than a few MeV, according to the recent doubly Λ hypernuclei experiments.[10,33] The lower bound for mass of the H particle is found to be 2224 MeV which favours our calculated value of 2291 MeV. Glendenning et al.[34] have proposed a vacuum mass of around 2.2 GeV for a condensate of H dibaryons found in neutron stars. Our estimate of 2.29 GeV is very close to this prediction. The mass of Ξ_6 is calculated to be 2.132 GeV which resembles the value 2.129 GeV of Λp given by Pigot et al.[31] Also, the mass of $\Sigma_6^{\pm}$ which is calculated as 2.291 GeV tallies well with that of Yokosawa[35] for the $\Lambda\Lambda$ state. The value of proton mass agrees favourably with that of Fleck et al.[36] However greater accuracy may be achieved by considering other effects like Pauli- blocking and quark-antiquark annihilation effects between the diquarks which are to be considered in our future work.It should be mentioned here that the maximum uncertainty lies in the estimation of the radius parameter which is not very well known.[28] Considering diquark radius to be the same as meson radius introduces an approximation in the estimation of diquark masses, which is reflected in the slight differences in calculated and experimental mass values for baryons. Moreover the constituent quark masses of the u and d quarks are usually chosen to lie between 300-350 MeV.[26] In the present work we have taken $m_u = m_d = 350MeV$,[20] which may introduce some difference into the results when compared with other estimates.

The quasi-particle model of diquark is found to be quite successful in reproducing the masses reasonably. It is interesting to note that the mass of the particles are obtained simply by adding the diquark masses which indicates the diquark is behaving as an independent entity without any interaction with the others and its mass get reduced simulating many body interaction.It may be mentioned that we come across such a scenario in the context of superfluidity where the phonons are behaving like noninteracting hypothetical particles where each phonon creates an environment around itself so that it behaves like a free particle. It may be suggested that the superfluid nature of the hadron vacuum may be represented by the quasi-particle nature of the diquark and it may be a fundamental entity for the configuration of a hadron. The predicted dibaryon masses are also in good agreement with others' works. However only future experiments could enlighten us on this aspect.

References

1. M. Gell-Mann, *Phys. Lett.* **8**, 214 (1964).
2. A. Bhattacharya *et al.*, *Int. J. Theor. Phys.* 10.1007/s10773-008-9684-3; A. Bhattacharya *et al.*, *Had. J.* **30**, 407 (2007).
3. A. Bialas, A. Bzdak, *Acta Phys. Polonica* **B38**, 159 (2007).
4. M. Oka, *Prog. Theor. Phys.* **112**, 1 (2004).
5. P. Maris, *Few Body Systems* **35**, 3 (2004).
6. B. Ram and V. Kriss, *Phys. Rev.* **D35**, 400 (1987).
7. A.S. de Castro *et al.*, *Z. Phys.* **C57**, 315 (1993).
8. R.L. Jaffe, *Phys. Rev. Lett.* **38**, 195 (1977).
9. R. Klingenberg, *J. Phys.* **G25**, R273 (1999).
10. J.K. Ahn *et al.*, *Phys. Rev. Lett.* **87**, 132504 (2001).
11. Z.Y. Zhang *et al.*, Phys. Rev. **C61**, 065204 (2000).
12. S.L. Zhu, *Phys. Rev.* **C70**, 045201 (2004).
13. C. Kuhn *et al.*, *J. Phys.* **G28**, 1707 (2002).
14. R.L. Ray *et al.*, *Phys. Rev.* **C62**, 024906 (2000).
15. B. Chakrabarti *et al.*, *Nucl. Phys. Proc. Supp.* **A782**, 392c (2007).
16. A. Haug. *Theoretical Solid State Physics*, Pergamon Press; p. 100 (1975).
17. S.N. Banerjee *et al.*, *Phys. Scr.* **37**, 201 (1988).
18. S.N. Banerjee *et al.*, *Int. J. Mod. Phys.* **A2**, 1829 (1987).
19. B. Chakrabarti, *Mod. Phys. Lett.* **A12**, 2133 (1997)
20. Particle Data Group, *Phys. Lett.* **B667**, 1-554 (2008); W.M. Yao *et al.*, *J. Phys.* **G33**, 1-510 (2006).
21. W. Lucha *et al.*, *Phys. Rep.* **200**, 127 (1991).
22. E.C. Reyes *et al.*, *Rev. Bras. Ensino Fis.* **25**, 1806 (2003).
23. V. Dmitrasinovic *et al.*, *Ann. Phys.(N.Y.)* **321**, 355 (2006).
24. L. D. Landau and E. M. Lifshitz, *Statistical Physics* Vol V/I p. 216, (1997).
25. J. Maiani *et al.*, *Phys. Rev.* **D71**, 014028 (2005); S.N. Banerjee *et al.*, *Int. J. Mod. Phys.* **17**, 4939 (2002).
26. R.K. Bhaduri, *Models of the Nucleon*, Addison-Wesley, p.45, (1988).
27. R. Jaffe and F. Wilczek, *Phys. Rev. Lett.* **91**, 232003 (2003).
28. H.P. Morsch *et al.*, *Phys. Rev. Lett.* **69**, 1336 (1992); *Z. Phys.* **A350**, 61 (1994).
29. P. Goyal *et al.*, *Prog. Theor. Phys.* **64**, 700 (1980).

30. H. Piekarz, *Nucl. Phys.* **A450**, 85c (1986).
31. C. Pigot *et al.*, *Nucl. Phys.* **B249**, 172 (1985).
32. S.Q. Xie, *J. Phys.* **G15**, 287 (1989).
33. H. Takahashi *et al.*, *Phys. Rev. Lett.* **87**, 212502 (2001).
34. K.N. Glendenning *et al.*; *Phys. Rev.* **C58**, 1298 (1998).
35. A. Yokosawa, *Proc. 19th Intl. Conf. HEP*, Tokyo, (Int. Acad. Pr.) 120, (1978)
36. S. Fleck *et al.*, *Phys. Rev.* **D38**, 1519 (1988).

AXIAL ANOMALY, DIRAC SEA, AND THE CHIRAL MAGNETIC EFFECT

DMITRI E. KHARZEEV

Department of Physics and Astronomy,
Stony Brook University, Stony Brook, New York 11794-3800, USA
Physics Department, Brookhaven National Laboratory,
Upton, New York 11973-5000, USA
Dmitri.Kharzeev@stonybrook.edu

Gribov viewed the axial anomaly as a manifestation of the collective motion of Dirac fermions with arbitrarily high momenta in the vacuum. In the presence of an external magnetic field and a chirality imbalance, this collective motion becomes directly observable in the form of the electric current – this is the chiral magnetic effect (CME). I give an elementary introduction into the physics of CME, and discuss the experimental status and recent developments.

Keywords: Axial anomaly; QCD; chiral magnetic effect.

1. Anomalies, as a manifestation of the high momentum collective motion in the vacuum

In the article entitled "Anomalies, as a manifestation of the high momentum collective motion in the vacuum," V.N. Gribov offered a deep insight into "one of the most beautiful and non-trivial phenomena in modern field theory"[1] – the axial[2,3] and scale[4–8] anomalies. According to Gribov, the source of anomalies can be traced back to the collective motion of particles with arbitrarily large momenta in the vacuum. This collective motion defies any UV cutoff that we may try to impose and "transfers the axial charge and the energy-momentum from the world with infinitely large momenta to our world of finite momenta".[1]

Let us illustrate this statement for the case of axial anomaly by considering the Dirac sea of massless fermions. In the absence of external fields (or parity-odd interactions), the chirality is conserved and there are two disconnected Fermi surfaces of left- and right-handed fermions. Now let us turn on an external classical field capable of changing the chirality of fermions – e.g. the parallel electric $\vec{E}$ and magnetic $\vec{B}$ fields. This field configuration will skew the balance between the Fermi surfaces of left- and right-handed fermions in the Dirac sea, transforming left-handed antiparticles into right-handed particles, or vice-versa, depending on the sign of the product $\vec{E} \cdot \vec{B}$.

The mechanism of the collective flow of chirality can be described as follows:[9,10] the presence of magnetic field B aligns the spins of the positive (negative) fermions in the direction parallel (anti-parallel) to $\vec{B}$. In the electric field E the positive fermions will experience the force eE and will move along $\vec{E}$; therefore their spin will have a positive projection on momentum, and we are dealing with the right fermions. Likewise, the negative fermions will be left-handed. After time t, the positive (right) fermions will increase their Fermi momentum to $p_R^F = eEt$, and the negative (left) will have their Fermi momentum decreased to $p_L^F = -p_R^F$. The one-dimensional density of states along the axis z that we choose parallel to the direction of fields $\vec{E}$ and $\vec{B}$ is given by $dN_R/dz = p_R^F/2\pi$. In the transverse direction, the motion of fermions is quantized as they populate Landau levels in the magnetic field. The transverse density of Landau levels is $d^2N_R/dxdy = eB/2\pi$. Therefore the density of right fermions increases per unit time as

$$\frac{d^4 N_R}{dt\ dV} = \frac{e^2}{(2\pi)^2}\ \vec{E} \cdot \vec{B}. \tag{1}$$

The density of left fermions decreases with the same rate, $d^4N_L/dt\ dV = -d^4N_R/dt\ dV$. The rate of chirality $Q_5 = N_R - N_L$ generation is thus

$$\frac{d^4 Q_5}{dt\ dV} = \frac{e^2}{2\pi^2}\ \vec{E} \cdot \vec{B}, \tag{2}$$

The quantity on the r.h.s. is the density of topological charge; its integral over four-dimensional space

$$q[A] = \frac{e^2}{8\pi^2} \int d^4x\ F^{\mu\nu} \tilde{F}_{\mu\nu}; \tag{3}$$

reveals the topological class to which the vector potential A belongs. It has to be integer, just as the difference between the numbers of right- and left-handed fermions. The relation (2) thus expresses the deep connection between the axial anomaly and the topology of classical gauge fields.

Having a classical field with an infinite number of quanta is important here since the picture described above involves changing the momenta of an infinite number of particles, and "a finite number of photons is not able to change the momenta of an infinite number of particles".[1] This feature of the anomaly gives an intuitive explanation of the absence of perturbative quantum corrections to the axial anomaly that can be established formally through the renormalization group arguments.[11–13] As will be discussed below, this property of the (electromagnetic) axial anomaly persists even when the coupling constant that determines the strength of (non-electromagnetic) interactions among the fermions becomes infinitely large.

The flow of chirality, as the derivation above reveals, is accompanied by the collective motion of particles at all momenta, including the momenta around the UV cutoff scale Λ_{UV} that we may attempt to introduce. Therefore our world of particles with finite momenta $p < \Lambda_{UV}$ cannot be isolated from the world of particles with arbitrarily high momenta, and this according to Gribov is the essence of quantum anomalies.

2. The chiral magnetic effect and Landau levels of Dirac fermions

Consider now the situation in which there exists an external magnetic field, and an imbalance between the Fermi momenta of left- and right-handed fermions. In the absence of an external electric field, this imbalance cannot be caused by electromagnetic interactions, but we can imagine that the imbalance may originate from other sources – e.g. from strong interactions of the fermions (quarks) with a non-Abelian gauge field configuration with non-trivial topological contents.

The presence of magnetic field aligns the spins of positive and negative fermions in opposite directions – along or against the direction of $\vec{B}$, respectively. Therefore being, say, right-handed means for the positive fermion to move along the direction of magnetic field, and for a negative fermion — to move against the direction of $\vec{B}$. Therefore, if the densities and Fermi-momenta of left- and right handed fermions are unequal in the presence of an external magnetic field, there should be an electric current and a separation of electric charge – this is the Chiral Magnetic Effect (CME).[14–18]

Let us introduce, in addition to magnetic field $\vec{B}$, an auxiliary electric field $\vec{E}$ and consider the energy balance of chirality generation. Changing chirality by one unit means transferring a massless fermion from the Fermi surface of left-handed particles to the one of the right-handed particles; this change costs an amount of energy equal to the difference of the corresponding Fermi-momenta $\mu_R - \mu_L = 2\mu_5$. If we multiply this energy by the rate of chirality change (2), we get the energy spent per unit time:

$$P = (\mu_R - \mu_L)\, \frac{e^2}{(2\pi)^2} \int d^3x\, \vec{E} \cdot \vec{B}. \tag{4}$$

As we argued above this energy powers the electric current, the power of which is given by

$$P = \int d^3x \; \vec{J} \cdot \vec{E}. \tag{5}$$

We can take $\vec{E}$ in the direction of $\vec{B}$ in this expression, and then get rid of the auxiliary electric field by taking the limit $\vec{E} \to 0$. This allows us to find the following expression for the density of CME current:[17]

$$\vec{J} = \frac{e^2 \mu_5}{2\pi^2} \vec{B}. \tag{6}$$

Note that this relation manifestly violates parity since magnetic field on the r.h.s. is a pseudo-vector whereas the electric current on the l.h.s. is a vector. Because of this, a static magnetic field with no curl cannot induce electric current in Maxwell electrodynamics (that is parity-even). In our case, the violation of parity is induced by the imbalance between the left- and right-handed fermions. Closely related phenomena have been discussed earlier in the physics of primordial electroweak plasma[19] and quantum wires.[20]

A more rigorous derivation[17] of (6) invokes the explicit sum over the contributions of all Landau levels of charged fermions. This sum is in general divergent, and one has to introduce a UV cutoff on the energy of Landau levels – this is the manifestation of the collective flow from the world of finite momenta to the world of infinite momenta discussed by Gribov. However, all excited Landau levels are degenerate in spin, and the opposite spin orientations give the contributions to the CME electric current that are opposite in sign and thus cancel each other. The lowest Landau level (LLL) of massless fermions is an exception since it is chiral, i.e. not degenerate in spin. Because of this, only the LLL contribution survives in the final expression (6) that does not contain any UV divergence.

3. The chiral magnetic effect and Maxwell-Chern-Simons electrodynamics

Let us now consider the CME in the effective theory of electromagnetism obtained by integrating the quarks out of the action.[18] Let us start from the QCD coupled to electromagnetism; the resulting theory possesses $SU(3) \times U(1)$ gauge symmetry:

$$\mathcal{L}_{\text{QCD+QED}} = -\frac{1}{4} G^{\mu\nu}_{\alpha} G_{\alpha\mu\nu} + \sum_f \bar{\psi}_f \left[i\gamma^\mu (\partial_\mu - igA_{\alpha\mu} t_\alpha - iq_f A_\mu) - m_f \right] \psi_f$$
$$- \frac{\theta}{32\pi^2} g^2 G^{\mu\nu}_{\alpha} \tilde{G}_{\alpha\mu\nu} - \frac{1}{4} F^{\mu\nu} F_{\mu\nu}, \tag{7}$$

where A_μ and $F_{\mu\nu}$ are the electromagnetic vector potential and the corresponding field strength tensor, and q_f are the electric charges of the quarks.

Let us discuss the electromagnetic sector of the theory (7). Electromagnetic fields will couple to the electromagnetic currents $J_\mu = \sum_f q_f \bar{\psi}_f \gamma_\mu \psi_f$. In addition, the θ-term in (7) will induce through the quark loop the coupling of $F\tilde{F}$ to the QCD topological charge. Let us introduce an effective pseudo-scalar field $\theta = \theta(\vec{x}, t)$ (playing the rôle of the axion[21–23] field, but without a kinetic term) and write down the resulting effective Lagrangian as

$$\mathcal{L}_{\mathrm{MCS}} = -\frac{1}{4} F^{\mu\nu} F_{\mu\nu} - A_\mu J^\mu - \frac{c}{4}\, \theta \tilde{F}^{\mu\nu} F_{\mu\nu}, \tag{8}$$

where

$$c = \sum_f q_f^2 e^2 / (2\pi^2). \tag{9}$$

This is the Lagrangian of Maxwell-Chern-Simons, or axion, electrodynamics that has been introduced previously in Refs. 24–26.

As we discussed above, the quantity $\tilde{F}^{\mu\nu} F_{\mu\nu}$ is the density of topological charge. Therefore the integral of this quantity over a four-dimensional volume should be an (integer) topological invariant sensitive only to the long distance, global properties of the gauge field. Such properties are determined by the asymptotic behavior of the field at the surface of the four-dimensional sphere, and thus the topological invariant has to be determined by the surface integral; Gauss theorem thus dictates that $\tilde{F^{\mu\nu}} F_{\mu\nu}$ has to be a full divergence:

$$\tilde{F}^{\mu\nu} F_{\mu\nu} = \partial_\mu J^\mu_{CS}; \tag{10}$$

the quantity J^μ_{CS} is the Chern-Simons current

$$J^\mu_{CS} = \epsilon^{\mu\nu\rho\sigma} A_\nu F_{\rho\sigma}, \tag{11}$$

that is a three-dimensional Chern-Simons form[27] promoted in four dimensions to a current by adding an extra index to the antisymmetric tensor. The Abelian three-dimensional Chern-Simons form

$$CS[A] = \int d^3x\, \epsilon^{\nu\rho\sigma} A_\nu F_{\rho\sigma} \tag{12}$$

is so-called magnetic helicity $\int d^3x\, \vec{A} \cdot \vec{B}$ measuring the linkage of the lines of magnetic flux.

If θ is a constant, then the entire last term in (8) represents a full divergence – therefore it does not affect the equations of motion and thus does not have any effect on the electrodynamics of charges. The situation is different if the effective field $\theta = \theta(\vec{x}, t)$ varies in space-time. Indeed, in this case we have

$$\theta \tilde{F^{\mu\nu}} F_{\mu\nu} = \theta \partial_\mu J^\mu_{CS} = \partial_\mu \left[\theta J^\mu_{CS}\right] - \partial_\mu \theta J^\mu_{CS}. \tag{13}$$

The first term on r.h.s. is again a full derivative and can be omitted; introducing notation

$$P_\mu = \partial_\mu \theta = (M, \vec{P}) \tag{14}$$

we can re-write the Lagrangian (8) in the following form:

$$\mathcal{L}_{\rm MCS} = -\frac{1}{4} F^{\mu\nu} F_{\mu\nu} - A_\mu J^\mu + \frac{c}{4}\, P_\mu J^\mu_{CS}. \tag{15}$$

Since θ is a pseudo-scalar field, P_μ is a pseudo-vector; as is clear from (15), it plays a rôle of the potential coupling to the Chern-Simons current (11). However, unlike the vector potential A_μ, P_μ is not a dynamical variable and is a pseudo-vector that is fixed by the dynamics of chiral charge – in our case, determined by the fluctuations of topological charge in QCD.

Let us write down the Euler-Lagrange equations of motion that follow from the Lagrangian (15), (11) (Maxwell-Chern-Simons equations):

$$\partial_\mu F^{\mu\nu} = J^\nu - P_\mu \tilde{F}^{\mu\nu}. \tag{16}$$

The first pair of Maxwell equations (which is a consequence of the fact that the fields are expressed through the vector potential) is not modified:

$$\partial_\mu \tilde{F}^{\mu\nu} = J^\nu. \tag{17}$$

It is convenient to write down these equations also in terms of the electric $\vec{E}$ and magnetic $\vec{B}$ fields:

$$\vec{\nabla} \times \vec{B} - \frac{\partial \vec{E}}{\partial t} = \vec{J} + c\left(M\vec{B} - \vec{P} \times \vec{E}\right), \tag{18}$$

$$\vec{\nabla} \cdot \vec{E} = \rho + c\vec{P} \cdot \vec{B}, \tag{19}$$

$$\vec{\nabla} \times \vec{E} + \frac{\partial \vec{B}}{\partial t} = 0, \tag{20}$$

$$\vec{\nabla} \cdot \vec{B} = 0, \tag{21}$$

where $(\rho, \vec{J})$ are the electric charge and current densities. One can see that the presence of Chern-Simons term leads to essential modifications of the Maxwell theory and induces, as we will see, the chiral magnetic effect.

Let us however start with a different phenomenon – the Witten effect:[28] magnetic monopoles at finite θ angle acquire electric charge and become "dyons". Consider, following Wilczek,[24] a magnetic monopole in the presence of a finite θ angle. In the core of the monopole $\theta = 0$, and away from the monopole θ acquires a finite non-zero value – therefore within a finite domain wall we have a non-zero $\vec{P} = \vec{\nabla}\theta$ pointing radially outwards from the monopole. According to (19), the domain wall thus acquires a non-zero charge density $c\vec{\nabla}\theta \cdot \vec{B}$. An integral along $\vec{P}$ (across the domain wall) yields $\int dl\ \partial\theta/\partial l = \theta$, and the integral over all directions of $\vec{P}$ yields the total magnetic flux Φ. By Gauss theorem, the flux is equal to the magnetic charge of the monopole g, and the total electric charge of the configuration is equal to

$$q = c\ \theta\ g = \frac{e^2}{2\pi^2}\ \theta\ g = \frac{e}{2\pi^2}\ \theta\ (eg) = e\ \frac{\theta}{2\pi}, \tag{22}$$

where we have used an explicit expression (9) for the coupling constant c, as well as the Dirac condition $ge = 4\pi \times$ integer.

Consider now a configuration where an external magnetic field $\vec{B}$ pierces a domain with $\theta \neq 0$ inside; outside $\theta = 0$. Let us assume first that the field θ is static, $\dot{\theta} = 0$. Assuming that the field $\vec{B}$ is perpendicular to the domain wall, we find from (19) that the upper domain wall acquires the charge density per unit area S of[15]

$$\left(\frac{Q}{S}\right)_{up} = +\ c\ \theta B \tag{23}$$

while the lower domain wall acquires the same in magnitude but opposite in sign charge density

$$\left(\frac{Q}{S}\right)_{down} = -\ c\ \theta B \tag{24}$$

Assuming that the domain walls are thin compared to the distance L between them, we find that the system possesses an electric dipole moment

$$d_e = c\,\theta\,(B\cdot S)\,L = \sum_f q_f^2 \left(e\,\frac{\theta}{\pi}\right)\left(\frac{eB\cdot S}{2\pi}\right)\,L; \tag{25}$$

for brevity of notations we put $\sum_f q_f^2 = 1$; it is easy to restore this factor in front of e^2 when needed. Static electric dipole moment is a signature of $\mathcal{P}$, $\mathcal{T}$ and $\mathcal{CP}$ violation (we assume that $\mathcal{CPT}$ invariance holds). The spatial separation of charge will induce the corresponding electric field $\vec{E} = c\,\theta\,\vec{B}$. The mixing of pseudo-vector magnetic field $\vec{B}$ and the vector electric field $\vec{E}$ signals violation of $\mathcal{P}$, $\mathcal{T}$ and $\mathcal{CP}$ invariances.

The formula (25) allows a simple interpretation: since $eB/2\pi$ is the transverse density of Landau levels of charged fermions in magnetic field B, the floor of the quantity $eB\cdot S/2\pi$ (i.e. the largest integer that is smaller than $eB\cdot S/2\pi$) is an integer number of fermions localized on the domain wall. Each fermion species contributes independently to this number as reflected by the factor N_f. Again we see that the electric dipole moment (25) arises from the electric charge $q \sim e\theta/\pi$ that is induced on the domain walls due to the gradient of the pseudo-scalar field θ.

If the domain is due to the fluctuation of topological charge in QCD vacuum, its size is on the order of QCD scale, $L \sim \Lambda_{\rm QCD}^{-1}$, $S \sim \Lambda_{\rm QCD}^{-2}$. This means that to observe an electric dipole moment in experiment we need an extremely strong magnetic field $eB \sim \Lambda_{\rm QCD}^2$. Fortunately, such fields exist during the early moments of a relativistic heavy ion collision.[16,29] Here we have assumed that the domain is static; this approximation requires the characteristic time of topological charge fluctuation $\tau \sim 1/\dot{\theta}$ be large on the time scale at which the magnetic field B varies. This assumption is only marginally satisfied in heavy ion collisions, and so we now need to consider also the case of $\dot{\theta} \neq 0$. Note however that if the medium produced in heavy ion collisions conducts electricity, then the decaying with time magnetic field will induce the circular electric current which in accord with Lenz's law will in turn produce a magnetic field.[30] The estimates[30] indicate that this mechanism can extend the lifetime of magnetic field in a very significant way.

Consider now the domain where $|\vec{P}| \ll \dot{\theta}$, i.e. the spatial dependence of $\theta(t,\vec{x})$ is much slower than the dependence on time.[16] Again, we will expose the domain to an external magnetic field $\vec{B}$ with $\vec{\nabla}\times\vec{B} = 0$, and assume that no external electric field is present. In this case we immediately get from (18) that there is an induced CME current[17]

$$\vec{J} = -c\ M\ \vec{B} = -\frac{e^2}{2\pi^2}\ \dot{\theta}\vec{B}. \tag{26}$$

4. Chiral magnetic and chiral vortical effects at strong coupling; relativistic hydrodynamics

Many of our arguments were based on the weak coupling picture, e.g. on the existence of Landau levels obtained by solving Dirac equation in the external magnetic field. One may worry that once the strong interactions among the quarks are turned on, this simple picture will break down. Nevertheless this does not happen – the essentially topological nature of the phenomenon protects it from being modified by quantum corrections, even at strong coupling. In particular, in holographic models (at infinite 't Hooft coupling) the magnitude of the chiral magnetic effect[31–33] appears the same as at weak coupling.[31,34,35]

The CME has been studied in lattice QCD coupled to electromagnetism, both in the quenched[36–38] and dynamical (domain wall) fermion[39] formulations; these simulations fully take account of strong interactions among the (anti)quarks. This suggests that the CME exists even when the coupling among the quarks is strong.

Quark-gluon plasma at strong coupling has been argued to behave as a nearly perfect fluid (for review, see Ref. 40), and an effective low-energy theory of strongly interacting fluids is well known – it is hydrodynamics. This invites a very interesting question about the role of axial anomaly in relativistic hydrodynamics that was addressed recently.[41] In a fluid, the role of magnetic field at finite baryon chemical potential μ_B can be played by vorticity $\vec{\omega}$ of the local fluid velocity $\vec{v}$:

$$\vec{\omega} = \vec{\nabla} \times \vec{v}. \tag{27}$$

This is quite natural since the rotating charged fluid generates an effective magnetic field $\mu_B\vec{\omega}$. As a result, the electric current can be induced by the rotation of the fluid with finite baryon and axial charge density even in the absence of an external magnetic field[15,41] – so-called "chiral vortical effect" (CVE). The topological origin of this phenomenon is manifest since the analog of Chern-Simons 3-form (12) in this case is the "kinetic helicity" $\int d^3x\ \vec{v} \cdot \vec{\omega}$. For a discussion of other vorticity-induced effects in heavy ion collisions, see e.g. Refs. 42–44.

A related effect — the emergence of a chiral current in a medium with finite baryon density, in an external magnetic field or in the presence of a vorticity the fluid — has been introduced in Refs. 45, 46, 41. The close connection between CME and the latter effect can be established for example by the method of dimensional reduction appropriate in the case of a strong magnetic field:[47] the simple relations $J_V^0 = J_A^1$, $J_A^0 = J_V^1$ between the vector J_V and axial J_A currents in the dimensionally reduced $(1+1)$ theory imply that the density of baryon charge must induce the axial current, and the density of axial charge must induce the charge current (CME).

5. Experimental status and a new test

Recently, STAR[48,49] and PHENIX[50,51] Collaborations at Relativistic Heavy Ion Collider reported experimental observation of charge asymmetry fluctuations. While the interpretation of the observed effect is still under intense discussion, the fluctuations in charge asymmetry have been predicted to occur in heavy ion collisions due to the Chiral Magnetic Effect (CME) in QCD coupled to electromagnetism.[14–18]

It is important to establish whether the CME explanation of charge asymmetry fluctuations is the correct one. First, it would be a direct observation of a topological effect in QCD. Second, the magnitude of this effect in the chirally broken phase is expected to be much smaller and hence the observation of the CME would manifest the restoration of chiral symmetry in the medium. The effort of quantifying the charge asymmetry fluctuations in QCD matter and of examining alternative explanations and backgrounds has already begun,[52–65] and there are plans to further study this effect at RHIC, LHC, NICA and FAIR.

Recently, a new test of the chiral magnetic and chiral vortical effects (CME and CVE) has been proposed.[66] The test relies only on the general properties of triangle anomalies. Consider anomalous hydrodynamics,[41] and suppose that the system under consideration has a chemical potential μ, coupled to a charge $\bar{q}\gamma^0 Bq$, where B is a flavor matrix, and an axial chemical potential μ_5, coupled to the axial charge $\bar{q}\gamma^0\gamma^5 Aq$, where A is another flavor matrix. For simplicity, we shall assume that both μ and μ_5 are much smaller than the temperature T (this assumption usually holds in relativistic heavy ion collisions). We also assume that electromagnetism couples to the current $\bar{q}\gamma^\mu Qq$, with Q being the charge matrix. If one measures a vector current $J^\mu = \bar{q}\gamma^\mu Vq$, then the result is

$$\vec{J} = \frac{N_c \mu_5}{2\pi^2}\left[\mathrm{tr}(VAQ)\,\vec{B} + \mathrm{tr}(VAB)\,2\mu\vec{\omega}\right] \tag{28}$$

where $\vec{B}$ and $\vec{\omega}$ are the external magnetic fields and the fluid vorticity respectively. The two parts of the current on the right hand side correspond to the CME and the CVE, respectively. The traces in the formula are related to the anomalous triangle diagram.

We shall consider two cases: $N_f = 3$, where u, d and s quarks are light, and $N_f = 2$ where only u and d quarks are light. In both cases, we assume A to be the unity matrix, $A = \mathbb{1}$ (which is expected if the chiral asymmetry is due to instanton events, which are flavor symmetric), and $B = (1/3)\mathbb{1}$. For $N_f = 3$, $Q = \text{diag}(2/3, -1/3, -1/3)$, and for $N_f = 2$, $Q = \text{diag}(2/3, -1/3)$. There are two currents that we will measure: the electromagnetic current J_E, corresponding to $V = Q$ and the baryon current J_B, corresponding to $V = B$. For CME, we get for the charge current (up to an overall factor of $N_c\ \mu_5 \vec{B}/(2\pi^2)$ which is common for both charge and baryon currents)

$$J_E^{CME} \sim \frac{2}{3}\ \ (N_f = 3)\ \ \text{or}\ \ \frac{5}{9}\ \ (N_f = 2) \tag{29}$$

and for the baryon current

$$J_B^{CME} = 0\ \ (N_f = 3)\ \ \text{or}\ \ \sim \frac{1}{9}\ \ (N_f = 2). \tag{30}$$

For CVE, the results are (up to the overall factor $N_c\ \mu_5 \mu \vec{\omega}/\pi^2$)

$$J_E^{CVE} = 0\ \ (N_f = 3)\ \ \text{or}\ \ \sim \frac{1}{3}\ \ (N_f = 2); \tag{31}$$

$$J_B^{CVE} \sim 1\ \ (N_f = 3)\ \ \text{or}\ \ \sim \frac{2}{3}\ \ (N_f = 2). \tag{32}$$

In the SU(3) case, the CME and CVE lead to completely different currents: the CME contributes only to the electromagnetic current and the CVE contributes only to the baryon current. In the SU(2) case, the separation is less clean, but the ratio of J_B/J_E still differs by a factor of ten.

Let us now discuss the implications for heavy ion collisions. It is known that the baryon chemical potential of the produced fireball depends on the collision energy: at smaller $\sqrt{s}$, μ is larger. Thus the CVE should be more important at lower energies. According to the computation above, J_B/J_E becomes larger as one lowers the energy of the collision. Moreover, since the symmetry arguments suggest that the magnetic field and the vorticity of the fluid have to be aligned, our results show that the two vectors $\vec{J}_B$ and $\vec{J}_E$ should point in the same direction. Therefore, in addition to the charge separation, there must be a baryon number separation. The two effects are positively correlated on the event-by-event basis, and the relative importance of baryon number separation increases as one lowers the energy of the collision.

6. Summary

The picture proposed by Gribov identifies the high-momentum collective motion in the Dirac sea as a source of axial anomaly. In a strong magnetic field, and in the presence of a local chirality imbalance, this collective motion becomes directly observable in the form of electric CME current. In heavy ion collisions, the local chirality imbalance can be readily supplied by the topological gluon field configurations in hot QCD matter, and a sufficiently strong magnetic field is delivered by the colliding ions. The effect thus can become observable, and there is an intriguing evidence from RHIC experiments for the expected charge asymmetry fluctuations.[49,50] Much remains to be done to verify (or disprove) the anomaly-related origin of the observed effect; however this work has already begun.

I am indebted to Julia Nyiri and Yuri Dokshitzer for their kind invitation to the Gribov-80 Conference. I thank my collaborators G. Başar, P. Buividovich, M. Chernodub, G. Dunne, K. Fukushima, L. McLerran, M. Polikarpov, D. Son, H. Warringa, H.-U. Yee, and A. Zhitnitsky for sharing their insights with me and numerous enjoyable discussions.

This work was supported in part by the U.S. Department of Energy under Contract No. DE-AC02-98CH10886.

References

1. V. N. Gribov, *"Anomalies, as a manifestation of the high momentum collective motion in the vacuum,"* KFKI-1981-66, Aug 1981;
   reprinted in V.N. Gribov *"Quark confinement and gauge theories"*, Phasis, Moscow (2001).
2. S. L. Adler, *Phys. Rev.* **177**, 2426 (1969).
3. J. S. Bell and R. Jackiw, *Nuovo Cim.* **A60**, 47 (1969).
4. C. G. Callan, *Phys. Rev.* **D2**, 1541 (1970).
5. K. Symanzik, *Commun. Math. Phys.* **18**, 227 (1970).
6. S. R. Coleman and R. Jackiw, *Annals Phys.* **67**, 552 (1971).
7. J. R. Ellis, *Nucl. Phys.* **B22**, 478 (1970).
8. J. C. Collins, A. Duncan and S. D. Joglekar, *Phys. Rev.* **D16**, 438 (1977).
9. H. B. Nielsen and M. Ninomiya, *Phys. Lett.* **B130**, 389 (1983).
10. E. Witten, *Nucl. Phys.* **B249**, 557 (1985).
11. A. Zee, *Phys. Rev. Lett.* **29**, 1198 (1972).
12. J. H. Lowenstein and B. Schroer, *Phys. Rev.* **D7**, 1929 (1973).
13. S. L. Adler, arXiv:hep-th/0405040.
14. D. Kharzeev, *Phys. Lett.* **B633**, 260 (2006), [arXiv:hep-ph/0406125].

15. D. Kharzeev and A. Zhitnitsky, *Nucl. Phys.* **A797**, 67 (2007), arXiv:0706.1026 [hep-ph].
16. D. E. Kharzeev, L. D. McLerran and H. J. Warringa, *Nucl. Phys.* **A803**, 227 (2008), arXiv:0711.0950 [hep-ph].
17. K. Fukushima, D. E. Kharzeev and H. J. Warringa, *Phys. Rev.* **D78**, 074033 (2008), arXiv:0808.3382 [hep-ph].
18. D. E. Kharzeev, *Annals Phys.* **325**, 205 (2010), arXiv:0911.3715 [hep-ph].
19. M. Giovannini and M. E. Shaposhnikov, *Phys. Rev. Lett.* **80**, 22 (1998), arXiv:hep-ph/9708303; *Phys. Rev.* **D57**, 2186 (1998), arXiv:hep-ph/9710234.
20. A.Yu. Alekseev, V.V. Cheianov and J. Frölich, *Phys. Rev. Lett.* **81**, 3503 (1998).
21. F. Wilczek, *Phys. Rev. Lett.* **40**, 279 (1978).
22. S. Weinberg, *Phys. Rev. Lett.* **40**, 223 (1978).
23. R. D. Peccei and H. R. Quinn, *Phys. Rev. Lett.* **38**, 1440 (1977).
24. F. Wilczek, *Phys. Rev. Lett.* **58**, 1799 (1987).
25. S. M. Carroll, G. B. Field and R. Jackiw, *Phys. Rev.* **D41**, 1231 (1990).
26. P. Sikivie, *Phys. Lett.* **B137**, 353 (1984).
27. S. S. Chern and J. Simons, *Annals Math.* **99**, 48 (1974).
28. E. Witten, *Phys. Lett.* **B86**, 283 (1979).
29. V. Skokov, A. Y. Illarionov and V. Toneev, *Int. J. Mod. Phys.* **A24**, 5925 (2009), arXiv:0907.1396 [nucl-th].
30. K. Tuchin, *Phys. Rev.* **C82**, 034904 (2010), arXiv:1006.3051 [nucl-th].
31. H. U. Yee, *JHEP* **0911**, 085 (2009), arXiv:0908.4189 [hep-th]; K. Y. Kim, B. Sahoo and H. U. Yee, arXiv:1007.1985 [hep-th].
32. A. Rebhan, A. Schmitt and S. A. Stricker, *JHEP* **1001**, 026 (2010), arXiv:0909.4782 [hep-th].
33. A. Gorsky, P. N. Kopnin and A. V. Zayakin, arXiv:1003.2293 [hep-ph].
34. V. A. Rubakov, arXiv:1005.1888 [hep-ph].
35. A. Gynther, K. Landsteiner, F. Pena-Benitez and A. Rebhan, arXiv: 1005.2587 [hep-th].
36. P. V. Buividovich, M. N. Chernodub, E. V. Luschevskaya and M. I. Polikarpov, *Phys. Rev.* **D80**, 054503 (2009), arXiv:0907.0494 [hep-lat].
37. P. V. Buividovich, E. V. Luschevskaya, M. I. Polikarpov and M. N. Chernodub, *JETP Lett.* **90**, 412 (2009), [*Pisma Zh. Eksp. Teor. Fiz.* **90**, 456 (2009)].
38. P. V. Buividovich, M. N. Chernodub, D. E. Kharzeev, T. Kalaydzhyan, E. V. Luschevskaya and M. I. Polikarpov, *Phys. Rev. Lett.* **105**, 132001 (2010), arXiv:1003.2180 [hep-lat].
39. M. Abramczyk, T. Blum, G. Petropoulos and R. Zhou, arXiv:0911.1348 [hep-lat].
40. T. Schäfer and D. Teaney, *Rept. Prog. Phys.* **72**, 126001 (2009), arXiv:0904.3107 [hep-ph].
41. D. T. Son and P. Surówka, *Phys. Rev. Lett.* **103**, 191601 (2009), arXiv:0906.5044 [hep-th].
42. Z. T. Liang and X. N. Wang, *Phys. Rev. Lett.* **94**, 102301 (2005), *Erratum-ibid.* **96**, 039901 (2006), arXiv:nucl-th/0410079.
43. B. Betz, M. Gyulassy and G. Torrieri, *Phys. Rev.* **C76**, 044901 (2007),

arXiv:0708.0035 [nucl-th].
44. F. Becattini, F. Piccinini and J. Rizzo, *Phys. Rev.* **C77**, 024906 (2008), arXiv:0711.1253 [nucl-th].
45. D. T. Son and A. R. Zhitnitsky, *Phys. Rev.* **D70**, 074018 (2004), arXiv:hep-ph/0405216.
46. M. A. Metlitski and A. R. Zhitnitsky, *Phys. Rev.* **D72**, 045011 (2005), arXiv:hep-ph/0505072.
47. G. Başar, G. V. Dunne and D. E. Kharzeev, *Phys. Rev. Lett.* **104**, 232301 (2010), arXiv:1003.3464 [hep-ph].
48. B. I. Abelev *et al.* (STAR Collaboration), *Phys. Rev. Lett.* **103**, 251601 (2009), arXiv:0909.1739 [nucl-ex].
49. B. I. Abelev *et al.* (STAR Collaboration), arXiv:0909.1717 [nucl-ex].
50. A. Ajitanand, S. Esumi and R. Lacey (PHENIX Collaboration), in: Proc. of the RBRC Workshops, vol. 96, 2010: "P- and CP-odd effects in hot and dense matter"; http://quark.phy.bnl.gov/ kharzeev/cpodd/
51. N. N. Ajitanand, R. A. Lacey, A. Taranenko and J. M. Alexander, arXiv:1009.5624 [nucl-ex].
52. A. Bzdak, V. Koch and J. Liao, *Phys. Rev.* **C81**, 031901 (2010), arXiv:0912.5050 [nucl-th]; arXiv:1005.5380 [nucl-th]; arXiv:1008.4919 [nucl-th].
53. S.-i. Nam, *Phys. Rev.* **D80**, 114025 (2009), arXiv:0911.0509 [hep-ph]; *Phys. Rev.* **D82**, 045017 (2010), arXiv:1004.3444 [hep-ph].
54. K. Fukushima, D. E. Kharzeev and H. J. Warringa, *Phys. Rev. Lett.* **104**, 212001 (2010), arXiv:1002.2495 [hep-ph]; *Nucl. Phys.* **A836**, 311 (2010), arXiv:0912.2961 [hep-ph].
55. A. Gorsky and M. B. Voloshin, arXiv:1006.5423 [hep-th].
56. B. Keren-Zur and Y. Oz, *JHEP* **1006**, 006 (2010), arXiv:1002.0804 [hep-ph].
57. W.-j. Fu, Y.-x. Liu and Y.-l. Wu, arXiv:1003.4169 [hep-ph].
58. S. Schlichting and S. Pratt, arXiv:1005.5341 [nucl-th]; arXiv:1009.4283 [nucl-th].
59. M. Asakawa, A. Majumder and B. Müller, *Phys. Rev.* **C81**, 064912 (2010), arXiv:1003.2436 [hep-ph].
60. S. A. Voloshin, arXiv:1006.1020 [nucl-th].
61. V. Orlovsky and V. Shevchenko, arXiv:1008.4977 [hep-ph].
62. A. R. Zhitnitsky, arXiv:1008.3598 [nucl-th].
63. B. Müller and A. Schäfer, arXiv:1009.1053 [hep-ph].
64. S. W. Mages, M. Aicher and A. Schäfer, arXiv:1009.1495 [hep-ph].
65. O. Rogachevsky, A. Sorin and O. Teryaev, arXiv:1006.1331 [hep-ph].
66. D. E. Kharzeev and D. T. Son, arXiv:1010.0038 [hep-ph].

$U_A(1)$ SYMMETRY RESTORATION FROM AN IN-MEDIUM η' MASS REDUCTION IN $\sqrt{s_{NN}} = 200$ GeV Au+Au COLLISIONS*

T. CSÖRGŐ

Dept. Physics, Harvard University,
17 Oxford Street, Cambridge, MA 02138, USA,
and
MTA KFKI RMKI,
H-1525 Budapest 114, P.O.Box 49, Hungary
csorgo@rmki.kfki.hu

R. VÉRTESI† and J. SZIKLAI‡

MTA KFKI RMKI,
H-1525 Budapest 114, P.O.Box 49, Hungary
†*vertesi@rmki.kfki.hu*
‡*sziklai@rmki.kfki.hu*

A reduction of the mass of the η' (958) meson may signal restoration of the $U_A(1)$ symmetry in a hot and dense hadronic matter, corresponding to the return of the 9th, "prodigal" Goldstone boson. We report on an analysis of a combined PHENIX and STAR data set on the intercept parameter of the two-pion Bose-Einstein correlation functions, as measuremed in $\sqrt{s_{NN}}$= 200 GeV Au+Au collisions at RHIC. To describe this combined PHENIX and STAR dataset, an in-medium η' mass reduction of at least 200 MeV is needed, at the 99.9% confidence level in a broad model class of resonance abundances.

1. Introduction

Although the quark model exhibits a $U(3)$ chiral symmetry in the limit of massless up, down and strange quarks, and in principle 9 massless Goldstone modes are expected to appear when this symmetry is broken, only 8 light pseudoscalar mesons are observed experimentally. This puzzling mystery is resolved by the Adler-Bell-Jackiw $U_A(1)$ anomaly: instantons tunneling between topologically different QCD vacuum states explicitely break the

*This work is supported by the Hungarian OTKA grant NK73143 and by HAESF, the Hungarian-American Enterprise Scholarship Fund.

$U_A(1)$ part of the $U(3)$ symmetry. Thus the 9th Goldstone boson is expected to be massive, and is associated with the η' meson, which has a mass of 958 MeV, approximately twice that of the other pseudoscalar mesons.

In high energy heavy ion collisions at RHIC, a hot and dense medium is created. Recent measurements of the direct photon spectrum in $\sqrt{s_{NN}}=$ 200 GeV Au+Au collisions indicate,[1] that the initial temperature in these reactions is at least 300 MeV, while hadrons as we know them may not exist above the Hagedorn temperature of $T_H \approx 170$ MeV.[2] Thus the matter created in heavy ion collisions at RHIC is hot enough to be a quark-gluon plasma.[1] Detailed analysis of the properties of this matter indicate that it flows like a perfect fluid,[3] and scaling properties of the elliptic flow indicate scaling with the number of constituent quarks,[4] hence this matter is sometimes referred to as a strongly interacting Quark-Gluon Plasma (sQGP),[5] or, in more direct terms, a perfect fluid of quarks.[3]

After this perfect fluid of quarks rehadronizes, a hot and dense hadronic matter may be created, where the $U_A(1)$ symmetry of the strong interactions may temporarily be restored.[6–8] Recent lattice QCD calculations indicate that such chirally symmetric but hadronic matter may exist below the critical temperature for quark deconfinement.[9] In such a medium, the mass of the $\eta'(958)$ mesons may be reduced to its quark model value of about 500 MeV, corresponding to the return of the "prodigal" 9th Goldstone boson.[7] Here we report on an indirect observation of such an in-medium η' mass modification based on a detailed analysis of PHENIX and STAR charged pion Bose-Einstein correlation (BEC) data.[10,11]

The abundance of the η' mesons with reduced mass may be increased at low p_{T}, by more than a factor of 10. One should emphasize that the η' (and η) mesons almost always decay after the surrounding hadronic matter has frozen out, due to their small annihilation and scattering cross sections, and their decay times that are much longer than the characteristic 5-10 fm/c decoupling times of the fireball created in high energy heavy ion collisions. Therefore one cannot expect a direct observation of the mass shift of the η' (or η) mesons: all detection possibilities of their in-medium mass modification have to rely on their enhanced production.

An enhancement of low transverse momentum η' mesons contributes to an enhanced production of soft charged pions mainly through the $\eta' \to \eta + \pi^+ + \pi^- \to (\pi^+ + \pi^0 + \pi^-) + \pi^+ + \pi^-$ decay chain and also through other, less prominent channels. As the η' decays far away from the fireball, the enhanced production of pions in the corresponding halo region will reduce the strength of the Bose-Einstein correlation between soft charged pions. The transverse mass ($m_T = \sqrt{m^2 + p_T^2}$) dependence of the extrapolated intercept parameter λ_* of the charged pion Bose-Einstein correlations was shown to be an observable that is sensitive to such an enhanced η' multiplicity, as pointed our first in Ref. 14 and discussed in Ref. 15. The predicted decrease of $\lambda_*(m_T)$ data at low transverse mass has been observed both by PHENIX[10] and STAR[11,12] at RHIC.

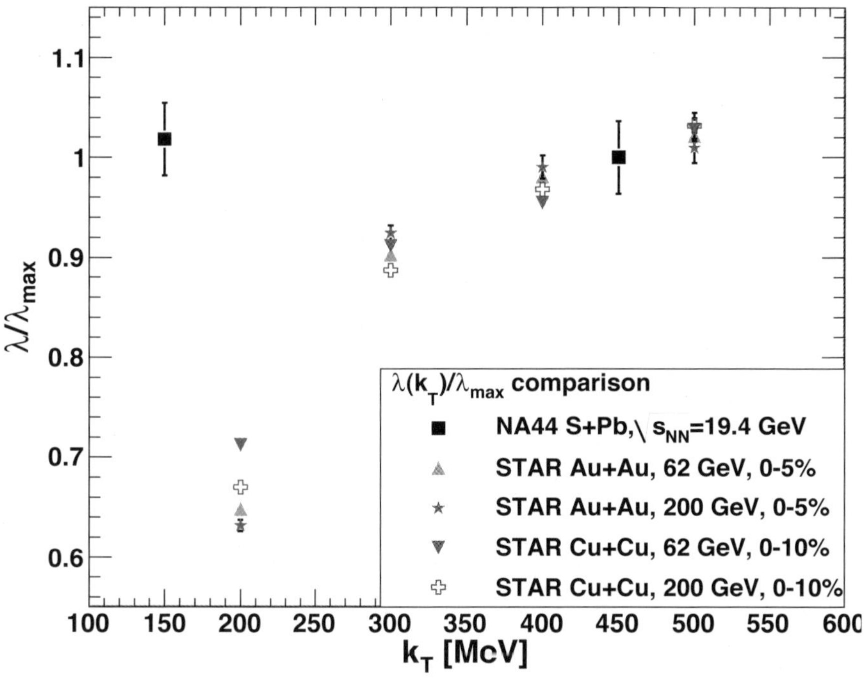

Fig. 1. The energy and system size dependence of the relative intercept parameter in the NA44 S+Pb and STAR Cu+Cu and Au+Au datasets.

Figure 1 indicates that a "hole" at low transverse mass, in the region characteristic to the $\eta' \to \eta + \pi^+ + \pi^- \to \eta' \to (\pi^+\pi^0\pi^-) + \pi^+ + \pi^-$ decay chain is not present in the S+Pb data set at CERN SPS at $\sqrt{s_{NN}}$= 19.4 GeV energy,[13] however, it is present and becomes slightly deeper as the system size and the colliding energy is increased from Cu+Cu to Au+Au and from $\sqrt{s_{NN}}$= 62.4 GeV to 200 GeV.[12]

2. Modeling and analysis method

Our main analysis tool was a Monte-Carlo simulation of the transverse mass dependence of the long lived resonance multiplicities including the possibility of an enhanced η' production at low transverse momentum, due to a partial in-medium $U_A(1)$ restoration and a related η' mass modification. This model and the related reduction of the effective intercept parameter of the two-pion Bose-Einstein correlation function was proposed first in Ref. 14 and detailed recently in Refs. 16–18.

In thermal models, the production cross sections of the light mesons are exponentially suppressed by the mass. Hence one expects about two orders of magnitude less η' mesons from the freeze-out than pions. This suppression, however, may be moderated as a consequence of a possible η' mass reduction, and the η' mesons may show up in an enhanced number. The number of in-medium η' mesons is calculated with an improved Hagedorn formula yielding the following η' enhancement factor:

$$f_{\eta'} = \left(\frac{m^*_{\eta'}}{m_{\eta'}}\right)^{\alpha} e^{-\frac{m_{\eta'} - m^*_{\eta'}}{T_{cond}}} . \tag{1}$$

This formula includes a prefactor with an expansion dynamics dependent exponent $\alpha \approx 1 - d/2$ for an expansion in d effective dimensions.[19] As a default value, $\alpha = 0$ was taken[14] and, for the systematic investigations, this parameter was varied between $-0.5 \leq \alpha \leq 0.5$. Other model parameters and their investigated ranges are described as follows: T_{cond} in the above formula corresponds to the temperature of the medium when the in-medium modified η' mesons are formed; its default value was taken to be $T_{cond} = 177$ MeV[14] and varied systematically between 140 and 220 MeV. Resonances with different masses were simulated with a mass dependent slope parameter $T_{eff} = T_{FO} + m\langle u_T \rangle^2$, where the default values of $T_{FO} = 177$ MeV and $\langle u_T \rangle = 0.48$[20] were utilized and systematically varied in the range of 100 MeV $\leq T_{FO} \leq$ 177 MeV and $0.40 \leq \langle u_T \rangle \leq 0.60$.

Once produced, the η' is expected to be decoupled from other hadronic matter, since its annihilation and scattering cross sections are very small.[7] If the η' mass is reduced in the medium, the observed η' spectrum will consist of two components. If the p_T of the η' is large enough, it can get on-shell and escape. This will produce a thermal component of the spectrum. Energy conservation at mid-rapidity implies ${m^*_{\eta'}}^2 + {p^*_{T,\eta'}}^2 = {m_{\eta'}}^2 + {p_{T,\eta'}}^2$. (In the latter equation the quantities marked with an asterisk denote the properties of the in-medium η', while the ones without an asterisk refer to the free η'.) On the other hand, η'-s with $p^*_{T,\eta'} \leq \sqrt{{m^*_{\eta'}}^2 - {m_{\eta'}}^2}$ will not be able to leave the hot and dense region through thermal fluctuation since they cannot compensate for the missing mass,[7,8] and thus will be trapped in the hot and dense region until it disappears. As the energy density of the medium is dissolved, the effect of QCD instantons increases and the trapped η' mesons regain their free mass and appear at low $p_{\rm T}$.

In our recent works of Refs. 16–18, we improved on earlier simulations of Ref. 14, that considered the trapped η' mesons to leave the dissolving medium with a negligible $p_{\rm T}$. That earlier approach resulted in a steep hole in the extrapolated intercept parameter $\lambda_*(m_{\rm T})$ at a characteristic transverse mass of $m_{\rm T} \leq 250$ MeV.[14,15,21] In that simplified scenario the only free parameter was the in-medium η' mass, determining the depth of the observed hole. In a recent analysis, summarized here, the η'-s from the decaying condensate were given a random transverse momentum, following Maxwell-Boltzmann statistics with an inverse slope parameter B^{-1}, which was necessary to obtain a quality description of the width and the slope of the $\lambda_*(m_{\rm T})$ data of PHENIX and STAR in the $m_{\rm T} \approx 300$ MeV region. Physically, B^{-1} is limited by T_{FO}, so the trapped η'-s may gain only moderate transverse momenta. Hence, the enhancement mostly appears at low p_T[6–8] just as in the first simulations. However, now the slope of "hole" of the $\lambda_*(m_{\rm T})$ curve is determined by B^{-1}, and, for certain values of the model parameters, the data can be reproduced quantitatively. (The λ_* values, actually used in the presented analysis, and their total errors are discussed in details in Ref. 18. Here $\lambda_*^{\rm max}$ is the $\lambda_*(m_{\rm T})$ value taken at $m_{\rm T} = 0.7$ GeV, with the exception of the STAR data, where the data point at the highest $m_{\rm T} = 0.55$ GeV is considered. Note that the $m_{\rm T}$ dependency of the $\lambda_*(m_{\rm T})$ measurements in the 0.5-0.7 GeV region is very weak.)

We have investigated a broad class of models of resonance production, including two different models that produce resonances without assuming local thermalization: FRITIOF[22] and UrQMD.[23] Resonance decays, including decay chains, were simulated with JETSET 7.4.[24]

The FRITIOF[22] Monte Carlo model, based on superposition of nucleon-nucleon collisions and the Lund string fragmentation model, cannot describe the behavior seen in $\lambda_*(m_{\rm T})/\lambda_*^{max}$ even when an arbitrary η' mass modification is considered. On the other hand, hadronic cascade based UrQMD,[23] as well as the quark coalescence model ALCOR[25] and the thermal resonance production models of Refs. 26–28, provide a successful fit in a certain range of the in-medium η' masses. The main difference between the thermal models that we utilized was in those resonance multiplicities that are not yet measured well: Ref. 26 predicts a factor of 1.6 more η-s and a factor of 3

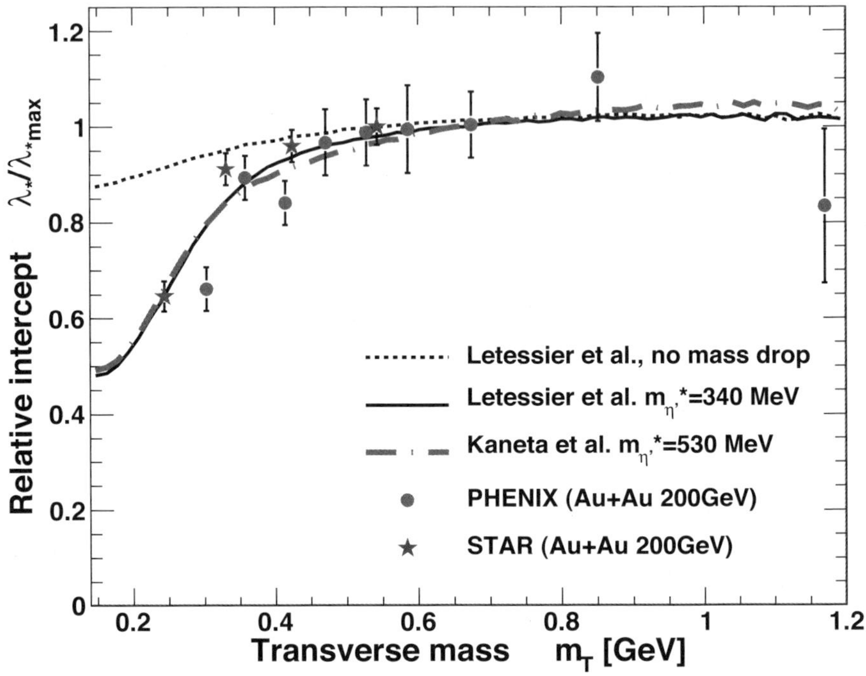

Fig. 2. The transverse mass dependence of the relative intercept parameter in the PHENIX and STAR dataset is reproduced with an in-medium mass modification of the η' mesons using two different resonance models as input. The same resonance models, but without in-medium mass modification, cannot explain these datasets.

more η'-s than the models of Refs. 27, 28. The relevant resonance fractions of these models are detailed in Table V of Ref. 18.

The dotted line in Fig. 2 indicates a scenario without an in-medium η' mass reduction, while the dot-dashed and solid lines show the enhancement required to describe the dip in the low $m_{\rm T}$ region of λ_* corresponding to the resonance multiplicities of Refs. 26, 27, respectively.

Based on extensive Monte-Carlo simulations, χ^2 of the fits to the data of Fig. 2 was computed as a function of $m^*_{\eta'}$ and B^{-1} for each resonance model and each fixed value of model parameters of α, T_{cond}, T_{FO} and $\langle u_{\rm T}\rangle$. The best values for the in-medium mass of η' mesons are in, or slightly below, the range $\sqrt{\frac{1}{3}(2m_{\rm K}^2 + m_\pi^2)} \leq m^*_{\eta'} \leq \sqrt{2m_{\rm K}^2 - m_\pi^2}$ predicted in Ref. 7, while all are above the lower limit of $m^*_{\eta'} \geq \sqrt{3}m_\pi$ given by Ref. 29. The $\lambda_*(m_{\rm T})/\lambda_*^{max}$ simulations for the best fits of two characteristic models are compared to the no-mass-drop scenario on Fig. 2, while the 1, 2 and 3-σ

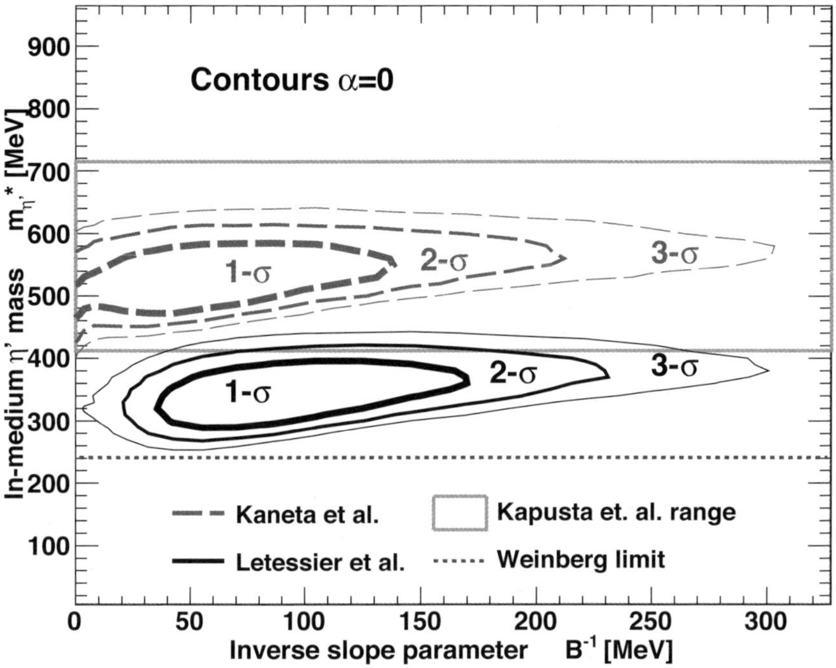

Fig. 3. Standard deviation contours on the (B^{-1}, $m^*_{\eta'}$) plain, obtained from $\lambda_*(m_\mathrm{T})/\lambda_*^{max}$ of Monte Carlo simulations based on particle multiplicities using two different models for hadronic resonances. The region between the horizontal solid lines indicates the theoretical range predicted by quark model considerations, while the dotted horizontal line stands for Weinberg's lower limit.

parameter boundaries are indicated in Fig. 3. Those models that describe both PHENIX and STAR $\lambda_*(m_\mathrm{T})/\lambda_*^{max}$ data in a statistically acceptable manner with the assumption of a sufficiently large in-medium η' mass reduction are all used for the estimation of systematics. The key parameters of the best fits are listed in Table 1.

3. Results

We have used different input models and setups to map the parameter space for a twofold goal: to determine, at least how big η' in-medium mass reduction is needed to be able to describe these datasets, and also to determine, what are the best values of the in-medium mass modification of the η' mesons. Utilizing our indirect method, we have also reconstracted the transverse mass dependent spectrum of these η' mesons.

3.1. *Lower limit on the in-medium η' mass reduction*

We excluded certain regions where a statistically acceptable fit to the data is not achievable, thus we can give a lower limit on the η' mass modification. At the 99.9 % confidence level, corresponding to a more than 5-σ effect, at least 200 MeV in-medium decrease of the mass of the $\eta'(958)$ meson was needed to describe both STAR 0-5 % central and PHENIX 0-30% central Au+Au data on $\lambda_*(m_{\rm T})/\lambda_*^{max}$ in $\sqrt{s_{NN}} = 200$ GeV Au+Au collisions at RHIC, in the considered model class.

3.2. *Best value of the in-medium η' mass reduction*

We have determined the best values and errors of the fitted $m^*_{\eta'}$ and B^{-1} parameters. The best simultaneous description of PHENIX[10] and STAR[11] relative intercept parameter data is achieved with an η' mass that is dramatically reduced in the medium created in central Au+Au collisions at RHIC from its vacuum value of 958 MeV to $340^{+50}_{-60}\,^{+280}_{-140} \pm 45$ MeV. The first error here is the statistical one determined by the 1-σ boundaries of the fit. The second error is from the choice of the resonance model and the parameters (α, T_{cond}, T_{FO} and $\langle u_{\rm T}\rangle$) of the simulation. The third error is the systematics resulting from slightly different PHENIX and STAR centrality ranges, particle identification and acceptance cuts. These effects have been estimated with Monte-Carlo simulations, detailed in Ref. 18, not to exceed 9.8%, 7% and 3% respectively. The main source of systematic errors is the choice of the resonance models. This is due to the unknown initial η' multiplicity, hence models like Ref. 26 with larger initial η' abundances require smaller in-medium η' mass modification, as compared to the models of Refs. 28, 27.

3.3. *Transverse mass spectrum of η' mesons*

In addition to the characterization of the in-medium η' mass modification, the transverse momentum spectra of the η and η' mesons have also been reported in Ref. 16. Figure 4 indicates the reconstructed spectrum of η' mesons in $\sqrt{s_{NN}}= 200$ GeV Au+Au collisions, for simulations based on resonance abundances of Refs. 27, 26. Normalization was carried out with respect to the η' multiplicity of the model described in Ref. 26. The spectrum of Fig. 4 features a characteristic low transverse momentum enhancement. Although PHENIX measured before the η spectrum in the $p_{\rm T} \geq 2$ GeV region,[30] as far as we know the spectrum of the η' particles has not been determined before in $\sqrt{s_{NN}}= 200$ GeV Au+Au collisions at RHIC.

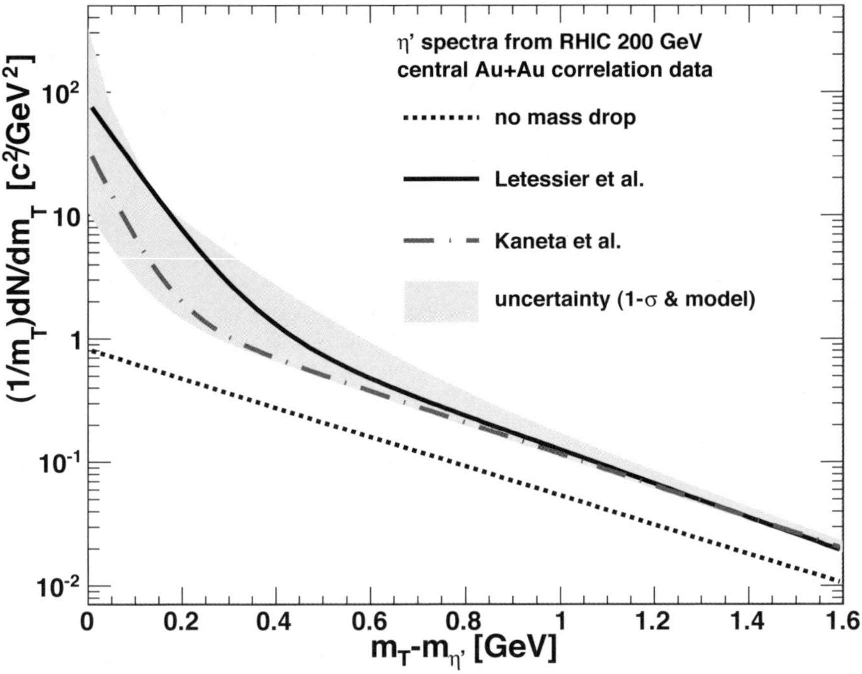

Fig. 4. The transverse mass dependent spectrum of the η' mesons, obtained using two different resonance models as input. The band indicates the systematic error, obtained from varying the resonance models as discussed in the text.

The restoration of the $U_A(1)$ symmetry and the symmetry between mass of the η and the η' mesons is illustrated on Fig. 5.

4. Discussion

Detailed analysis of the STAR and PHENIX $\lambda_*(m_\mathrm{T})/\lambda_*^{max}$ dataset recorded at 7.7, 9.2, 11.5, 39 and 62.4 GeV during 2010 has just been started,[31] marking the beginning of the RHIC energy scan program.

At present, detailed data are available from the NA44 collaboration at $\sqrt{s_{NN}}$ = 19.4 GeV[13] as well as from the STAR collaboration at $\sqrt{s_{NN}}$ = 62.4 and 200 GeV Cu+Cu and Au+Au collisions, the latter at different centrality classes within the 0%–80% range.[12] The NA44 data at $\sqrt{s_{NN}}$ = 19.4 GeV does not feature an η' mass drop effect. A positive sign of the η' mass modification is apparent in each case of the STAR datasets, indicating that the mass modification effect is nearly at maximum in $\sqrt{s_{NN}}$ = 200 GeV Au+Au collisions and reduces with decreasing cen-

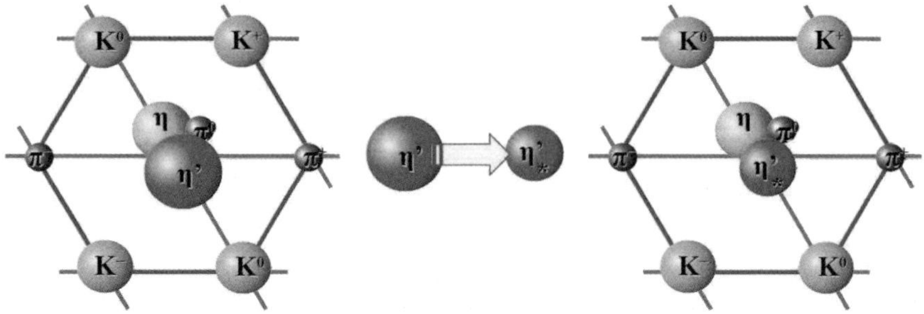

Fig. 5. The left hand side plot indicates the 9 pseudoscalar mesons, the size of the peebles is proportional to the mass, respectively. The central plot indicates that the η' meson reduces its mass in a hot and dense hadronic medium created in Au+Au collisions at RHIC. The right-hand side plot indicates that after the $U_A(1)$ symmetry is effectively restored, the mass of the η and the η' mesons will be similar.

Table 1. Best fits of $m^*_{\eta'}$ and B^{-1} for different resonance multiplicity models, followed by χ^2/NDF and the corresponding confidence level (CL). The integrated η' and η enhancement factors $f_{\eta'}$ and f_η are followed by the 5-σ limits of maximum in-medium masses. Errors on $m^*_{\eta'}$ values represent 1-σ boundaries, while the 5-σ limits include systematic errors too.

Resonance model	$m^*_{\eta'}$ (MeV)	B^{-1} (MeV)	$\frac{\chi^2}{NDF}$ (CL %)	$f_{\eta'}$	f_η	5-σ limit $m^*_{\eta'}$ (MeV)
ALCOR[25]	490^{+60}_{-50}	42	20.2/11 (4.29)	43.4	5.25	≤ 700
Kaneta[26]	530^{+50}_{-50}	55	22.8/11 (4.12)	25.6	3.48	≤ 730
Letessier[27]	340^{+50}_{-60}	86	18.9/11 (6.35)	67.6	4.75	≤ 570
Stachel[28]	340^{+50}_{-60}	86	18.9/11 (6.38)	67.6	4.97	≤ 570
UrQMD[23]	400^{+50}_{-40}	86	18.9/11 (6.14)	45.0	7.49	≤ 660

trality, colliding energy and system size. We have estimated the magnitude of the system size and energy dependence between 62.4 GeV Cu+Cu and 200 GeV Au+Au collisions to be not larger than 15%, which is substantially less than the dominant systematic error coming from the choice of the resonance model.

The *dilepton spectrum* has been measured recently in minimum bias Au+Au collisions at $\sqrt{s_{NN}} = 200$ GeV, and a large enhancement was observed in the low invariant mass region $m_{ee} < 1$ GeV.[32] Low transverse mass enhancement of the η' and η production results in dilepton enhancement just in this kinematic range.[7] Estimations using the enhancement factors in Table 1 indicate that the observed in-medium η' mass drop is indeed a promising candidate to explain this dilepton excess.

PHENIX recently reported a two-component transverse momentum spectrum in dilepton channel *direct photon measurements*,[32] which provides an additional testing possibility to constrain the two component structure of the η' spectra reported here.

5. Summary

Our report presents a statistically significant, indirect observation of an in-medium mass modification of the η' mesons in $\sqrt{s_{NN}} = 200$ GeV Au+Au collisions at RHIC. These results were recently published in Refs. 16–18. A similar search for in-medium η' mass modification provided negative result in S+Pb reactions at CERN SPS energies.[14] More detailed studies of the excitation function, the centrality and system size dependence of the $\lambda_*(m_\mathrm{T})/\lambda_*^{max}$ could provide important additional information about the onset and saturation of the partial $U_A(1)$ symmetry restoration in hot and dense hadronic matter. Studies of the low-mass dilepton spectrum and measurements of other decay channels of the η' meson may shed more light on the reported magnitude of the low p_T η' enhancement and the related $U_A(1)$ symmetry restoration in high energy heavy ion collisions.

Acknowledgments

We thank the Organizers of the Gribov 80 Memorial Workshop for creating an inspiring scientific atmosphere and providing an excellent setting for scientific discussions. We also would like to thank to professors R. J. Glauber and Gy. Wolf for inspiring and clarifying discussions. T. Cs. is grateful to R. J. Glauber for his kind hospitality at the Harvard University. Our research was supported by Hungarian OTKA grant NK 73143. T. Cs. has also been supported by a Senior Leader and Scholar Fellowship by the Hungarian American Enterprise Scholarship Fund (HAESF).

References

1. A. Adare *et al.* (PHENIX Collaboration), *Phys. Rev. Lett.* **104**, 132301 (2010), arXiv:0804.4168 [nucl-ex].
2. R. Hagedorn, *Nuovo Cim. Suppl.* **3**, 147 (1965).
3. K. Adcox *et al.* (PHENIX Collaboration), *Nucl. Phys.* **A757**, 184 (2005), arXiv:nucl-ex/0410003.
4. A. Adare *et al.* (PHENIX Collaboration), *Phys. Rev. Lett.* **98**, 162301 (2007), arXiv:nucl-ex/0608033.

5. E. Shuryak, *Nucl. Phys.* **A774**, 387 (2006), arXiv:hep-ph/0510123.
6. T. Kunihiro, *Phys. Lett.* **B219**, 363 (1989); *ibid.* **B245** 687(E) (1990).
7. J. I. Kapusta, D. Kharzeev and L. D. McLerran, *Phys. Rev.* **D53**, 5028 (1996).
8. Z. Huang and X. N. Wang, *Phys. Rev.* **D53**, 5034 (1996).
9. Z. Fodor and S. D. Katz, arXiv:0908.3341 [hep-ph].
10. S. S. Adler *et al.*, *Phys. Rev. Lett.* **93**, 152302 (2004).
11. J. Adams *et al.*, *Phys. Rev.* **C71**, 044906 (2005).
12. B. I. Abelev *et al.*, *Phys. Rev.* **C80**, 024905 (2009).
13. H. Beker *et al.*, *Phys. Rev. Lett.* **74**, 3340 (1995).
14. S. E. Vance, T. Csörgő and D. Kharzeev, *Phys. Rev. Lett.* **81**, 2205 (1998).
15. T. Csörgő, *Heavy Ion Phys.* **15**, 1 (2002), arXiv:hep-ph/0001233.
16. T. Csörgő, R. Vértesi and J. Sziklai, *Phys. Rev. Lett.* **105**, 182301 (2010), arXiv:0912.5526 [nucl-ex].
17. R. Vértesi, T. Csörgő and J. Sziklai, *Nucl. Phys.* **A830**, 631C (2009).
18. R. Vértesi, T. Csörgő and J. Sziklai, arXiv:0912.0258 [nucl-ex].
19. T. Csörgő and B. Lörstad, *Phys. Rev.* **C54**, 1390 (1996).
20. S. S. Adler *et al.*, *Phys. Rev.* **C69**, 034909 (2004).
21. M. Csanád for PHENIX Collaboration, *Nucl. Phys.* **A774**, 611 (2006).
22. B. Anderson *et al.*, *Nucl. Phys.* **B281**, 289 (1987).
23. M. Bleicher *et al.*, *J. Phys.* **G25**, 1859 (1999).
24. T. Sjöstrand, *Comp. Phys. Commun.* **82**, 74 (1994).
25. T. S. Biró, P. Lévai and J. Zimányi, *Phys. Lett.* **B347**, 6 (1995).
26. M. Kaneta and N. Xu, arXiv:nucl-th/0405068.
27. J. Letessier and J. Rafelski, *Eur. Phys. J.* **A35**, 221 (2008).
28. S. A. Bass *et al.*, *Nucl. Phys.* **A661**, 205 (1999).
29. S. Weinberg, *Phys. Rev.* **D11**, 3583 (1975).
30. S. S. Adler *et al.*, *Phys. Rev.* **C75**, 024909 (2007).
31. B. I. Abelev *et al.*, *Phys. Rev.* **C81**, 024911 (2010).
32. A. Adare *et al.*, *Phys. Rev.* **C81**, 034911 (2010).

INITIAL TEMPERATURE OF THE STRONGLY INTERACTING QUARK GLUON PLASMA CREATED AT RHIC

M. CSANÁD

Department of Atomic Physics, Eötvös University,
Budapest, Pázmány P. s. 1/a, H-1117, Hungary
csanad@elte.hu

A 1+3 dimensional solution of relativistic hydrodynamics is analyzed in this paper. Momentum distribution and other observables are calculated from the solution and compared to hadronic measurements from the Relativistic Heavy Ion Collider (RHIC). The solution is compatible with the data, but only the freeze-out point of the evolution is determined. Many equation of states and initial states (initial temperatures) are valid with the same freeze-out distribution, thus the same hadronic observables. The observable that would distinguish between these initial temperatures is momentum distribution of photons, as photons are created throughout the evolution of the fireball created in RHIC collisions. The PHENIX experiment at RHIC measures such data via low invariant mass e^+e^- pairs. Average temperature from this data is $T = 221 \pm 23$(stat)± 18(sys) MeV, while a model calculation with initial temperature $T_{init} = 370$ MeV agree with the data.

Keywords: Heavy ion collisions; hydrodynamics; equation of state; temperature; hadron spectra; photon spectra.

1. Perfect fluid hydrodynamics

In the last several years it has been revealed that the strongly interacting Quark Gluon Plasma produced[1] in the collisions of the Relativistic Heavy Ion Collider (RHIC) is a nearly perfect fluid,[2] i.e. it can be described with perfect fluid hydrodynamics.

Perfect fluid hydrodynamics is based on local conservation of entropy or number density (n), energy-momentum density ($T^{\mu\nu}$). The fluid is perfect if the energy-momentum tensor is diagonal in the local rest frame, i.e. viscosity and heat conduction are negligible. This can be assured if $T^{\mu\nu}$ is chosen as $T^{\mu\nu} = (\epsilon + p)u^\mu u^\nu - pg^{\mu\nu}$, where u^μ is the flow field in the fluid, ϵ is energy density, p is pressure and $g^{\mu\nu}$ is the metric tensor, diag$(1, -1, -1, -1)$.

The conservation equations are closed by the equation of state, which gives the relationship between ϵ and p. Typically $\epsilon = \kappa p$ is chosen, where the proportionality "constant" κ may depend on temperature T, which in turn is connected to the density n and pressure p via $p = nT$.

The exact, analytic result for hydrodynamic solutions is, that the hadronic observables do not depend on the initial state or the dynamical equations separately, just through the final state.[3,4] Thus if we fix the final state from the data, the equation of state can be anything that is compatible with the particular solution. This is the framework of several hydro solutions as detailed in the next paragraph.

Many solve the above equations numerically, but there are only a few exact solutions. Historically the first is the implicit 1+1 dimensional accelerating solution of Landau and Khalatnikov.[5–7] Another renowned 1+1 dimensional solution of relativistic hydrodynamics was found by Hwa and Bjorken:[8–10] it is simple, explicit and exact, but accelerationless.

Important are solutions[11,12] which are explicit and describe a relativistic acceleration, i.e. combine the properties of the Landau-Khalatnikow and the Hwa-Bjorken solutions. With these one can have an advanced estimate on the energy density,[13] but investigation of transverse dynamics is not possible by these solutions.

The only exact 1+3 dimensional relativistic solution, from which observables like momentum distribution, correlation function and elliptic flow were calculated[4] is the one in Ref. 14. Observables from this solution were computed and compared to data in Ref. 4.

2. The analyzed solution

The analyzed solution[14] describes an ellipsoidally symmetric expansion. The ellipsoids are given by constant values of the scale variable s:

$$s = \frac{r_x^2}{X(t)^2} + \frac{r_y^2}{Y(t)^2} + \frac{r_z^2}{Z(t)^2}, \tag{1}$$

here $X(t)$, $Y(t)$, and $Z(t)$ are time dependent scale parameters (axes of the $s = 1$ ellipsoid), only depending on the time t. Spatial coordinates are r_x, r_y, and r_z. The velocity-field is described by a Hubble-type expansion:

$$u^\mu(x) = \gamma \left(1, \frac{\dot{X}(t)}{X(t)} r_x, \frac{\dot{Y}(t)}{Y(t)} r_y, \frac{\dot{Z}(t)}{Z(t)} r_z\right), \tag{2}$$

where x means the four-vector (t, r_x, r_y, r_z), and $\dot{X}(t) = dX(t)/dt$, similarly for Y and Z. The $\dot{X}(t) = \dot{X}_0$, $\dot{Y}(t) = \dot{Y}_0$, $\dot{Z}(t) = \dot{Z}_0$ (i.e. all are constant) criteria must be fulfilled, ie. the solution is accelerationless. This is one of the drawbacks of this solution.

The temperature $T(x)$ and number density $n(x)$ are:

$$n(x) = n_0 \left(\frac{\tau_0}{\tau}\right)^3 \nu(s), \tag{3}$$

$$T(x) = T_0 \left(\frac{\tau_0}{\tau}\right)^{3/\kappa} \frac{1}{\nu(s)}, \tag{4}$$

$$p(x) = p_0 \left(\frac{\tau_0}{\tau}\right)^{3(\kappa+1)/\kappa}, \tag{5}$$

where τ is the proper time, s is the above scaling variable, $\nu(s)$ is an arbitrary function, while $n_0 = n|_{s=0,\tau=\tau_0}$, $T_0 = T|_{s=0,\tau=\tau_0}$ and $p_0 = p|_{s=0,\tau=\tau_0}$ with $p_0 = n_0 T_0$ (hence p does not depend on the spatial coordinates only τ). Furthermore, τ_0 is the time of the freeze-out, thus T_0 is the central freeze-out temperature. The parameter κ is arbitrary, i.e. any value of κ yields a solution. The function $\nu(s)$ is chosen as:

$$\nu(s) = e^{-bs/2}, \tag{6}$$

where b is then the temperature gradient. If the fireball is the hottest in the center, then $b < 0$. An example time evolution of the temperature distribution is shown in Fig. 1.

3. Hadronic observables

The picture widely used in hydro models is that the pre freeze-out (FO) medium is described by hydrodynamics, and the post FO medium is that of observed hadrons. In our framework we assume that the freeze-out can happen at any proper time, e.g. in case of a self-quenching effect or if the phase space evolution is that of a collisionless gas. See details in Ref. 4. The hadronic observables can be extracted from the solution via the phase-space distribution at the FO. This will correspond to the hadronic final state or source distribution $S(x, p)$.

We do not need to fix a special equation of state, because the same final state can be achieved with different equations of state or initial conditions.[3] Thus the hadronic observables do not restrict the value of κ.

The first calculated observable is invariant transverse momentum distribution $N_1(p_t)$ of a particle with mass m:[4]

$$N_1(p_t) = 2\pi \overline{N}\, \overline{V} \left(m_t - \frac{p_t^2(T_{\text{eff}} - T_0)}{m_t T_{\text{eff}}}\right) \exp\left[-\frac{m_t^2 + m^2}{2m_t T_0} - \frac{p_t^2}{2m_t T_{\text{eff}}}\right],$$

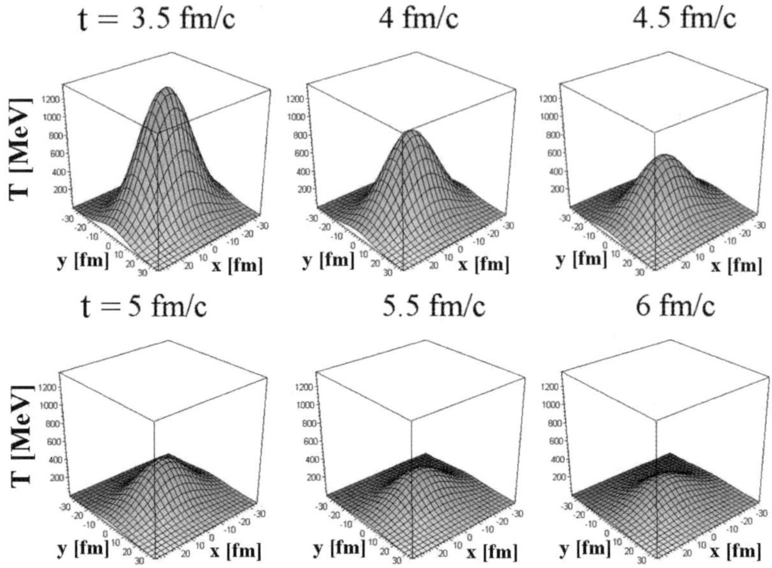

Fig. 1. Temperature distribution in the transverse plane (x-y) is shown for various times with an example parameter set. The fireball is the hottest in the center and it cools down as time elapses.

with the following auxiliary quantities:

$$\overline{N} = \mathcal{N} n_0 \left(\frac{2T_0 \tau_0^2 \pi}{m_t} \right)^{3/2}, \tag{7}$$

$$\overline{V} = \sqrt{\left(1 - \frac{T_0}{T_x}\right)\left(1 - \frac{T_0}{T_y}\right)\left(1 - \frac{T_0}{T_z}\right)}, \tag{8}$$

$$\frac{1}{T_{\text{eff}}} = \frac{1}{2}\left(\frac{1}{T_x} + \frac{1}{T_y}\right). \tag{9}$$

Furthermore, T_x ,T_y, T_z are the effective temperatures, i.e. inverse logarithmic slopes of the distribution:

$$T_x = T_0 + \frac{m_t T_0 \dot{X}_0^2}{b(T_0 - m_t)}, \tag{10}$$

$$T_y = T_0 + \frac{m_t T_0 \dot{Y}_0^2}{b(T_0 - m_t)}, \tag{11}$$

$$T_z = T_0 + \frac{m_t T_0 \dot{Z}_0^2}{b(T_0 - m_t)}, \tag{12}$$

where $\dot{X}_0$, $\dot{Y}_0$ and $\dot{Z}_0$ are the (constant) expansion rates of the fireball, T_0 its central temperature at FO and b the temperature gradient.

We also calculate the elliptic flow, which describes the azimuthal asymmetry of the momentum distribution:[4]

$$v_2(p_t) = \frac{I_1(w)}{I_0(w)} \tag{13}$$

where I_0, and I_1 are the modified Bessel functions while

$$w = \frac{p_t^2}{4m_t}\left(\frac{1}{T_y} - \frac{1}{T_x}\right). \tag{14}$$

See details of the calculation in Ref. 4. The formula for v_2 gives back previously found formulas of non-relativistic solutions[15] and relativistic solutions.[16,17] Also the formula for $N_1(p_t)$ is similar to results of the previously mentioned papers.

Third observable we calculate is the two-particle Bose-Einstein (HBT) correlation radii of identical bosons:[18]

$$R_x^2 = \frac{T_0\tau_0^2(T_x - T_0)}{M_t T_x}, \tag{15}$$

$$R_y^2 = \frac{T_0\tau_0^2(T_y - T_0)}{M_t T_y}, \tag{16}$$

$$R_z^2 = \frac{T_0\tau_0^2(T_z - T_0)}{M_t T_z}, \tag{17}$$

where M_t is the transverse mass belonging to the average momentum $K = 0.5(p_1 + p_2)$ of the pair, which is (at mid-rapidity) $M_t = 0.5\,(m_{t,1} + m_{t,2}\,)$. The $m_{t,1}$, and $m_{t,2}$ quantities are the transverse masses, the T_x, T_y, and T_z are the effective temperatures belonging to the average momentum (i.e. here $T_x = T_x|_{M_t}$). The calculations are detailed in ref.[1] To compare the HBT radii with the data the Bertsch-Pratt[19] frame is to be used. It has three axes: the *out* is the direction of the average transverse momentum of the pair, the *long* direction is equal to the direction z, and the *side* direction is orthogonal to both of them. The result for $R_{\rm out}$, $R_{\rm side}$ and $R_{\rm long}$ is:

$$R_{\rm out}^2 = R_{\rm side}^2 = \frac{R_x^2 + R_y^2}{2}, \tag{18}$$

$$R_{\rm long}^2 = R_z^2. \tag{19}$$

Clearly in this solution the out and side radii are equal. This can be attributed to the instantaneous freeze-out; a non-zero freeze-out duration would make $R_{\rm out}^2$ bigger by a term of $\Delta\tau^2 p_t^2/E^2$. Supported by the data, we use the $\Delta\tau = 0$ approximation in our solution, which corresponds to instantaneous freeze-out.

An important consequence of the above results is that neither spectra nor elliptic flow nor correlation radii depend on the EoS itself, only through the final state parameters. If we determine for example T_0, the freeze-out central (at the center means here $r_x = r_y = r_z = 0$) temperature, κ or the initial temperature T_{initial} still cannot be calculated. We only know that they are connected through the consistency condition $T_{\text{initial}} = T_0(\tau_0/\tau_{\text{initial}})^{3/\kappa}$, see Eq. (4), i.e. they can be co-varied (softer EoS requires smaller initial temperature for a given freeze-out proper-time). Thus κ or T_{initial} has to be determined from another measurement, e.g. the spectrum of thermal photons.

4. Comparing the hadronic observables to RHIC data

The above results were compared in Ref. 4 to PHENIX data of 200 GeV Au+Au collisions. Above formulas were fitted to describe spectra and HBT positive pion data[20,21] (0-30% centrality) and elliptic flow data[22] for $\pi^\pm$, $K^\pm$, p and $\overline{\text{p}}$ particles (0-92% centrality).

The fit results are shown in Fig. 2, see details of the fit in Ref. 4. Important are the following: central freeze-out temperature T_0 is around 200 MeV for both datasets (with an error of 7 MeV), and the fireball is colder away from the center. The expansion eccentricity is positive, it tells us that the expansion is faster in-plane. Because of the Hubble-flow this means that the source is in-plane elongated, similarly to the result of Ref. 23. The freeze-out happens at a proper-time of $\tau_0 = 7.7 \pm 0.8$ fm/c.

Our fit parameters describe the fireball at the freeze-out. However, the solution is time-dependent, most importantly the temperature depends on time as described by Eq. (4). We plotted the time-dependence of the central temperature in Fig. 3 for several values of κ, i.e. several EoS'. From this, assuming for example an average κ of 10, see Ref. 24, one can also calculate the initial central temperature of the fireball based on Eq. (4):

$$T_{\text{initial}} = T_0 \left(\frac{\tau_0}{\tau_{\text{initial}}} \right)^{3/\kappa} \tag{20}$$

This yields 370 MeV at t_{initial}=1 fm/c (note that $t = \tau$ at the center).

One still would like to determine the value for κ and the initial temperature from experimental observables. The key are thermal photon invariant momentum distributions, because photons are not suppressed by the medium, they are produced according to the local temperature at all times. If one could measure the yield of thermal (direct) photons, one could compare it to time-integrated hydrodynamic results and determine EoS and initial temperature. In the next section thus experimental results on direct photons are reviewed.

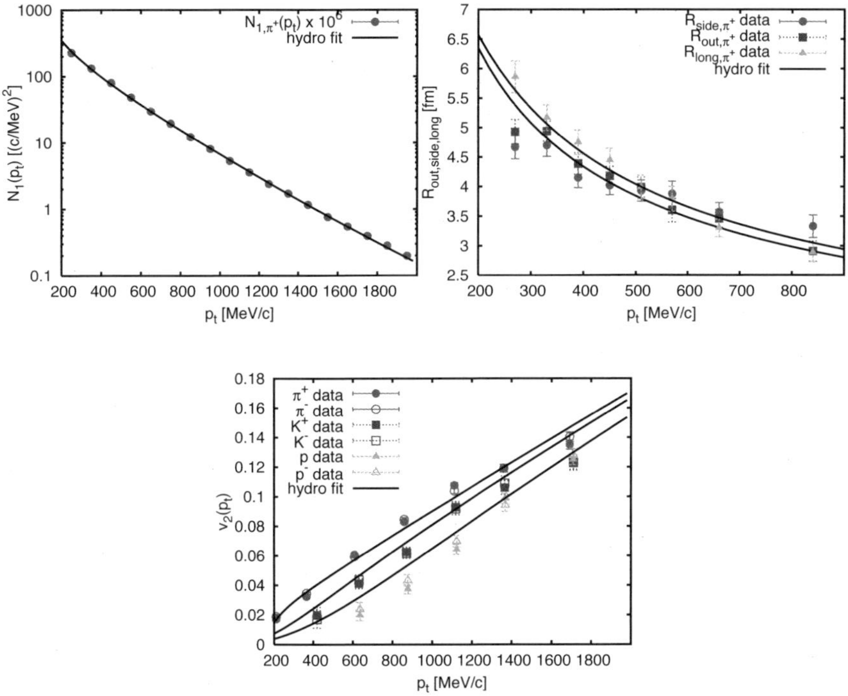

Fig. 2. Fits to 0–30% centrality PHENIX Au+Au spectra[20] (top left) HBT radii[22] (top right) and 0-92% centrality PHENIX Au+Au elliptic flow[21] (bottom). See details of the fit in Ref. 4.

5. Direct photons at the PHENIX experiment at RHIC

Thermal photons from the partonic phase are predicted to be the dominant source of direct photons for $1 < p_t < 3$ GeV/c in Au+Au collisions at the Relativistic Heavy Ion Collider (RHIC).[26] The measurement of direct photons in this kinematic domain is however very difficult due to the background from hadronic decay photons. At PHENIX, an alternative approach is used.[25] The idea is that any source of high energy photons can also emit virtual photons which then convert to e^+e^- pairs.

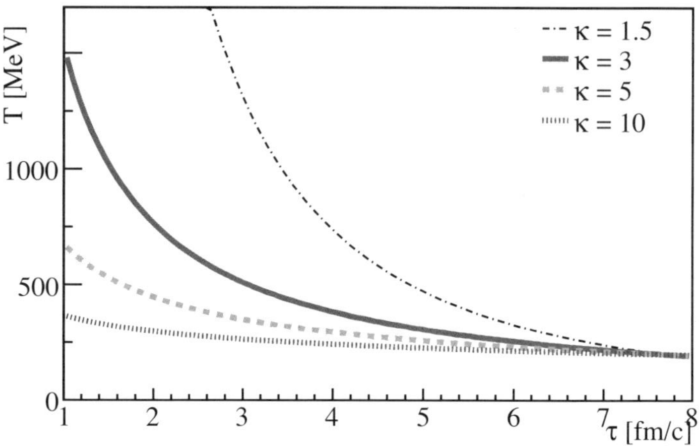

Fig. 3. Time dependence of the central temperature of the fireball, from Eq. (4) is shown for different κ values. In reality κ may change with time, we show here the curves only for fixed κ values. Assuming an average of $\kappa = 10^{24}$ one gets an initial temperature of 370 MeV at t_{initial}=1 fm/c, in agreement with PHENIX measurements.[25]

The relation between photon production and the associated e^+e^- pair production is then[25,27]

$$\frac{d^2 n_{ee}}{dm} = \frac{2\alpha}{3\pi m}\sqrt{1 - \frac{4m_e^2}{m^2}}\left(1 + \frac{2m_e^2}{m^2}\right) S dn_\gamma \tag{21}$$

Here α is the fine structure constant, m_e and m are the masses of the electron and the e^+e^- pair respectively, and S is a process dependent factor.

Invariant mass distribution of e^+e^- pairs was measured thus in PHENIX[25,27] for $m_{ee} < 300$ MeV/c^2 and for $1 < p_z < 5$ GeV/c in Au+Au and $p + p$ collisions at $\sqrt{s_{NN}} = 200$ GeV. Top panel of Fig. 4 shows the measured[25] mass spectra of e^+e^- pairs in $p + p$ and Au+Au collisions for different ranges of p_t, comparing them to expected yields from dielectron decays of hadrons, calculated using a Monte Carlo hadron decay generator based on meson production as measured by PHENIX.[28] The Au+Au data show a relatively high excess above the hadronic background, which indicates an the production of virtual photons in Au+Au collisions. PHENIX assumed that the excess is entirely due to internal conversion of direct photons and deduce the real direct photon yield from the e^+e^- pair yield using Eq. (21).[28]

In order to calculate direct photon yields the $f(m) = (1 - r)f_c(m) + rf_{dir}(m)$ fit function was used to the mass distribution in each p_t bin separately, where $f_c(m)$ is the mass distribution from the hadronic decays

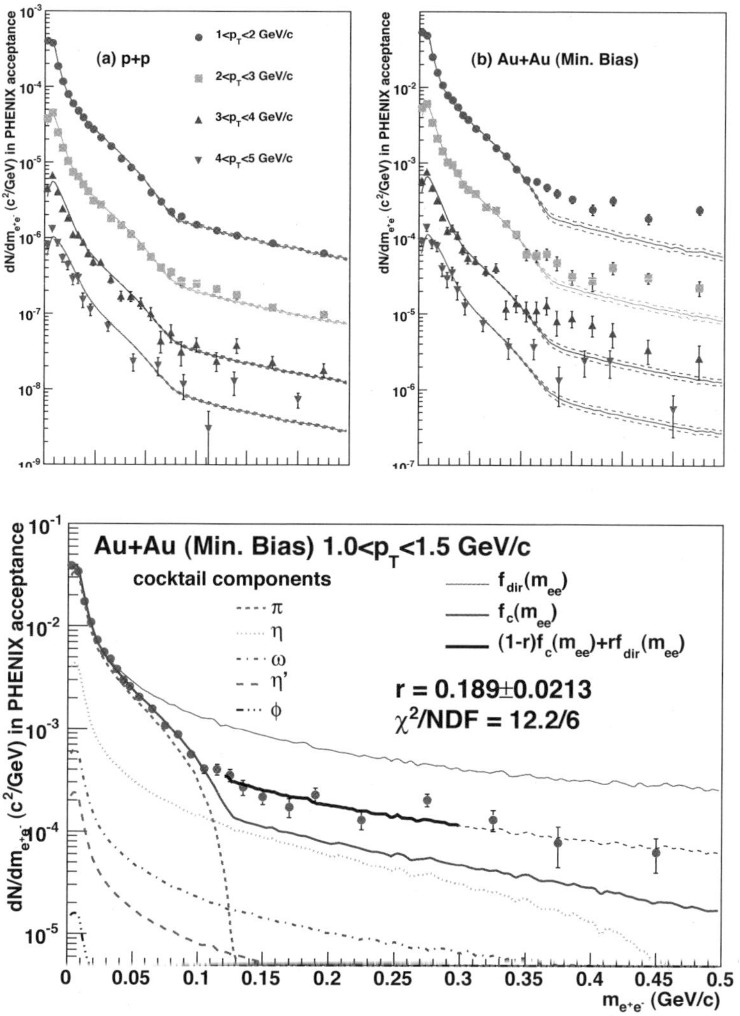

Fig. 4. The e^+e^- pair invariant mass distributions in $p+p$ and minimum bias Au+Au collisions from Ref. 25 is shown in the top panel. Mass distribution of electron pairs measured in Au+Au minimum bias events[25] is shown for $1.0 < p_t < 1.5$ GeV/c in the bottom panel. The fit is explained in the text and originally in Ref. 25.

(shown in the top panel of Fig. 4 with solid lines), and $f_{dir}(m)$ is the expected shape of the direct photon internal conversion, and r is the fit parameter. Such a fit is shown in the bottom panel of Fig. 4 for one p_t bin. The direct photon yield can be calculated then using the direct photon fraction r as $dN^{\text{direct}}(p_t) = r \times dN^{\text{inclusive}}(p_t)$ where $dN^{\text{inclusive}}(p_t)$ is the yield of all photons (direct and decay inclusive). The inclusive photon yield can be calculated as $dN_\gamma^{\rm inclusive} = N_{ee}^{\rm data} \times (dN_\gamma^{\rm cocktail}/N_{ee}^{\rm cocktail})$, where $N_{ee}^{\rm data}$ and $N_{ee}^{\rm cocktail}$ are the measured and cocktail e^+e^- pair yields and $dN_\gamma^{\rm cocktail}$ is the yield of photons from the cocktail.

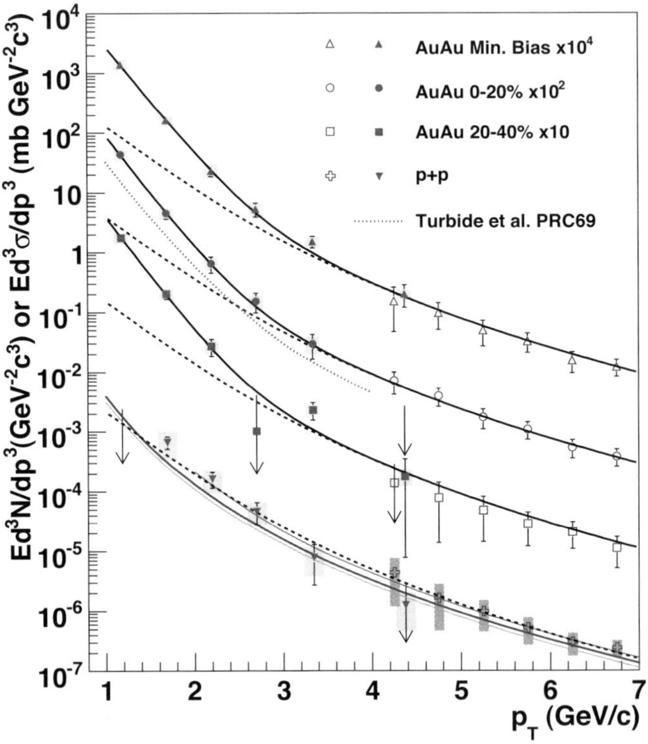

Fig. 5. Invariant cross section ($p + p$) and invariant yield (Au+Au) of direct photons is shown, taken from Ref. 25. The red points are from the analysis presented in Ref. 25 and blue points are from Refs. 29, 30. Dashed black curves are modified power-law fits to the $p+p$ data,[25] while the solid black curves are exponential plus modified power-law fits.[25] The red dashed curve is the direct photon spectrum in central Au+Au collisions in Ref. 26.

The obtained direct photon spectra[25] are shown in Fig. 5 and compared to similar data of Refs. 29, 30. The direct photon yields are consistent with a NLO pQCD calculation in $p + p$. The shape of the direct photon spectra above a binary collisions scaled $p+p$ spectrum is exponential in p_t, with an inverse slope $T = 221 \pm 23(\text{stat})\pm 18(\text{sys})$ MeV in central Au+Au collisions.[25] This temperature can be regarded as a time-average of the temperature of the fireball, represents thus an experimental lower limit to the initial temperature. The shape of the thermal photon spectrum from a model calculation with initial temperature $T_{init} = 370$ MeV agrees with the data.[25]

6. Summary

Exact parametric solutions of perfect hydrodynamics were long searched for in order to describe the matter produced in heavy ion collisions at RHIC. We extracted hadronic observables from the relativistic, 1+3 dimensional, ellipsoidally symmetric, exact solution of Ref. 14. We calculated momentum distribution, elliptic flow and Bose-Einstein correlation radii from the solution. We compared the results to 200 GeV Au+Au PHENIX data.[20–22] The solution is compatible with the data. If using an experimentally determined average EoS of $\kappa \approx 10$,[24] our results yield approximately 370 MeV at $\tau_{\text{initial}}=1$ fm/c, in agreement with recent PHENIX photon measurements.[25] The time-average of the temperature in these collisions is $T = 221 \pm 23(\text{stat})\pm 18(\text{sys})$ MeV,[25] which is a lower limit of the initial temperature. From detailed comparisons with hydrodynamical models of direct photon emission, PHENIX concluded that $T_{init} = 300$ MeV is the lowest possible initial temperature at 1 fm/c, that is consistent with such an average slope parameter. In the future we will compare hadron and photon observables simultaneously to the analyzed analytic hydrodynamic solution.

Acknowledgments

The author is thankful for the support of the organizers of the Gribov-80 Memorial Workshop, it was a pleasure to participate in a conference of so many well-known scientists, in such a pleasent and friendly atmosphere. The author also gratefully acknowledges the support of the Hungarian OTKA grant NK 73143.

References

1. K. Adcox *et al.*, *Nucl. Phys.* **A757**, 184 (2005).
2. R. A. Lacey *et al.*, *Phys. Rev. Lett.* **98**, 092301 (2007).
3. M. Csanád, *Acta Phys. Polon.* **B40**, 1193 (2009).
4. M. Csanád and M. Vargyas, *Eur. Phys. J.* **A44**, 473 (2010).
5. L. D. Landau, *Izv. Akad. Nauk SSSR Ser. Fiz.* **17**, 51 (1953).

6. I. M. Khalatnikov, *Zhur. Eksp. Teor. Fiz.* **27**, 529 (1954).
7. S. Z. Belenkij and L. D. Landau, *Nuovo Cim. Suppl.* **3S10**, 15 (1956).
8. R. C. Hwa, *Phys. Rev.* **D10**, 2260 (1974).
9. C. B. Chiu, E. C. G. Sudarshan and K.-H. Wang, *Phys. Rev.* **D12**, 902 (1975).
10. J. D. Bjorken, *Phys. Rev.* **D27**, 140 (1983).
11. T. Csörgő, M. I. Nagy and M. Csanád, *Phys. Lett.* **B663**, 306 (2008).
12. A. Bialas, R. A. Janik and R. B. Peschanski, *Phys. Rev.* **C76**, 054901 (2007).
13. M. Csanád, M. I. Nagy and T. Csörgő, *Eur. Phys. J. ST* **155**, 19 (2008).
14. T. Csörgő, L. P. Csernai, Y. Hama and T. Kodama, *Heavy Ion Phys.* **A21**, 73 (2004).
15. T. Csörgő, S. V. Akkelin, Y. Hama, B. Lukacs and Y. M. Sinyukov, *Phys. Rev.* **C67**, 034904 (2003).
16. M. Csanád, T. Csörgő and B. Lörstad, *Nucl. Phys.* **A742**, 80 (2004).
17. M. Csanád *et al.*, *Eur. Phys. J.* **A38**, 363 (2008).
18. R. M. Weiner, *Introduction to Bose-Einstein correlations and subatomic interferometry* (Wiley and Sons, 2000).
19. S. Pratt, *Phys. Rev.* **D33**, 1314 (1986).
20. S. S. Adler *et al.*, *Phys. Rev.* **C69**, 034909 (2004).
21. S. S. Adler *et al.*, *Phys. Rev. Lett.* **93**, 152302 (2004).
22. S. S. Adler *et al.*, *Phys. Rev. Lett.* **91**, 182301 (2003).
23. M. Csanád, B. Tomášik and T. Csörgő, *Eur. Phys. J.* **A37**, 111 (2008).
24. R. A. Lacey and A. Taranenko, *PoS* **CFRNC2006**, 021 (2006).
25. A. Adare *et al.*, *Phys. Rev. Lett.* **104**, 132301 (2010).
26. S. Turbide, R. Rapp and C. Gale, *Phys. Rev.* **C69**, 014903 (2004).
27. A. Adare *et al.*, *Phys. Rev.* **C81**, 034911 (2010).
28. A. Adare *et al.*, *Phys. Lett.* **B670**, 313 (2009).
29. S. S. Adler *et al.*, *Phys. Rev. Lett.* **94**, 232301 (2005).
30. S. S. Adler *et al.*, *Phys. Rev. Lett.* **98**, 012002 (2007).

EXACT SOLUTIONS OF PERFECT FLUID HYDRODYNAMICS: THEORY AND APPLICATIONS IN HIGH–ENERGY EXPERIMENTS*

MÁRTON I. NAGY

MTA KFKI RMKI, H-1525 Budapest 114, P.O. Box 49, Hungary
nagymarci@rmki.kfki.hu

Some exact and explicit solutions of the relativistic hydrodynamical equations are reviewed in context of high-energy heavy ion physics. Hydrodynamics is a powerful tool to describe the experimentally observed features (*e.g.* scaling laws) in the final state of high-energy collisions. Hydrodynamical solutions (both exact, analytic and numerical) can then be used to draw consequences on the initial state or the equation of state of the created matter. These findings can help to reach the ultimate goal of heavy-ion physics, the mapping of the phase structure of the strong interaction. In this paper we discuss exact and phenomenologically relevant solutions of hydrodynamics, and hint at some applications of them. We detail one important application: the estimation of the initial energy density from the observed final state.

1. Introduction

The goal of heavy-ion physics is to map the phase structure of the strong force governing nuclear interactions. Experimentally, colliding heavy nuclei with high enough energies create conditions favorable to the formation of an expected new state of strongly interacting matter, the Quark Gluon Plasma. The precise determination of the location of the phase-transition line and investigation of the properties of this new matter needs theoretical models able to interpret the experimental results. In the experiments, one measures the final state of the reaction, and to dynamically connect this to the investigated initial state, hydrodynamics is a natural choice. It relies only on local energy-momentum conservation and on the concept of local thermal equilibrium.

*This work was supported by the Hungarian OTKA grants T038406 and T049466.

Hydrodynamical modeling was the most successful to describe the collective phenomena observed at the heavy-ion experiments at the RHIC accelerator, such as scaling of the azimuthal anisotropy of momentum distribution (*elliptic flow*), transverse momentum dependence of the characteristic size of two-particle correlation functions (HBT-radii). Because of this, the general consensus has arisen that the matter created at RHIC is a strongly coupled liquid of quarks, rather than a gas-like state. It also turned out that the fluid is a nearly perfect one (its sheer viscosity to entropy density ratio, η/s being close to the conjectured quantum limit[1] of $1/4\pi$). These facts give the justification of hydrodynamical modeling.

The ground idea of hydrodynamical modeling stems from Fermi[2] and Landau,[3] that is, to treat the soft (non-perturbative) particle production as a result of some thermal distribution. The initial state and the final state is connected by a hydrodynamical solution, either numerical or analytical, and knowing the final state, one can infer the initial state (given the proper equation of state), or if one assumes a model-initial state, one can constrain the equation of state based on experimental results. In this paper the focus is on the analytic solutions. The drawback of this approach is obviously the difficulty of finding solutions of a highly coupled set of nonlinear partial differential equations. But once a proper solution is found, many general facts of the dynamics are revealed.

The task is thus to find such solutions to the hydrodynamical equations that can characterize the space-time evolution of the strongly interacting deconfined matter created in heavy-ion collisions. In the following we highlight some historic and some recent development in this field, and show some applications of the solutions, and discuss what can be learned with them from the experimental results.

2. Perfect fluid hydrodynamics

Notations: We use the following notations: the space-time coordinate is $x^\mu \equiv (t, \mathbf{r})$, the metric tensor is $g^{\mu\nu} \equiv diag\,(1, -1, \ldots)$. (We use units where the speed of light is equal to 1.) We investigate $1+1$ and $1+3$ dimensional flows as well, the notation D will stand for the dimensionality of the space: $g^\mu_\mu = D+1$, $\delta_{kk} = D$. (Greek letters denote Lorentz indices, Latin letters denote three-vector indices.) For $1+1$ dimensional solutions, and spherical solutions, the so-called Rindler coordinates τ (proper-time) and η (space-time pseudorapidity) are often useful: their definition is $t = \tau \cosh \eta$, $r = \tau \sinh \eta$, here r is a radial coordinate (in case of spherical symmetry) or the only spatial coordinate (in case of $1+1$ dimensions). The inverse formulas are $\tau = \sqrt{t^2 - r^2} = \sqrt{x^+ x^-}$, $\eta = 1/2 \ln(x^+/x^-)$, where $x^\pm \equiv t \pm r$. The four-velocity of the fluid is $u^\mu \equiv \gamma(1, \mathbf{v})$, where $\gamma = \sqrt{1 - v^2}$, the fluid rapidity Ω defined as $v = \tanh \Omega$ will also be used. T is the temperature, ε is the energy density, p is the pressure, and σ is the entropy density.

Basic equations: As mentioned, the equations of relativistic hydrodynamics stem only from the requirement of local thermal equilibrium (that's why it is possible to speak of pressure and temperature fields), and local energy-momentum conservation: $\partial_\nu T^{\mu\nu} = 0$, where $T^{\mu\nu}$ is the (symmetric) stress-energy-momentum tensor of the flowing matter. Its form is[3] $T^{\mu\nu} = (\varepsilon + p) u^\mu u^\nu - p g^{\mu\nu}$ for a perfect fluid, a form due to Landau.

The relativistic Euler and energy conservation equations are thus

$$(\varepsilon + p) u^\nu \partial_\nu u^\mu = (g^{\mu\rho} - u^\mu u^\rho) \partial_\rho p \quad , \quad (\varepsilon + p) \partial_\rho u^\rho + u^\rho \partial_\rho \varepsilon = 0. \tag{1}$$

It can be shown that this form of energy conservation equation is equivalent to entropy conservation (valid indeed for perfect fluid): $\partial_\mu(\sigma u^\mu) = 0$.

The much simpler non-relativistic equations are obtained as limiting cases of the previous equations. While taking this limit, one has to assume that the velocities are much smaller than 1, and also that the fluid consists of individual particles (with conserved number density n, mass m_0 and corresponding chemical potential μ), and that the energy density is mainly from the rest mass: if $\varepsilon_0 \equiv \varepsilon - n m_0 \ll n m_0$. The non-relativistic Euler and the energy conservation equations are then

$$n m_0 \left(\frac{\partial \mathbf{v}}{\partial t} + (\mathbf{v} \nabla) \mathbf{v} \right) = -\nabla p \quad , \quad \frac{\partial \varepsilon_0}{\partial t} + \nabla(\varepsilon_0 \mathbf{v}) = -p(\nabla \mathbf{v}), \tag{2}$$

and the conservation equation for n is $\frac{\partial n}{\partial t} + \nabla(n\mathbf{v}) = 0$ in the non-relativistic case, while it is $\partial_\nu(n u^\nu) = 0$ in the relativistic case.

The Equation of State (EoS) of the matter connects ε with p and T. A simple EoS is used in most known exact solutions of hydrodynamics. It is $\varepsilon = \kappa p$ in the relativistic case, (corresponding to an ideal gas of massless particles), or $\varepsilon_0 = \kappa p$ in the non-relativistic case (corresponding to a non-relativistic ideal gas.) This EoS closes the set of equations. We can include the $p = nT$ condition for the EoS, characteristic to an ideal gas.

3. Exact solutions

In this section we present some important hydrodynamical solutions. For the detailed derivations, see the respective references.

Historical solutions: The first and one of the historically most important analytic solutions of relativistic hydrodynamics were the Landau-Khalatnikov solution,[4,5] which is an exact but implicit, $1+1$ dimensional solution. It was used for decades in the description of cosmic ray reactions. One of the important observables in high-energy experiments is the rapidity distribution of the particles $\frac{\mathrm{d}n}{\mathrm{d}y}$, *i.e.* the distribution with respect to the rapidity $y \equiv 1/2 \ln\left((E+p_z)/(E-p_z)\right)$ (the z direction is the direction of the collision). One can calculate this distribution from the Landau-Khalatnikov solution, it turns out to be a nearly-Gaussian distribution, which is in close agreement with the observations. The Landau-Khalatnikov solution is, however, an extremely complicated implicit solution, thus it is not easy to even write it down. The key point is that one expresses the space-time coordinates through the fluid rapidity and the temperature (here $\theta = \ln T$) from a potential $\Phi(x_+, x_-)$, with $\partial_+\Phi = Te^{\Omega}$, $\partial_-\Phi = Te^{-\Omega}$. Φ can be calculated from its Legendre transform $\chi\left(Te^{\Omega}, Te^{-\Omega}\right) \equiv \Phi - x_+Te^{\Omega} - x_-Te^{-\Omega}$, which obeys the Khalatnikov equation:

$$\partial_\theta^2 \chi\left(\theta, \Omega\right) + \left(\kappa - 1\right) \partial_\theta \chi\left(\theta, \Omega\right) - \kappa \partial_\Omega^2 \chi\left(\theta, \Omega\right) = 0, \qquad (3)$$

The solution of this equation can be written up with integral-formulas using the Green's function formalism;[4] the defining characteristic of the Landau-Khalatnikov solution is its initial condition: a finite piece of matter at rest.

Another important historical solution is the Hwa-Bjorken solution,[6,7] which is an over-simplistic boost-invariant $1+1$ dimensional accelerationless solution. The boost invariance (which is a reasonable approximation for collisions of infinite bombarding energy) tells that σ (thus T) depends only on τ, not η. The following forms give the Hwa-Bjorken solution:

$$\Omega = \eta \quad \Longleftrightarrow \quad v = r/t \quad , \quad \sigma_0/\sigma = \tau_0/\tau. \qquad (4)$$

The expression of T depends on the actual value of κ in the EoS. The rapidity distribution corresponding to this solution is constant, and thus it is applicable only the central rapidity region, where indeed a flat "plateaux" is present in the experimental $\frac{\mathrm{d}n}{\mathrm{d}y}$ measurements. Nevertheless, it allowed Bjorken to obtain a simple estimation of the initial energy density of the reaction; this estimation is still used in the interpretation of the experimental results. This application is what rendered the Hwa-Bjorken solution a unique status.

Non-relativistic solutions: The non-relativistic hydrodynamical equations are much simpler and easier to solve, and (maybe surprisingly) have relevance to the description of the experimental data. One of the interesting exact results is a family of ellipsoidally expanding solutions, discovered step-by-step by many authors.[8–12] The most general solution (containing most of the other solutions as special cases) in this class describes self-similar ellipsoidal expansion,[10] with an arbitrary scaling function $\nu(S)$ as follows:

$$\mathbf{v} = \left(\frac{\dot{A}(t)}{A(t)}x, \frac{\dot{B}(t)}{B(t)}y, \frac{\dot{C}(t)}{C(t)}z\right), \qquad S = \frac{x^2}{A^2(t)} + \frac{y^2}{B^2(t)} + \frac{z^2}{C^2(t)},$$

$$T = T_0\left(\frac{A_0B_0C_0}{ABC}\right)^{\frac{1}{\kappa}}\tau(S), \qquad n = n_0\frac{A_0B_0C_0}{ABC}\nu(S), \qquad p = nT. \quad (5)$$

(The free choice of one of the $\tau(S)$ and $\nu(S)$ functions fixes the form of the other one.[10]) This is a solution if the time-dependent principal axes $A(t)$, $B(t)$, and $C(t)$ obey an *ordinary* set of differential equations. This feature is what called as *parametric solution.* Once one assumes that such an expanding solution is valid for the description of the dynamics after the collision, one can parametrize the hadronic final state, and one can follow the time evolution back to the initial state.

This solution serves as a base of the Buda-Lund model, which is successful in describing particle spectra, correlations, and their scalings.[13,14] Also, this is the only one (as far as I know) which can be generalized as a parametric solution for arbitrary $\kappa(T)$, *i.e.* arbitrary temperature-dependent speed of sound EoS (with some restriction on the fairly general density and temperature profile: one has to assume Gaussian density and spatially homogeneous temperature distribution[8]). This result — though non-relativistic — is unique, and makes possible to use any QCD–inspired EoS. One would very much look for relativistic solutions with this property.

Relativistic accelerationless solutions: The generalization of the Hwa-Bjorken solution to arbitrary number of spatial dimensions is also a solution which falls into the category of the parametric solutions mentioned above:[15] these solutions are the relativistic equivalents of them. They have an accelerationless, spherically symmetric velocity profile. The solution is

$$\mathbf{v} = \frac{\mathbf{r}}{t}, \qquad S = \frac{1}{t^2}\left(\frac{x^2}{A^2} + \frac{y^2}{B^2} + \frac{z^2}{C^2}\right)$$

$$n = n_0\left(\frac{\tau_0}{\tau}\right)^3\frac{1}{\mathcal{T}(S)}, \qquad T = T_0\left(\frac{\tau_0}{\tau}\right)^{3\frac{1}{\kappa}}\mathcal{T}(S), \qquad p = nT, \quad (6)$$

where $\mathcal{T}$ is an arbitrary function of the ellipsoidal scaling variable S, with principal axes A, B, and C in the directions x, y, and z. Other generalizations, *e.g.* to hyperbolic profiles,[15] and more general, ellipsoidally symmetric (but still accelerationless) velocity fields,[16] were also found, besides other important accelerationless solutions.[17]

Accelerating solutions: A recent approach toward new solutions was a generalization of the Bjorken ansatz (the boost invariance) to a harmonic ansatz: this in 1+1 dimensions read $\frac{\partial}{\partial x^-}\frac{\partial}{\partial x^+}\Omega = 0$. A new class of solutions is obtained when substituted into the hydrodynamical equations:[18]

$$p = p_0 \exp\left\{-\frac{(1+\kappa)^2}{4\kappa}\left(l_+^2 + l_-^2\right) + \frac{\kappa^2-1}{2\kappa} l_+ l_-\right\}, \qquad \Omega = \frac{1}{2}\left(l_+^2 - l_-^2\right). \tag{7}$$

The notations are

$$l_\pm(x_\pm) = \sqrt{\ln F_\pm}, \qquad x_\pm = h\int^{F_\pm} \frac{dx}{\sqrt{\ln x}}, \tag{8}$$

where h is an arbitrary constant. The later formula gives the F function implicitly. This solution, although not fully explicit, is very interesting, since it interpolates between the Landau and the Bjorken pictures (fixed h, $l_\pm \to \infty$, and $h \to 0$, respectively). If one calculates the entropy density per unit rapidity, which is proportional to the observable particle distribution, it depends on the assumed freeze-out surface, but in general it is approximately Gaussian.[18] More general expressions for the entropy flow $\frac{dS}{dn}$, based partially on the Khalatnikov method, were discussed recently in Ref. 19.

A different recent approach to find new and relevant solutions was to generalize the Hwa-Bjorken flow in the following way:[20,21]

$$v = \tanh(\lambda\eta), \qquad p = p_0\left(\tau_0/\tau\right)^{\lambda D(\kappa+1)/\kappa} \cosh^{-(d-1)\Phi_\lambda}(\eta/2) \tag{9}$$

expressions are indeed explicit, exact solutions of the hydrodynamical equations, for certain values of the parameters λ, D, Φ_λ, and κ. (λ is a parameter of the solution.) For $\lambda = 1$ the Hwa-Bjorken solution is recovered, for $\lambda \neq 1$ the solution is accelerating. The Φ_λ parameter is introduced because for some choices of λ the pressure depends on η as well. There are five allowed parameter-sets listed in Table I of Ref. 20. (See this for a detailed discussion.) Important cases are: for $D = 1$ and $\kappa = 1$, $\Phi_\lambda = 0$, but λ can take arbitrary values. This solution is the basis for an advanced estimate of the initial energy density of the reaction, discussed briefly below. (We note that for $\kappa = 1$, the general explicit solution of the hydrodynamical problem is found[20] because the equations can be cast in a linear form. For three-dimensional flows, this idea resulted in a broad class of new general solutions.[22] It is not clear how these results could be generalized for any other κ values, thus to the case of some physically more relevant EoS.)

Another important case of Eq. (9) is the $\lambda = 2$, $\kappa = D$ case, it is valid in three-dimensions for a realistic EoS. This is a solution,[20] which describes an uniformly accelerating explosion from an initial pointlike, infinitely dense domain. Generalizations of this solution were recently found,[23] where the initial condition is a „smeared" point (with finite temperature and total energy).

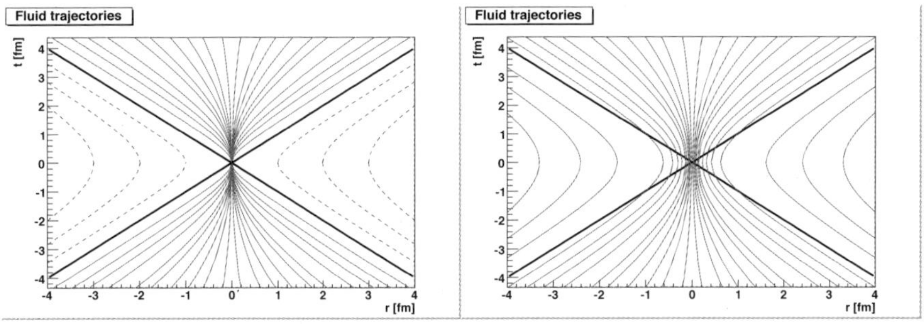

Fig. 1. Left: fluid trajectories of the $\lambda = 2$ case accelerating solution of Eq. (9), inside (solid) and outside the lightcone (dashed). Right: trajectories of the „smeared" generalization of it, for $\rho > 0$ (and $\mathbf{B} = 0$) in Eq. (10).

Also, for the first time, rotating solutions were found: the initial fireball can also rotate. This generalization is written up as

$$\mathbf{v} = \frac{2t\mathbf{r} + \mathbf{B} \times \mathbf{r}}{t^2 + r^2 + \rho} \quad \text{and} \quad T = \frac{T_0 \tau_0^2}{\sqrt{(\tau^2 + \rho)^2 + 4\rho r^2 - (\mathbf{B} \times \mathbf{r})^2}}, \tag{10}$$

with $p = p_0 \left(T/T_0\right)^{\kappa+1}$. Here ρ is an arbitrary constant, and $\mathbf{B}$ is a constant three-vector. The detailed phenomenological implications of these solutions (as the calculation of observable single-particle spectra, two-particle correlations) are not yet explored, but the rotating solutions described here deserve thorough investigation: for non-central heavy-ion collisions, rotation of the initial fireball is definitely present.

This rotating solution, with more general than spherical (or $1+1$ dimensional) symmetry may open the quest for other solutions of even more general symmetry properties. One such solution is the recently found class of conformally symmetric solutions with generalized boost-invariance.[26] In the following we return to Eq. (9) and delineate an application of that solution with far-reaching consequences.

Energy density estimation: The $\kappa = 1$, $D = 1$ accelerating solutions discussed above allow to an easy approximate calculation of the $\frac{dn}{dy}$ rapidity distribution.[20] These distributions qualitatively agree with the observed peaked (Gaussian) structure, and the λ parameter can be extracted from a fit to measured data, and from that one can to improve Bjorken's original estimate[7] of the initial energy density: the work done by the fluid (because of acceleration) and the shift in the estimated origin of the trajectories caused by the presence of acceleration leads to the conclusion that the Bjorken estimate needs to be corrected by a factor greater than 1: for $\sqrt{s_{NN}} = 200$ GeV Au+Au collisions, from rapidity distributions measured by BRAHMS,[24] one gets a factor of 2.0 ± 0.1 correction. (Taking the softness of the EoS, ie. the fact that $\kappa > 1$) into account, a conjecture can be made of a correction factor of 2.9 ± 0.2[21]). This result is in accordance with recent PHENIX findings about the initial temperature of the collision zones, based on the measurement of the direct photon spectrum.[25] The knowledge of the initial energy density is important in the interpretation of experimental data, since one has to be able to tell at what actual point on the QCD phase diagram a measurement is made. It seems thus that the energy densities expected originally for LHC are realized already at RHIC collisions, and at LHC one can expect even higher initial energy densities.

4. Summary

We have reviewed the concept of hydrodynamical modeling in heavy-ion physics, with focus on exact analytic solutions. We have presented some historic results and some more recent ones. We tried to demonstrate the phenomenological importance: if one has a solution with a few adjustable fit parameters, it gives insight into the dynamics of the created matter in an analytical way. We displayed an application of one of the recent accelerating solutions: the advanced energy density estimate of the heavy-ion reaction, which corrects the famous Bjorken estimate. It turns out that the based on this estimate (which is in very close proximity of that suggested by direct photon spectral measurements) the energy density reached in top energy RHIC collisions approach those expected only for LHC.

The topic of applying hydrodynamics, in particular exact solutions of hydrodynamics in heavy-ion physics is a flourishing field. However, exact solutions to the hydrodynamical equations are hard to find. For instance, there is no known solution in more than one spatial dimensions with a little bit more general equation of state, and there are very few solutions that exhibit acceleration. The quest for such solutions (e.g. for an ellipsoidally symmetric one) would lead to a more accurate description of the observables, and thus through the transport properties, and the equation of state one can test the models of the deconfined QCD matter.

Acknowledgments

The author is grateful to the organizers of the Gribov-80 Memorial Workshop at Trieste for their kindness and support. The author acknowledges the support of the Hungarian OTKA grants T038406 and T049466.

References

1. P. Kovtun, D. T. Son and A. O. Starinets, *Phys. Rev. Lett.* **94**, 111601 (2005).
2. E. Fermi, *Prog. Theor. Phys.* **5**, 570 (1950).
3. L. D. Landau, *Izv. Akad. Nauk Ser. Fiz.* **17**, 51 (1953).
4. I.M. Khalatnikov, *Zhur. Eksp. Teor. Fiz.* **27**, 529 (1954); see also Ref. 3.
5. S. Z. Belenkij and L. D. Landau, *Nuovo Cim. Suppl.* **3S10**, 15 (1956) [*Usp. Fiz. Nauk* **56**, 309 (1955)].
6. R. C. Hwa, *Phys. Rev.* **D10**, 2260 (1974).
7. J. D. Bjorken, *Phys. Rev.* **D27**, 140 (1983).
8. T. Csörgő, S. V. Akkelin, Y. Hama, B. Lukács and Yu. M. Sinyukov, *Phys. Rev.* **C67**, 034904 (2003), arXiv:hep-ph/0108067.
9. S. V. Akkelin, T. Csörgő, B. Lukács, Yu. M. Sinyukov and M. Weiner, *Phys. Lett.* **B505**, 64 (2001), arXiv:hep-ph/0012127.
10. T. Csörgő, *Acta Phys. Polon.* **B37**, 483 (2006).
11. J. Bondorf, S. Garpman and J. Zimányi, *Nucl. Phys.* **A296**, 320 (1978).
12. P. Csizmadia, T. Csörgő and B. Lukács, *Phys. Lett.* **B443**, 21 (1998).
13. M. Csanád, T. Csörgő, B. Lörstad and A. Ster, *J. Phys.* **G30** (2004) S1079.
14. T. Csörgő and B. Lorstad, *Phys. Rev.* **C54**, 1390 (1996).
15. T. Csörgő, L. P. Csernai, Y. Hama and T. Kodama, *Heavy Ion Phys.* **A21**, 73 (2004).
16. Y.M. Sinyukov and I.A. Karpenko, nucl-th/0506002.
17. T. S. Biró, *Phys. Lett.* **B487**, 133 (2000).
18. A. Bialas, R. A. Janik and R. Peschanski, arXiv:0706.2108 [nucl-th].
19. G. Beuf, R. Peschanski and E. N. Saridakis, *Phys. Rev.* **C78**, 064909 (2008).
20. M. I. Nagy, T. Csörgő and M. Csanád, *Phys. Rev.* **C77**, 024908 (2008).
21. T. Csörgő, M. I. Nagy and M. Csanád, *Phys. Lett.* **B663**, 306 (2008).
22. M. S. Borshch and V. I. Zhdanov, *SIGMA* **3**, 116 (2007).
23. M. I. Nagy, arXiv:0909.4285 [nucl-th].
24. I. G. Bearden *et al.*, *Phys. Rev. Lett.* **94**, 162301 (2005).
25. A. Adare *et al.*, *Phys. Rev. Lett.* **104**, 132301 (2010).
26. S. S. Gubser, arXiv:1006.0006 [hep-th].

WEAK INTERACTION CONTRIBUTION TO THE INCLUSIVE HADRON-HADRON SCATTERING CROSS SECTIONS AT HIGH p_T

B. L. IOFFE

A.I. Alikhanov Institute of Theoretical and Experimental Physics,
B. Cheremushkinskaya 25, 117218, Moscow, Russia

It is demonstrated that the strong power-like scaling violation in the transverse momentum distribution of inclusive hadron production, observed by the CDF Collaboration in $\bar{p}p$ collisions at the Tevatron is caused by the contribution of weak interaction. The contribution of the weak interaction is increasing with energy at high energies.

The CDF Collaboration have measured the inclusive cross sections of charged hadron production at high transverse momentum p_T at $\bar{p}p$ collisions at c.m. energy 1.96 TeV.[1] Surprisingly, the strong power-like scaling violation was observed at $p_T > 30$ GeV: at $p_T \approx 100$ GeV the data indicate that the scaling law $Ed\sigma/d^3p \sim 1/p_T^4$ is violated more than by one order of magnitude. The scaling law $Ed\sigma/d^3p \sim 1/p_T^4$ for inclusive hadron production in hadron-hadron scattering was proved using a very general basis – the light-cone dominance of hard processes in strong interactions.[2] Therefore, the observation of the violation of the scaling law resulted in a serious confusion. Theoretically the observed phenomenon was discussed in the paper by Albino et al.[3] The authors of Ref. 3 addressed the scaling violation to factorization breaking at high transverse momentum charged hadron production. Such an explanation is not satisfactory: QCD has no scale parameters besides Λ_{QCD} and inclusive cross sections are infra-red stable in QCD. In principle there is the dimensional parameter in the problem in view – the energy of the collision. But, as it is well known,[2,4] the energy is related to the longitudinal size of the collision region, but not to the transverse size, which determines the cross section. In a recent paper[5] an attempt was made to construct a model, describing the data, but as well as in Ref. 3 no success was achieved. At the same time the measurements of inclusive jet production[6,7] demonstrate a good agreement with the scaling law and the theoretical expectations.

In this paper it is shown, that the scaling law violation in inclusive cross sections of charged hadron production at high p_T, observed by CDF Collaboration, is described by contribution of weak interaction. The idea is that weak interaction has the scale parameters – the masses of W and Z bosons. At high p_T the contribution of weak interaction to the inclusive cross section is strongly enhanced by the presence of W and Z resonances in comparison with strong interaction contribution which falls steeply with p_T. Weak interaction contribution has a peak at $p_T = m_W/2$. Due to these circumstances the weak interaction contribution becomes compatible with strong ones at $p_T \gtrsim 30$ GeV.

The weak interaction Lagrangian is that of the Standard Model:

$$
\begin{aligned}
L = \frac{g}{\sqrt{2}} \Bigg\{ & \left[W_\mu^+ \bar{u} \gamma_\mu \frac{1}{2}(1+\gamma_5) d + W_\mu^- \bar{d} \gamma_\mu \frac{1}{2}(1+\gamma_5) u \right] \\
& + Z_\mu \frac{1}{\cos\theta_W} \left[\bar{u} \gamma_\mu \frac{1}{2}(1+\gamma_5) \left(\frac{1}{2} - \frac{2}{3}\sin^2\theta_W \right) u \right. \\
& + \bar{u} \gamma_\mu \frac{1}{2}(1-\gamma_5) \left(-\frac{2}{3}\sin^2\theta_W \right) u + \bar{d} \gamma_\mu \frac{1}{2}(1+\gamma_5) \left(-\frac{1}{2} + \frac{1}{3}\sin^2\theta_W \right) d \\
& \left. + \bar{d} \gamma_\mu \frac{1}{2}(1+\gamma_5) \left(\frac{1}{3}\sin^2\theta_W \right) d \right] + (u \to c,\ d \to s) \Bigg\} . \qquad (1)
\end{aligned}
$$

Here u and d are the fields of u and d quarks, θ_W is the Weinberg angle, $\sin^2\theta_W \approx 0.230$. The coupling constant g is

$$
g^2 = \frac{e^2}{\sin^2\theta_W}, \quad e^2 = \frac{1}{137} \qquad (2)
$$

The matrix element of weak interaction contribution to the inclusive cross section in $\bar{p}p$ collision is represented by the diagram of Fig. 1.

There are also the diagrams, where $\bar{p}$ fragments into $u, \bar{s}, c$ and p – into $\bar{d}, c, \bar{s}$, correspondingly, as well as the diagrams with W^- and Z in annihilation channel. (The contribution of W and Z exchange in the t-channel is negligible.) In order to compare the results with CDF data, let us calculate the inclusive cross section integrated over pseudorapidity

$$
\eta = \frac{1}{2} \ln \frac{E' + p'_{\parallel}}{E' - p'_{\parallel}}, \qquad (3)
$$

where $E' = \sqrt{p'^2_{\parallel} + p'^2_T}$ is the energy of the detected charged particle,

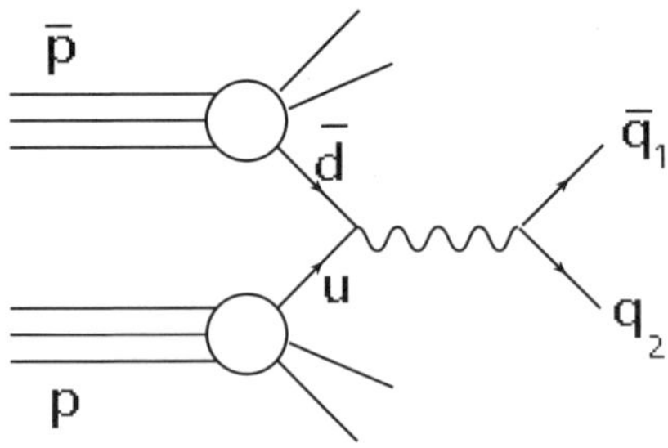

Fig. 1. The diagram, describing the quark pair production at high p_T in case of W^+ exchange in the annihilation channel.

$p' = (p'_{\|}, \mathbf{p}'_T)$, with $p'_{\|}$ and $\mathbf{p}'_T$ being projections of its momenta parallel and perpendicular to the beam direction. The contribution to the inclusive cross section of the diagram of Fig. 1 is equal

$$E' \int \frac{d\sigma_{weak}}{d^3p'} d\eta' \Bigg|_{|\eta'|<\eta}$$
$$= \frac{9}{8} \frac{g^4}{(2\pi)^2} \frac{p_T}{E} \int \frac{dx_1 dx_2 dx_3}{x_1 x_2} \frac{(x_1^2 + x_2^2)(1 + th^2\eta)}{(4E^2 x_1 x_2 - m_W^2)^2 + m_W^2 \Gamma_W^2}$$
$$\times \frac{F_u(x_1) F_d(x_2)}{(x_1 + x_2) x_3 sech\ \eta - \frac{p_T}{E}} \sum_i \left[D_u^i(x_3) + D_d^i(x_3) + D_s^i(x_3) + D_c^i(x_3) \right]. \tag{4}$$

Here E is the proton or antiproton energy in c.m.s., m_W and Γ_W are the W mass and width, $F_u(x_1), F_d(x_2)$ – are u and d-quark distributions in proton, $D_u^i, D_d^i, D_s^i, D_c^i$ are the fragmentation functions of u, d, s, c quarks into the ith charged particle, the sum is performed over all charged particles. The integration domain in variables x_1, x_2, x_3 is restricted by

$$x_1 x_2 > \left(\frac{p_T}{E} \right)^2$$

$$(x_1 + x_2) x_3 sech\ \eta > \frac{p_T}{E} (1 + x_3 sech\ \eta) \tag{5}$$

As it is well known (see e.g. Ref. 8 and references herein), in case of production of narrow vector resonances (like $\omega, \varphi, J/\psi, \Upsilon$) in e^+e^--annihilation $e^+e^- \to V$ the radiative effects due to the emission of real or virtual photons by initial e^+ and e^- are very important. The resonance curves are widened and the resonance maxima are suppressed. The cross section of the process $e^+e^- \to V$ without radiative effects is described by the Breit-Wigner formula

$$\sigma(e^+e^- \to V) = \frac{12\pi}{s}\frac{\Gamma^V_{e^+e^-}}{M}\mathrm{Im}f_0(\sqrt{s}), \tag{6}$$

where $\Gamma^V_{e^+e^-}$ is the electron width of V-resonance and M is its mass,

$$f_0(\sqrt{s}) = \frac{(1/2)M}{-\sqrt{s}+M-i\Gamma/2}, \tag{7}$$

$\sqrt{s}$ is the total energy of an e^+e^- pair in their c.m.s. and Γ is the V total width. In Ref. 8 it was shown that the account of radiative effects results in the substitution

$$f_0(\sqrt{s}) \to f(\sqrt{s}) = [\ f_0(\sqrt{s})\ ]^{-\beta_{QED}} \tag{8}$$

where

$$\beta_{QED} = \frac{4\alpha}{\pi}\left[\ln\frac{\sqrt{s}}{m_e} - \frac{1}{2}\right], \tag{9}$$

$\alpha = e^2 = 1/137$ and m_e is the electron mass. A similar situation takes place in the production of W or Z bosons by the annihilation of a quark pair. The diagrams corresponding to gluon corrections of the first order are shown in Fig. 2.

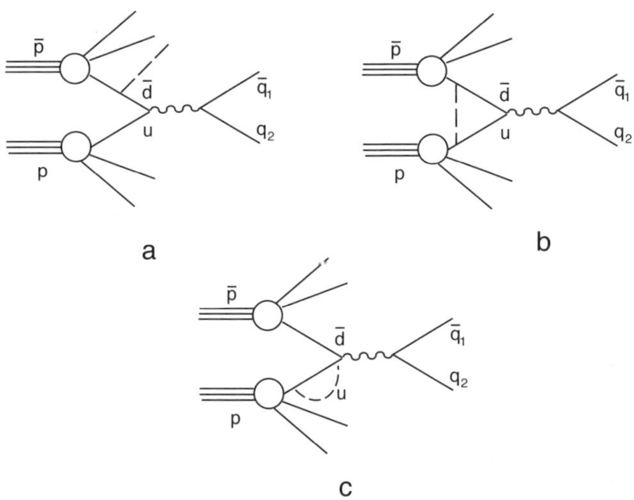

Fig. 2. Gluon corrections to the process $\bar{q}q \to W$. Dashed lines correspond to gluons.

The diagrams Figs. 2,b,c contribute through interference with the diagram of Fig. 1. The gluon emission by final quarks $\bar{q}_1, q_2$ can be neglected, since these quarks are carrying large p_T. In analogy with (8), (9) the account of gluon corrections results in the substitution in Eq. (4):

$$\frac{1}{(4x_1x_2E^2 - m_W^2)^2 + m_W^2\Gamma^2} \rightarrow \frac{1}{m_W^3\Gamma_W}\left[\frac{m_W^3\Gamma_W}{(4x_1x_2E^2 - m_W^2)^2 + m_W^2\Gamma_W^2}\right]^{1-\beta_{QCD}} \tag{10}$$

where

$$\beta_{QCD} = \frac{8}{3}\frac{\alpha_s(s)}{\pi}\left[\ln\frac{s}{M_{char}^2} - 1\right], \qquad s = 4x_1x_2E^2, \tag{11}$$

and M_{char} is the characteristic mass of strong interaction, $M_{char} \sim 1$ GeV. In the derivation of () it was assumed that $\alpha_s(2p_T)$ is small, $\alpha_s(s) \ll 1$, the $\ln[s/m_{char}^2]$ is large, the product $\alpha_s(s)\ln[s/M_{char}^2]$ is of the order of 1 and the terms $\sim (\alpha_s \ln[s/M_{char}^2])^n$ are summed.

Taking into account the gluon corrections we have instead of (4):

$$E'\int \frac{d\sigma_{weak}}{d^3p'}d\eta'\bigg|_{|\eta'|<\eta} = \frac{9}{8}\frac{g^2}{(2\pi)^2}\frac{p_T}{E}\int \frac{dx, dx_2, dx_3}{x_1x_2}\frac{(x_1^2 + x_2^2)(1 + th^2\eta)}{m_W^3\Gamma_W}$$

$$\times\left[\frac{m_W^3\Gamma_W}{(4x_1x_2E^2 - m_W^2)^2 + m_W^2\Gamma_W^2}\right]^{1-\beta_{QCD}} \frac{F_u(x_1)F_d(x_2)}{(x_1 + x_2)x_3 sech\,\eta - p_T/E}$$

$$\times\sum_1 [D_u^i(x_3) + D_d^i(x_3) + D_s^i(x_3) + D_c^i(x_3)] \tag{12}$$

At $p_T \approx 40 - 100$ GeV $\alpha_s(2p_T) = 0.12 - 0.10$ and $\beta_{QCD} \approx 0.6 - 0.8$. So, the account of gluon corrections drastically changes the results. At small x the quark distributions and fragmentation functions behave as $(1/x)^\gamma$, where γ is equal or larger than 1 and small x_1, x_3 or x_2, x_3 are dominating in (12). The consequence of this fact is that the cross section (12) increases with beam energy E and decreases with p_T more slowly, than $1/p_T^4$. Therefore the measurements of inclusive cross sections at high p_T at LHC are very promising. The detailed calculation of weak contribution to the inclusive cross section at high p_T, using quark distributions found by MSTW2008[9] and CTEQ 6.6.M[10] and available information on distribution functions will be presented in a separate publication.

Finally, let us now explain, why the scaling violation is not observed experimentally in inclusive jet production.[6,7] In this case in the strong interaction mechanism the main role is plaid by the diagrams with gluon exchange, like those presented in Fig. 3, and the contribution of weak interaction is small in comparison with them.

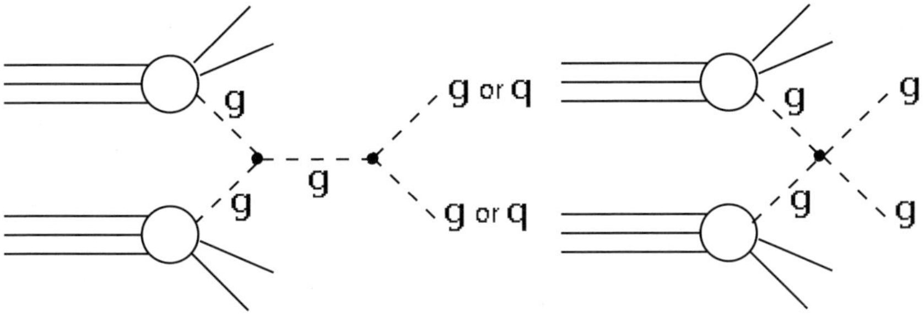

Fig. 3. The diagrams representing the contributions to inclusive jets production due to gluon exchange.

Acknowledgments

I am thankful to A.B. Kaidalov for useful remarks and to A.G. Oganesian who found a few errors in my calculations.

This work was supported in parts by RFBR grant 09-02-00732 and by CRDF Cooperative Program grant RUP2-2961-MO-09. The support of the European Community – Research Infrastructure Integrating Activity "Study of Strongly Interacting Matter" under Seventh Framework Program of EU is acknowledged.

References

1. T. Aaltonen *et al.* (CDF Collaboration), *Phys. Rev* **D79**, 112005 (2009).
2. B.L. Ioffe, *Phys. Lett.* **30B**, 123 (1969).
3. S. Albino, B.A. Kniehl and G. Kramer, *Phys. Rev. Lett.* **104**, 242001 (2010), arXiv: 1003.2127 [hep-ph].
4. V.N. Gribov, B.L. Ioffe and I.Ya. Pomeranchuk, *Yad. Phiz.* **2**, 768 (1965); *Sov. J. Nucl. Phys.* **2**, 549 (1966).
5. A.S. Yoon, E. Wenger, and G. Roland, arXiv: 1003.5928 [hep-ph].
6. T. Aaltonen *et al.* (CDF Collaboration), *Phys. Rev.* **D78**, 052006 (2008); *ibid.* **D79**, 119902(E) (2009), arXiv: 0807.2204 [hep-ph] .
7. V.M. Abazov *et al.* (D0 Collaboration), *Phys. Rev. Lett.* **101**, 06001 (2008), arXiv: 0802.2400 [hep-ph].
8. B.L. Ioffe, V.A. Khoze and L.N. Lipatov, *Hard Processes*, North Holland, 1984, Chapter 2, Sec. 2.11.2.
9. A.D. Martin, W.J. Stirling, R.S. Thorne and G. Watt, *Eur. Phys. J.* **C63**, 189 (2009), arXiv: 0901.0002 [hep-ph].
10. P.M. Nadolsky *et al.* (CTEQ Collaboration), *Phys. Rev.* **D78**, 013004 (2008), arXiv: 0802.0007 [hep-ph].

ARE THE PROTON-PROTON AND PROTON-ANTIPROTON INTERACTIONS DIFFERENT AT VERY HIGH ENERGIES?

V.A. ABRAMOVSKY* and N.V. RADCHENKO

Novgorod State University, Russia
* *Victor.Abramovsky@novsu.ru*

There is an additional subprocess of hadron production in three quark strings in $p\bar{p}$ scattering. Its contribution is proportional to the logarithm of full energy squared so this subprocess is important at very high energies. It leads to a difference in multiplicity distributions in pp and $p\bar{p}$ interactions. Due to this effect the inclusive cross sections of $p\bar{p}$ scattering are larger than the inclusive cross sections of pp scattering. The difference in inclusive cross sections might be very evident in events with high multiplicities, and it is shown for energy $\sqrt{s} = 900$ GeV.

1. Introduction

It is generally accepted that total cross sections of pp and $p\bar{p}$ interactions at high energies $\sqrt{s} \to \infty$ ($\sqrt{s}$ – full energy in center-of-mass system) are the same $\sigma_{tot}^{pp} \equiv \sigma_{tot}^{p\bar{p}}$. Also elastic cross sections are equal $\sigma_{el}^{pp} \equiv \sigma_{el}^{p\bar{p}}$ and differential elastic cross sections $d\sigma_{el}^{pp}/dt \equiv d\sigma_{el}^{p\bar{p}}/dt$ (t – four-momentum transfer squared). It follows from the Pomeranchuk[1] theorem, which was proved for constant total cross sections. This theorem was generalized by Iden[2] for increasing total cross sections, which fulfill the Froissart theorem.

It is also generally accepted both by experimentalists and majority of theorists that multiplicity properties of pp and $p\bar{p}$ interactions such as inclusive cross sections, multiplicity distributions and mean multiplicities are the same.

However, we insist that the last statement is incorrect. In the present report we argue that in principle these multiplicity properties are different for pp and $p\bar{p}$ interactions.

We are based on the Low Constituents Number Model (LCNM), which was proposed in Refs. 3, 4 and then concretized in Ref. 5. The model can be formulated as follows.

1. On the first step before the collision there is small number of constituents in initial hadrons. In every hadron this is component either with only valence quarks or with valence quarks and one additional gluon. These gluons fill the whole spectrum in rapidity space.

2. On the second step the hadrons interaction is carried out by gluon exchange between the valence quarks and initial gluons. The hadrons gain the color charge.

3. On the third step after interaction the colored hadrons move apart and when the distance between them becomes larger than the confinement radius, the lines of color electric field gather into the string. This string breaks out into secondary hadrons.

LCNM gives good description of various experimental data. For example, the behavior of total cross sections of pp and $p\bar{p}$ interactions in LCNM is given in Fig. 1. It is described by formulae

$$\begin{aligned} \sigma_{tot}^{p(\bar{p})p} &= 63.52 s^{-0.358} \mp 35.43 s^{-0.56} + \sigma_0^{pp} + \sigma_1^{pp} \ln s + \sigma_2^{pp} (\ln s)^2, \\ \sigma_0^{pp} &= 20.08 \pm 0.42, \quad \sigma_1^{pp} = 1.14 \pm 0.13, \quad \sigma_2^{pp} = 0.16 \pm 0.01. \end{aligned} \tag{1}$$

All parameters are given in mb. The first two terms present non vacuum contributions and were taken from Ref. 6. The term σ_1^{pp} presents contribution from one gluon exchange between hadrons components, which contain only valence quarks. The value $\sigma_1^{pp} \ln s$ gives contribution from one gluon exchange between component of one hadron, which contains only valence quarks, and component of another hadron, which contains valence quarks and one additional gluon. The value $\sigma_2^{pp}(\ln s)^2$ gives contribution from one gluon exchange between hadrons components, which contain one additional gluon besides valence quarks. In the similar way it is possible to obtain in LCNM good description of total cross sections of $\pi^{\pm}p$, $K^{\pm}p$, γp and $\gamma\gamma$ interactions.[5]

Differential cross sections of elastic scattering at high energies are shown in Fig. 2, they are described by formulae

$$\begin{aligned} \frac{d\sigma^{el}}{dt} &= \frac{1}{16\pi} \left[\sigma_0 + \sigma_1 \ln s + \sigma_2 (\ln s)^2\right]^2 \left(1 + \rho^2\right) \exp\{-B(s)|t|\}, \\ B(s) &= 7.12 + 0.34 \ln s + 0.02 \ln^2 s, \end{aligned} \tag{2}$$

with the same values of σ_0^{pp}, σ_1^{pp}, σ_2^{pp}; here contributions from non vacuum reggeons are neglected. Ratio of the real to imaginary parts of elastic scattering amplitude is $\rho = 0.125$. Value of diffractive slope cone was fitted from experimental data.

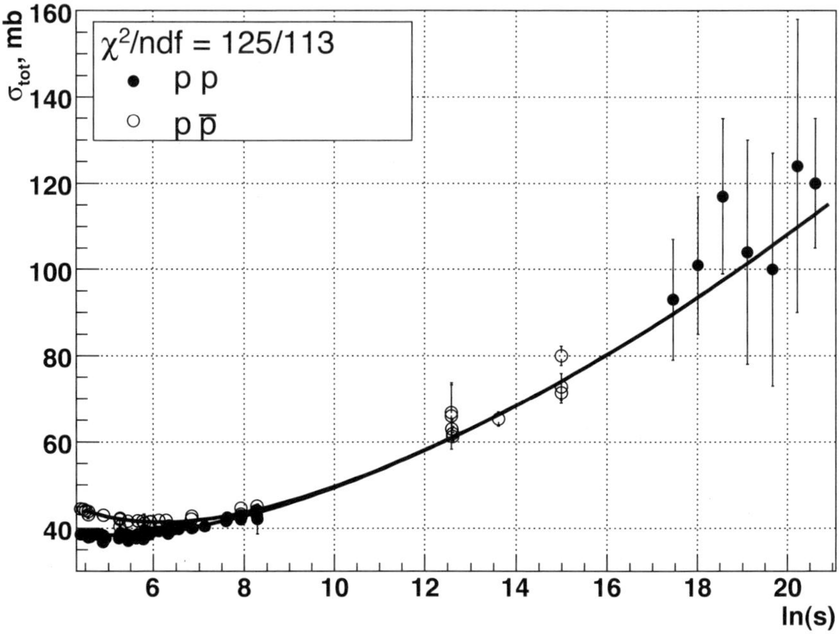

Fig. 1. Fitting of total cross sections of pp and $p\bar{p}$ collisions in LCNM.

Such an agreement with experimental data gives a possibility to expect that LCNM can adequately describe the picture of hadronic interactions at high energies.

2. Expansion of inelastic processes by topological subprocesses

In Regge and parton approaches, in which partons are considered as colorless particles, inelastic processes correspond to sum of subprocesses, so-called pomeron or reggeon showers.[7] Relative weights of these subprocesses are defined by the AGK theorem.[8] The capability of this topological expansion is well justified and caused by possibility to separate leading classes of diagrams, neglecting the others.

In case of QCD situation with partons – quarks and gluons, is much more difficult. The first reason is in very complicated color structure that have to circulate by lines of diagram. The second reason is irrelevance of

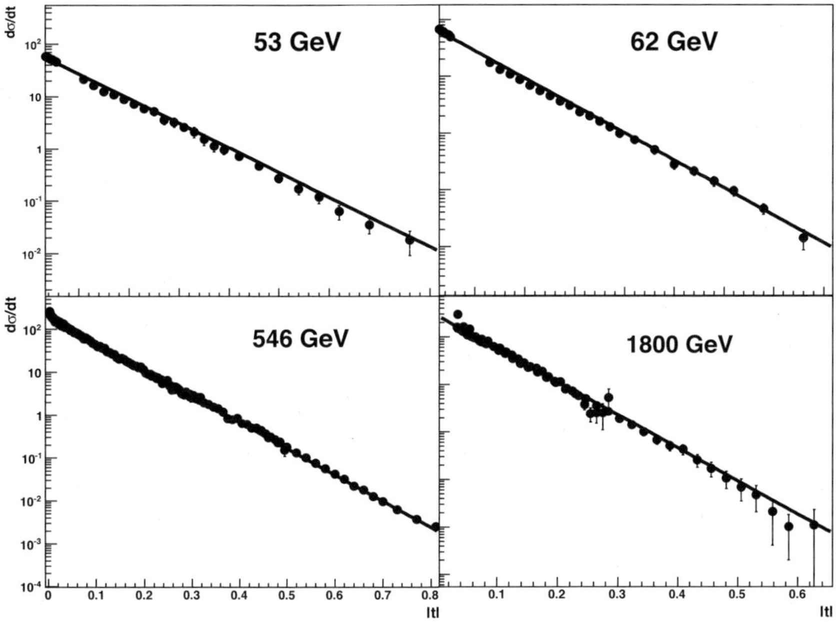

Fig. 2. Fitting of differential elastic cross sections in LCNM.

QCD diagram technique in region of soft physic because of absence of any small coupling constant. There is no distinct causes to neglect some classes of diagrams. The third reason is in unavoidable occurrence of interaction vertices with large number of quasi-particles. It leads to significant combinatoric complexity.

In this regard the realization of dual topological model,[9] where pomerons are cylinders and pomeron showers are cut cylinders with two quark strings, presumably turns up a hypothesis. What is more, all basic formulaes are deduced from the Regge theory.

As an alternative hypothesis we accept LCNM, in which only three types of inelastic processes exist.[10,11] Our basic assumptions are the following.

1. Color field string with octet charges at its endpoints arises from one gluon exchange between hadrons components, which contain only valence quarks, both in pp and $p\bar{p}$ interactions. This string we call a gluon string. It can be argued that when gluon string decays to secondary hadrons, their multiplicity distribution obeys normal distribution. This subprocess we call a shower from gluon string (Fig. 3, subprocess 1).

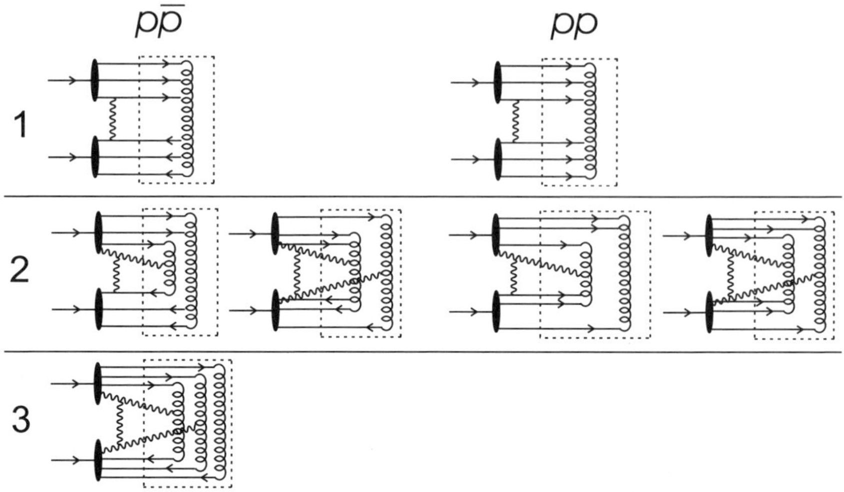

Fig. 3. Types of inelastic processes in $p\bar{p}$ and pp collisions.

2. Color field string with quark and antiquark (or quark and diquark, or diquark and antidiquark) at its endpoints arises from one gluon exchange between hadrons components, which contain additional gluons. These strings we call a quark strings. Both for pp and $p\bar{p}$ two quark strings are formed in this case. Decay of these two strings lead to production of secondary hadrons. This subprocess we call a shower from two quark strings (Fig. 3, subprocess 2).

3. Combination with three quark strings with quarks from protons and antiquarks from antiprotons at their endpoints arises in $p\bar{p}$ interaction from one gluon exchange between hadrons components, which contain one additional gluon besides valence quarks. There is no such capability in pp collisions. Decay of these three strings leads to production of secondary hadrons. This subprocess we call a shower from three quark strings (Fig. 3, subprocess 3). The difference in multiplicity properties of pp and $p\bar{p}$ interactions is determined by this very subprocess.

3. Weights of inelastic subprocesses in pp and $p\bar{p}$ collisions

Experimental charged multiplicities are normalized by non single diffraction cross sections $\sigma_{nsd} = \sigma_{tot} - \sigma_{el} - \sigma_{sd}$. Vacuum contribution to this cross sections are the same as for total cross sections and is defined by formulae

$$\sigma_{vac}^{nsd} = \sigma_0^{nsd} + \sigma_1^{nsd} \ln s + \sigma_2^{nsd} (\ln s)^2 = \sigma_0^{nsd}(1 + \delta_1^{nsd} \ln s + \delta_2^{nsd} \ln^2 s), \quad (3)$$

We obtained coefficients $\delta_1^{nsd} = 0.075 \pm 0.011$ and $\delta_2^{nsd} = 0.007 \pm 0.001$.[11] The value of $\sigma_0^{nsd} = 13.3$ was fixed taking into account constant part of total cross section, constant part of elastic cross section and constant part of single diffraction cross section. The last value is taken from Ref. 13.

Using values of parameters δ_1^{nsd} and δ_2^{nsd} we got weights of inelastic processes for pp and $p\bar{p}$.

Weights of inelastic processes for $p\bar{p}$ collisions are the following.

Weight of shower from gluon string is equal to

$$\frac{1}{1 + \delta_1^{nsd} \ln s + \delta_2^{nsd} (\ln s)^2}.$$

Weight of shower from two quark strings is equal to

$$\frac{\delta_1^{nsd} \ln s + (1 - c)\, \delta_2^{nsd} (\ln s)^2}{1 + \delta_1^{nsd} \ln s + \delta_2^{nsd} (\ln s)^2}.$$

Weight of shower from three quark strings is equal to

$$\frac{c\, \delta_2^{nsd} (\ln s)^2}{1 + \delta_1^{nsd} \ln s + \delta_2^{nsd} (\ln s)^2}.$$

In order to explain the meaning of parameter c, let us go back to Fig. 3. In case of $p\bar{p}$ interaction the configuration with two additional gluons gives both showers from two and three quark strings. The weights of these showers are $(1 - c)$ and c correspondingly. We chose the value $c = 0.75$ from the estimation of ATLAS Coll. results,[14] where the dependance of inclusive cross sections from transverse momentum was obtained in pseudorapidity range $|\eta| < 2.5$. We believe that this result is correct.

There is no shower from three quark strings in case of pp collisions. Therefore we have the following weights.

Weight of shower from gluon string is equal to

$$\frac{1}{1 + \delta_1^{nsd} \ln s + \delta_2^{nsd} (\ln s)^2}.$$

Weight of shower from two quark strings is equal to

$$\frac{\delta_1^{nsd} \ln s + \delta_2^{nsd} (\ln s)^2}{1 + \delta_1^{nsd} \ln s + \delta_2^{nsd} (\ln s)^2}.$$

4. Inclusive pseudorapidity distributions in pp and $p\bar{p}$ interactions

Invariant inclusive cross section of production of one charged particle in event with n charged particles can be defined as following:

$$(2\pi)^3 2E_1 \frac{\mathrm{d}^3\sigma_n^{incl}}{\mathrm{d}^3 p_1} = \frac{1}{(n-1)!}$$
$$\times \sum_{m=0}^{\infty} \frac{1}{m!} \int \mathrm{d}\tau_{n-1+m} \left| A_{2\to n+m}\left(\vec{p}_1; \vec{p}_2, \ldots, \vec{p}_n, \vec{q}_{n+1}, \ldots, \vec{q}_{n+m}\right)\right|^2, \quad (4)$$

where $A_{2\to n+m}\left(\vec{p}_1, \ldots, \vec{p}_n, \vec{q}_{n+1}, \ldots, \vec{q}_{n+m}\right)$ – amplitude of production of n charged and m neutral particles with corresponding momenta $\vec{p}_i$, $\vec{q}_j$; $\mathrm{d}\tau_{n+m}$ – corresponding phase volume. Integration in $\mathrm{d}\tau_{n-1+m}$ is performed by all momenta starting from $\vec{p}_2, \ldots$.

Invariant inclusive cross section can be written as:

$$(2\pi)^3 2E_1 \frac{\mathrm{d}^3\sigma^{incl}}{\mathrm{d}^3 p_1} = \sum_{n=1}^{\infty} (2\pi)^3 2E_1 \frac{\mathrm{d}^3\sigma_n^{incl}}{\mathrm{d}^3 p_1}. \quad (5)$$

Inclusive cross section is normalized by mean multiplicity of corresponding cross section of inelastic process, here we use non single diffraction cross section σ^{nsd}.

$$\int \mathrm{d}^3 p_1 \frac{\mathrm{d}^3\sigma^{incl}}{\mathrm{d}^3 p_1} = \langle n\rangle\, \sigma^{nsd} \quad (6)$$

At the same time cross section $\mathrm{d}^3\sigma_n^{incl}/\mathrm{d}^3 p_1$ is normalized by the following relation:

$$\int \mathrm{d}^3 p_1 \frac{\mathrm{d}^3\sigma_n^{incl}}{\mathrm{d}^3 p_1} = n\,\sigma_n. \quad (7)$$

where n – number of charged particles in event and σ_n – corresponding topological cross section.

We can obtain expressions for inclusive cross sections $\mathrm{d}\sigma^{incl}/\mathrm{d}\eta$ for pseudorapidity η (or $\mathrm{d}\sigma^{incl}/\mathrm{d}y$ for rapidity y) by using integral of (6) and (7) of transverse components of momentum $\vec{p}_1$.

Normalization of these cross sections is obvious.

$$\int \mathrm{d}\eta \frac{\mathrm{d}\sigma^{incl}}{\mathrm{d}\eta} = \langle n\rangle\, \sigma^{nsd} \quad (8)$$

$$\int \mathrm{d}\eta \frac{\mathrm{d}\sigma_n^{incl}}{\mathrm{d}\eta} = n\,\sigma_n \quad (9)$$

We can rewrite (9) as

$$\frac{1}{\sigma^{nsd}}\int \mathrm{d}\eta \frac{\mathrm{d}\sigma_n^{incl}}{\mathrm{d}\eta} = n\,\frac{\sigma_n}{\sigma^{nsd}} = n\,P_n. \tag{10}$$

We think that inclusive cross sections $\mathrm{d}\sigma^{incl}/\mathrm{d}\eta$ ($\mathrm{d}\sigma^{incl}/\mathrm{d}y$) are the most informative. UA5 Collaboration gave data[15] on inclusive cross sections in nine bins depending on number of charged particles ($2 \leqslant n \leqslant 10$, $12 \leqslant n \leqslant 20$, …, $n \geqslant 82$).

We define the following notations:

$$\sigma^{(1)} = \sum_{n=2}^{10} \sigma_n, \quad \sigma^{(2)} = \sum_{n=12}^{20} \sigma_n, \quad \ldots \quad \sigma^{(9)} = \sum_{n=82}^{\infty} \sigma_n, \tag{11}$$

$$\sum_{i=1}^{9} \sigma^{(i)} = \sigma^{nsd}. \tag{12}$$

Also we define

$$\frac{\mathrm{d}\sigma^{(1)incl}}{\mathrm{d}\eta} = \sum_{n=2}^{10} \frac{\mathrm{d}\sigma_n^{incl}}{\mathrm{d}\eta}, \quad \ldots \quad \frac{\mathrm{d}\sigma^{(9)incl}}{\mathrm{d}\eta} = \sum_{n=82}^{\infty} \frac{\mathrm{d}\sigma_n^{incl}}{\mathrm{d}\eta}, \tag{13}$$

$$\sum_{i=1}^{9} \frac{\mathrm{d}\sigma^{(i)incl}}{\mathrm{d}\eta} = \frac{\mathrm{d}\sigma^{incl}}{\mathrm{d}\eta}. \tag{14}$$

Data of UA5 Collaboration are given in format

$$\frac{1}{\sigma^{(i)}}\frac{\mathrm{d}\sigma^{(i)incl}}{\mathrm{d}\eta}. \tag{15}$$

We integrated the expression $\dfrac{1}{\sigma^{nsd}}\dfrac{\mathrm{d}\sigma^{(i)incl}}{\mathrm{d}\eta}$ over pseudorapidity space using (9) and obtained

$$\frac{1}{\sigma^{nsd}}\int \mathrm{d}\eta \frac{\mathrm{d}\sigma^{(i)incl}}{\mathrm{d}\eta} = \bar{n}^{(i)}, \tag{16}$$

where $\bar{n}^{(i)}$ is defined as

$$\bar{n}^{(1)} = \sum_{n=2}^{10} n\,P_n, \quad \bar{n}^{(2)} = \sum_{n=12}^{20} n\,P_n, \quad \ldots \quad \bar{n}^{(9)} = \sum_{n=82}^{\infty} n\,P_n.$$

Non single diffraction cross sections are the same for pp and $p\bar{p}$ interactions because of Pomeranchuk theorem. But shapes of multiplicity distribution curves are different for pp and $p\bar{p}$ since underlying elementary subprocesses are different. Therefore values of $\bar{n}^{(i)}$ are different for pp and $p\bar{p}$ collisions, we denote them $\bar{n}^{(i)}_{pp}$ and $\bar{n}^{(i)}_{p\bar{p}}$ correspondingly.

From relation (16) it follows

$$\int \mathrm{d}\eta \frac{\mathrm{d}\sigma^{(i)incl}_{pp}}{\mathrm{d}\eta} = \frac{\bar{n}^{(i)}_{pp}}{\bar{n}^{(i)}_{p\bar{p}}} \int \mathrm{d}\eta \frac{\mathrm{d}\sigma^{(i)incl}_{p\bar{p}}}{\mathrm{d}\eta}. \tag{17}$$

One of solutions of the integral equation (17) can be written as

$$\frac{\mathrm{d}\sigma^{(i)incl}_{pp}}{\mathrm{d}\eta} = \frac{\bar{n}^{(i)}_{pp}}{\bar{n}^{(i)}_{p\bar{p}}} \frac{\mathrm{d}\sigma^{(i)incl}_{p\bar{p}}}{\mathrm{d}\eta}. \tag{18}$$

From LCNM and fitting of multiplicity distributions for pp and $p\bar{p}$ at different energies[11] we obtained prediction for multiplicity distribution in pp scattering at $\sqrt{s} = 900$ GeV.

Table 1. Coefficients for inclusive pseudorapidity distributions.

	$2 \leqslant n \leqslant 10$	$12 \leqslant n \leqslant 20$	$22 \leqslant n \leqslant 30$
$\bar{n}^{(i)}_{p\bar{p}}/\bar{n}^{(i)}_{pp}$	0.76 ± 0.01	0.86 ± 0.01	0.99 ± 0.01
	$32 \leqslant n \leqslant 40$	$42 \leqslant n \leqslant 50$	$52 \leqslant n \leqslant 60$
$\bar{n}^{(i)}_{p\bar{p}}/\bar{n}^{(i)}_{pp}$	1.09 ± 0.01	1.10 ± 0.01	1.18 ± 0.01
	$62 \leqslant n \leqslant 70$	$72 \leqslant n \leqslant 80$	$n \geqslant 82$
$\bar{n}^{(i)}_{p\bar{p}}/\bar{n}^{(i)}_{pp}$	1.35 ± 0.02	1.45 ± 0.02	1.26 ± 0.02

We calculated the values of coefficients $\bar{n}^{(i)}_{pp}/\bar{n}^{(i)}_{p\bar{p}}$ for nine bins of multiplicity to estimate the difference in inclusive cross sections for pp and $p\bar{p}$ with (18), Table 1. Values of probabilities P_n for pp we took from our prediction in case when 75% of two gluons give three quark strings in $p\bar{p}$, values of P_n for $p\bar{p}$ we took from UA5 experiment.[16] Inclusive pseudorapidity distributions (without normalization) for bins $42 \leqslant n \leqslant 50$ and $52 \leqslant n \leqslant 60$ of charged multiplicity are shown in Figs. 4 and 5.

5. Conclusion

As we have shown above, there is possible difference in inclusive cross sections of pp and $p\bar{p}$ interactions. The result of ATLAS Coll., where dependance of inclusive cross sections from transverse momentum was obtained in pseudorapidity range $|\eta| < 2.5$ at energy $\sqrt{s} = 900$ GeV, can be a suggestion to this difference. The values reported by ATLAS Coll. are 1.2 times

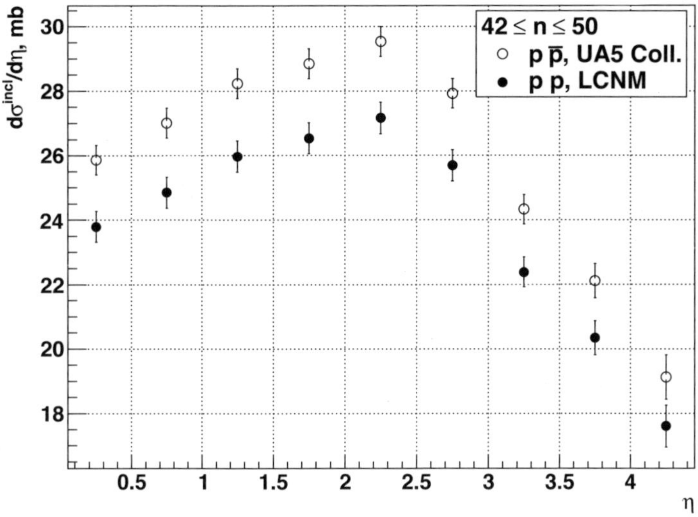

Fig. 4. Inclusive pseudorapidity distributions for $42 \leqslant n \leqslant 50$.

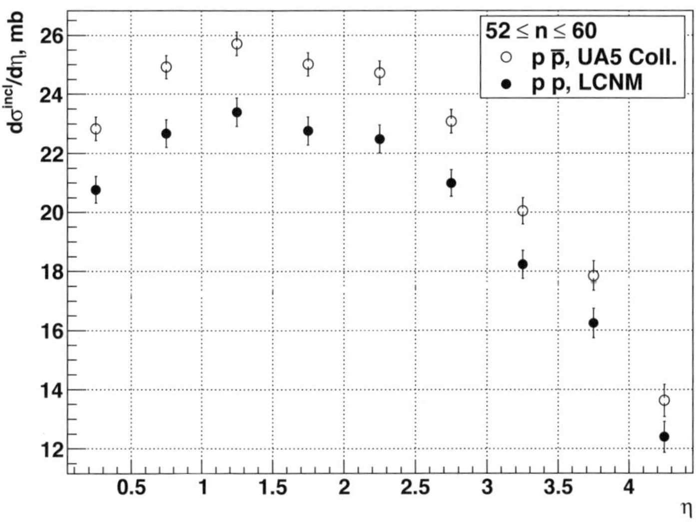

Fig. 5. Inclusive pseudorapidity distributions for $52 \leqslant n \leqslant 60$.

lower than corresponding values, obtained by UA5 Coll. for $p\bar{p}$. This effect is observed in large range of transverse momentum, up to $p_\perp = 2$ GeV/c. We think that ATLAS Coll. discovered the new effect – difference in multiple production in pp and $p\bar{p}$ interactions.

We propose to experimentalists to analyze inclusive cross sections in bins with high multiplicities, where the difference will be the most evident.

Acknowledgement

For one of the authors (V.A.A.) Vladimir Naumovich Gribov always was and still remains a teacher, to whom he is indebted and grateful all his life.

V.A.A. is thankful to the Organizing Committee, Julia Nyíri, Yuri Dokshitzer, Péter Lévai, Luciano Bertocchi and Daniele Treleani for the invitation and to the Organizing Committee for partial financial support.

References

1. I.Ya. Pomeranchuk, *Zh. Eksp. Teor. Fiz.* **34**, 725 (1958).
2. R.J. Iden, *High Energy Collisions of Elementry Particles*, Cambridge, 1967.
3. V.A. Abramovsky and O.V. Kancheli, *Pisma Zh. Eksp. Teor. Fiz.* **31**, 566 (1980).
4. V.A. Abramovsky and O.V. Kancheli, *Pisma Zh. Eksp. Teor. Fiz.* **32**, 498 (1980).
5. V.A. Abramovsky and N.V. Radchenko, *Particles and Nuclei, Letters* **6**, 607 (2009).
6. J.R. Cudell *et al.*, *Phys. Rev.* **D61**, 034019 (2000).
7. V.A. Abramovsky and O.V. Kancheli, *Pisma Zh. Eksp. Teor. Fiz.* **15**, 559 (1972).
8. V.A. Abramovsky, V.N. Gribov and O.V. Kancheli, *Yad. Fiz.* **18**, 595 (1973) [*Sov. J. Nucl. Phys.* **18**, 308 (1974)].
9. A.B. Kaidalov and K.A. Ter-Martirosyan, *Sov. J. Nucl. Phys.* **40**, 135 (1984).
10. V.A. Abramovsky and N.V. Radchenko, arXiv:0912.1041v1 (2009).
11. V.A. Abramovsky and N.V. Radchenko, *Particles and Nuclei, Letters* **6**, 717 (2009).
12. G.C. Rossi and G. Veneziano, *Nucl. Phys.* **B123**, 507 (1977).
13. B.Z. Kopelovich and B.G. Zakharov, preprint Dubna E2-87-911 (1987).
14. ATLAS Coll., *Phys. Lett.* **B688**, 21 (2010).
15. G.J. Alner *et al.*, *Z. Phys.* **C33**, 1 (1986).
16. R.E. Ansorge *et al.*, *Z. Phys.* **C43**, 357 (1989).

TSALLIS – PARETO-LIKE DISTRIBUTIONS IN HADRON-HADRON COLLISIONS*

G. G. BARNAFÖLDI† and T.S. BIRÓ

KFKI Research Institute for Particle and Nuclear Physics of the HAS
29-33 Konkoly-Thege Miklós Str.
H-1121 Budapest, Hungary
† bgergely@rmki.kfki.hu

K. ÜRMÖSSY and G. KALMÁR

KFKI Research Institute for Particle and Nuclear Physics of the HAS
29-33 Konkoly-Thege Miklós Str.
H-1121 Budapest, Hungary
and
Eötvös University
1/A. Pázmány Péter Sétány
H-1117 Budapest, Hungary

Non-extensive thermodynamics is a novel approach in the field of high energy physics. The low and intermediate transverse momentum spectra are extremely well reproduced by a Tsallis – Pareto distribution. However, the derivation of the Tsallis parameter is still an unsolved question. Understanding hadronization processes in heavy-ion collisions requires more detailed tests, especially at high transverse momenta, where we are far from the thermal equilibrium state. Here we focus on parton fragmentation processes at the final state, we analyze whether standard power-law-tailed fragmentation functions do overlap with Tsallis – Pareto energy distributions, reproducing experimental data. Moreover, we investigated a possible QCD evolution of the parameters.

1. Introduction

Strong correlations between particles can rule out the Boltzmann – Gibbs and rather lead to non-extensive statistics.[1] This case is typical for heavy ion physics which can be a possible testbed for non-extensive thermodynam-

*This work was supported by: Hungarian National Research Found (OTKA) PD73596, NK77816 and K68108.

†Work was partially supported by ICTP and the János Bolyai Research Scholarship of the Hungarian Academy of Sciences.

ical phenomena, due to the fact that high-energy nuclear reactions involve strongly correlated processes.

Non-extensive thermodynamics is a novel approach to investigate high-energy heavy ion collisions, furthermore it is particularly successful for describing the intermediate and high transverse momentum regimes. Based on the idea in Refs. 2, 3 we found that the hadron production in $AuAu$ collisions at RHIC energies can be well fitted by the Tsallis – Pareto distribution.[4]

Recent experimental data measured at the LHC in proton-proton collisions[5,6] have also shown Tsallis – Pareto-like distributions, but the fitted parameters are not yet interpreted. Furthermore, fits for earlier Tevatron data[7] seem to deviate from the parameterization at the highest transverse momentum region, as it was pointed out in Refs. 8, 9.

In this short contribution we present the re-parameterized fragmentation functions for charge-averaged pion yields by a Tsallis – Pareto-like distribution. We apply a suggested *ansatz* based on Ref. 10 in order to satisfy the Dokshitzer – Gribov – Lipatov – Altarelli – Parisi (DGLAP) scale evolution equation.[11] The inclusion of this new fragmentation model leads to a slightly modified curve in the pQCD improved parton model result, supporting the applicability of our model.

2. Tsallis – Pareto distribution for hadronization

Based on an earlier test the Tsallis – Pareto-like distribution were found to be a good fit to various experimentally measured p_T spectra in proton-proton collisions in a wide energy range.[10] Parallel to this, we found that an inclusion of a $\sim \ln\left(\ln\left(Q^2\right)\right)$ term leads to parameters scaling with energy correctly. Relating these findings to a pQCD based parton model picture, the origin of the Tsallis – Pareto-distribution may arise from the final state, where strongly correlated multiparton states are hadronizing. Our assumption: the hadronization (or fragmentation) should be the best candidate for the non-extensive processes yielding Tsallis – Pareto-distributions.

To test our assumption within the pQCD based parton model framework, we parameterized a fragmentation function for charge-averaged pions ($\pi = \pi^+ + \pi^-$) using Tsallis – Pareto-like distribution. We used the form for the parameterization as:

$$D^{\pi}_{p_i}\left(z\right) \sim \left(1 + (q_i - 1) \cdot \frac{z}{T_i}\right)^{-1/(q_i - 1)} , \tag{1}$$

where z is the momentum fraction of the hadron h originated from a parton

p_i for all cases namely, $\{u, d, s, c, b, \bar{u}, \bar{d}, \bar{s}, \bar{c}, \bar{b}, g\}$. We fitted parameters q_i and T_i using existing fragmentation function parameterizations.[12,13]

Since the above Eq. (1) has no scale evolution by itself, we used an *ansatz* similar to Ref. 10 for each parameter:

$$T_i \longrightarrow T_i(Q^2) = T_{i1} \cdot \ln\left(\ln(Q^2)\right) + T_{i0} \ ,$$

$$q_i \longrightarrow q_i(Q^2) = q_{i1} \cdot \ln\left(\ln(Q^2)\right) + q_{i0} \ . \tag{2}$$

Here T_{i0}, T_{i1}, q_{i0} and q_{i1} are those parameters, which we have fitted.[a] They are listed in Table 1 for all partonic channels.

Table 1. Parameters for the Tsallis–Pareto distribution in all partonic channels for charge-averaged pions.

parton, i	T_{i1}	T_{i0}	q_{i1}	q_{i0}
$u, \bar{u}, d, \bar{d}$	−0.057753	0.239825	0.124000	0.860351
$s, \bar{s}$	−0.093988	0.343175	0.265042	0.384453
$c, \bar{c}$	−0.048170	0.205408	0.134546	0.762087
$b, \bar{b}$	−0.033599	0.156249	0.103565	0.803255
g	−0.118556	0.394749	0.318477	0.253205

These parameters, for all important partonic channels $(u, d, s, g, \bar{u}, \bar{d}, \bar{s})$ to fragment into pions, can be found in Figure 1 as a function of the scale parameter Q^2. We plotted parameters for the light constituent quark contributions $q_{u,d}$ and $T_{u,d}$ by the *solid red line* and the *dashed red line*, respectively. For strange quarks q_s and T_s are plotted with *dotted black* and *dense-dotted black* lines, finally the gluon contributions are indicated by *long dash-dotted blue* and *short dash-dotted blue* lines

Due to the symmetry of the π state quarks and anti-quarks fragment equally. Moreover, isospin symmetry leads to identical u and d channel contributions. The asymptotic scaling values at large Q^2, are the same for the constituent, $q_{u,d}(Q^2)$ and for sea quark or gluon contributions, $q_{s,g}(Q^2)$ which is $\approx 1.20 - 1.25$. Herewith the QCD scaling behavior is clearly seen. The effective temperatures, T_i, show a monotonic decrease to 0, while the q_i indicate a monotonic increase for all partonic channels.

The temperatures $T_i(Q^2)$ and $q_i(Q^2)$ are fitted so that they approach the same value at the highest Q^2. This agrees with the assumption of a state complying with a general requirement of the thermodynamics.

[a]Note, that also the normalizations were fitted for each partonic channel.

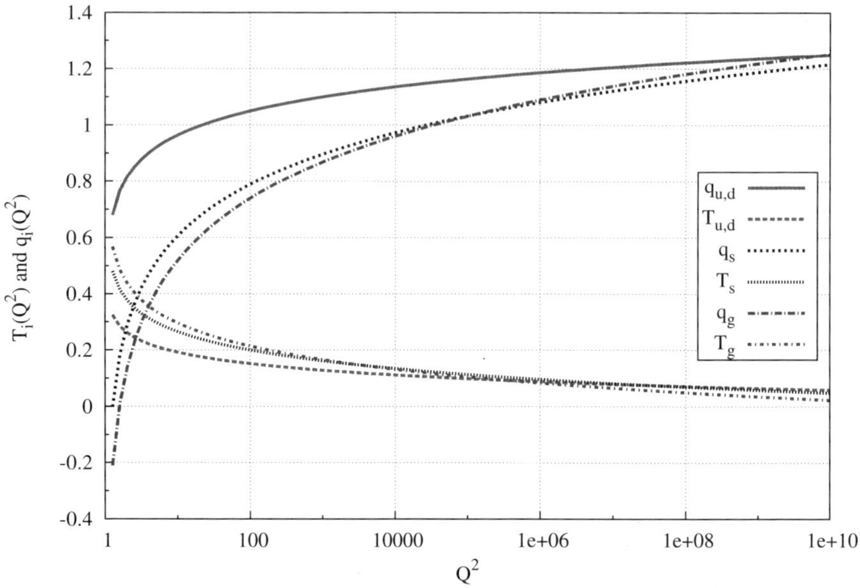

Fig. 1. The scaling parameters T_i, and q_i of Tsallis – Pareto distribution for various parton-hadron channels. (Color online.)

3. Parameter space for the fitted channels

We plotted the parameter space for the fitted distribution, Eq. (1) on Fig. 2. Since the obtained $q_i(Q^2)$ and $T_i(Q^2)$ parameters are connected via the scale parameter Q^2 in Eq. (2), they generally populate branches in the parameter space. For the case of the π hadronization, two remarkable branches appear: one for constituent quarks, (u and d) and another branches for sea contributions, like s and c quarks[b] and for the gluon, g. For constituents quarks in the π, these regions are $0.7 < q_{u,d} < 1.25$ and 0 GeV $< T_{u,d} < 0.3$ GeV, meanwhile for s, c, and g $0 < q_{s,c,b,g} < 1.25$ and 0 GeV $< T_{s,c,b,g} < 0.2 - 0.4$ GeV.

In both cases, at intermediate values of the parameters a huge dispersion can be inspected, these points are spread in the parameter space. The extreme parameter values form the above mentioned branches.

Comparing the parameter space with the fits of our earlier work,[10] we find overlapping regions with the earlier results. However, especially for the

[b]Parameter region for b quarks is almost overlapping with c.

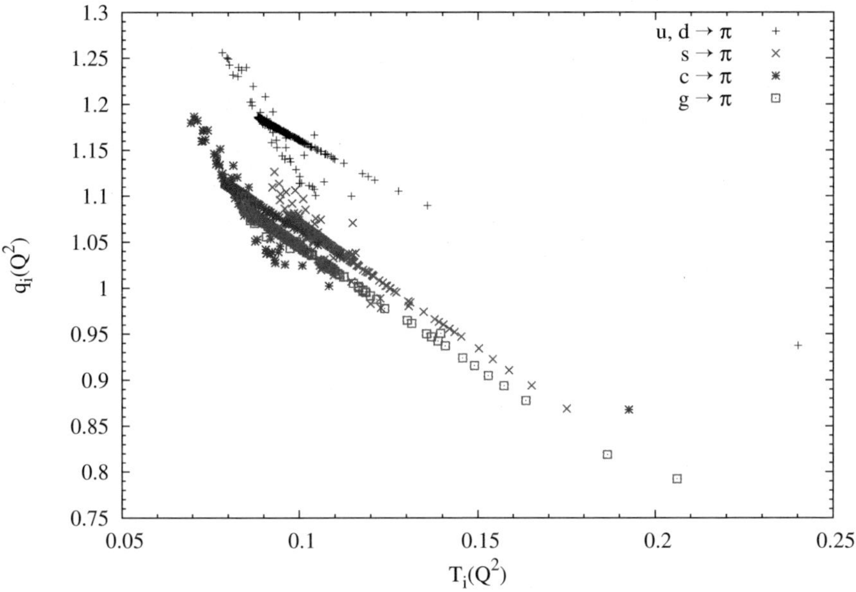

Fig. 2. Parameter space for q and T parameters of the fitted Tsallis – Pareto distributions for various parton-hadron channels. (Color online.)

light constituent quark channels, parameters are going far below the values were fitted by the experimental hadron transverse momentum spectra.

4. An application within pQCD based parton model

In the previous sections we parameterized a Tallis – Pareto-like fragmentation function. We found, that it satisfies the DGLAP scale evolution by construction, and is a good candidate for an easy parameterization.

As a simple test of our re-parameterized fragmentation functions, we implemented them into a next-to-leading order pQCD improved parton model framework.[14]

On Fig. 3 we plotted the inclusive spectrum of a minimum bias π^+ production in $\sqrt{s} = 7$ TeV proton-proton collision. Calculations were carried out in a next-to-leading order pQCD based parton model framework, with MRST-(cg) parton distribution.[15] For comparison we plotted two curves: one with AKK fragmentation function indicated by the *dashed blue* curve and the re-parameterized Tsallis – Pareto-like hadronization plotted with the *solid red* curve.

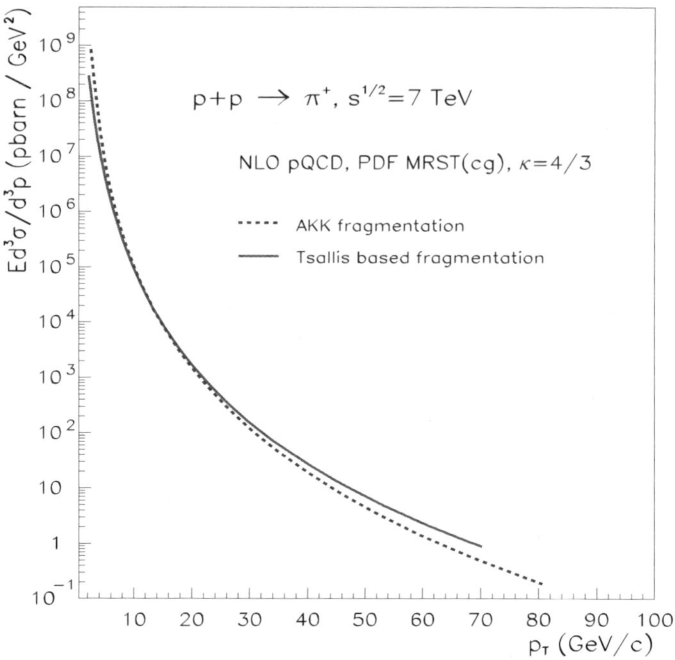

Fig. 3. Application of the Tsallis – Pareto-like hadronization in a pQCD based parton model for $p + p \longrightarrow \pi^+ + X$ reaction. (Color online.)

As Fig. 3 presents, our result overlaps with the original AKK-based calculations, especially at the low momentum fraction values. On the other hand at high p_T the new distribution gives a slightly higher yield. The validity of our model might be clarified by future data from Large Hadron Collider's experiments.

5. Conclusions and outlook

In this short contribution we presented re-parameterized fragmentation functions for the charge-averaged pion using Tsallis – Pareto-like distributions. Based on a suggested *ansatz*,[10] we fitted AKK parameterization to Eqs. (1) and (2). The proposed form of the *ansatz*, $\sim \ln\left(\ln\left(Q^2\right)\right)$, looks promising as a scaling extension of the Tsallis – Pareto-distribution for pQCD application, with its correct asymptotical behavior at high Q^2. Furthermore, any physically acceptable fragmentation model has to deal with the Dokshitzer – Gribov – Lipatov – Altarelli – Parisi (DGLAP) scale evolution equation.

The obtained parameters for π were found to be in the following regions: for channels u and d in $0.7 < q_{u,d} < 1.25$ and in 0 GeV $< T_{u,d} < 0.3$ GeV, meanwhile for channels s, c, b and g were in $0 < q_{s,c,b,g} < 1.25$ and in 0 GeV $< T_{s,c,b,g} < 0.2 - 0.4$ GeV. On the other hand the population in the parameter space features two main branches, one for constituents and one sea and gluon contributions.

The inclusion of the Tsallis – Pareto-based fragmentation leads to a slightly modified curve in a pQCD improved parton model result, presenting the applicability of our model.

References

1. C. Tsallis, *J. Stat. Phys.* **52**, 479 (1988); *Eur. Phys. J.* **A40**, 257 (2009).
2. T. S. Biró, *Europhys. Lett.* **84**, 56003 (2008).
3. G. Wilk and Z. Wlodarczyk, *Phys. Rev.* **C79**, 054903 (2009).
4. T. S. Biró, K. Ürmössy and G. G. Barnaföldi, *J. Phys.* **G35**, 044012 (2008).
5. V. Khachatryan *et al.* (CMS Collaboration), *JHEP* **1002**, 041 (2010).
6. V. Khachatryan *et al.* (CMS Collaboration), CSM-QCD-10-006, CERN-PH-EP-2010-009, FERMILAB-PUB-10-170-CMS, *arXiv:1005.3299*
7. T. Aaltonen *et al.* (CDF Collaboration), *Phys. Rev.* **D79**, 112005 (2009).
8. R. Ichou and D. d'Enterria, *Phys. Rev.* **D82**, 014015 (2010).
9. F. Arleo, D. d'Enterria and A. S. Yoon, *JHEP* **1006**, 035 (2010).
10. G. G. Barnaföldi, K. Ürmössy and T. S. Biró: *in print* Proceedings of the Hot Quarks 2010 (2010).
11. V. N. Gribov and L. N. Lipatov, *Sov. J. Nucl. Phys.* **15**, 438 (1972); G. Altarelli and G. Parisi, *Nucl.Phys* **B126**, 298 (1977); Yu. L. Dokshitzer, *Sov. Phys. JETP* **46**, 641 (1977).
12. S. Albino, B. A. Kniehl and G. Kramer, *Nucl. Phys.* **B725**, 181 (2005).
13. S. Kretzer, *Phys. Rev.* **D62**, 054001 (2000).
14. Y. Zhang *et al*, *Phys. Rev.* **C65**, 034903 (2002).
15. A. D. Martin, R. G. Roberts, W. J. Stirling and R. S. Thorne, *Eur. Phys. J.* **C23**, 73 (2002)

THE PHOTON-JET ANGULAR DISTRIBUTION IN $p\bar{p}$ COLLISIONS AT $\sqrt{s} = 1.96$ TEV

ZOUINA BELGHOBSI

Laboratoire de Physique Théorique (LPTh),
Département de Physique, Université de Jijel,
PB.98 Ouled Aissa, DZ-18000 Jijel, Algeria
belghob@univ-jijel.dz
belghob@yahoo.fr

In this work, we study the production of a large-p_T photon in association with a jet in proton-proton collisions. We assess the sensitivity of various photon+ jet correlation observables to the photon fragmentation functions, especially the photon-jet angular distribution.

1. Introduction

The phenomenology of prompt photons is very rich and interesting, as the photon can be considered as a pointlike particle described by QED, leading to clean expermental signatures.

At the LHC, photon +jet final states will be important for jet calibration and parton distribution function (pdf) studies.[1,2] An accurate knowledge of the photon+jet rate, in particular, is required to estimate and control the reducible background (which comes from photon-jet and jet-jet events) to the Higgs boson search in the diphoton channel.[3] The production of prompt photons in hadronic collisions may be schematically seen as originating from either of two mechanisms. In the first one, which may be called "direct" (D), the photon behaves as a high p_T colorless parton, i.e., it takes part in the hard subprocess, and is most likely well separated from any hadronic environnment see Figure 1. In the other one which may be called "fragmentation" (F), the photon behaves hadronlike, i.e., it results from the collinear fragmentation of colored high p_T parton (see Figure 2). In the latter case, it is most probably accompanied by hadrons — unless the photon carries away most of the transverse momentum of the fragmenting parton.

From a technical point of view, (F) emerges from the calculation of the higher order corrections in the perturbative expansion inpowers of the strong coupling α_s. The higher order corrections to the cross section can be split into (1) a contribution free from these final state collinear singularities, to be added to the Born term so as to build (D), and (2) a contribution (F) involving these singularities together with accompagnying large collinear logarithms.

In this article, (D) is defined as the Born term plus the fraction of the highest order corrections from which final state collinear singularities and accompanying collinear logarithms have been subtracted according to the $\overline{MS}$ factorization scheme, whereas (F) is defined as the contribution involving a fragmentation function of any parton into a photon defined in the $\overline{MS}$ scheme. The partonic cross section can thus be written

$$d\sigma^\gamma = d\sigma^{(D)}(\mu^2, M^2, M_F^2) + \sum_{k=q,\bar{q},g} d\sigma_k^{(F)}(\mu^2, M^2, M_F^2) \otimes D_{\gamma/k}(M_F^2), \tag{1}$$

where μ, M, M_F are respectively, the (arbitrary) renormalization, initial state factorization, and final state fragmentation scales, and "$\otimes$" stands for a convolution over the fragmentation variable.

The fragmentation component represents a fraction of the inclusive prompt photon signal which grows with the center-of-mass energy of the collision. It becomes dominant at collider energies, in particular, the TeV collider experiments CDF and D0 at the Tevatron, ATLAS and CMS at the LHC- do not measure inclusive photons, because at these energies the inclusive prompt photon signal would be swamped by a large background of secondary photons from decays of fast neutral mesons (mainly π^0, as well as η, etc.). Instead these experiments impose isolation criteria on the hadronic final states of photon candidate events, requiring that the photon be not accompanied by more than a prescribed amount of hadronic transverse energy in some given cone about the photon. The isolation cuts do not only supress the background, they also substantially reduce the (F) component. Yet some fraction of the (F) component may survive and affect shapes of various tails of distributions, especially for correlation observables. Of course the sensitivity is even larger when loose isolation cuts are applied. We wish to stress the interest of photon-jet correlation observables, in particular, in constraining the photon fragmentation functions, which requires to go beyond the lowest order. Similar studies have been performed in Ref. 4. In these works the (F) component was calculated at lowest order (LO) only.

Fig. 1. Direct mechanism in which photon takes part in the hard process.

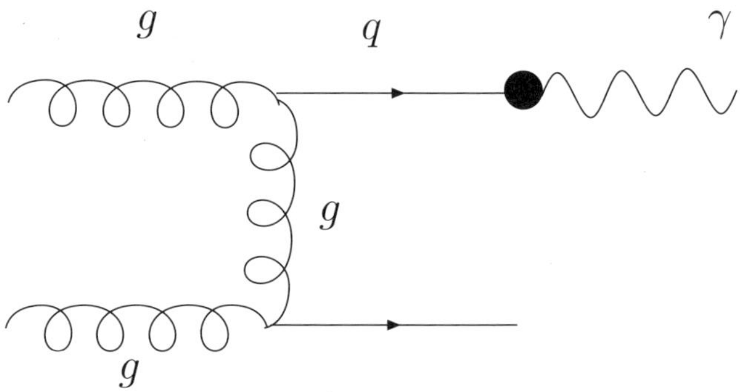

Fig. 2. Fragmentation mechanism in which photon results from the collinear fragmentation of parton.

2. Photon-jet angular distribution

An observable expected to receive a distinctive contribution from the (F) component is the photon-jet angular distribution which has been measured by the CDF Collaborations[5,6] and is defined as follows.

At LO, corresponding to $2 \to 2$ kinematics, $\cos\theta^*$ is the cosine of the angle between the photon direction and the beam axis in the center-of-mass system of the partonic subprocess. It also coincides with $\cos\theta^* = \tanh y^*$ where $y^* = (y_\gamma - y_{jet})/2$. This angular distribution is expected to receive a dominant contribution from the (F) component when $\cos\theta^*$ becomes close to 1. Indeed, at lowest order, the (D) component proceeds via a t-channel quark-exchange yielding a behavior $\sim 1/(1-\cos\theta^*)$ for the partonic amplitude squared, whereas the (F) component involves also gluon exchange in the t-channel, yielding a behavior $\sim 1/(1-\cos\theta^*)^2$. On this ground, one thus expects (F) to take over for $\cos\theta^*$ values close enough to 1.

Depending on the cuts applied, the $\cos\theta^*$ dependences coming from the partonic transition matrix element squared may be blurred by an extra dependence coming through the parton luminosity. Focusing on the direct (D) contribution, and parametrizing the LO phase space as

$$d(p_T^\gamma)^2 dy_\gamma dy_{jet} = \frac{1}{2} d(p^*)^2 dy_B d\cos\theta^* \tag{2}$$

in terms of the variables $\cos\theta^* = \tanh y^*$, $y_B = (y_\gamma + y_{jet})/2$, and $p^* = p_T^\gamma \cosh y^*$, the LO distribution in $\cos\theta^*$ reads

$$\frac{d\sigma}{d\cos\theta^*} = \sum_{i,j} \int dy_B dp^* G_{i/P}(x_i) G_{j/P}(x_j) \frac{d\hat{\sigma}_{ij}}{d\cos\theta^* dy_B dp^*} \tag{3}$$

with

$$x_{i,j} = \frac{2p^*}{\sqrt{S}} e^{\pm y_B} \tag{4}$$

In particular, one of the two pdfs involved in the distribution in $\cos\theta^*$ has an argument x which grows with y^* i.e. with $\cos\theta^*$ at fixed p_T^γ, so that this pdf decreases (towards zero) if $\cos\theta^*$ increases (towards one). In the absence of extra cuts, this decrease actually takes over the growth of the partonic cross section with growing $\cos\theta^*$ over the whole range. At LO, this can be neutralized by imposing cuts on y_B and p^* independent from y^*,[5,6] so that the integration over y_B and p^* in Eq. (3) yields $\cos\theta^*$ independent factors.

Beyond LO the definition of $\cos\theta^*$ has to be extended. This extension is not unique, and various definitions can be found in the literature. Here we take*

$$\cos\theta^* = \tanh y^*, \tag{5}$$

where

$$y^* = \frac{1}{2}(y_\gamma - y_{leadingjet}), \tag{6}$$

$y_\gamma - y_{leadingjet}$ being the difference† between the rapidity of the photon and the rapidity of the *leading jet*, i.e. the jet of highest transverse energy.

*An alternative possibility is the one used in Ref. 5, which, in short, combines several jets of multijet final state into one so-called superjet recoiling against the photon, in order to stick to a 2 → 2 kinmatics as close as possible. See Ref. 5 for more details.

†Note that the definition of $\cos\theta^*$ given by Eqs. (5) and (6) refers to a quantity which is invariant under longitudinal boosts along the beam axis.

Furthermore, the higher order contributions to the angular distribution involve an extra convolution smearing over kinematical configurations, so that the interpretation of this distribution beyond LO is less transparent.

Besides, measurements at colliders most often involve isolated photons, in which case the (F) component is quite reduces. Namely, when the hadronic transverse energy accompanying the photon is required to be smaller than $E_{T_{max}}$, the (F) contribution is roughly proportional to $(1 - z_c) \simeq E_{T_{max}}/p_T^\gamma$, the width of the support $[z_c, 1]$ of the convolution with the photon fragmentation functions. If $E_{T_{max}}$ is chosen such that this ratio is always small, the dominance of the t-channel gluon exchange from the (F) component is never effective; yet one might still expect a sizable distorsion of the angular distribution for $\cos\theta^*$ close enough to 1.

2.1. *CDF Collaboration analysis*

In a preliminary study[6] subsequent to the analysis published in Ref. 5 and based on a data set with larger statistics and extended towars lower values of p_T^γ, the CDF Collaboration found a discrepancy between the measured $\cos\theta^*$ distribution and the theoretical prediction of Ref. 4. This subsequent preliminary CDF analysis concluded that extra dijetlike contributions involving t-channel gluon exchange would be necessary to bridge the gap, and that these extra contributions might come from NLO contributions to the (F) component. Since the prediction of Ref. 4 involves an account of (F) at LO only, we have revisited this observable and computed the effects of accounting for (F) at NLO. The CDF Collaboration adopted a procedure to patch together the contributions from data corresponding to two regions that were distinct in p^* and y_B though overlapping in $\cos\theta^*$, in order to maximize the range in $\cos\theta^*$ displayed on one and the same plot. In particular the distribution was normalized to 1 in the bin farthest from $\cos\theta^* = 1$, and the data sets from the two regions were normalized to each other in one overlapping bin in $\cos\theta^*$.

We understand this approximate procedure to have been dictated by the use of limited statistics and precision of Run I data but we did not follow the same procedure in our study for several reasons. First, the normalization to 1 in the bin farthest from $\cos\theta^* = 1$ aims at getting rid of the numerical factor coming from the integration of partonic luminosity. This is fine as long as only one partonic subprocess contributes- or at least, when one yields a much greater contribution than all the others. However, in the

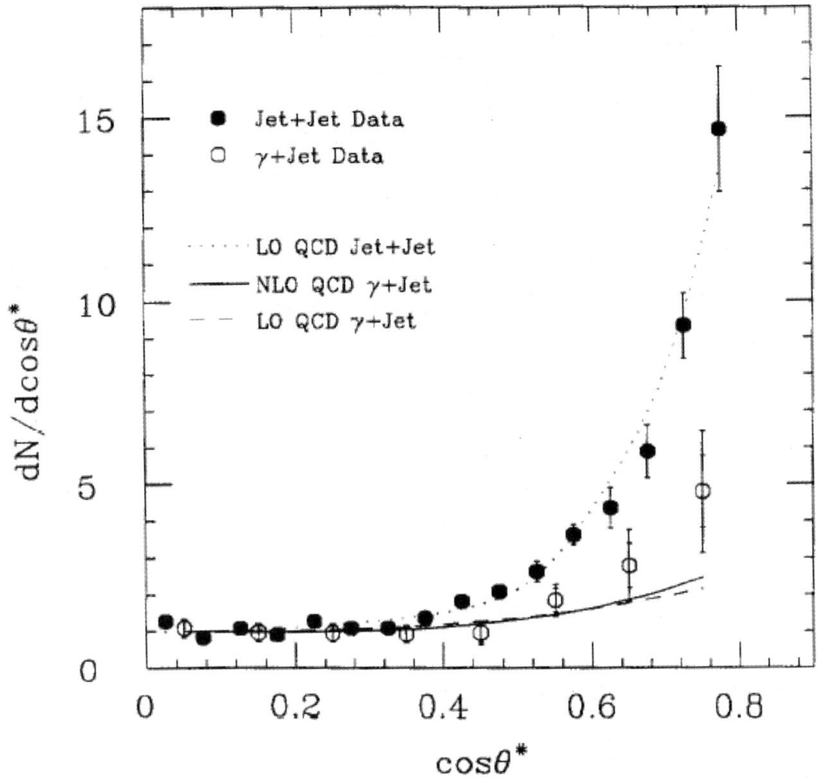

Fig. 3. Prompt photon $dN/d\cos\theta^*$ after background subtraction. The photon data (open circles) are compared to LO-QCD (dashes) and NLO-QCD (solid) shown are previously published dijet data (solid circles) and theory curves for LO dijet tree level diagrams (dots). The data and theory curves are normalized to an area of 3.0 in the region $|\cos\theta^*| < 0.3$.

present case, the distribution is of the form

$$\frac{d\sigma}{d\cos\theta^*} = \sum_s \mathcal{L}^{(s)} \frac{d\hat{\sigma}^{(s)}}{d\cos\theta^*}, \tag{7}$$

i.e. a linear combination of contributions coming from several subprocesses s. In particular, considering (D) only, gq (or $g\bar{q}$) initiated and $q\bar{q}$ initiated processes which contribute at LO have distinct functional dependences on $\cos\theta^*$. Second, the integrated partonic luminosity factors $\mathcal{L}^{(s)}$ depend not only on the subprocess s but also on the integration regions in phase space. The relative weights of the subprocesses from the contributions (D) and

(F) thus differ in the two regions defined by CDF as "1" and "2". The normalization enforced by the matching procedure of CDF is not harmless on the impact of the NLO correction to the (F) contribution, and might *reduce* the impact of thos correction.

Therefore we did not stick to the CDF study. We focused on a study of the magnitude of the (F) contribution, without making a direct comparison with the CDF Run I data. Our study has been made for $\sqrt{S} = 1.96$ TeV, with the following definitions and kinematic cuts: $| y_\gamma| \leq 0.9$, $p_T \geq 30$ GeV, $p_T^{leading\ jet} \geq 25$ GeV. The jets were defined according to the D0 algorithm,[7] with cone aperture $R_c = 0.7$. The photon isolation required that in a cone of aperture $R = 0.4$ in rapidity and azimuthal angle around the photon direction, the fraction of maximal hadronic transverse energy $E_{T_{max}}/p_T^\gamma$ be less than a prescribed value ϵ, which we varied from 0.05 to 0.3. The further cut 45 GeV $\leq p^* \leq$ 55 GeV was imposed. The pdf set CTEQ 6.1 was used together with BFG[8] set II for the fragmentation functions, with the scale choice $\mu = M_F = M = p_T^\gamma/2$.

We have considered three ingredients which may affect the size of the contribution of the (F) component. One is the account for the NLO corrections to the many subprocesses; another one concerns the uncertainty on the fragmentation functions; yet another one deals with a possible mismatch between the implementation of isolation at partonic vs hadronic level.

Let us first consider the impact of the NLO corrections to the (F) component. For the standard scale choice $\mu = M_F = M = p_T^\gamma/2$, the effect is to multiply the component (F) at LO by about a factor of 2. From the top-right see Figure 4, the impact on the total (D)+ (F) both at NLO amounts to an increase by 4 per cent in the upper $\cos\theta^*$ range with respect to (w.r.t) (D) at NLO + (F) only at LO for $\epsilon = 0.05$.

Is it possible to increase the (F) contribution by modifying the fragmentation functions? When a stringent isolation cut is required on the photon candidates as in the CDF experiment, the (F) contribution involves the photon fragmentation function at $z \geq z_c$ i.e. rather close to 1. In this region the fragmentation functions are dominated by their so-called anomalous parts predicted by perturbative QCD. Their poorly known nonperturbative parts, which would be the only adjustable ingredients, play no role: thus the (F) contribution to $\cos\theta^*$ distibution is rather tightly constrained.

We have tackled the issue of the account of isolation at the partonic vs hadronic level by varying the value of the isolation parameter ϵ from 0.05 to 0.3. As already mentioned above and as can be seen on see Figure 4, the separate (D) and (F) contributions do depend strongly on ϵ at NLO; yet

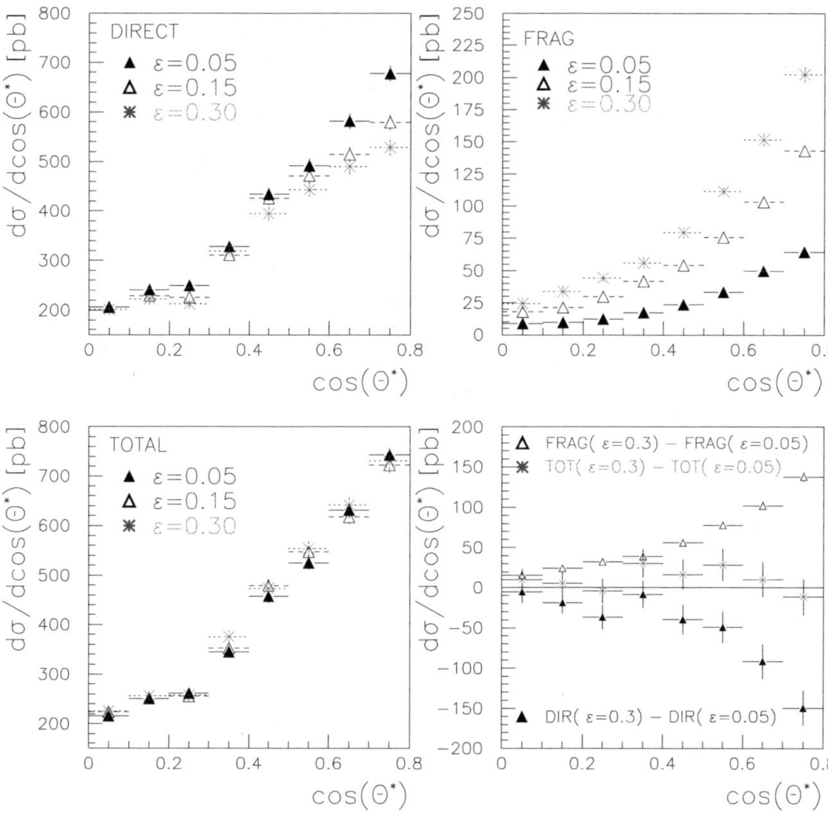

Fig. 4. (Color online). Sensitivity of the distribution of $\cos\theta^*$ in photon + jet at NLO to the isolation parameter $\epsilon = E_{T_{max}}/p_T^\gamma$. Top left: Direct (D) contribution only. Top right: Fragmentation (F) contribution only. Bottom left: total (D) + (F) contribution. Bottom right: differences in (D), (F) and total (D) + (F) between $\epsilon = 0.3$ and 0.05.

strong cancellations turn out to occur between (D) and (F) so that the total (D) + (F) depends on ϵ only very mildly, at least as long as the infrared sensitive term $\alpha_s \ln \epsilon$ in (D) does not become large — otherwise the fixed order prediction becomes unreliable. Therefore, changing the calorimetric isolation parameter by as a factor of 6 does not modify the total contribution to the $\cos\theta^*$ distribution significantly.

3. Conclusion

In this work, we have studied photon-jet correlations in hadronic collisions based on the NLO program JETPHOX[‡], which is a Monte Carlo program of partonic event generator type which incorporates NLO corrections to both direct photons and photons from fragmentation. The program is flexible to account for user-defined kinematic cuts and photon isolation parameters. We sudied the photon-jet angular distribution $\cos\theta^*$ in view of possible constraints on the parton distribution functions in the proton, in particular, the gluon.

To summarize, the results of our calculations show that the idea of playing with the fragmentation component as suggested in the CDF analysis turns out to be ineffective in the conditions which we have considered. It would be worthwhile to perform a quantitative analysis of the much larger statistics data set gathered in Run II, without relying on the questionable matching procedure used in the CDF Run I analysis.On the other hand, the *verry small* fraction of hadronic events which pass the isolation cuts corresponds to the tail of fragmentation at large z which is not constrained by the data. These background events yield, namely, dijet-type contributions involving t-channel gluon exchange, which might explain part of the discrepancy observed, and the distribution in $\cos\theta^*$ at $\cos\theta^* \to 1$ might provide an enhanced sensitivity to this contamination w.r.t other prompt photon observables.

Acknowledgments

I want to thank the organizers for inviting me to this Gribov-80 Memorial Workshop on Quantum Chromodynamics and Beyond. I want also to thank all my collaborators: J.-Ph. Guillet, E. Pilon, M. Werlen (LAPTH, Annecy), M. Fontannaz (LPT, Orsay) and G. Heinrich (IPPP, Durham).

References

1. G.L. Bayatian *et al.* (CMS Collaboration), *J. Phys.* **G 34**, 995 (2007); R. Adolphi *et al.* (CMS Collaboration), *JINST* **3**, S08004 (2008).
2. G. Aad *et al.* (ATLAS Collaboration), arXiv:0901.0512 (2009); G. Aad *et al.* (ATLAS Collaboration), *JINST* **3**, S08003 (2008).
3. Estimates using JETPHOX can be found in Ref. 2, Table 1, p. 1214. See also: Y. Fang, CERN, Report No. CERN-THESIS-2008-073, p. 78, Table 5.3.

[‡]Code available at the following URL address:
http://*lappweb.in2p3.fr/lapth/PHOX*$_F$*AMILY/main.html*.

4. H. Baer, J. Ohnemus and J.F. Owens, *Phys. Lett.* **B234**, 127 (1990).
5. F. Abe *et al.* (CDF Collaboration), *Phys. Rev. Lett.* **71**, 679 (1993).
6. J.I. Lamoureux, *AIP Conf. Proc.* **357**, 548 (1996).
7. G.C. Blazey *et al.*, in Batavia 1999, *Weak Boson Physics in Run II*, edited by U. Baur, R.K. Ellis and D. Zeppenfeld (Fermilab, Batavia, 2000) (Fermilab Report No. FERMILAB-PUB-00-297), p. 47.
8. L. Bourhis, M. Fontannaz and J.Ph. Guillet, *Eur. Phys. J.* **C2**, 529 (1998).

RECENT RESULTS IN GENERAL FIELD THEORY AND GRAVITY

NEUTRINO OSCILLATIONS IN QUANTUM MECHANICS AND QUANTUM FIELD THEORY

E. AKHMEDOV

Max-Planck Institut für Kernphysik,
Saupfercheckweg 1, 69117 Heidelberg, Germany
and
NRC Kurchatov Institute, Moscow, Russia
akhmedov@mpi-hd.mpg.de

Quantum mechanical and quantum field theoretical approaches to neutrino oscillations are discussed and compared.

Keywords: Neutrino oscillations; quantum mechanics; quantum field theory.

1. Introduction

The title of this workshop is "QCD and beyond", and my talk is certainly in the second category. Still, I believe that its topic is fully relevant to this meeting dedicated to the 80th anniversary of birthday of Vladimir Naumovich Gribov. I would like to remind you the 1969 paper[1] of Gribov and Pontecorvo, which was an important contribution to the theory of neutrino oscillations. In particular, in this paper oscillations of Majorana neutrinos were considered for the first time, and important observations on possible oscillations of solar neutrinos were made.

By now, neutrino oscillations have been firmly established in experiments with solar, atmospheric, reactor and accelerator neutrinos (see, *e.g.*, Refs. 2, 3 for reviews). The theory of this phenomenon has been actively developed since the 1960s, and is now quite mature. Still, it cannot be considered as finished or complete. Although neutrino oscillations appear to be a simple quantum mechanical phenomenon, a closer look at them reveals a number of subtle and even paradoxical issues. It is probably for this reason that debates on the fundamentals of the oscillation theory and on the correctness of different theoretical approaches to neutrino oscillations do not cease in the literature, with many papers published in the recent years. The questions that are currently being debated include

- Do different neutrino mass eigenstates composing a flavour eigenstate have same energy or same momentum?
- Do neutrino states evolve in space or in time?
- What is the role of QM uncertainty relations in ν oscillations?
- Is wave packet description necessary?
- What determines the size of the neutrino wave packets?
- Under what conditions can the oscillations be observed?
- When are the oscillations described by a universal probability?
- Is the standard oscillation formula correct?
- Is the oscillation probability Lorentz invariant?

It is well known by now that neutrino oscillations can be consistently described either in the quantum-mechanical (QM) wave packet approach, or within a quantum field theoretic (QFT) framework.[a] In this talk I will review and compare these two approaches, which will allow me to discuss the answers to some of the questions listed above. The results presented here are mainly based on the work done in collaboration with Alexei Smirnov and Joachim Kopp.[4–6]

2. Same energies or same momenta?

Neutrinos emitted or absorbed in charged-current weak-interaction processes are flavour eigenstates ν_α ($\alpha = e, \mu, \tau$), which are linear superpositions of mass eigenstates ν_i ($i = 1, 2, 3$). In the basis where the mass matrix of charged leptons has been diagonalized the fields describing the massive neutrinos ν_i and flavour-eigenstate neutrinos ν_α and the corresponding states $|\nu_i\rangle$ and $|\nu_\alpha\rangle$ are related by

$$\nu_\alpha = \sum_i U_{\alpha i}\, \nu_i \,, \qquad\qquad |\nu_\alpha\rangle \;=\; \sum_i U_{\alpha i}^*\, |\nu_i\rangle \,, \tag{1}$$

where U is the leptonic mixing matrix. Since neutrinos propagate macroscopic distances between their source and detector, the mass-eigenstate components ν_i are actually on the mass shell, *i.e.* their energies, momenta and masses satisfy the relation $E_i^2 = \vec{p}_i^{\,2} + m_i^2$. It then immediately follows that different mass eigenstates cannot have the same energy and the same momentum simultaneously. Therefore, in most derivations of the standard formula for the probability of neutrino oscillations in vacuum it is usually

[a]Although in a number of sources a plane wave approach to neutrino oscillations is employed, it is actually marred by inconsistencies and, if applied correctly, does not lead to neutrino oscillations at all.

assumed that different ν_i composing a given flavour eigenstate either have the same momentum or the same energy. The derivation typically proceeds as follows.

If one assumes that all the mass eigenstates composing the initially produced flavour state $|\nu(0)\rangle = |\nu_\alpha\rangle$ have the same momentum, then, after time t has elapsed, the ith mass eigenstate will simply pick up the phase factor $\exp(-iE_i t)$, and the evolved state $|\nu(t)\rangle$ will be given by

$$|\nu(t)\rangle = \sum_i U_{\alpha i}^* \, e^{-iE_i t} |\nu_i\rangle \,. \tag{2}$$

Projecting this state onto the flavour state $|\nu_\beta\rangle$ and taking the squared modulus of the resulting transition amplitude, one gets the probability of the neutrino flavour transition $\nu_\alpha \to \nu_\beta$ after the time interval t:

$$P(\nu_\alpha \to \nu_\beta; t) = \left| \sum_i U_{\beta i} \, e^{-iE_i t} \, U_{\alpha i}^* \right|^2 . \tag{3}$$

Since for ultra-relativistic neutrinos $E_i \simeq p + \frac{m_i^2}{2p}$, and for relativistic point-like particles the distance L they propagate during the time interval t satisfies $L \simeq t$, one finally finds

$$P(\nu_\alpha \to \nu_\beta; L) = \left| \sum_i U_{\beta i} \, e^{-i\frac{\Delta m_{ij}^2}{2p} L} \, U_{\alpha i}^* \right|^2 , \tag{4}$$

where $\Delta m_{ij}^2 = m_i^2 - m_j^2$ and the index j corresponds to any of the mass eigenstates. This is the standard formula describing neutrino oscillations in vacuum. Note that in this approach neutrino states actually evolve only in time (see Eq. (3)); the usual coordinate dependence of the oscillation probability (4) is only obtained by invoking the additional "time-to-space conversion" assumption $L \simeq t$. Without this conversion, one would have come to a paradoxical conclusion that neutrino oscillations could be observed by just putting the neutrino detector immediately next to the source and waiting long enough.

Alternatively, one could assume that all the mass-eigenstate neutrinos composing the initially produced flavour state $|\nu_a\rangle$ have the same energy. Using the fact that the spatial propagation of the ith mass eigenstate is described by the phase factor $e^{i\mathbf{p}_i\mathbf{x}}$ and that for a relativistic neutrino of mass m_i and energy E one has $p_i \simeq E - m_i^2/2E$, one again comes to the same standard formula (4) for the oscillation probability. Note that in this case the neutrino flavour evolution occurs in space and it is not necessary to invoke the "time-to-space conversion" relation to obtain the standard oscillation formula.

The above two alternative derivations of the oscillation probability are very simple and transparent, and they allow one to arrive very quickly at the desired result. The trouble with them is that they are both wrong. In general, there is no reason whatsoever to assume that different mass eigenstates composing a flavour neutrino state emitted or absorbed in a weak-interaction process have either the same energy or the same momentum. Indeed, the energies and momenta of particles emitted in any process are dictated by the kinematics of the process and by the experimental conditions. The direct analysis of, *e.g.*, 2-body decays with simple kinematics, such as pion decay, allows one to find the 4-momenta of the emitted particles and shows that neither energies nor momenta of the different neutrino mass eigenstates composing the produced flavour state are the same.

3. QM approach, energy and momentum uncertainties and energy-momentum conservation

How can one resolve this problem? It is sufficient to remember that in reality energies and momenta of all particles have intrinsic quantum-mechanical uncertainties. This is a consequence of the fact that the processes in which these particles are produced or detected are always confined to finite space-time intervals.

Since energy and momentum uncertainties of particles are usually completely negligible compared to their energies and momenta themselves, in most situations these uncertainties can be safely neglected. This is, however, not justified when neutrino oscillations are considered, since the neutrino energy and momentum uncertainties, as tiny as they are, are crucially important for the oscillation phenomenon – in their absence the oscillations just would not occur. Indeed, if neutrinos were produced, for example, with no momentum uncertainty, this would mean that their source was completely delocalized in space, and therefore neutrino oscillations as a function of baseline L would be unobservable. Similar arguments apply to the neutrino energy uncertainty (see[4,6] for more detailed discussions).

From the above it is clear that in theoretical analyses of neutrino oscillations within the QM approach neutrinos should be described by wave packets rather than by plane waves (later on I will discuss an alternative possibility, realized in the QFT framework, in which neutrinos are described by propagators).

Obviously, energy and momentum uncertainties are in no contradiction with the fact that energy-momentum conservation is an exact law of nature. However, the expectation values of energies and momenta of the particles participating in a given localized process satisfy only approximate conservation laws.

Let me now very briefly review the QM wave packet approach to neutrino oscillations. The state vector of the neutrino state, after time t from its production and at the distance x from the production point, is

$$|\nu_\alpha(t,\vec{x})\rangle \;=\; \sum_i U_{\alpha i}^*\,|\nu_i(t,\vec{x})\rangle \;=\; \sum_i U_{\alpha i}^*\,\Psi_i^P(t,\vec{x})|\nu_i\rangle\,. \tag{5}$$

Here the coordinate-space wave function (wave packet) of the ith mass eigenstate is

$$\Psi_i^P(t,\vec{x}) = \int \frac{d^3p}{(2\pi)^3}\, f_i^P(\vec{p})\, e^{i\vec{p}\vec{x}-iE_i(p)t}\,, \tag{6}$$

where $f_i^P(\vec{p})$ is the momentum distribution function, which is assumed to have a sharp maximum at some momentum $\vec{P}$, *i.e.* the width of the peak $\sigma_{pP} \ll P$. Likewise, for the detected state (centered at $\vec{x}=\vec{L}$) one writes

$$|\nu_\beta(\vec{x})\rangle = \sum_k U_{\beta k}^*\,\Psi_k^D(\vec{x})|\nu_i\rangle\,, \qquad \Psi_k^D(\vec{x}) = \int \frac{d^3p}{(2\pi)^3}\, f_k^D(\vec{p})\, e^{i\vec{p}(\vec{x}-\vec{L})}\,. \tag{7}$$

The momentum distribution function $f_i^D(p)$ has a peak of the width σ_{pD} at $\vec{p}=\vec{P}'$. The coordinate-space wave packets of the produced and detected neutrino states have the spatial widths $\sigma_{xP} \sim 1/\sigma_{pP}$ and $\sigma_{xD} \sim 1/\sigma_{pD}$, respectively.[b] The transition amplitude is given by

$$\mathcal{A}_{\alpha\beta}(T,\vec{L}) = \langle\nu_\beta|\nu_\alpha(T,\vec{L})\rangle = \sum_i U_{\alpha i}^* U_{\beta i}\,\mathcal{A}_i(T,\vec{L})\,, \tag{8}$$

where $\mathcal{A}_i(T,\vec{L})$ is the contribution of the ith neutrino mass eigenstate:

$$\mathcal{A}_i(T,\vec{L}) = \int \frac{d^3p}{(2\pi)^3}\, f_i^P(\vec{p})\, f_i^{D*}(\vec{p})\, e^{-iE_i(p)T+i\vec{p}\vec{L}}\,. \tag{9}$$

It can be shown that this amplitude is strongly suppressed unless the condition $|\vec{L}-\vec{v}_iT| \lesssim \sigma_x$ is satisfied, where v_i is the group velocity of the wave packet and $\sigma_x \sim \max\{\sigma_{xP},\sigma_{xD}\}$. The quantity $\mathcal{A}_i(T,\vec{L})$ represents an effective wave packet, which depends on the characteristics of both the neutrino production and detection processes, with the center at $\vec{L}=\vec{v}_iT$ and the spatial width σ_x. For example, for Gaussian wave packets

$$\mathcal{A}_i(T,\vec{L}) \propto \exp\left[-\frac{(\vec{L}-\vec{v}_iT)^2}{4\sigma_x^2}\right], \qquad \sigma_x^2 \;\equiv\; \sigma_{xP}^2 + \sigma_{xD}^2\,. \tag{10}$$

[b]Usually the Gaussian form is assumed for the wave packets of the produced and detected neutrino states. Here I will try to keep the discussion as general as possible and will not specify the exact form of the neutrino wave packets.

We can now answer the question of how two completely different and wrong assumptions (same energy and same momentum) lead to exactly the same and correct result for the oscillation probability $P_{\alpha\beta}$. In calculating this probability, one has to compute the oscillation phase, which is the difference of the phases acquired by two different neutrino mass eigenstates:

$$\Delta\phi = \Delta E \cdot T - \Delta p \cdot L \,. \tag{11}$$

Here I omitted the indices corresponding to the neutrino mass eigenstates and for simplicity considered 1-dimensional case, assuming that $\vec{p}$ and $\vec{L}$ are parallel (this is a very good approximation when the distance between the neutrino source and detector is large compared to their spatial extensions).

Consider the case $\Delta E \ll E$ (relativistic or quasi-degenerate neutrinos). Then

$$\Delta E \simeq \frac{\partial E}{\partial p}\Delta p + \frac{\partial E}{\partial m^2}\Delta m^2 = v\Delta p + \frac{1}{2E}\,\Delta m^2 \,, \tag{12}$$

and one finds from (11)

$$\Delta\phi = (v\Delta p + \frac{1}{2E}\Delta m^2)\, T - \Delta p \cdot L \tag{13}$$

$$= -(L - vT)\,\Delta p + \frac{\Delta m^2}{2E}\,T \,. \tag{14}$$

First, we note that under the "same momentum" assumption $\Delta p = 0$ the first term in (14) vanishes. If, in addition, one adopts the "time-to-space conversion" assumption $T = L/v$, Eq. (14) gives just the standard oscillation phase $\frac{\Delta m^2}{2p}\,L$.

Notice, however, that the first term in (14) vanishes not only when $\Delta p = 0$, but also at the center of the neutrino wave packet, where $L = vT$. Away from this point, $L \neq vT$, but $|L - vT|$ never exceeds substantially the spatial width of the wave packet σ_x. Thus, the standard oscillation phase is obtained also in the case $\Delta p\, \sigma_x \sim \Delta p/\sigma_p \ll 1$.

Likewise, expanding Δp in powers of ΔE and Δm^2 one can express the oscillation phase (11) in the form equivalent to (14):

$$\Delta\phi = -\frac{1}{v}(L \;-\; vT)\Delta E \;+\; \frac{\Delta m^2}{2p} L \,. \tag{15}$$

Under the "same energy" assumption $\Delta E = 0$ the first term in (15) vanishes, and the standard oscillation phase results (this time without invoking the "time-to-space conversion" relation). However, this term is negligible even if $\Delta E \neq 0$ provided that $\Delta E \sigma_x / v \ll 1$. Note that for on-shell neutrinos their energy and momentum uncertainties satisfy $\sigma_E \simeq v\sigma_p$, and therefore the condition $E\sigma_x/v \ll 1$ is equivalent to $\Delta E \ll \sigma_E$.

Thus, the standard oscillation phase results without any unjustified assumptions like "same energy" or "same momentum" provided that

- Neutrinos are relativistic or quasi-degenerate in mass, *i.e.* $\Delta E \ll E$
- Energy and momentum differences of different neutrino mass eigenstates ΔE and Δp are small compared to, respectively, energy and momentum uncertainties of the neutrino state σ_E and σ_p.

The energy and momentum uncertainties σ_E and σ_p that I discuss here are actually the effective uncertainties, which depend on the corresponding uncertainties both at neutrino production and detection. The energy uncertainty σ_E is dominated by the smaller between the production and detection energy uncertainties, and similarly for the momentum uncertainty σ_p. Note that the conditions $\Delta E \ll \sigma_E$ and $\Delta p \ll \sigma_p$ are actually the coherence conditions for neutrino production and detection, which are necessary conditions for the observability of neutrino oscillations: Indeed, too accurate neutrino energy or momentum measurement at production or detection would allow one to determine the neutrino mass and so would kill the oscillations.

Normally, the fulfilment of $\Delta E \ll \sigma_E$ also implies that the condition $\Delta p \ll \sigma_p$ is satisfied (see sec. 5.3 of Ref. 4).

3.1. *Neutrino oscillations and coherence*

What are the general conditions for the observability of neutrino oscillations? The keyword here is *coherence*. Actually, in addition to coherence of neutrino production and detection which I already discussed, the coherent propagation condition must also be satisfied. Neutrinos can lose coherence if they propagate very long distances because different mass-eigenstate components of the flavour states propagate with different group velocities and therefore separate. If the separation exceeds the spatial width of the wave packets to such an extent that different mass eigenstates can no longer interfere in the detector, the coherence of the neutrino state is lost. The coherent propagation condition is therefore

$$L \ll l_{\rm coh} = \frac{v}{\Delta v}\sigma_x \,, \qquad (16)$$

where Δv is the difference of the group velocities of different neutrino mass eigenstates.

The coherent production/detection conditions $\Delta E \ll \sigma_E$ and $\Delta p \ll \sigma_p$ are actually equivalent to the localization conditions, which require that the localization regions of the individual neutrino emitter and recipient l_S and l_D be small compared to the neutrino oscillation length $l_{\rm osc}$. In reality, one has to impose also the conditions that the macroscopic sizes of the neutrino source and detector L_S and L_D be small compared to $l_{\rm osc}$ – otherwise the integration over the neutrino production or detection coordinate would average the oscillations out. These conditions may be more stringent than the requirement $l_{S,D} \ll l_{\rm osc}$.

The coherent neutrino production/detection conditions are satisfied for most processes of practical interest. The coherent propagation condition is satisfied in all known cases except for neutrinos of astrophysical and cosmological origin.

It is interesting to note that even the non-observation of neutrino oscillations at baselines that are much shorter than the oscillation length is a consequence of and a firm evidence for coherence of the neutrino emission and detection processes: if it were broken (*i.e.* if the different neutrino mass eigenstates were emitted and absorbed incoherently), the survival probability of neutrinos of a given flavour, instead of being practically equal to one, would correspond to averaged neutrino oscillations.

3.2. *The oscillation probability*

Let me now discuss the expression for the oscillation probability in the QM wave packet approach. It is obtained by taking the square modulus of the transition amplitude (8). Since the neutrino emission and detection times are not measured (or not accurately measured) in most experiments, one usually integrates the obtained expression over T (a rigorous justification of this procedure is given in the QFT approach[5]). This gives

$$P(\nu_\alpha \to \nu_\beta; L) = \sum_{i,k} U_{\alpha i}^* U_{\beta i} U_{\alpha k} U_{\beta k}^* \, e^{-i\frac{\Delta m_{ik}^2}{2P} L} \, F_{ik} \,, \tag{17}$$

where

$$\begin{aligned} F_{ik} = \int \frac{dq}{2\pi v} f_i^P(r_k q - \Delta E_{ik}/2v + P_i) f_i^{D*}(r_k q - \Delta E_{ik}/2v + P_i) \\ \times f_k^{P*}(r_i q + \Delta E_{ik}/2v + P_k) f_k^D(r_i q + \Delta E_{ik}/2v + P_k)\, e^{i\frac{\Delta v}{v} qL} . \end{aligned} \tag{18}$$

Here the following notation is used:

$$v \equiv \frac{v_i + v_k}{2} \,, \qquad \Delta v \equiv v_k - v_i \,, \qquad r_{i,k} \equiv \frac{v_{i,k}}{v} \,. \tag{19}$$

Note that the standard probability for neutrino oscillations in vacuum obtains from (17) when $F_{ik} = 1$. The integration variable q in (18) is essentially the deviation of the neutrino momentum from the peak momentum P, therefore the main contribution to the integral comes from the region $|q| \lesssim \sigma_p$. From Eq. (18) one can then readily see that

- For $(\Delta v/v)\sigma_p L \ll 1$ (*i.e.* $L \ll l_{\rm coh} = (v/\Delta v)\sigma_x$) the quantity F_{ik} is approximately independent of L; in the opposite case it is strongly suppressed.
- F_{ik} is also strongly suppressed unless $\Delta E_{ik}/v \ll \sigma_p$, *i.e.* unless $\Delta E_{ik} \ll \sigma_E$. If the latter condition is violated, there will be little overlap between the factors $f_i^{P,D}$ and $f_k^{P,D}$ in the integrand, leading to a suppression of F_{ik}.

It is easy to see that these two conditions are actually the coherent propagation and coherent production/detection conditions discussed above.

If all the coherence conditions are satisfied we find an unsuppressed E- and L-independent quantity $F_{ik} = const.$, but the oscillation probability (17) is not automatically correctly normalized. It can still be normalized "by hand" by imposing the unitarity condition:

$$\sum_\beta P_{\alpha\beta}(L) = 1\,. \tag{20}$$

This gives[c]

$$F_{ii} = \int \frac{dp}{2\pi v} |f_i^P(p)|^2 \, |f_i^D(p)|^2 = 1\,. \tag{21}$$

This can be achieved by properly normalizing $f_i^P(p)$ and/or $f_i^D(p)$. However, this would require a nonstandard normalization of these momentum distribution functions; moreover, since the integral in (21) depends on the overlap of $f_i^P(p)$ and $f_i^D(p)$, no *independent* normalization of the produced and detected neutrino wave function would do the job. Contrary to this, in the QFT approach the correctly normalized $P_{\alpha\beta}(L)$ is obtained automatically (and the meaning of the normalization procedure adopted in the QM wave packet approach is actually clarified).[5]

Once F_{ik} is normalized according to (21), Eq. (17) yields the standard oscillation probability (4) provided that the coherent production/detection and coherent propagation conditions are satisfied.

[c]The normalization condition in (21) turns out to be crucial for proving Lorentz invariance of the oscillation probability in the QM approach.[4] This is non-trivial because the QM formalism, unlike the QFT one, is not manifestly Lorentz invariant.

To summarize, the QM wave packet formalism provides a consistent approach for describing neutrino oscillations. Its shortcomings are:
(1) the form of the neutrino wave packets is postulated rather than derived;
(2) the neutrino production and detection processes are not considered;
(3) an *ad hoc* normalization procedure is necessary, which is not properly justified within the QM wave packet framework.

The advantage of this approach is its relative simplicity.

4. QFT approach to neutrino oscillations

In QFT, the asymptotic *in-* and *out-* states are considered to be on the mass shell, therefore the neutrino flavour eigenstates, which do not have a definite mass, cannot be treated as asymptotic particles. The way out of this is to include the neutrino production and detection processes into the consideration; neutrinos are then treated as intermediate particles and are described by propagators rather than by wave functions. In this approach neutrino production, propagation and detection is treated as a single process, which is described by the Feynman diagram of Fig. 1. [d] The fact that

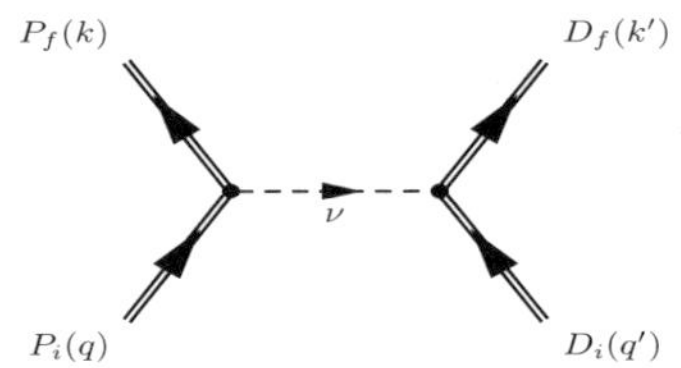

Fig. 1. Feynman diagram describing neutrino production, propagation and detection.

flavour eigenstates are coherent superpositions of different mass eigenstates translates into the requirement that the diagrams with exchanges of different mass eigenstate neutrinos add coherently. This approach was pioneered in Ref. 7 and subsequently developed by many authors; for a comprehensive review see Ref. 8, more recent results can be found in Ref. 5.

In the QFT approach to neutrino oscillations the external particles, *i.e.* the particles that accompany neutrino production and detection, are assumed to be described by localized wave packets, which leads to the localization of the production and detection regions. For simplicity, I will

[d]After my talk Arkady Vainshtein remarked that in his opinion this approach is also QM rather than QFT. His point of view is that the true QFT is where the loops are. We agreed, however, that this is largely a matter of semantics.

assume that the production process involves only one initial state and one final state particle (besides the neutrino), and similarly for the detection process. The generalization to the case of an arbitrary number or participating particles is straightforward. Direct application of the Feynman rules leads to the following expression for the transition amplitude:

$$i\mathcal{A}_{\alpha\beta} = \sum_j U^*_{\alpha j} U_{\beta j} \int \frac{d^3p}{(2\pi)^3}\, \Phi_{jP}(E_j(\vec{p}), \vec{p}) \Phi_{jD}(E_j(\vec{p}), \vec{p})\, e^{-iE_j(\vec{p})T + i\vec{p}\vec{L}}, \tag{22}$$

where

$$\Phi_{jP} = \int d^4x_1 e^{ipx_1} \int [dq] \int [dk] f_{Pi}(\vec{q}) f^*_{Pf}(\vec{k}) e^{-i(q-k)x_1} M_{jP}(q,k)\,,$$

$$\Phi_{jD} = \int d^4x_2 e^{-ipx_2} \int [dq'] \int [dk'] f_{Di}(\vec{q}') f^*_{Df}(\vec{k}') e^{-i(q'-k')x_2} M_{jD}(q',k'). \tag{23}$$

Here $M_{jP}(q,k)$ and $M_{jD}(q',k')$ are the matrix elements of the production and detection processes for the neutrino ν_j, the quantities f in the integrands in (23) are the momentum distribution functions characterizing the wave packets of the external particles, and I used the shorthand notation $[dq] \equiv d^3q/[(2\pi)^3\sqrt{2E_{Pi}(\vec{q})}]$, *etc.*

4.1. *General properties of the neutrino wave packets*

Comparing (22) with Eqs. (8) and (9), we see that the transition amplitudes obtained within the QM and QFT formalisms coincide if one identifies

$$f_j^P(\vec{p}) = \Phi_{jP}(E_j(\vec{p}), \vec{p})\,, \qquad f_j^D(\vec{p}) = \Phi^*_{jD}(E_j(\vec{p}), \vec{p})\,. \tag{24}$$

This is actually easy to understand: from the definition (23) of $\Phi_{jP}(E_j(p), \vec{p})$ it follows that it is the probability amplitude for the production process in which the jth mass eigenstate neutrino ν_j is emitted with momentum $\vec{p}$; but this is nothing but the momentum-state neutrino wave packet $f_j^P(\vec{p})$. A similar argument applies to the neutrino detection process and $f_{jD}(\vec{p})$.

Thus, even though the QFT approach does not require neutrino wave functions, it allows one to find them. If the wave functions of the external particles are known, the neutrino wave packets can be found explicitly rather than being postulated, as they are in the QM formalism.

Equation (23) allows one to study general properties of the wave packets of the produced and detected neutrino states. In the momentum space they are given by the matrix elements of neutrino production or detection multiplied by "smeared δ-functions" representing approximate conservation laws for the energy and momentum expectation values of the particles involved in the neutrino production and detection. How exactly this smearing occurs will depend on the form of the wave packets of the external particles.

Can we match the expressions for the neutrino wave packets obtained in QFT with those usually postulated in the QM approach? Consider *e.g.* the produced neutrino states. Using the Gaussian wave packets of the external particles as an example, one can show[5] that such a matching is possible provided that one applies the following changes to the QM results:

(*i*) The momentum uncertainties of the neutrino mass eigenstates are replaced by the effective ones, which are anisotropic. The longitudinal momentum uncertainty (*i.e.* the one in the direction of the neutrino momentum) is

$$\frac{1}{(\sigma^{||}_{pP\,\mathrm{eff}})^2} = \frac{1}{\sigma^2_{pP}} + \frac{(\vec{v}_j - \vec{v}_P)^2}{\sigma^2_{eP}}\,, \qquad (25)$$

where σ_{eP} and σ_{pP} are the energy and momentum uncertainties inherent in the neutrino production process, $\vec{v}_j$ is the velocity of ν_j and $\vec{v}_P$ is an effective velocity of the neutrino source. At the same time, the transverse momentum uncertainties satisfy $\sigma^{\perp}_{p\,\mathrm{eff}} = \sigma_{pP}$.

(*ii*) The mean momentum $\vec{P}$ is shifted according to $\vec{P} \to \vec{P}_{\mathrm{eff}} = \vec{P} + \vec{\delta}$, where $\vec{\delta}$ is a calculable (typically small) quantity.

(*iii*) The wave packet of each neutrino mass eigenstate gets an extra factor $N_j = \exp[-\tilde{\gamma}_j]$, where $\tilde{\gamma}_j \sim (E_j - E_P)^2/4\sigma^2_{eP}$ with E_j and E_P being the energy of the jth neutrino mass eigenstates and the mean neutrino energy, respectively.

As follows from Eq. (25), the effective momentum uncertainties that should be used to describe the wave packets of emitted neutrinos in the QM formalism are not just equal to the true momentum uncertainty at production σ_{pP}, as naively expected, but also depend on the energy uncertainty σ_{eP}, which is an independent parameter, as well as on the neutrino velocity $\vec{v}_j$ and on the velocity of the neutrino source $\vec{v}_P$. It is this effective momentum uncertainty that determines the spatial width of the neutrino wave packet in the direction of its motion: $\sigma_x \sim (\sigma^{||}_{pP\,\mathrm{eff}})^{-1}$. It is known that the energy and momentum uncertainties at neutrino production satisfy $\sigma_{eP} \le \sigma_{pP}$.[4] Therefore, except for $\vec{v}_j \approx \vec{v}_P$, the effective longitudinal momentum uncertainty is dominated by σ_{eP}. In the limit $\sigma_{eP}, \vec{v}_P \to 0$, which corresponds to a stationary neutrino source approximation,[8] the effective longitudinal momentum uncertainty $\sigma^{||}_{pP\,\mathrm{eff}}$ vanishes, even though the true momentum uncertainty σ_{pP} is nonzero. As follows from Eq. (16), this implies an infinite coherence length, in accordance with the well known result for the stationary case.

From the above point (*iii*) one can see that if $|E_i - E_j| \ll \sigma_{eP}$, the factors N_j are essentially the same for all neutrino mass eigenstates and can be included in their common normalization factor. In the opposite case the coherence of different neutrino mass eigenstates is lost.

4.2. *Oscillation probability and its normalization*

What is actually calculated in the QFT approach is the probability of the overall neutrino production-propagation-detection process rather than the oscillation probability $P_{\alpha\beta}(L)$. How can one then find $P_{\alpha\beta}(L)$ and is this always possible?

To answer these questions, let me first recall the operational definition of $P_{\alpha\beta}(L)$. In a detection process sensitive to ν_β the detection rate is

$$\Gamma_\beta^{\rm det} = \int dE\, j_\beta(E)\sigma_\beta(E)\,, \tag{26}$$

where $\sigma_\beta(E)$ is the detection cross section and $j_\beta(E)$ is the energy density (spectrum) of the ν_β flux at the detector. If an isotropic source at a distance L from the detector emits ν_α with the energy spectrum $d\Gamma_\alpha^{\rm prod}(E)/dE$, the energy density of the ν_β flux at the detector is

$$j_\beta(E) = \frac{1}{4\pi L^2}\frac{d\Gamma_\alpha^{\rm prod}(E)}{dE}P_{\alpha\beta}(L,E)\,. \tag{27}$$

Substituting this into (26), we obtain the detection rate for the overall process

$$\Gamma_{\alpha\beta}^{\rm tot} \equiv \int dE\,\frac{d\Gamma_{\alpha\beta}^{\rm tot}(E)}{dE} = \frac{1}{4\pi L^2}\int dE\,\frac{d\Gamma_\alpha^{\rm prod}(E)}{dE}\,P_{\alpha\beta}(L,E)\,\sigma_\beta(E)\,, \tag{28}$$

from which one can find the oscillation probability:

$$P_{\alpha\beta}(L,E) = \frac{d\Gamma_{\alpha\beta}^{\rm tot}(E)/dE}{\frac{1}{4\pi L^2}\,[d\Gamma_\alpha^{\rm prod}(E)/dE]\,\sigma_\beta(E)}\,. \tag{29}$$

An important ingredient in this argument is the assumption that, at a fixed neutrino energy E, the overall rate factorizes into the production rate, propagation (oscillation) probability and detection cross section. If this does not hold, the very notion of the oscillation probability would lose its sense and one would have to deal instead with the overall rate of neutrino production, propagation and detection.

Let us now come back to the QFT-based treatment of neutrino oscillations and try to cast the rate of the overall process in the form (28). If we succeed, this would allow us to find the oscillation probability.

Let us first find some ingredients necessary for such a calculation. Application of the standard rules of QFT yields for the neutrino production probability

$$P_\alpha^{\rm prod} = \sum_j |U_{\alpha j}|^2 \frac{1}{8\pi^2}\int dE\,\big|\Phi_P(E,p_j)\big|^2 4Ep_j\,, \tag{30}$$

whereas the detection probability is given by

$$P_\beta^{\rm det}(E) = \sum_k |U_{\beta k}|^2 |\Phi_D(E,p_k)|^2 \frac{1}{V}\,, \tag{31}$$

where V is the normalization volume. Here I have taken into account that on the mass shell the quantities $\Phi_{jP}(E,p_j)$ and $\Phi_{jD}(E,p_j)$ defined in (23) depend on the index j only through the neutrino momentum p_j and denoted

$$\Phi_{jP}(E,p_j) \equiv \Phi_P(E,p_j)\,, \qquad\qquad \Phi_{jD}(E,p_j) \equiv \Phi_D(E,p_j)\,. \tag{32}$$

to simplify the notation.

The next step is to express the overall rate of the process as an integral over the neutrino energy. I skip the details (which can be found in[5]) and go directly to the result. If the expression for the overall rate of the process is factorizable, the oscillation probability should be obtainable from Eq. (29), which yields

$$\text{“}P_{\alpha\beta}(L,E)\text{”} = \frac{\sum_{j,k} B_{jk}^{\alpha\beta}\Phi_P(E,p_j)\Phi_D(E,p_j)\Phi_P^*(E,p_k)\Phi_D^*(E,p_k)e^{i(p_j-p_k)L}}{\sum_j |U_{\alpha j}|^2\,|\Phi_P(E,p_j)|^2\,p_j\,\sum_k |U_{\beta k}|^2\,|\Phi_D(E,p_k)|^2 p_k^{-1}}. \tag{33}$$

Here I have used the shorthand notation

$$B_{jk}^{\alpha\beta} = U_{\alpha j}^* U_{\beta j} U_{\alpha k} U_{\beta k}^*\,. \tag{34}$$

The quotation marks in (33) are to remind us that we yet have to prove that this quantity can indeed be interpreted as the oscillation probability, i.e. that the factorization of the QFT-derived $d\Gamma_{\alpha\beta}^{\rm tot}(E)/dE$ into $d\Gamma_\alpha^{\rm prod}(E)/dE$, $P_{\alpha\beta}(L,E)$ and $\sigma_\beta(E)$ holds. It is not difficult to show[5] that this happens when $|p_j - p_k| \ll p_j, p_k$, i.e. for ultra-relativistic or quasi-degenerate in mass neutrinos. Then the oscillation probability $P_{\alpha\beta}(E,L)$ can be defined as a sensible quantity and is given by

$$P_{\alpha\beta}(L,E) = \frac{\sum_{j,k} B_{jk}^{\alpha\beta}\Phi_P(E,p_j)\Phi_D(E,p_j)\Phi_P^*(E,p_k)\Phi_D^*(E,p_k)e^{-i\frac{\Delta m_{jk}^2}{2p}L}}{|\Phi_P(E,p)|^2\,|\Phi_D(E,p)|^2}. \tag{35}$$

One can readily make sure that this expression is properly normalized and automatically satisfies the unitarity condition (20). If $|p_j - p_k| \ll \sigma_p$, *i.e.* if the production/detection coherence condition is satisfied, the factors $\Phi_P(E, p_{j,k})$ and $\Phi_D(E, p_{j,k})$ can be be pulled out of the sums in the numerator of (35), and the standard oscillation probability (4) results. Thus, the oscillation probability exists and is independent of the neutrino production and detection processes provided that neutrinos are ultra-relativistic or quasi-degenerate in mass and the coherence conditions are satisfied.

The above considerations also allow us to shed some light on the meaning of the normalization condition imposed on the oscillation probability in the QM wave packet approach, which looked rather arbitrary. As I already discussed, the QM and QFT approaches can be matched if the QM quantities f_{jP} and f_{jD} are identified with the QFT functions $\Phi_{jP}(E_j, \vec{p})$ and $\Phi^*_{jD}(E_j, \vec{p})$, respectively. The latter, however, bear information not only on the properties of the emitted and absorbed neutrinos, but also on the production and detection processes. The QM normalization procedure can then be easily seen to be equivalent to the division of the overall rate of the process by the production rate and detection cross section, as in Eq. (35).

We have seen that the quantum mechanical and quantum field theoretic approaches to neutrino oscillations are closely related; however, the QFT framework gives a more consistent and general theoretical description of this phenomenon. Apart from providing expressions for the oscillation probabilities and event rates in its own right, it can be used to derive the input parameters required for the QM approach and to elucidate some QM procedures which were not properly justified or fully understood within that approach.

References

1. V. N. Gribov and B. Pontecorvo, *Phys. Lett.* **B28**, 493 (1969).
2. A. Strumia and F. Vissani, arXiv:hep-ph/0606054.
3. M. C. Gonzalez-Garcia and M. Maltoni, *Phys. Rept.* **460**, 1 (2008).
4. E. K. Akhmedov and A. Y. Smirnov, *Phys. Atom. Nucl.* **72**, 1363 (2009) (shortened version). The full version can be found in arXiv:0905.1903 [hep-ph].
5. E. K. Akhmedov and J. Kopp, *JHEP* **1004**, 008 (2010).
6. E. K. Akhmedov and A. Y. Smirnov, arXiv:1008.2077 [hep-ph].
7. I. Y. Kobzarev, B. V. Martemyanov, L. B. Okun and M. G. Shchepkin, *Sov. J. Nucl. Phys.* **32**, 823 (1980) [*Yad. Fiz.* **32** (1980) 1590].
8. M. Beuthe, *Phys. Rept.* **375**, 105 (2003).

PRODUCTION AND PROPAGATION OF AN ENTANGLED LEPTON-NEUTRINO PAIR

B. MESZÉNA

Institute of Physics, Eötvös University,
Budapest, H-1117, Hungary
kredenc@fazekas.hu

A. PATKÓS

Department of Atomic Physics, Eötvös University
Research Group for Biological and Statistical Physics
Budapest, H-1117, Hungary
patkos@hector.elte.hu

The evolution of the entangled muon-neutrino system emerging from charged pion decay is explored both in vacuum and in matter. The study is based on a Weisskopf-Wigner type wave-packet description. Explicit formulae are derived displaying modulation and attenuation of the oscillations due to additional time scales characterising the production process. The case of neutrinos disentangled due to the detection of the muon is also considered.

Keywords: Neutrino oscillation; quantum coherence.

1. Introduction

The analysis of the neutrino oscillation phenomena necessarily includes beyond the propagation also the production and detection stages, since the finiteness of the space-time distance between the starting and the final events is an essential part of the phenomenon. This feature leads to the smearing of the energy and the momentum of the particles and requires a wave packet description. Further time scales characterising the production and the detection appear, which modulate and damp the fundamental oscillations. More than 40 years after the fundamental paper of Gribov and Pontecorvo[1] these aspects are still constantly discussed in the literature of the past decades.[2–6] Terrestrial observability of decoherence effects was estimated very recently by Kayser and Kopp.[7] This is the latest development in the ongoing attempts to extract more information from the oscillation experiments beyond the oscillation length.[8,9]

Recent investigations focus on the entanglement aspects of the lepton-neutrino pair propagation in different "experimental" situations.[10] Most recently Wu *et al.*[11] analysed the dependence of the density matrix on the detection time of the accompanying lepton when it is detected in addition to the propagation time of the neutrino. Their analysis of a two-body decay producing an entangled lepton-neutrino pair relies on first order perturbation theory which is strictly valid only for times shorter than the lifetime of the mother-particle. This approach seems to be more relevant for long-lived sources (for example for beta decay of long-lived radiactive isotopes). Decohering effects reflecting the finite width of a Gaussian wave packet were analyzed in Ref. 12. Decoherence resulting from the finite lifetime of the mother-particle was systematically investigated in Refs. 13, 14. Both studies considered disentangled neutrinos.

Our goal in this contribution is to derive simple expressions governing the coherence of the system in a unified treatment which includes the effects of all relevant timescales and relies on an approach which makes use of wave packets. In our discussion we make use of two-body decays and the description is valid for times much larger than the source's lifetime. An important process of this type is the decay of charged pions $\pi \to \mu + \nu_\mu$, to which we shall mostly refer in this paper. The treatment will be extended also to the propagation through matter where density dependent resonant increase of the coherence length is observed, which reaches even infinity for some specific value of the lepton density. Consequences of muon detection will be carefully explored and compared to the oscillations displayed by an entangled neutrino.

The paper starts by an overview of the results of the entangled wave-packet treatment of the 2-flavor muon+neutrino propagation following Ref. 4. Simple approximate formulae which describe the modulation and decoherence of the basic oscillations both in vacuum and matter represent the main results of our note. In the second part we repeat the analysis for the case, when the muon is detected near the decay tube of the pion. We achieve also in this case a unified picture of the effects of finite lifetime and the period of the oscillations observed due to the finite propagation time.

2. Wave-packet evolution without observing the entangled muon

The state vector of an entangled collinearly outflying muon-neutrino pair in the Weisskopf-Wigner approximation reads as follows:[4]

$$|\Psi(x_\nu, x_\mu, t)\rangle = (\psi_1(x_\nu, x_\mu, t)\cos\Theta|m_1\rangle + \psi_2(x_\nu, x_\mu, t)\sin\Theta|m_2\rangle)|\mu\rangle,$$

$$\psi_i(x_\nu, x_\mu, t) = \mathcal{N}\int dp_\nu \int dp_\mu f_\pi(p_\nu + p_\mu) e^{i(p_\nu x_\nu + p_\mu x_\mu) - i(E_{\nu_i} + E_\mu)t} \times \frac{1}{E_{\nu_i} + E_\mu - E_\pi + iM_\pi\Gamma/2E_\pi}, \tag{1}$$

where $E^2_{\nu_i} = m_i^2 + p_\nu^2$, $E_\pi^2 = M_\pi^2 + (p_\nu + p_\mu)^2$ and the state vector of the muon-type neutrino created from the decay is $|\nu_\mu\rangle = \cos\Theta|m_1\rangle + \sin\Theta|m_2\rangle$. The label m_i refers to the mass of the i-th mass eigenstate, $f_\pi(p)$ is the momentum profile, Γ is the energy width of the decaying pion, which will be later assumed to have zero momentum on the average. $\mathcal{N}$ is an appropriate normalisation constant which ensures that $\int dx_\nu \int dx_\mu |\psi_i|^2 = 1$ is fulfilled. Eq.(1) arises from the complete Weisskopf-Wigner expression when $t >> \Gamma^{-1}$. (It should be contrasted with the treatment of Ref. 11 which is valid oly for early times $t << \Gamma^{-1}$.)

The survival probability of ν_μ can be studied by projecting (1) on the initial (muon + muon-neutrino) state. The spatial propagation of the μ-type neutrino profile is tracked when the muon is not detected, e.g. one integrates the absolute square of the projection over x_μ:

$$Prob(x_\nu, t) = \int dx_\mu \left(\cos^4\Theta|\psi_1|^2 + \sin^4\Theta|\psi_2|^2 + \frac{1}{2}\sin^2(2\Theta)\mathrm{Re}(\psi_1^*\psi_2)\right). \tag{2}$$

Even simpler is the question about the survival of ν_μ without its localisation, e.g. integrating also over x_ν. The oscillating part of this probability is proportional to the real part of the following integral:

$$I_{12} = \int dx_\nu \int dx_\mu \psi_1^*\psi_2 = \int dp_\nu \int dp_\mu |f(p_\nu + p_\mu)|^2 e^{i(E_{\nu_1} - E_{\nu_2})t} \times \frac{1}{(E_{\nu_1} + E_\mu - E_\pi + iM_\pi\Gamma/2E_\pi)(E_{\nu_2} + E_\mu - E_\pi - iM_\pi\Gamma/2E_\pi)}. \tag{3}$$

Another way to recognize its interest is to realise that it determines the off-diagonal element of the flavor density matrix in the mass eigenbasis, and characterizes this way the coherence of the two-state system:

$$\rho_{12}^{(m)}(t) = \frac{1}{2} I_{12}\sin(2\Theta). \tag{4}$$

Oscillation phenomena actually are governed by this component. The experimentally observed oscillation in the diagonal matrix elements in the flavor or interaction eigenbase ($\rho^{(f)}$) is just the consequence of the orthogonal transformation connecting the two bases.

A simple and numerically accurate approximate representation of I_{12} can be derived following[4] by expanding all factors of the integrand of (3) to linear order around some sort of "mean" momenta (denoted by capital letters): P_ν, P_μ, and P_π. They are determined from the requirement of energy and momentum conservation upon neglecting the mass of the neutrino. After the expansion one demonstrates with an appropriate change of variables that the two-variable integral factorizes and one integration even can be performed via Cauchy's theorem. The final result is of the following form:

$$I_{12} = \exp\left(i\frac{\Delta m^2}{2P_\nu}t\right)\frac{2\pi}{1-i\frac{\Delta m^2}{2P_\nu}\frac{E_\pi}{M_\pi\Gamma}}F(t)\exp\left(-\frac{\Delta m^2}{4P_\nu^2}\frac{M_\pi\Gamma}{E_\pi(1-v_\mu)}t\right),$$

$$F(t) = \int dp|f(p)|^2\exp\left(-i\frac{\Delta m^2}{2P_\nu^2}\frac{v-v_\mu}{1-v_\mu}(p-P_\pi)t\right). \qquad (5)$$

Here Δm^2 is the squared mass difference of the two neutrinos, and $v = P_\pi/E_\pi, v_\mu = P_\mu/(P_\mu^2+m_\mu^2)^{1/2}$. The second factor in this expression suppresses the amplitude of the oscillations only when the lifetime of the mother-particle is much larger than the oscillation time, which is not the case of the pion decay. Clearly, the integral $F(t)$ represents the Fourier transform of the wave function profile of the pion, and implies the decay of the oscillating term with characteristic size $1/d$. The last factor provides attenuation of the oscillation with some 300 years characteristic decay time when one makes use of realistic parameters.[14] The coherence length emerging from the concurrence of the last two factors is determined by $max(\Gamma, d)$. Below we shall assume $P_\pi = 0$, since then the formulae look much simpler.

Modified formulae reflecting medium effects can be written in an analogous form. Starting with a Hamilton operator which contains the Fermi interaction of the e-type neutrinos with the constituents of the medium[15] one solves the Schrödinger equation of the density matrix. For the leading effect the mass difference in the denominator can be neglected (in both factors we write for the neutrino energy its value calculated at P_ν). Then

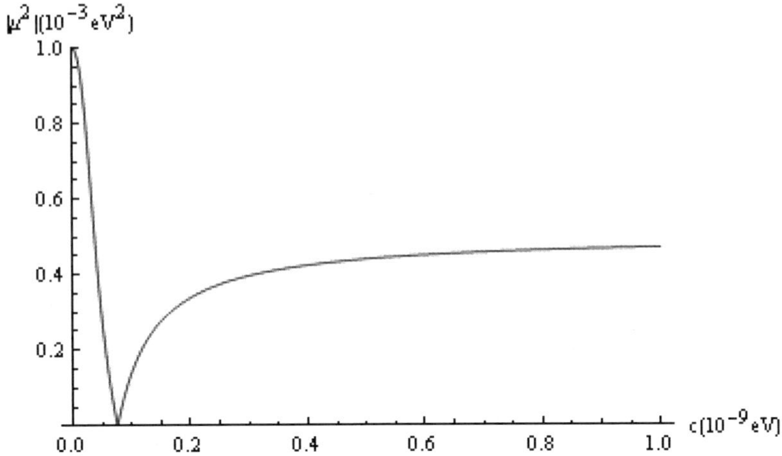

Fig. 1. The parameter μ^2 which determines the coherence damping in matter as a function of the parameter c. (For the figure the parameter values $\Delta m^2 = 10^{-3}$ eV$^2, M = 100$ MeV$, \Theta = \pi/6$ are used.)

for the probability of the flavor flip one finds

$$P(\nu_\mu \to \nu_e, t) = (2\pi N)^2 \int dp_\nu \int dp_\mu \left(\frac{\Delta m^2}{\Delta m^2_{eff}(p_\nu)}\right)^2 |f(p_\nu + p_\mu)|^2 \times \frac{(1 - \cos[(\Delta m^2_{eff}(p_\nu)/2p_\nu)t]) \sin^2(2\Theta)/2}{|E_{P_\nu} + E_\mu - E_\pi - i\Gamma/2|^2}. \quad (6)$$

Here we have introduced the medium and momentum dependent effective squared mass splitting

$$\Delta m^2_{eff}(p) = \sqrt{(\Delta m^2)^2 + c^2p^2 - 2cp\Delta m^2 \cos(2\Theta)}, \qquad c = 2\sqrt{2}NG_F \quad (7)$$

(G_F is the Fermi constant, N the density of the medium).

Assuming slow variation of the effective mass splitting over the momentum range of the neutrino, the double integral can be evaluated with the same approximate technique as explained for (5). It leads to the following factorized form:

$$P(\nu_\mu \to \nu_e, t) = \left(\frac{\Delta m^2}{2\Delta m^2_{eff}(P_\nu)}\right)^2 \sin^2(2\Theta)(1 - \mathrm{Re} I_{matter}),$$

$$I_{matter} = 2\pi \exp\left(i\frac{\Delta m^2_{eff}(P_\nu)}{2P_\nu}t\right) F_{matter}(t) \exp\left(-\frac{|\mu^2|}{4P_\nu^2}\frac{\Gamma}{1 - v_\mu}t\right). \quad (8)$$

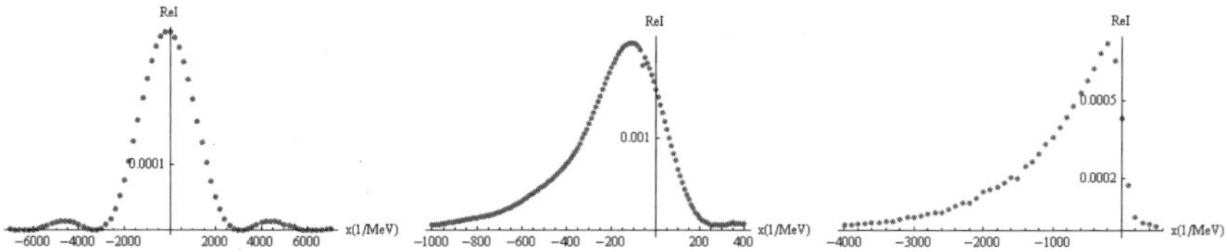

Fig. 2. The starting ($t = 0$) probability distribution of a rightward propagating neutrino wave-packet. The inverse lengths characteristic for the figures from the left to the right are the following (measured in MeV): $(\Gamma, d) = (0.01, 0.001), (0.005, 0.01), (0.001, 0.01)$, respectively.

The absence of the second factor of the analogous expression (5) tells that here we discuss the case when the lifetime of the pion is much shorter than the oscillation period. F_{matter} has the same form as in (5) with a substitution $\Delta m^2 \to \mu^2$, where

$$\mu^2 = \frac{\Delta m^2(\Delta m^2 - cP_\nu \cos(2\Theta))}{\Delta m^2_{eff}(P_\nu)}. \tag{9}$$

As a function of c (e.g. the matter density) μ^2 is nonmonotonic and with realistic parameters for $N \approx 2 \times 10^2 N_A/cm^3$ it reaches zero (N_A is the Avogadro number). Its variation with c, proportional to the electron density in the medium is displayed in Fig.1. It is worthwhile to emphasize that the quantities influencing the variation of the oscillation time and of the coherence length in matter are different.

Although with the integration over x_ν information was lost on the location of the neutrino, one might estimate it intuitively with the identification $\bar{x}_\nu \approx t$. More detailed information can be obtained by evaluating (2), which we are going to discuss next.

In order to ease the triple momentum integral in its last term occuring after the x_μ integration, we have assumed that $f(p)$ is just a window function of size $2d$. The result still reflects all generic features not depending on the specific form of the pion profile. In this specific case by careful mapping of the range of variation the p_μ integration can be performed analytically. Denoting its (not-quite-transparent) result by $G(p_\nu, q_\nu)$ one reduces the integral to the form

$$I(x_\nu, t) = 2\pi N^2 \int\int_{|p_\nu - q_\nu| < 2d} dp_\nu dq_\nu G(p_\nu, q_\nu) \times \exp[i(q_\nu - p_\nu)x_\nu - i(E_{\nu 1}(q_\nu) - E_{\nu 2}(p_\nu))t]. \tag{10}$$

The remaining two integrations were performed numerically. The shape of the profile of the muon-type neutrino signal is obtained by adding to this

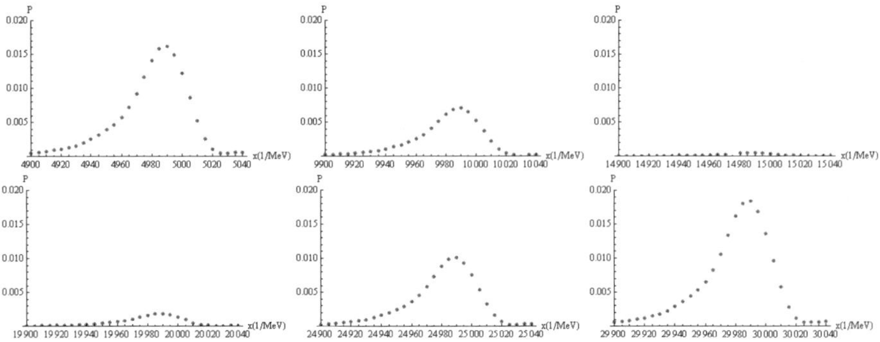

Fig. 3. Propagation of the ν_μ-neutrino profile with oscillating amplitude. Time increases from left to right and top to down

term also the diagonal contributions. The resulting probability distribution at $t = 0$ depends on the relation of Γ and d, as can be seen from Fig. 2. The shapes propagate with oscillating amplitude and the forms displayed in the Figure will reappear periodically with some changes reflecting decoherence. This propagation with oscillating amplitude is represented by the sequence of shapes appearing in Fig.3. The time coordinate values appearing on the abscissas shows the progress of the ν_μ signal. The parameters coincide with those characterising the middle figure of Fig.2. For such "early" times the decoherence due to the different velocities of the different Fourier-components is not noticeable.

When $d << \Gamma$ the profile is symmetric to some central point which can be identified with the actual position of the neutrino. When one goes over to the opposite situation where $d >> \Gamma$, the time-profile becomes asymmetric. Its forward slope becomes very steep of size $\sim 1/d$, while the size of the back-tail is determined by Γ^{-1}. Unfortunately the resolution of the neutrino detectors will not allow for a long time the detailed mapping of the neutrino signal. Still this part of our analysis shows that the detection rate might be influenced by the way the wave function of the detector particle overlaps with the neutrino profile.

3. Effects of muon detection on the neutrino propagation

In realistic experiments the detection of the accompanying muons usually provides the estimate for the neutrino flux generated in the decay tube of the pions. This disentangling event has an impact on the observations of the neutrino oscillations which we shall discuss below.

Our crude picture of muon detection assumes that the detection consists of measuring the momentum of the muon by a spatially extended detector. Let its measured value be Q and the measurement happens at time t_μ. This disentangling event produces the following one-particle wave function for the neutrino:

$$|\nu_\mu\rangle = \phi_1 \cos\Theta |m_1\rangle + \phi_2 \sin\Theta |m_2\rangle,$$

$$\phi_i(x_\nu, t) = n_i(Q) N \int dp_\nu \frac{f(p_\nu + Q) e^{iE_\mu(Q)t_\mu}}{E_{\nu,i} + E_\mu(Q) - E_\pi + i\Gamma/2} e^{i(p_\nu x_\nu - E_{\nu,i} t)}. \quad (11)$$

Here $n_i(Q)$ is an appropriate normalisation factor for the one-particle wave function of the neutrino. The exponent of the last factor is constant, therefore can be omitted.

We wish to compare the oscillating part of the survival probability in this case to (3), therefore we integrate over x_ν:

$$I_{12}^{dis}(t) = 2\pi n_1(Q) n_2(Q) N^2$$
$$\times \int dp_\nu \frac{|f(p_\nu + Q)|^2 e^{i(E_{\nu 1} - E_{\nu 2})t}}{(E_{\nu 1} + E_\mu(Q) - E_\pi - i\Gamma/2)(E_{\nu 2} + E_\mu(Q) - E_\pi + i\Gamma/2)}. \quad (12)$$

This integral can be estimated analytically in two limiting cases: $d << \Gamma$ and $d >> \Gamma$. When $d << \Gamma$, the momentum uncertainty is smaller therefore p_ν varies around $P_\nu = -Q$. The variation of the numerator is faster, therefore the denominator can be taken out as a constant from the integration. The requirement of the normalisation of I_{12}^{dis} to unity when $\Delta m^2 = 0$ leads to the cancellation of these factors with $n_1 n_2$. In the opposite limiting case $d >> \Gamma$ the uncertainty of the energy is small therefore the "mean" value of p_ν is calculated from the energy conservation: $P_\nu = M_\pi - \sqrt{m_\nu^2 + Q^2}$. The remaining integral is performed via Cauchy's theorem. The results of the integration in the two limiting cases are as follows:

$$I_{12}^{dis}(d << \Gamma) \approx 2\pi \exp\left(-i\frac{\Delta m^2}{2Q} t\right) \int dp |f(p)|^2 \exp\left(-i\frac{\Delta m^2}{2Q^2} pt\right),$$

$$I_{12}^{dis}(d >> \Gamma) \approx \exp\left(-i\frac{\Delta m^2}{2Q} t\right) \frac{1}{1 + i\frac{\Delta m^2}{2Q\Gamma}} \exp\left(-\frac{\Delta m^2 \Gamma}{4Q^2} t\right). \quad (13)$$

It follows that the coherence length in case of the disentangled neutrino is determined by $min(\Gamma, d)$. One can easily find that physically this corresponds to the case when the coherence is suppressed by the vanishing overlap of the different mass-components of the neutrino wave function. Its characteristic time is determined by the difference of the group velocities of the two components.

In conclusion we summarize our findings:

- In case of the entangled muon+neutrino pair if Γ and d are of different order of magnitude then the coherence time/length is always shorter than the separation time of the two eigen-mass modes. In the disentangled case, however, this latter determines the coherence of the oscillations.
- The dependence of the coherence length on the muon momentum and velocity is slightly different in the entangled and the disentangled cases (cf. (5) and (13).

In an experimental situation only some fraction of the muons is detected because of the finite efficiency of the muon-detector. Therefore the neutrino beam realistically is a mixture of entangled and disentangled neutrinos, which will necessitate some refinement of the analysis when the experiments become sensitive to such details. The measurement of the coherence length of the oscillations might extend the conventional oscillation paradigm in the not-too-distant future. Since there is no universal characterisation of the decoherence, future measurements of the decoherence length might bring informations on the actual wave-packet profile and also on the degree of entanglement of the neutrino beam.

Acknowledgments

This research was supported by the Hungarian Research Fund Grants K77534 and T068108.

References

1. V.N. Gribov and B. Pontecorvo, *Phys. Lett.* **B28**, 493 (1969).
2. B. Kayser, *Phys. Rev.* **D24**, 110 (1981).
3. H. Lipkin, *Phys. Lett.* **B348**, 604 (1995).
4. M. Nauenberg, *Phys. Lett.* **B447**, 23 (1999); Erratum: *ibid.* **B452**, 434 (1999).
5. E.Kh. Akhmedov and J. Kopp, *JHEP10*, 008 (2010).
6. E.Kh. Akhmedov and A.Yu. Smirnov, arXiv:1008.2077.
7. B. Kayser and J. Kopp, arXiv:1005.408.
8. E.Lisi, A. Marrone and D. Montanino, *Phys. Rev. Lett.* **85**, 1166 (2000).
9. M. Blennow, T.Ohlsson and W. Winter, *JHEP* **0506**, 049 (2005).
10. A.G. Cohen, S.L. Glashow and Z. Ligeti, *Phys. Lett.* **B678**, 191 (2009).
11. J. Wu, J.A. Hutasoit, D. Boyanovsky and R. Holman, *Phys. Rev.* **D82**, 013006 (2010).
12. C. Giunti and C.W. Kim, *Phys. Rev.* **D58**, 017301 (1998).
13. W. Grimus and P. Stockinger, *Phys. Rev.* **D54**, 3414 (1996).
14. W. Grimus, P. Stockinger and S. Mohanty, *Phys. Rev.* **D59**, 013011 (1998).
15. G. Sigl and G. Raffelt, *Nucl. Phys.* **B406**, 423 (1993).

UNIVERSALITY OF LEADING RELATIVISTIC CORRECTIONS TO BOUND STATE GYROMAGNETIC RATIOS

MICHAEL I. EIDES* and TIMOTHY J. S. MARTIN

Department of Physics and Astronomy, University of Kentucky, Lexington, KY 40506, USA
and Petersburg Nuclear Physics Institute, Gatchina, St.Petersburg 188300, Russia
** eides@pa.uky.edu*

The leading relativistic (nonrecoil and recoil) corrections to bound state g-factors of particles with an arbitrary spin are discussed. These corrections are universal for any spin and depend only on the free particle gyromagnetic ratios. The physical reasons behind this universality are explained.

Keywords: Gyromagnetic ratio; relativistic corrections; electromagnetically bound systems; arbitrary spins.

1. Introduction

Gyromagnetic ratios of particles in hydrogenlike bound states have become in the last ten-fifteen years an active field of experimental and theoretical research. The gyromagnetic ratio of a bound electron is proportional to the ratio of the spin flip and cyclotron frequencies of a hydrogenlike ion and to the electron-ion mass ratio. The experimental uncertainties of the ratio of the spin flip and cyclotron frequencies of the hydrogenlike carbon $^{12}C^{5+}$ and oxygen $^{16}O^{7+}$ were reduced to $5-7$ parts in 10^{10}, see Refs. 1, 2 and review in Ref. 3. The theoretical expression for the bound state g-factor was also greatly improved recently (see, e.g. Refs. 4, 5 and references in Ref. 3), and the theoretical uncertainty was reduced to $1.5-5.5$ parts in 10^{11}. As a result measurements of the bound electron g-factor became the best source for precise values of the electron mass in atomic units.[3] This bright picture is marred by the discrepancy on the magnitude of the leading relativistic corrections to bound state g-factors existing in the literature. Leading relativistic (both nonrecoil and recoil) corrections to g-factors of loosely bound spin one half particles were calculated a long time ago.[6–8] These corrections in the case of loosely bound particles with arbitrary spins were calculated in Ref. 9. It turned out that the leading relativistic and recoil corrections to bound state g-factors are universal, they do not depend on the spin of the constituents but only on their free g-factors. Among other results, this feature allowed the authors of Ref. 9 to sum all nonrecoil and recoil corrections of order $\alpha^n(Z\alpha)^2$ to bound state g-factors in hydrogenlike ions. Universality of leading binding corrections to bound state g-factors was challenged in Refs. 10 and 11. The results of Refs. 10 and 11 contained the contributions explicitly depending on spin, both for nonrecoil and recoil corrections. These terms shift the theoretical value of the bound state g-factors of the hydrogenlike carbon $^{12}C^{5+}$ and oxygen $^{16}O^{7+}$ by about $2-3$ parts in 10^{11}. The discrepancy will become even more phenomenologically relevant if proposed improvement[12] of the experimental accuracy by two orders of magnitude is achieved. Later the universal results of Ref. 9 were confirmed in Refs. 13 and 14.

Below we derive an effective nonrelativistic QED (NRQED) Hamiltonian for charged particles with arbitrary spins and calculate the leading relativistic and recoil corrections to the bound state g-factors in loosely bound two-particle systems. We show that these corrections are universal for all spins, and explain the physical reasons behind this universality.

2. NRQED Lagrangian

A loosely bound two-particle system is effectively nonrelativistic, with characteristic velocities of constituents of order $Z\alpha$. We are looking for the leading nonrecoil and recoil corrections to bound state g-factors of order $(Z\alpha)^2$. NRQED is a natural framework for calculation of these corrections. The NRQED Lagrangian sufficient for calculation of these corrections should include terms in nonrelativistic expansion up to and including terms of order v^2. The well known (see, e.g., Ref. 15) NRQED Lagrangian for the spin one half case is constructed from the covariant derivatives $\partial_0 + ieA_0$ and $\boldsymbol{D} = \boldsymbol{\nabla} - ie\boldsymbol{A} = i(\boldsymbol{p} - e\boldsymbol{A})$, electric and magnetic fields $\boldsymbol{E}$ and $\boldsymbol{B}$, and the spin operator $\boldsymbol{S}$. Notice that in a loosely bound two-particle system the scalar potentialA_0 is of order v^2, $\langle eA_0\rangle \sim (Z\alpha)^2$. For higher spin particles we should, besides the spin, include also higher irreducible intrinsic multipole moments as the building blocks of the NRQED Lagrangian. Technically these multipole moments are polynomials in the components of the spin operator that for higher spins do not reduce to numerical tensors and operators linear in spin. The general NRQED Lagrangian has the form

(compare with Ref. 15 for spin one half case)

$$\mathcal{L} = \phi^+ \Bigg\{ i(\partial_0 + ieA_0) + \frac{\boldsymbol{D}^2}{2m} + \frac{\boldsymbol{D}^4}{8m^3} + c_F \frac{e\boldsymbol{S}\cdot\boldsymbol{B}}{2m} + c_D \frac{e(\boldsymbol{D}\cdot\boldsymbol{E} - \boldsymbol{E}\cdot\boldsymbol{D})}{8m^2}$$
$$+ c_Q \frac{eQ_{ij}(D_iE_j - E_iD_j)}{8m^2} + c_S \frac{ie\boldsymbol{S}\cdot(\boldsymbol{D}\times\boldsymbol{E} - \boldsymbol{E}\times\boldsymbol{D})}{8m^2}$$
$$+ c_{W1} \frac{e[\boldsymbol{D}^2(\boldsymbol{S}\cdot\boldsymbol{B}) + (\boldsymbol{S}\cdot\boldsymbol{B})\boldsymbol{D}^2]}{8m^3} + c_{W2} \frac{-eD^i(\boldsymbol{S}\cdot\boldsymbol{B})D^i}{4m^3}$$
$$+ c_{p'p} \frac{e[(\boldsymbol{S}\cdot\boldsymbol{D})(\boldsymbol{B}\cdot\boldsymbol{D}) + (\boldsymbol{D}\cdot\boldsymbol{B})(\boldsymbol{S}\cdot\boldsymbol{D})]}{8m^3} \Bigg\} \phi, \tag{1}$$

where $Q_{ij} = S_iS_j + S_jS_i - (2/3)\boldsymbol{S}^2\delta_{ij}$ is proportional to the electric quadrupole moment operator ($Q_{ij} \equiv 0$ for spin one half), and ϕ is a $2s+1$-component spinor field for a particle with spin s. We included in the Lagrangian in Eq. (1) operators of dimensions not higher than four, but omitted some terms (like the terms with derivatives of magnetic field[a]) that are irrelevant for calculation of the leading recoil corrections. Let us mention that gauge invariant bilinears in $\boldsymbol{E}$ and $\boldsymbol{B}$ are of too high order to generate leading relativistic contributions of order $(Z\alpha)^2$ to bound state g-factors.

The coefficients in the NRQED Lagrangian for spin one half charged particles are determined from comparison of the scattering amplitudes in nonrelativistic theory and in relativistic QED. We would like to follow the same path for the case of arbitrary spin, but renormalizable QED for charged particles with high spin does not exist. However, the rules for calculation of all one-photon interactions of charged particles with arbitrary spin were constructed some time ago in Refs. 16 and 17. This construction uses only Lorentz invariance and local current conservation, and is valid for charged particles of arbitrary spin. The interaction vertex that includes all higher multipole moments in the approach of Refs. 16 and 17 is a direct generalization of the ordinary spin one half vertex

$$\Gamma_\mu = e\frac{(p_1+p_2)_\mu}{2m}F_e(q^2,\tau) - F_m(q^2,\tau)\frac{e\Sigma_{\mu\nu}q^\nu}{2m}, \tag{2}$$

where $q = p_2 - p_1$, $\Sigma_{\mu\nu}$ is the generalization of ordinary spin one half $\sigma_{u\nu}$, S_μ is a covariant spin four-vector, $\tau = (q\cdot S)^2$, and $F_e(0,0) = 1$,

[a]See Ref. 13 for the explicit form of the Hamiltonian that includes all operators with dimensions not higher than four.

$F_m(0,0) = g/2$. The wave functions are spinors with dotted and undotted indices that are symmetrized among themselves (for more details see Refs. 16, 17, and 18). The form of the vertex in Eq. (2) is uniquely fixed by the requirements of Lorentz invariance, C, P and charge conservation. Charged particles with higher spins automatically carry higher multipole moments that arise as coefficients in expansion of the form factors F_e and F_m over τ.[16,17] These intrinsic electric and magnetic multipole moments are treated phenomenologically, and we do not try to calculate them. The phenomenological approach to multipole moments is an advantage for our purposes because we would like to describe how g-factors of not necessarily electromagnetic origin (for example the g-factor of a spin one deuteron) change in a loosely bound electrodynamic system.

We find coefficients in Eq. (1) by comparing one-photon scattering amplitudes in NRQED and in the relativistic formalism of Refs. 16 and 17 with vertex Eq. (2). Although some terms in Eq. (1) are bilinear in electromagnetic fields $\boldsymbol{A}$ and $\boldsymbol{E}$ they still can be restored from one-photon terms due to gauge invariance. Therefore the one-photon relativistic vertex in Eq. (2) is sufficient for calculation of all the coefficients in Eq. (1). We calculated scattering amplitudes off an external electromagnetic field using the nonrelativistic Lagrangian in Eq. (1) and using the relativistic one-photon vertex in Eq. (2) at $q^2 = 0$ and $\tau = 0$. In the relativistic calculation we used noncovariantly normalized particle spinors in the generalized standard representation, which is necessary for consistency with the respective nonrelativistic results. Diagrammatically this choice of spinors and representation corresponds to the Foldy-Wouthuysen transformation (for more details, see, e.g., Ref. 18). After nonrelativistic expansion we compared results of the relativistic calculation with the nonrelativistic ones and obtained values of all constants in the Lagrangian in Eq. (1)

$$c_F = \frac{g}{2}, \quad c_D = (g-1)\frac{\boldsymbol{\Sigma}^2}{3}, \quad c_S = g-1, \quad c_Q = -2\lambda(g-1),$$
$$c_{W1} = \frac{g+2}{4}, \quad c_{W2} = \frac{g-2}{4}, \quad c_{p'p} = \frac{g-2}{2}, \tag{3}$$

where $\boldsymbol{\Sigma}^2 = 4S$, $\lambda = 1/(2S-1)$ for integer spin and $\boldsymbol{\Sigma}^2 = 4S+1$, $\lambda = 1/(2S)$ for half integer spin. Dependence on the magnitude of charged particle spin arose in the coefficients before the Darwin term and the induced electric quadrupole interaction. The g-factor in Eq. (3) is the gyromagnetic ratio defined by the magnetic form factor $F_m(0,0) = g/2$ in Eq. (2). In the spin one half case g reduces to a sum of the QED perturbation series if the charged particle is subject only to electromagnetic interactions. The coefficients in Eq. (3) in the spin one half case coincide with the respective coefficients in Ref. 15, if the phenomenological g-factor is substituted in the expressions in Ref. 15 instead of the perturbative $g = 2(1 + \alpha/2\pi)$. As an independent test of the effective Lagrangian in Eq. (1) we considered the charged $W^{\pm}$-boson sector of the Glashow-Weinberg-Salam Electroweak Theory amended by the anomalous magnetic moment term. We derived the effective NRQED Lagrangian for the $W^{\pm}$ bosons. This Lagrangian coincides with the Lagrangian in Eq. (1) for spin one charged particles.

The coefficients in Eq. (3) are calculated ignoring all loop diagrams and q^2 and τ dependence of the form factors in Eq. (2). Both the loop diagrams in relativistic QED and multipole expansion of the form factors would generate further corrections to the coefficients in Eq. (3). In ordinary renormalizable spin one half QED (as well as in the renormalizable QED of spin one $W^{\pm}$ vector bosons) all diagrams, besides those that give contributions only to the free particle g-factors, generate corrections to the coefficients that are additionally suppressed by powers of $Z\alpha$. We expect the same effect in any reasonable theory for higher spin particles. It is also obvious that accounting for q^2 and τ dependent terms in the form factors in Eq. (2) generates terms suppressed by additional powers of $Z\alpha$. Notice that we ignored the internal electric quadrupole moment implicit in the form factor $F_e(q^2, \tau)$ in Eq. (2), but still a term with an induced electric quadrupole moment arose in Eq. (1) with a coefficient dependent on the magnitude of spin. This immediately means that electric quadrupole interaction with external electric field depends on the magnitude of spin. Something like this mechanism could in principle make the interaction of the magnetic dipole moment with an external magnetic field spin-dependent, and lead to dependence of bound state g-factor on the magnitude of spin. But this happens neither with the coefficient before the term $\boldsymbol{S} \cdot \boldsymbol{B}$, nor with the coefficients before the other terms in Eq. (1) that give contributions to leading binding correction to the bound state g-factors.

Formally two-photon relativistic Compton effect diagrams are needed to obtain coefficients before all terms in the NRQED Lagrangian bilinear in electromagnetic fields. However, as mentioned above, all gauge noninvariant terms in Eq. (1) bilinear $\boldsymbol{A}$ and $\boldsymbol{E}$ can be restored from one-photon diagrams with the help of gauge invariance. Any gauge invariant terms connected with the two-photon diagrams are of too high order in $Z\alpha$ to contribute to the leading relativistic corrections of order $(Z\alpha)^2$. This allows us to avoid consideration of the relativistic two-photon diagrams.

The Lagrangian in Eq. (1) with the coefficients from Eq. (3) is sufficient for calculation of the leading relativistic corrections to the bound g-factor in the nonrecoil case.

3. Effective Two-Particle Hamiltonian. Center of Mass Motion

Next we construct the effective nonrelativistic quantum mechanical Hamiltonian for a loosely bound electrodynamic system of two particles needed to calculate both nonrecoil and recoil corrections of order $(Z\alpha)^2$ to the bound state g-factors. The interaction between two charged particles with accuracy up to $(Z\alpha)^2$ is described by the one photon exchange which generates Coulomb and Breit interactions. We calculated the two-particle scattering amplitude for particles with arbitrary spins and magnetic moments using the vertices from Eq. (2) and the photon propagator in the Coulomb gauge. After nonrelativistic expansion we obtained the interaction potential (see also Refs. 13 and 19)

$$
\begin{aligned}
V_{int}(\boldsymbol{p}_1,\boldsymbol{p}_2,\boldsymbol{r}) = e_1e_2\Bigg[&\frac{1}{4\pi r} - (g_1-1)\frac{1}{8m_1^2}\frac{\boldsymbol{\Sigma}_1^2}{3}\delta(\boldsymbol{r}) - (g_1-1)\frac{3\lambda_1}{\pi}\frac{r^ir^jQ_{ij}^{(1)}}{16m_1^2r^5} \\
&- (g_2-1)\frac{1}{8m_2^2}\frac{\boldsymbol{\Sigma}_2^2}{3}\delta(\boldsymbol{r}) - (g_2-1)\frac{3\lambda_2}{\pi}\frac{r^ir^jQ_{ij}^{(2)}}{16m_2^2r^5} - \frac{\boldsymbol{r}(\boldsymbol{r}\cdot\boldsymbol{p}_1)\cdot\boldsymbol{p}_2}{8\pi m_1m_2r^3} \\
&- \frac{\boldsymbol{p}_1\cdot\boldsymbol{p}_2}{8\pi m_1m_2r} - (g_1-1)\frac{2\boldsymbol{S}_1\cdot(\boldsymbol{r}\times\boldsymbol{p}_1)}{16\pi m_1^2r^3} + g_1\frac{2\boldsymbol{S}_1\cdot(\boldsymbol{r}\times\boldsymbol{p}_2)}{16\pi m_1m_2r^3} \\
&+ (g_2-1)\frac{2\boldsymbol{S}_2\cdot(\boldsymbol{r}\times\boldsymbol{p}_2)}{16\pi m_2^2r^3} - g_2\frac{2\boldsymbol{S}_2\cdot(\boldsymbol{r}\times\boldsymbol{p}_1)}{16\pi m_1m_2r^3} \\
&+ \frac{g_1g_2}{16\pi m_1m_2}\left(\frac{\boldsymbol{S}_1\cdot\boldsymbol{S}_2}{r^3} - \frac{3(\boldsymbol{S}_1\cdot\boldsymbol{r})(\boldsymbol{S}_2\cdot\boldsymbol{r})}{r^5} - \frac{8\pi}{3}\boldsymbol{S}_1\cdot\boldsymbol{S}_2\delta(\boldsymbol{r})\right)\Bigg],
\end{aligned}
\tag{4}
$$

where $\boldsymbol{r}_{1(2)}$, $\boldsymbol{p}_{1(2)}$, $\boldsymbol{S}_{1(2)}$, $m_{1(2)}$, $g_{1(2)}$, and $Q_{ij}^{(1(2))}$ are the coordinate, momentum, spin, mass, gyromagnetic ratio, and induced quadrupole moment of the first (second) particle, and $\boldsymbol{r} = \boldsymbol{r}_1 - \boldsymbol{r}_2$ is the relative coordinate.

This interaction is a natural generalization of the spin one half one-photon potential (see, e.g., Ref. 18). The only difference is that like in the Lagrangian in Eq. (1) the coefficients in Eq. (4) before the Darwin terms depend on the magnitude of particles' spins, and new terms with induced electric quadrupole moments arise (we again ignore intrinsic electric quadrupole moments and other intrinsic multipole moments not relevant for calculation of the leading corrections to bound state g factors). The interaction in Eq. (4) is calculated in the absence of a small uniform external magnetic field that is present in the g-factor problem. This drawback is easily repaired by the minimal substitution $\boldsymbol{p}_i \to \boldsymbol{p}_i - e_i\boldsymbol{A}_i$, $\boldsymbol{A}_i = \boldsymbol{B}\times\boldsymbol{r}_i/2$.

Combining the nonrecoil Lagrangian in Eq. (1) and the one-photon potential (after minimal substitution) in Eq. (4) we obtain a total effective nonrelativistic two-particle quantum mechanical Hamiltonian for electromagnetically interacting particles with arbitrary spins (we preserve below only the terms relevant for calculation of the g-factor contributions)

$$H = H_1 + H_2 + H_{int}, \tag{5}$$

where

$$\begin{aligned} H_1 = {} & \frac{(\boldsymbol{p}_1 - e_1\boldsymbol{A}_1)^2}{2m_1} - g_1\frac{e_1}{2m_1}(\boldsymbol{S}_1\cdot\boldsymbol{B})(1-\frac{\boldsymbol{p}_1^2}{2m_1^2}) \\ & - (g_1-2)\frac{e_1}{2m_1}(\boldsymbol{S}_1\cdot\boldsymbol{B})\frac{\boldsymbol{p}_1^2}{2m_1^2} + (g_1-2)\frac{e_1}{2m_1}\frac{(\boldsymbol{p}_1\cdot\boldsymbol{B})(\boldsymbol{S}_1\cdot\boldsymbol{p}_1)}{2m_1^2}, \end{aligned} \tag{6}$$

$$\begin{aligned} H_{int} = {} & \frac{e_1e_2}{4\pi r} + e_1e_2\Bigg[-(g_1-1)\frac{2\boldsymbol{S}_1\cdot(\boldsymbol{r}\times(\boldsymbol{p}_1 - e_1\boldsymbol{A}_1))}{16\pi m_1^2r^3} \\ & + g_1\frac{2\boldsymbol{S}_1\cdot(\boldsymbol{r}\times(\boldsymbol{p}_2 - e_2\boldsymbol{A}_2))}{16\pi m_1m_2r^3} + (g_2-1)\frac{2\boldsymbol{S}_2\cdot(\boldsymbol{r}\times(\boldsymbol{p}_2 - e_2\boldsymbol{A}_2))}{16\pi m_2^2r^3} \\ & - g_2\frac{2\boldsymbol{S}_2\cdot(\boldsymbol{r}\times(\boldsymbol{p}_1 - e_1\boldsymbol{A}_1))}{16\pi m_1m_2r^3}\Bigg], \end{aligned} \tag{7}$$

and H_2 is obtained from H_1 by the substitution $1 \to 2$.

The nonrelativistic effective two-particle Hamiltonian in Eq. (5) describes all (nonrecoil and recoil) leading relativistic corrections to bound state g-factors of each of the constituents. The coefficients before all terms in the Hamiltonian in Eq. (5) do not depend on the magnitude of spin, and the corrections to the bound state g-factors are connected with the terms in the Hamiltonian that are linear in external magnetic field. Already at this stage we see that these corrections are universal and do not depend on the magnitude of spin.

Actual calculation of the recoil corrections requires separation of the effects connected with the motion of the bound system as a whole from the internal effects. This task is not quite trivial because the center of mass variables do not separate in the presence of external field. For the current case of a small external magnetic field a solution was suggested in

Refs. 6 and 9. The main idea is to impose the condition that the center mass of a loosely bound system moves in a small external field exactly in the same way as a respective elementary particle with the same mass and charge. Neither canonical nor kinetic momentum of a charged particle are conserved in external uniform magnetic field. Instead, in the symmetric gauge $\boldsymbol{A}(\boldsymbol{r}) = \boldsymbol{B} \times \boldsymbol{r}/2$, the pseudomomentum $\boldsymbol{p} + q\boldsymbol{A}(\boldsymbol{r})$ is conserved,[6]

$$[H, \boldsymbol{p} + q\boldsymbol{A}(\boldsymbol{r})] = 0, \tag{8}$$

where $\boldsymbol{p}$ and $\boldsymbol{r}$ are the canonical momentum and coordinate of the charged particle, q is its charge[b], and the Hamiltonian has the standard form $H = (\boldsymbol{p} - q\boldsymbol{A}(\boldsymbol{r}))^2/(2m)$. Classically conservation of pseudomomentum means that the center of the Larmor orbit remains at rest. A naive transition to the standard center of mass coordinates $\boldsymbol{r} = \boldsymbol{r}_1 - \boldsymbol{r}_2$, $\boldsymbol{R} = \mu_1\boldsymbol{r}_1 + \mu_2\boldsymbol{r}_2$ ($\mu_i = m_i/(m_1 + m_2)$) does not secure conservation of total pseudomomentum $\boldsymbol{P} + (e_1 + e_2)\boldsymbol{A}(\boldsymbol{R})$, where $\boldsymbol{P} = \boldsymbol{p}_1 + \boldsymbol{p}_2$. To satisfy the transparent physical requirement of total pseudomomentum conservation (conservation of the position of the Larmor orbit center) transition to the center of mass coordinates should be accompanied by the unitary transformation of the Hamiltonian $H \to U^{-1}HU$, where $U = e^{i(e_1\mu_2 - e_2\mu_1)A(\boldsymbol{R})\cdot\boldsymbol{r}}$. After this transformation the internal Hamiltonian acquires the form

$$H = \frac{\boldsymbol{p}^2}{2m_r} + \frac{e_1e_2}{4\pi r} + H^{(1)}_{spin} + H^{(2)}_{spin} + H_r, \tag{9}$$

where $m_r = m_1m_2/(m_1+m_2)$ is the reduced mass, $H^{(1)}_{spin}$ ($H^{(2)}_{spin}$) describes interaction of the first (second) spin with the external field, and H_r includes all other terms in the Hamiltonian. Explicitly the Hamiltonian for the first particle spin interaction with the external magnetic field is

$$H^{(1)}_{spin} = -\frac{e_1}{2m_1}(\boldsymbol{S}_1 \cdot \boldsymbol{B})\Bigg\{g_1\left[(1 - \frac{\boldsymbol{p}^2}{2m_1^2}) - \frac{e_2[e_1 - (e_1 + e_2)\mu_1^2]}{24\pi m_1 r} - \frac{e_2[e_2 - (e_1 + e_2)\mu_2^2]}{12\pi m_2 r}\right] + (g_1 - 2)\left[\frac{\boldsymbol{p}^2}{3m_1^2} - \frac{e_2[e_1 - (e_1 + e_2)\mu_1^2]}{24\pi m_1 r}\right]\Bigg\}. \tag{10}$$

The Hamiltonian $H^{(2)}_{spin}$ for the second constituent spin has a similar form. The leading binding correction to the g-factor is completely described by the Hamiltonian in Eq. (10). We calculate its matrix element with the help of the first order perturbation theory between the Schrödinger-Coulomb wave

[b]All charges in this paper carry sign, so, for example, for an electron $e = -|e|$.

functions that are eigenfunctions of the unperturbed internal Hamiltonian. After simple calculations we obtain the bound state g-factors with account of the leading relativistic corrections of order $(Z\alpha)^2$ for s-states with the principal quantum number n

$$g_1^{bound} = g_1\Bigg[(1 - \frac{\mu_2^2 e_1^2 e_2^2}{2(4\pi)^2 n^2}) + \frac{\mu_2 e_1 e_2^2[e_1 - (e_1+e_2)\mu_1^2]}{6(4\pi)^2 n^2} + \frac{\mu_1 e_1 e_2^2[e_2 - (e_1+e_2)\mu_2^2]}{3(4\pi)^2 n^2}\Bigg] + (g_1 - 2)\Bigg[\frac{\mu_2^2 e_1^2 e_2^2}{3(4\pi)^2 n^2} + \frac{\mu_2 e_1 e_2^2[e_1 - (e_1+e_2)\mu_1^2]}{6(4\pi)^2 n^2}\Bigg], \quad (11)$$

$$g_2^{bound} = g_2\Bigg[(1 - \frac{\mu_1^2 e_1^2 e_2^2}{2(4\pi)^2 n^2}) + \frac{\mu_1 e_1^2 e_2[e_2 - (e_1+e_2)\mu_2^2]}{6(4\pi)^2 n^2} + \frac{\mu_2 e_1^2 e_2[e_1 - (e_1+e_2)\mu_1^2]}{3(4\pi)^2 n^2}\Bigg] + (g_2 - 2)\Bigg[\frac{\mu_1^2 e_1^2 e_2^2}{3(4\pi)^2 n^2} + \frac{\mu_1 e_1^2 e_2[e_2 - (e_1+e_2)\mu_2^2]}{6(4\pi)^2 n^2}\Bigg]. \quad (12)$$

These results resolve the discrepancy mentioned in the Introduction in favor of the results in Ref. 9 (see also Ref. 13). The remarkable property of the expressions in Eq. (11) and Eq. (12) is that they are universal for particles of any spin; they depend only on the g-factors of free charged particle, not on the magnitude of their spins. Technically this happened because no terms in the effective two-particle NRQED Hamiltonian in Eq. (5) relevant for calculation of the leading relativistic corrections contain spin-dependent coefficients λ_i, $\mathbf{\Sigma}_i^2$.

4. Universality and the Bargmann-Michel-Telegdi Equation

Universality of the $(Z\alpha)^2$ corrections to the bound state g-factors in Eq. (11) and Eq. (12) requires physical explanation. We already noticed above that simple analysis of the dimensions and spin structure of all terms in the NRQED Lagrangian in Eq. (1) leads to the conclusion that terms with derivatives of electric fields do not generate contributions to the leading relativistic corrections to the bound state g-factors. Omission of the field derivatives is the basic assumption for validity of the Bargmann-Michel-Telegdi (BMT) equation.[18,20] Hence, one can use the BMT equation for motion of spin in external electromagnetic field for derivation of the leading corrections to the bound state g-factors as was suggested in Ref. 9. Then the leading relativistic corrections are universal if the BMT equation is universal for all spins. To explain universality of the coefficients in the BMT equation let us recall the main steps in its derivation. The basic idea is to use Lorentz invariance to generalize the nonrelativistic precession equation

$$\dot{\boldsymbol{S}} = g\frac{e}{2mc}\boldsymbol{S} \times \boldsymbol{B} \tag{13}$$

to a relativistically invariant equation for motion of spin in external field. Following Refs. 20 and 18 we introduce a four-pseudovector a^μ ($a^2 = -\boldsymbol{S}^2$) to describe spin. In the particle rest frame $a^\mu = (0, \boldsymbol{S})$. In the relativistic generalization of Eq. (13) the derivative on the LHS should be over proper time $d\tau = \gamma dt$ ($\gamma = 1/\sqrt{1-v^2}$). On the RHS we are looking for a vector that is linear and homogenous in the gauge invariant external EM field $F_{\mu\nu}$ and in the spin pseudovector a_μ. Then the most general relativistically invariant equation for spin motion is

$$\frac{da^\mu}{d\tau} = \alpha F^\mu{}_\nu a^\nu + \beta u^\mu F^{\nu\lambda} u_\nu a_\lambda, \tag{14}$$

where u^μ is the four-velocity, $u^\mu = (\gamma, \gamma\boldsymbol{v})$, $u^\mu u_\mu = 1$, and α and β are unknown constants. The values of these constants are uniquely restored using the nonrelativistic precession equation Eq. (13) and the classical relativistic equation of motion for a charged particle in external field. We obtain

$$\beta = \frac{e}{m} - \alpha = -(g-2)\frac{e}{2m}. \tag{15}$$

Returning to ordinary time t and noncovariant spin vector $\boldsymbol{S}$ we can write the BMT equation in the form

$$\begin{aligned}\frac{d\boldsymbol{S}}{dt} &= \frac{e}{2m}\boldsymbol{S} \times \left\{\left(g - 2 + \frac{2}{\gamma}\right)\boldsymbol{B}\right.\\ &\left. - \frac{(g-2)\gamma}{1+\gamma}\boldsymbol{v}\cdot\boldsymbol{B}\boldsymbol{v} + \left(g - \frac{2\gamma}{1+\gamma}\right)[\boldsymbol{E}\times\boldsymbol{v}]\right\}.\end{aligned} \tag{16}$$

We see that all coefficients in the BMT equation are universal for any spin, and this universality follows from the universality of nonrelativistic spin precession and Lorentz invariance.

Next we sketch the derivation of the leading relativistic corrections to bound state g-factors based on the BMT equation,[9] in order to demonstrate how the universality of the BMT equation leads to universality of the leading corrections to the bound state g-factors. At the first step we represent

Eq. (16) as the Heisenberg equation of motion for the spin operator

$$i\frac{d\boldsymbol{S}}{dt} = [\boldsymbol{S}, H], \tag{17}$$

with the Hamiltonian

$$\begin{aligned} H = -\frac{e\hbar}{2m}\boldsymbol{S}\cdot\Bigg\{&\left(g-2+\frac{2}{\gamma}\right)\boldsymbol{B} \\ &-\frac{(g-2)\gamma}{1+\gamma}\boldsymbol{v}\cdot\boldsymbol{B}\boldsymbol{v}+\left(g-\frac{2\gamma}{1+\gamma}\right)[\boldsymbol{E}\times\boldsymbol{v}]\Bigg\}. \end{aligned} \tag{18}$$

To calculate the leading relativistic corrections to the bound state g-factor we expand the Hamiltonian in Eq. (18) up to quadratic terms in v^2, $\gamma \approx 1+\boldsymbol{v}^2/2$. We obtain the nonrelativistic Hamiltonian

$$\begin{aligned} H \approx -\frac{e}{2m}\Bigg\{&\left(g-\frac{(\boldsymbol{p}-e\boldsymbol{A})^2}{m^2}\right)\boldsymbol{S}\cdot\boldsymbol{B}-(g-2)\frac{((\boldsymbol{p}-e\boldsymbol{A})\cdot\boldsymbol{B})(\boldsymbol{S}\cdot(\boldsymbol{p}-e\boldsymbol{A}))}{2m^2} \\ &+(g-1)\,\frac{\boldsymbol{S}\cdot[\boldsymbol{E}\times(\boldsymbol{p}-e\boldsymbol{A})]}{m}\Bigg\}. \end{aligned} \tag{19}$$

where we made the substitution $\boldsymbol{v} = (\boldsymbol{p}-e\boldsymbol{A})/m$ required by gauge invariance. The Hamiltonian in Eq. (19) coincides with the one in Eq. (6), and explains why the latter Hamiltonian has universal coefficients.

Besides the Hamiltonian in Eq. (6) calculation of the recoil corrections to bound state g-factors also requires knowledge of the interaction Hamiltonian in Eq. (7). Above we derived this interaction Hamiltonian from the relativistic one-photon exchange, but it is easy to see that it includes only the ordinary nonrelativistic spin-orbit and spin-other orbit interactions. It is well known that these interactions are universal for any spin and we could write the Hamiltonian in Eq. (7) using only nonrelativistic quantum mechanical expressions. This finally explains universality of the corrections of order $(Z\alpha)^2$ to the bound state g-factors.

Acknowledgments

We are deeply grateful to Peter Mohr and Barry Taylor who attracted our attention to the problem of spin dependence. We greatly appreciate the help, advice, and input of Howard Grotch who participated at the initial stage of this project. We thank Krzysztof Pachucki for informing us about his results.[13] This work was supported by the NSF grant PHY–0757928.

References

1. H. Häffner, T. Beier, S. Djekić *et al.*, *Eur. Phys. J.* **D22**, 163 (2003).
2. J. Verdú, S. Djekić, T. Valenzuela *et al.*, *Phys. Rev. Lett.* **92**, 093002 (2004).
3. P. J. Mohr, B. N. Taylor and D. B. Newell, *Rev. Mod. Phys.* **80**, 633 (2008).
4. K. Pachucki, A. Czarnecki, U. D. Jentschura and V. A. Yerokhin, *Phys. Rev.* **A72**, 022108 (2005).
5. U. D. Jentschura, *Phys. Rev.* **A79**, 044501 (2009).
6. H. Grotch and R. A. Hegstrom, *Phys. Rev.* **A4**, 59 (1971).
7. R. N. Faustov, *Phys. Lett.* **B33**, 422 (1970).
8. F. E. Close and H. Osborn, *Phys. Lett.* **B34**, 400 (1971).
9. M. I. Eides and H. Grotch, *Ann. Phys. (N.Y.)* **260**, 191 (1997).
10. A. P. Martynenko and R. N. Faustov, *JETP* **93**, 471 (2001) [*ZhETF* **120**, 539 (2001)].
11. R. N. Faustov and A. P. Martynenko, *Phys. Atom. Nucl.* **65**, 271 (2002) [*Yad. Fiz.* **65**, 297 (2002)].
12. W. Quint, B. Nikoolabakht and U. D. Jentschura, *JETP Letters* **87**, 30 (2008) [*Pis'ma v ZhETF* **87**, 36 (2008)].
13. K. Pachucki, *Phys. Rev.* **78**, 012504 (2008).
14. M. I. Eides and T. J. S. Martin, *Phys. Rev. Lett.* **105**, 100402 (2010).
15. T. Kinoshita and M. Nio, *Phys. Rev.* **D53**, 4909 (1996).
16. I. B. Khriplovich, A. I. Milstein and R. A. Sen'kov, *Phys. Lett.* **A221**, 370 (1996).
17. I. B. Khriplovich, A. I. Milstein and R. A. Sen'kov, *JETP* **84**, 1054 (1997) [*ZhETF* **111**, 1935 (1997)].
18. V. B. Berestetskii, E. M. Lifshitz and L. P. Pitaevskii, *Quantum electrodynamics*, 2nd Edition, Pergamon Press, Oxford, 1982.
19. R. N. Lee, A. I. Milstein and M. Schumacher, *Phys. Rev.* **A64**, 032507 (2001).
20. V. Bargmann, L. Michel and V. Telegdi, *Phys. Rev. Lett.* **2**, 435 (1959).

ON YANGIAN SYMMETRY IN PLANAR $N = 4$ SYM

NIKLAS BEISERT

Max-Planck-Institut für Gravitationsphysik
Albert-Einstein-Institut
Am Mühlenberg 1, 14476 Potsdam, Germany
nbeisert@aei.mpg.de

Planar $\mathcal{N} = 4$ supersymmetric Yang–Mills theory appears to be perturbatively integrable. This work reviews integrability in terms of a Yangian algebra and compares the application to the problems of anomalous dimensions and scattering amplitudes.

Keywords: Integrability; Yang-Mills; local operators; anomalous dimensions; scattering matrix.

1. Introduction

Integrability is a very useful feature of selected physical models. It allows one to rely on certain algebraic properties to solve them exactly and to determine physical observables efficiently. Unfortunately, in general integrability is restricted to at most two-dimensional models. These can be discrete, e.g. spin chains, statistical physics models, or continuous, e.g. sigma models such as two-dimensional (super)gravity and worldsheet models string theory.

Despite this severe restriction, signs of integrability have been discovered in four-dimensional gauge theories: Lipatov noticed that the BFKL Hamiltonian[1–5] describing the evolution of reggeized gluons in QCD high-energy scattering is integrable[6] and closely related to the Heisenberg spin chain[7,8] (see also Ref. 9 for a recent account and additional references). The crucial additional assumption which enables integrability in this four-dimensional model is the 't Hooft large-N_c or *planar limit*.[10] In this limit the gauge group dynamics reduces to two-dimensional surfaces on which the integrable structure lives.

Another instance of integrability in large-N_c gauge theory is deep inelastic scattering where anomalous dimensions of local operators are responsible for scaling violations. The anomalous dimensions of local operators can be described by the DGLAP evolution equation which was initiated by Gribov and Lipatov.[11–13] It was noticed that also these evolution equations are integrable to some extent.[14–18]

In 2002 a new line of developments started for a particular four-dimensional gauge theory, namely $\mathcal{N} = 4$ maximally supersymmetric Yang–Mills ($\mathcal{N} = 4$ SYM). This model, consisting of a $\mathrm{U}(N_\mathrm{c})$ gauge field, 4 flavours of massless adjoint fermions and 6 flavours of massless adjoint scalars, is relevant to the AdS/CFT string/gauge duality. Integrability was shown to apply to all leading-order planar anomalous dimensions.[19,20] Unlike in the analogous problem in QCD, integrability was moreover demonstrated to survive in higher-order quantum corrections[21,22] hinting at complete integrability of the planar sector of the theory.

In this paper we review integrability of planar $\mathcal{N} = 4$ SYM in the guise of Yangian symmetry. We shall focus on the problems of anomalous dimensions of local operators (Sec. 2) and the spacetime scattering matrix (Sec. 3) in order to reveal the close similarities between them (Sec. 4).

2. Anomalous Dimensions of Local Operators

Scaling dimensions of local operators represent a key set of observables in a conformal field theory. They determine to a large extent the spacetime dependence of correlation functions. The spectrum of scaling dimensions can be viewed as the conformal analog of the mass spectrum of composite particles in a non-conformal field theory. In $\mathcal{N} = 4$ SYM the planar spectrum turned out to be governed by an integrable system with an underlying Yangian algebra. In the following we shall review local operators and the role the Yangian algebra.

2.1. *Framework*

Local operators are local, gauge-invariant combinations of the scalars $\mathcal{B}$, fermions $\mathcal{C}$ and gauge field strengths $\mathcal{F}$ as well as their covariant derivatives $\mathcal{D}$. Gauge invariant combinations are constructed as traces of products of covariant fields, e.g.

$$\begin{aligned} \bar{\mathcal{O}}_1(x) &= \mathrm{Tr}\big(\mathcal{B}_m(x)\,\mathcal{B}_m(x)\big), \\ \bar{\mathcal{O}}_2(x) &= \mathrm{Tr}\big(\mathcal{D}^\mu\mathcal{B}_m(x)\,\mathcal{D}_\mu\mathcal{B}_n(x)\big), \\ \bar{\mathcal{O}}_3(x) &= \dots. \end{aligned} \qquad (1)$$

One can also construct multi-trace operators, such as $\bar{\mathcal{O}}_1(x)\bar{\mathcal{O}}_2(x)$, but in the planar limit these decouple and can be safely ignored, see Fig. 1.

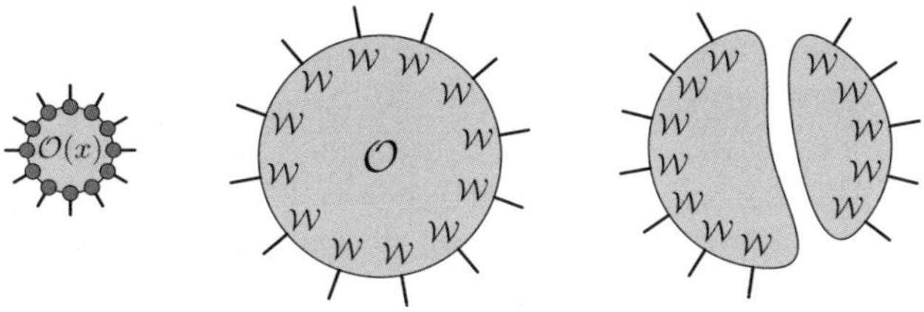

Fig. 1. Local operators as gauge invariant combination of fields at a common point in spacetime (left). Focusing on the $\mathrm{U}(N_c)$ gauge group structure alone, operators are classified by the number of traces: single-trace (middle) and multi-trace (right) operators.

In a perturbative QFT on flat Minkowski spacetime the correlator of two such operators takes the generic form $\langle\bar{\mathcal{O}}_A(x)\bar{\mathcal{O}}_B(y)\rangle = F_{AB}(x-y, g, \mu, \epsilon)$ due to Poincaré symmetry. Here μ is the regularisation scale and ϵ is the parameter of dimensional regularisation. Importantly, the result is generically divergent as one removes the regulator, i.e. at $\epsilon \to 0$. This also applies to $\mathcal{N}=4$ SYM. Superficially, it contradicts the fact that $\mathcal{N}=4$ SYM is a finite CFT where two-point functions take a particular form which depends only on the *scaling dimensions* D_A of the local operators $\tilde{\mathcal{O}}_A$ [a]

$$\langle\tilde{\mathcal{O}}_A(x)\,\tilde{\mathcal{O}}_B(y)\rangle = \frac{\delta_{AB}}{|x-y|^{2D_A}}\,. \tag{2}$$

To recover this form one has to find the right linear combinations $\tilde{\mathcal{O}}_A$ of the bare operators $\bar{\mathcal{O}}_A$, see e.g. Ref. 23. This is usually done in two steps: First, renormalisation absorbs the divergencies into the definition of the operators $\bar{\mathcal{O}}_A = Z_A{}^B\mathcal{O}_B$. Then the operators are diagonalised to achieve the above form by means of another linear transformation $\mathcal{O}_A \to \tilde{\mathcal{O}}_A$. Of course the composition of the two maps is yet another linear map, but it still makes sense to distinguish the two steps: Renormalisation can be performed abstractly on a basis of states as in (1) to the end that one can enumerate renormalised operators $\mathcal{O}_B$ in an equivalent basis. Conversely, diagonalisation requires the precise knowledge of the set of operators one is interested in. Moreover it requires to solve algebraic equations, potentially of very high degree. It should be noted though that the splitting remains somewhat ambiguous to the extent that Z is uniquely determined by the

[a]This expression applies to scalar operators; spinning operators have a different, yet uniquely determined and x-dependent structure in the numerator.

model only modulo some transformations (which are eventually compensated by the diagonalisation).

Next we establish a useful basis of single-trace local operators in $\mathcal{N} = 4$ SYM

$$\mathcal{O} = \mathrm{Tr}\big(\mathcal{W}_1 \mathcal{W}_2 \mathcal{W}_3 \dots \mathcal{W}_n\big), \qquad \mathcal{W}_k \in \{\mathcal{D}^j \mathcal{B}, \mathcal{D}^j \mathcal{C}, \mathcal{D}^j \mathcal{F}\}. \tag{3}$$

The matrices $\mathcal{W}_k$ represent the scalars $\mathcal{B}$, the fermions $\mathcal{C}$, the gauge field strengths $\mathcal{F}$ or their (multiple) covariant derivatives $\mathcal{D}$ (we hide the spacetime and internal indices). All fields are evaluated at a common point in spacetime which we need not specify further for the enumeration. Due to the trace, the definition of the local operators is invariant w.r.t. (graded) cyclic shifts $\mathcal{W}_k \to \mathcal{W}_{k+1}$, $\mathcal{W}_n \to \mathcal{W}_1$.

In enumerating the local operators one should take the (quantum) equations of motions of the fields into account. For example, $\mathcal{D}^2 \mathcal{B}$ can be expressed through products of the fields such as $\mathcal{B}^3$ or $\mathcal{C}^2$. Such combinations are already accounted for in (3), so we can discard the term $\mathcal{D}^2 \mathcal{B}$ (irrespectively of the precise form of the quantum equation of motion). Similarly, the terms $\mathcal{D} \cdot \mathcal{C}$ and $\mathcal{D} \cdot \mathcal{F}$, as well as $\mathcal{D} \wedge \mathcal{F}$ and $\mathcal{D} \wedge \mathcal{D}$ can be dropped. A minimal basis for the fields $\mathcal{W}$ can be expressed most conveniently using spinor indices $\alpha, \beta, \ldots = 1, 2$ and $\dot{\alpha}, \dot{\beta} = 1, 2$ for the Lorentz algebra $\mathfrak{so}(3,1) = \mathfrak{sl}(2, \mathbb{C})$ as well as spinor indices $a, b, \ldots = 1, 2, 3, 4$ for the the internal algebra $\mathfrak{so}(6) = \mathfrak{su}(4)$. It turns out that in our basis Lorentz spinor indices are totally symmetric while internal spinor indices are totally antisymmetric. Such a basis can be represented through states of a supersymmetric oscillator[24] with two plus two bosonic operators $\mathrm{a}^{\dagger\alpha}, \mathrm{b}^{\dagger\dot{\alpha}}$ and four fermionic operators $\mathrm{d}^{\dagger a}$. Then the various fields of $\mathcal{N} = 4$ SYM decompose as follows

$$\begin{aligned} &\mathcal{F} \sim \mathrm{b}^\dagger \mathrm{b}^\dagger, \ \mathcal{C} \sim \mathrm{b}^\dagger \mathrm{d}^\dagger, \qquad \mathcal{B} \sim \mathrm{d}^\dagger \mathrm{d}^\dagger, \\ &\mathcal{D} \sim \mathrm{a}^\dagger \mathrm{b}^\dagger, \ \bar{\mathcal{C}} \sim \mathrm{a}^\dagger \mathrm{d}^\dagger \mathrm{d}^\dagger \mathrm{d}^\dagger, \ \bar{\mathcal{F}} \sim \mathrm{a}^\dagger \mathrm{a}^\dagger \mathrm{d}^\dagger \mathrm{d}^\dagger \mathrm{d}^\dagger \mathrm{d}^\dagger, \end{aligned} \tag{4}$$

where we have suppressed the indices. Note that all physical fields in (3) are uncharged w.r.t. the operator

$$\mathrm{C} = 2 + N_{\mathrm{a}} - N_{\mathrm{b}} - N_{\mathrm{d}}, \tag{5}$$

where the $N_{\mathrm{a,b,d}}$ measure the occupation numbers of the oscillators $\mathrm{a}, \mathrm{b}, \mathrm{d}$. For local operators one introduces further indices for the sites, e.g.

$$\mathrm{Tr} \mathcal{B}^{ab} \mathcal{B}^{cd} \sim \mathrm{d}_1^{\dagger a} \mathrm{d}_1^{\dagger b} \mathrm{d}_2^{\dagger c} \mathrm{d}_2^{\dagger d} |0\rangle. \tag{6}$$

Note that on the r.h.s. cyclicity is automatic while on the l.h.s. it must be imposed by hand. Altogether we have seen that local operators can be expressed through states of a supersymmetric harmonic oscillator subject to a charge and a cyclicity constraint.

2.2. *One-Loop Hamiltonian*

The anomalous dimensions of local operators originate from the divergent contributions to their two-point functions. They are therefore captured by the renormalisation matrix Z. More precisely, the matrix of anomalous dimensions for the $\mathcal{O}_A$ is given by the logarithmic derivative of Z w.r.t. the logarithm of the renormalisation scale μ

$$\delta \mathrm{D} \sim Z^{-1} \frac{\mu \, dZ}{d\mu} \,. \tag{7}$$

The eigenvalues of the matrix $\delta\mathrm{D}$ represent the quantum corrections δD_A in the scaling dimensions D_A of the eigen-operators $\tilde{\mathcal{O}}_A$. The matrix can be interpreted as a Hamiltonian of a quantum mechanical system: It acts on the states in a systematic fashion determined by connected Feynman diagrams attached to the fields constituting the local operators, see Fig. 2. The planar limit suppresses crossing lines in Feynman diagrams, therefore $\delta\mathrm{D}$ acts on a set of adjacent fields along the single-trace state (3).

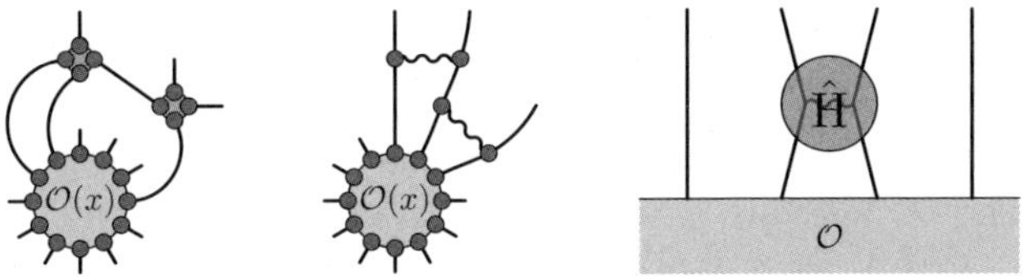

Fig. 2. Non-planar (left) and planar (middle) gluing of interactions to a local operator $\mathcal{O}$. The planar Hamiltonian $\hat{\mathrm{H}}$ acts on a pair of nearest neighbours (right) when zooming into the trace structure of $\mathcal{O}$.

The number of fields involved in the action of $\delta\mathrm{D}$ increases with the loop order. At the leading one-loop order the action is between nearest neighbours, cf. Fig. 2

$$\delta \mathrm{D}^{(1)} = \hat{\mathrm{H}} = \sum_{k=1}^{n} \hat{\mathrm{H}}_{k,k+1} \,, \tag{8}$$

and it can be interpreted as the Hamiltonian of a quantum spin chain.[19] The one-loop Hamiltonian is invariant under the free superconformal symmetries. The representation of the latter on the fields can be expressed conveniently in the oscillator framework using bilinears in the operators $(\mathrm{a}^\dagger, \mathrm{b}, \mathrm{d})$ and $(\mathrm{a}, \mathrm{b}^\dagger, \mathrm{d}^\dagger)$[24]

$$\begin{aligned}
\mathrm{L}^\alpha{}_\beta &= \mathrm{a}^{\dagger\alpha}\,\mathrm{a}_\beta - \tfrac{1}{2}\,\delta^\alpha_\beta \mathrm{a}^{\dagger\gamma}\,\mathrm{a}_\gamma, & \bar{\mathrm{L}}^{\dot\alpha}{}_{\dot\beta} &= \mathrm{b}^{\dagger\dot\alpha}\,\mathrm{b}_{\dot\beta} - \tfrac{1}{2}\,\delta^{\dot\alpha}_{\dot\beta}\mathrm{b}^{\dagger\dot\gamma}\,\mathrm{b}_{\dot\gamma},\\
\mathrm{D} &= \tfrac{1}{2}\mathrm{a}^{\dagger\gamma}\,\mathrm{a}_\gamma + \tfrac{1}{2}\mathrm{b}_{\dot\gamma}\,\mathrm{b}^{\dagger\dot\gamma}, & \mathrm{R}^a{}_b &= \mathrm{d}^{\dagger a}\mathrm{d}_b - \tfrac{1}{4}\,\delta^a_b\mathrm{d}^{\dagger c}\,\mathrm{d}_c,\\
\bar{\mathrm{Q}}^{\dot\alpha}_b &= \mathrm{b}^{\dagger\dot\alpha}\,\mathrm{d}_b, & \bar{\mathrm{S}}^b_{\dot\alpha} &= \mathrm{d}^{\dagger b}\,\mathrm{b}_{\dot\alpha},\\
\mathrm{Q}^{\beta a} &= \mathrm{a}^{\dagger\beta}\,\mathrm{d}^{\dagger a}, & \mathrm{S}_{\beta a} &= \mathrm{a}_\beta\,\mathrm{d}_a,\\
\mathrm{P}^{\beta\dot\alpha} &= \mathrm{a}^{\dagger\beta}\,\mathrm{b}^{\dagger\dot\alpha}, & \mathrm{K}_{\beta\dot\alpha} &= \mathrm{a}_\beta\,\mathrm{b}_{\dot\alpha}.
\end{aligned} \tag{9}$$

Single-trace states (3) transform in tensor product representations of the above.

Invariance under free superconformal symmetry imposes strong constraints on $\hat{\mathrm{H}}$. The crucial observation is that the tensor product of two field representations decomposes into a sequence of irreducible representations distinguished by their overall superconformal spin j. The latter can be measured using the quadratic Casimir of $\mathfrak{psu}(2,2|4)$ in analogy to the total spin of $\mathfrak{su}(2)$. Symmetry demands that that the Hamiltonian has a common eigenvalue for all components of an irreducible multiplet.[b] Hence it suffices to specify the eigenvalues, and we can write the nearest-neighbour Hamiltonian as[25]

$$\hat{\mathrm{H}}_{k,k+1} = \sum_{j=0}^{\infty} c_j \hat{\mathrm{P}}_{k,k+1;j}. \tag{10}$$

Here the operator $\hat{\mathrm{P}}_{k,k+1;j}$ projects a two-particle state to its components with superconformal spin j. Now there are several ways to determine the unspecified eigenvalues c_j: Direct calculation in the one-loop quantum field theory shows that the coefficients are given by the elements of the harmonic series[25–28]

$$c_j \sim h(j) = \sum_{k=1}^{j} \frac{1}{k} = \Psi(j+1) - \Psi(1), \qquad \Psi(z) = \frac{\Gamma'(z)}{\Gamma(z)}. \tag{11}$$

Two other methods of determining the coefficients purely algebraically are described in the following two subsections.

The analog of the above Hamiltonian for quasi-partonic operators in QCD is very similar, and it has a particular feature which was noticed in Refs. 15, 17, 18, see Ref. 29 for a review. Namely, the appearance of the digamma function Ψ hints at *integrability*, cf. Ref. 30. In particular,

[b]This holds for multiplets of multiplicity 1; for higher multiplicity n, invariance allows an action equivalent to a $n \times n$ matrix.

Lipatov realised in Refs. 14, 16 that for $\mathcal{N} = 4$ SYM the Hamiltonian is particularly simple, and also made the prophetic connection to the newly proposed AdS/CFT correspondence to strings on $AdS_5 \times S^5$. Several years later and in a different context, integrability of the one-loop Hamiltonian was rediscovered in Refs. 19, 21. Most importantly, it was also put to use by establishing a set of Bethe equations to determine the spectrum of planar one-loop anomalous dimensions very efficiently. In particular, the thermodynamic limit of long chains, $n \to \infty$, became accessible[19,31,32] and could be compared to results from string theory,[33–35c] see Refs. 37, 38, 39 for reviews.

2.3. *Leading-Order Yangian*

Integrable spin chains with manifest Lie algebra symmetry $\mathfrak{g}$ typically have a Yangian algebra Y underlying their structure.[40] The Yangian is a quantum algebra based on (half of) the affine extension of the Lie algebra. That is to say, next to the Lie generators J^A, there are level-one Yangian generators $\widehat{\mathrm{J}}^A$. These obey similar commutation relations as the Lie generators, namely[d]

$$[\mathrm{J}^A, \mathrm{J}^B] = F^{AB}_C \mathrm{J}^C, \qquad [\mathrm{J}^A, \widehat{\mathrm{J}}^B] = F^{AB}_C \widehat{\mathrm{J}}^C, \tag{12}$$

from which two sets of Jacobi-identities follow. However, a third Jacobi-identity involving two Yangian generators is quantum-deformed to the following Serre relation

$$\begin{aligned} &[[\mathrm{J}^A, \widehat{\mathrm{J}}^B], \widehat{\mathrm{J}}^C] + [[\mathrm{J}^B, \widehat{\mathrm{J}}^C], \widehat{\mathrm{J}}^A] + [[\mathrm{J}^C, \widehat{\mathrm{J}}^A], \widehat{\mathrm{J}}^B] \\ &\quad = F^{AG}_D F^{BH}_E F^{CI}_F F_{GHI} \mathrm{J}^{\{D} \mathrm{J}^E \mathrm{J}^{F\}}. \end{aligned} \tag{13}$$

A representation of a Lie algebra can sometimes be lifted to an evaluation representation of the corresponding Yangian. For these, J acts as in the Lie algebra and $\widehat{\mathrm{J}} \simeq u\mathrm{J}$ with u the spectral parameter of the evaluation representation. Clearly, the two commutation relations (12) are satisfied automatically, but in addition the r.h.s. of the the Serre relation (13) must vanish. This is true for the above superconformal representation,[41] consequently the spin chain transforms in a representation of the Yangian. Due to homogeneity of the spin chain, the spectral parameters of all sites should be equal.

---

[c] It turned out only later that the matching was more of a coincidence than a confirmation for AdS/CFT due to an order of limits issue, see Ref. 36.

[d] For reasons of clarity we treat all generators to be bosonic, the generalisation to superalgebras by insertion of appropriate sign factors is straight-forward.

In addition to multiplication, a quantum algebra has a comultiplication operation $\Delta : \mathrm{Y} \to \mathrm{Y} \otimes \mathrm{Y}$ with

$$\Delta(\mathrm{J}^A) = \mathrm{J}^A \otimes 1 + 1 \otimes \mathrm{J}^A, \qquad \Delta(\widehat{\mathrm{J}}^A) = \widehat{\mathrm{J}}^A \otimes 1 + 1 \otimes \widehat{\mathrm{J}}^A + F^A_{BC} \mathrm{J}^B \otimes \mathrm{J}^C. \quad (14)$$

It is compatible with the multiplication, in particular with the Serre relation (13). Its main purpose is to define tensor products of representations, i.e. it determines how the algebra acts on the spin chain. The action on the tensor product of n fields is determined by $\Delta^{n-1}(\mathrm{J})$

$$\Delta^{n-1}(\mathrm{J}^A) = \sum_{k=1}^{n} \mathrm{J}^A_k, \qquad \Delta^{n-1}(\widehat{\mathrm{J}}^A) = \sum_{k=1}^{n} \widehat{\mathrm{J}}^A_k + F^A_{BC} \sum_{j<k=1}^{n} \mathrm{J}^B_j \mathrm{J}^C_k. \quad (15)$$

When using an evaluation representation with homogeneous evaluation parameter u, we see that the first term in the action of $\widehat{\mathrm{J}}^A$ equals the superconformal action $u\mathrm{J}^A$; therefore nothing is lost by fixing u to a particular value, e.g. $u = 0$. Note that while the representation of the Lie generators J follows the usual pattern for tensor products, the representation of Yangian generators $\widehat{\mathrm{J}}$ non-trivially combines the various sites of the chain. The action of J^A and $\widehat{\mathrm{J}}^A$ is depicted in Fig. 3.

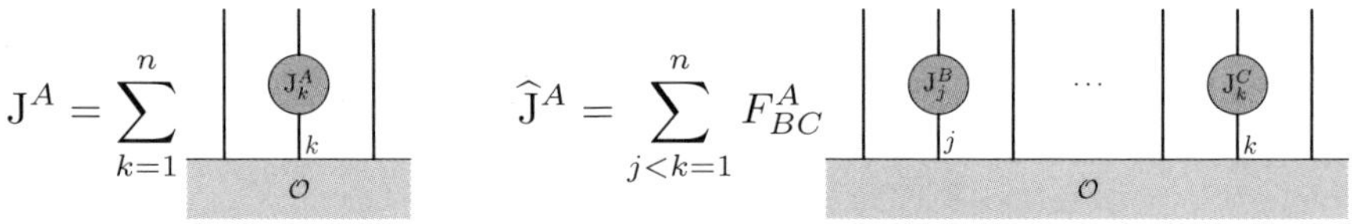

Fig. 3. Action of the free superconformal J^A and Yangian $\widehat{\mathrm{J}}^A$ generators on spin chain state alias a local operators $\mathcal{O}$.

An integrable Hamiltonian $\hat{\mathrm{H}}$ is invariant under the Lie symmetries J, but it is typically not exactly invariant under the Yangian generators $\widehat{\mathrm{J}}$. It commutes up to a difference of two terms[42]

$$[\Delta(\mathrm{J}^A), \hat{\mathrm{H}}_{12}] = 0, \qquad [\Delta(\widehat{\mathrm{J}}^A), \hat{\mathrm{H}}_{12}] \sim \mathrm{J}^A_1 - \mathrm{J}^A_2, \quad (16)$$

On a chain with n sites the commutator yields only boundary terms $\mathrm{J}_1 - \mathrm{J}_n$ essentially because periodic boundary conditions are not compatible with the definition of the Yangian. This means that the spectrum of $\hat{\mathrm{H}}$ does not organise into multiplets of the Yangian, but merely of the Lie algebra. Nevertheless one can consider the Yangian to be a symmetry of the (bulk) Hamiltonian, because commutation (up to boundary terms) does yield non-trivial constraints on $\hat{\mathrm{H}}$ which guarantee its integrability. In particular, commutation requires the following recursion relation for the coefficients c_j of the Hamiltonian[42]

$$c_j = c_{j-1} + \frac{1}{j}\,. \tag{17}$$

This relation is precisely satisfied by the coefficients from field theory (11), and hence planar one-loop $\mathcal{N} = 4$ SYM is integrable.

2.4. *Higher Loops*

Going to higher loops the above picture changes although integrability apparently remains valid. The symmetry generators as well as the Hamiltonian receive corrections in the coupling constant

$$\mathrm{J}(g) = \sum_{k=0}^{\infty} g^k \mathrm{J}^{(k/2)}, \quad \widehat{\mathrm{J}}(g) = \sum_{k=0}^{\infty} g^k \widehat{\mathrm{J}}^{(k/2)}, \quad \hat{\mathrm{H}}(g) = \sum_{k=0}^{\infty} g^k \hat{\mathrm{H}}^{(k/2)}. \tag{18}$$

The structure of the operators must remain compatible with planar Feynman diagrams, therefore an operator at $\mathcal{O}(g^k)$ involves at most $k+2$ ingoing plus outgoing fields, see Fig. 4. In particular, the number of sites of the chain is allowed to fluctuate.

$\mathrm{J}(g) =$ 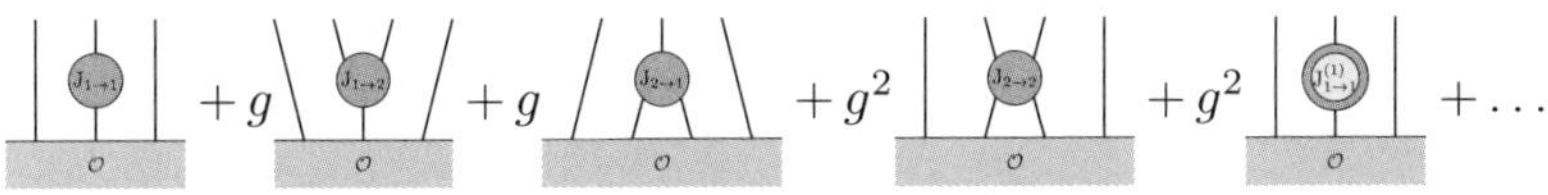

Fig. 4. Perturbative action of a superconformal generator on a spin chain state. The deformations involve long-range and dynamic interactions.

Despite the above deformations of the representations, the algebra relations (12) and (13) should remain unchanged. Generally this involves cancellations between products of terms at various orders. These cancellations leave some space for ambiguities, and unfortunately the deformations cannot be defined uniquely. It turns out that the ambiguities correspond to perturbative similarity transformations $\mathrm{J} \to \mathcal{X}\mathrm{J}\mathcal{X}^{-1}$ of the generators which leave all algebra relations invariant.[e] Only at low orders the set of permissible similarity transformations is empty and the algebra becomes unique.

[e]Ambiguities (e.g. of ordering) are a generic problem of quantum algebras, which is also the reason why the Serre relation (13) is not formulated in the form of $[\widehat{\mathrm{J}}^A, \widehat{\mathrm{J}}^B] = \ldots$ analogously to (12). Hence quantum algebras are typically defined modulo certain types of deformations.

We have seen that symmetry determines the one-loop Hamiltonian (10) up to a sequence of coefficients c_j. It turns out that the higher-loop corrections impose even stronger constraints: The point is that the Hilbert space of the spin chain decomposes into irreducible multiplets which are distinguished by their scaling dimension (among other quantum numbers). For the free superconformal algebra, the multiplets can be of short/atypical or of long/typical type.[43] Short multiplets must have (half) integral superconformal scaling dimension while long multiplets can have irrational scaling dimensions. However, the Hamiltonian attributes anomalous dimensions to almost all irreducible multiplets, long or short. Considering the spin chain Hamiltonian as the radiative correction $\delta\mathrm{D}$ to the dilatation generator D seemingly leads to a paradox. It is resolved if the right combination of short multiplets join to form a long multiplet.[f] This can only work if the short multiplets have coincident one-loop anomalous dimensions, which thus puts constraints on $\hat{\mathrm{H}}$. On the level of the algebra, the joining of short multiplets into a long one is achieved through deformations of the superconformal generators $\mathrm{Q}, \bar{\mathrm{Q}}, \mathrm{P}$ and $\mathrm{S}, \bar{\mathrm{S}}, \mathrm{K}$ at order $\mathcal{O}(g)$. These map one site to two or vice versa, cf. Fig. 5. The algebra turns out to completely determine them. In-

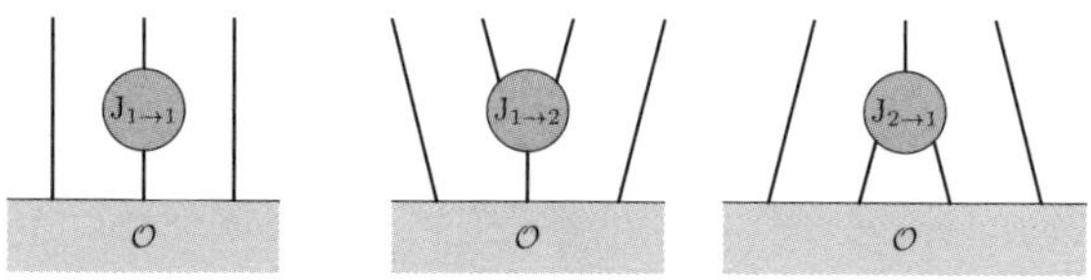

Fig. 5. Free action of a generic superconformal generator J (left) and leading-order corrections to $\mathrm{Q}, \bar{\mathrm{Q}}, \mathrm{P}$ (middle) and $\mathrm{S}, \bar{\mathrm{S}}, \mathrm{K}$ (right).

variance of the Hamiltonian then fixes the coefficients to the values of field theory (11)[37]

$$c_j \sim \sum_{k=1}^{j} \frac{1}{k} \,. \tag{19}$$

It is curious to see that integrability as well as higher-loop consistency lead to precisely the same constraints of the one-loop Hamiltonian. On the one hand, one can view it as a consistency condition of higher-loop integrability, but on the other hand, the semantic relation between the two approaches remains somewhat obscure.

[f] A similar mechanism is required for the Higgs effect where a massless vector and a massless scalar combine into a massive vector.

The leading-order deformation of the generators was established explicitly in Refs. 22, 44 for closed sectors of the full theory. In these sectors the construction was also continued by a few more perturbative orders. Knowledge of the higher-loop Hamiltonian revealed first strong hints that integrability survives[20] in perturbation theory. It was later shown that also the action of the Yangian can be deformed appropriately,[45–49] which establishes integrability rigorously at a given perturbative order. We note that the structure of the deformed Yangian action always follows a pattern analogous to the coproduct rule (14): It consists of a bi-local combination of superconformal representations (properly expanded at each order of perturbation theory) and a local contribution which can be viewed as a short-distance regularisation of the bi-local term, see Fig. 6.

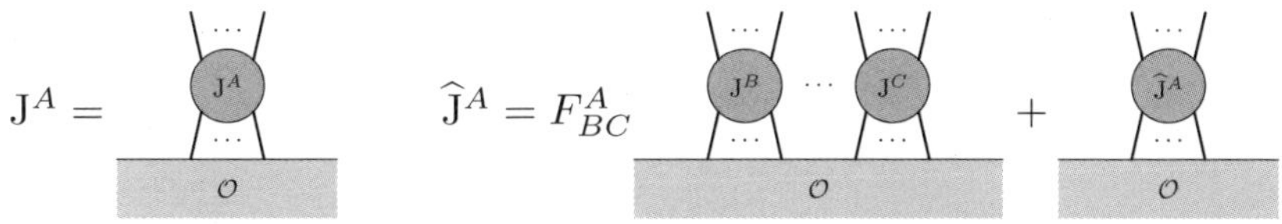

Fig. 6. Structure of the deformed representation of the superconformal and Yangian algebra. The bilocal contributions to the Yangian are determined by the superconformal generators while the local contributions can be viewed as a short-distance regularisation thereof.

3. Scattering Amplitudes

A different type of observable which plays an important role in quantum field theories is the scattering matrix. Integrability has also been observed for $\mathcal{N} = 4$ SYM in this context, and apparently it leads to substantial simplifications in their structure. We now review scattering amplitudes and their Yangian symmetry.

3.1. *Framework*

A scattering amplitude of n particles is a function of the particle momenta p_k, spins or helicities, flavours and gauge degrees of freedom A_k. Statistics requires that this function is (graded) symmetric under the simultaneous interchange of all quantum numbers associated to any pair of particles. This symmetry can be enforced by summing over all (graded) permutations of particles with the associated quantum numbers

$$\mathcal{A}^{\text{full}}_{1\ldots n} = \sum_{\pi\in S_n} \mathcal{A}^{\text{ordered}}_{\pi(1)\ldots\pi(n)}. \tag{20}$$

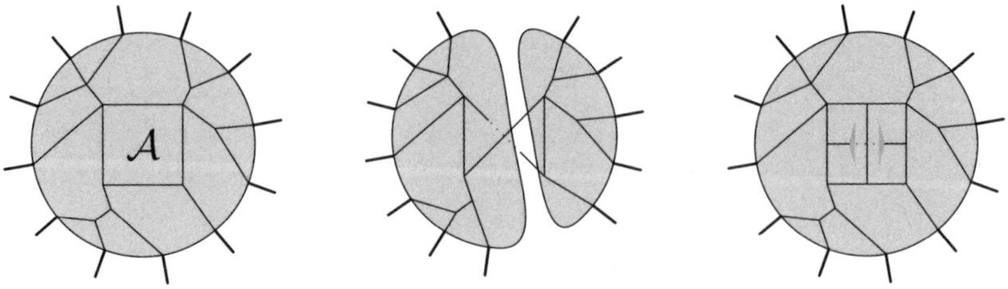

Fig. 7. Colour-ordered scattering amplitudes expanded into $\mathrm{U}(N_\mathrm{c})$ traces: single-trace (left), multi-trace (middle) and non-planar (right) contributions.

In $\mathcal{N} = 4$ SYM all particles transform in the adjoint representation of the gauge group. The gauge indices for a $\mathrm{U}(N_\mathrm{c})$ gauge group can be expanded in a basis of traces of $\mathrm{U}(N_\mathrm{c})$ generators $T_k := T^{A_k}$ in the fundamental representation, see Fig. 7

$$\begin{aligned}\mathcal{A}^{\text{ordered}}_{1\ldots n} &= \frac{1}{n}\operatorname{Tr}(T_1 \ldots T_n)\mathcal{A}^{\text{single-trace}}_{1\ldots n} \\ &+ \sum_{k=1}^{n} \frac{1}{2k(n-k)} \operatorname{Tr}(T_1 \ldots T_k) \operatorname{Tr}(T_{k+1} \ldots T_n)\, \mathcal{A}^{\text{double-trace}}_{1\ldots k|k+1\ldots n} \\ &+ \ldots . \end{aligned} \tag{21}$$

The prefactors of $1/k$ and $1/2$ are the appropriate symmetry factors for $\mathbb{Z}_k$ cyclic and S_2 permutation symmetry. Note that the various multi-trace or *colour-ordered* contributions to the amplitude now just depend on the particle momenta, spins/helicities and flavours, but not on the gauge structure anymore.

Next, all particles in $\mathcal{N} = 4$ SYM are massless. The on-shell momenta p_k are light-like and can be represented as bilinear combinations of bosonic spinors λ^β and $\tilde{\lambda}^{\dot\alpha}$ [50]

$$p^{\beta\dot\alpha} = \sigma^{\beta\dot\alpha}_\mu p^\mu = \lambda^\beta \tilde{\lambda}^{\dot\alpha}. \tag{22}$$

The two spinors are complex conjugates, $\tilde{\lambda}^{\dot\alpha} = \pm(\lambda^\alpha)^*$, where the sign determines the sign of the particle energy. Furthermore, all flavours of on-shell particles – scalars Φ, fermions Ψ and gluons Γ – can be conveniently combined into a field on superspace[51]

$$\Omega(\lambda, \tilde{\lambda}, \bar{\eta}) = \Gamma + \bar{\eta}^a \Psi_a + \tfrac{1}{2}\bar{\eta}^a\bar{\eta}^b\Phi_{ab} + \tfrac{1}{6}\varepsilon_{abcd}\bar{\eta}^a\bar{\eta}^b\bar{\eta}^c\bar{\Psi}^d + \tfrac{1}{24}\varepsilon_{abcd}\bar{\eta}^a\bar{\eta}^b\bar{\eta}^c\bar{\eta}^d\bar{\Gamma}. \tag{23}$$

This is useful because we can now scatter the superfield Ω instead of the individual fields: The colour-ordered amplitudes then turn into plain function on the configuration superspace parametrised by λ_k, $\tilde{\lambda}_k$ and $\bar{\eta}_k$ or, collectively, Λ_k

$$\mathcal{A}^{\text{colour-ordered}}(\Lambda_1, \ldots, \Lambda_n). \tag{24}$$

In particular, the flavour indices have been traded in completely for expansion coefficients in the $\bar{\eta}_k$. Also the helicity is determined by the flavour in $\mathcal{N} = 4$ SYM, so that no indices remain. Such colour-ordered amplitudes on superspace will be the standard objects we shall consider.

Note that the multiplication of λ^α by a complex phase does not change the momentum p^μ. Such a multiplication is equivalent to a rotation about the particle momentum, and thus the amplitude must transform according to the particle helicity. In effect, the amplitude is constrained by

$$\mathcal{A}(\ldots, \Lambda_k, \ldots) = e^{2i\varphi}\mathcal{A}(\ldots, e^{i\varphi}\Lambda_k, \ldots), \quad e^{i\varphi}(\lambda, \tilde{\lambda}, \eta) := (e^{i\varphi}\lambda, e^{-i\varphi}\tilde{\lambda}, e^{i\varphi}\eta). \tag{25}$$

Put differently, the differential operator

$$\mathrm{C}_k = 2 + \lambda_k^\alpha \frac{\partial}{\partial \lambda_k^\alpha} - \tilde{\lambda}_k^\alpha \frac{\partial}{\partial \tilde{\lambda}_k^{\dot{\alpha}}} - \bar{\eta}_k^a \frac{\partial}{\partial \bar{\eta}_k^a} \tag{26}$$

acting on any leg k annihilates the amplitude.

Furthermore, one can classify amplitudes by an operator B which effectively measures the overall helicity of the particles

$$\mathcal{A}_n = \sum_{k=2}^{n-2} \mathcal{A}_{n,k}, \qquad \mathrm{B}\mathcal{A}_{n,k} = 4k\mathcal{A}_{n,k}, \qquad \mathrm{B} = \sum_{k=1}^{n} \bar{\eta}_k^a \frac{\partial}{\partial \bar{\eta}_k^a} = N_{\bar{\eta}}\,. \tag{27}$$

Note that due to $\mathfrak{su}(4)$-invariance of the amplitude, the $\bar{\eta}$'s can only appear in groups of four and due to supersymmetry there must be between 8 and $4(n-2)$ of them. The amplitude with the minimum number of eight $\bar{\eta}$'s is called MHV and it is typically the simplest among those with the same number of legs.

3.2. *Tree-Level Amplitudes*

The MHV amplitudes $\mathcal{A}_n^{\text{MHV}} := \mathcal{A}_{n,2}$ have particularly simple expressions[50,52]

$$\mathcal{A}_n^{\text{MHV}} = \frac{\delta^4(P_n)\,\delta^8(Q_n)}{\langle 1,2\rangle \cdots \langle n,1\rangle} \tag{28}$$

with the overall momentum P_n and its fermionic partner Q_n

$$P_n^{\beta\dot{\alpha}} = \sum_{k=1}^{n} \lambda_k^\beta \tilde{\lambda}_k^{\dot{\alpha}}, \qquad Q_n^{\beta a} = \sum_{k=1}^{n} \lambda_k^\beta \bar{\eta}_k^a. \tag{29}$$

Furthermore invariants of the spinors are obtained by contraction with the antisymmetric invariant tensor $\varepsilon_{\alpha\beta}$ or $\varepsilon_{\dot\alpha\dot\beta}$

$$\langle \lambda, \mu\rangle := \varepsilon_{\alpha\beta}\lambda^\alpha\mu^\beta, \qquad [\tilde\lambda, \tilde\mu] := \varepsilon_{\dot\alpha\dot\beta}\tilde\lambda^{\dot\alpha}\tilde\mu^{\dot\beta}. \tag{30}$$

We abbreviate $\langle\lambda_j, \lambda_k\rangle$ as $\langle jk\rangle$. We can use the above expression $\mathcal{A}^{\mathrm{MHV}}$ to confirm $\mathcal{N} = 4$ superconformal invariance of scattering amplitudes. The generators of $\mathfrak{psu}(2,2|4)$ acting on a single free field take particularly simple expressions using the spinor helicity superspace variables $\lambda, \tilde\lambda, \bar\eta$[53]

$$\begin{aligned}
\mathrm{L}^\alpha{}_\beta &= \lambda^\alpha\partial_\beta - \tfrac{1}{2}\delta^\alpha_\beta\lambda^\gamma\partial_\gamma, & \bar{\mathrm{L}}^{\dot\alpha}{}_{\dot\beta} &= \tilde\lambda^{\dot\alpha}\tilde\partial_{\dot\beta} - \tfrac{1}{2}\delta^{\dot\alpha}_{\dot\beta}\tilde\lambda^{\dot\gamma}\tilde\partial_{\dot\gamma},\\
\mathrm{D} &= \tfrac{1}{2}\partial_\gamma\lambda^\gamma + \tfrac{1}{2}\tilde\lambda^{\dot\gamma}\tilde\partial_{\dot\gamma}, & \mathrm{R}^a{}_b &= \bar\eta^a\bar\partial_b - \tfrac{1}{4}\delta^a_b\bar\eta^c\bar\partial_c,\\
\bar{\mathrm{Q}}^{\dot\alpha}_b &= \tilde\lambda^{\dot\alpha}\bar\partial_b, & \bar{\mathrm{S}}^b_{\dot\alpha} &= \bar\eta^b\tilde\partial_{\dot\alpha},\\
\mathrm{Q}^{\beta a} &= \lambda^\beta\bar\eta^a, & \mathrm{S}_{\beta a} &= \partial_\beta\bar\partial_a,\\
\mathrm{P}^{\beta\dot\alpha} &= \lambda^\beta\tilde\lambda^{\dot\alpha}, & \mathrm{K}_{\beta\dot\alpha} &= \partial_\beta\tilde\partial_{\dot\alpha}.
\end{aligned} \tag{31}$$

The action on the amplitude is given by the standard tensor product rule as the sum over all fields

$$\mathrm{J}^A = \sum_{k=1}^{n} \mathrm{J}^A_k. \tag{32}$$

Invariance under the Lorentz $\mathrm{L}, \bar{\mathrm{L}}$ and internal R rotations is manifest because the amplitude (28) is constructed only from scalar combinations.

The dilatation generator D counts the number of λ's and $\tilde\lambda$'s. For scaling invariance the overall number must equal $-2n$. Using the degrees of homogeneity of the three components (28)

$$\delta^4(P) \sim \lambda^{-4}\tilde\lambda^{-4}, \qquad \delta^8(Q) \sim \bar\eta^8\lambda^8, \qquad \frac{1}{\langle 12\rangle\ldots\langle n1\rangle} \sim \lambda^{-2n}. \tag{33}$$

invariance follows straight-forwardly. Furthermore, it is clear that each C_k annihilates the amplitude as desired because both delta-functions are invariant, and the denominator contributes λ_k^{-2} for each k.

Next, the translations P and the supertranslations Q annihilate the amplitude due to the two delta-functions $\delta^4(P_n)$ and $\delta^8(Q_n)$ because P acts through multiplication by the overall momentum P_n (analogously for Q and Q_n).

Invariance under the conjugate supertranslation is less obvious. As it contains a derivative w.r.t. $\bar\eta$, it acts non-trivially only on the fermionic delta-function $\delta^8(Q_n)$

$$\bar{\mathrm{Q}}^{\dot\alpha}_b\delta^8(Q_n) = \sum_{k=1}^{n}\tilde\lambda^{\dot\alpha}_k\bar\partial_{b,k}\delta^8(Q_n) = \sum_{k=1}^{n}\lambda^\gamma_k\tilde\lambda^{\dot\alpha}_k\frac{\partial\delta^8(Q_n)}{\partial Q^{\gamma b}_n} = P^{\gamma\dot\alpha}_n\frac{\partial\delta^8(Q_n)}{\partial Q^{\gamma b}_n}. \tag{34}$$

Due to the presence of the bosonic delta-function $\delta^4(P_n)$ the conjugate supermomentum annihilates the amplitude. The derivation for invariance under the conjugate superboost $\bar{\mathrm{S}}$ is analogous, but there are important subtleties to be discussed in Sec. 3.4.

To show invariance under the conformal boost K and superboost S takes the largest number of steps. The two derivations are analogous and we consider only the superboost S. It contains a derivative w.r.t. $\bar{\eta}$ which again only acts on the fermionic delta-function. By a sequence of transformation we can recombine the terms into useful combinations

$$\begin{aligned}\mathrm{S}_{\alpha b}\delta^8(Q_n) &= \sum_{k=1}^{n} \partial_{\alpha,k}\lambda_k^\gamma \frac{\partial\delta^8(Q_n)}{\partial Q_n^{\gamma b}} = \left(\mathrm{L}^\gamma{}_\alpha + \tfrac{1}{2}\delta_\alpha^\gamma \sum_{k=1}^{n}(\lambda_k^\delta\partial_{\delta,k}+2)\right)\frac{\partial\delta^8(Q_n)}{\partial Q_n^{\gamma b}} \\ &= \frac{\partial\delta^8(Q_n)}{\partial Q_n^{\gamma b}}\mathrm{L}^\gamma{}_\alpha + \frac{1}{2}\frac{\partial\delta^8(Q_n)}{\partial Q_n^{\alpha b}}\left(\sum_{k=1}^{n}\lambda_k^\delta\partial_{\delta,k} - 3 + 7 + 2n\right).\end{aligned} \tag{35}$$

The first step consists in rewriting $\partial_\alpha\lambda^\gamma$ as a Lorentz generator $\mathrm{L}^\gamma{}_\alpha$. In the next step these generators are commuted past the fermionic delta-function. This picks up $-\frac{3}{2}$ from the Lorentz generator and $\frac{7}{2}$ from the weight in λ. The point is then that the remaining denominator and bosonic delta-function in $\mathcal{A}_n^{\mathrm{MHV}}$ (28) are Lorentz invariant have overall weight λ^{-4-2n} according to (33). Hence the amplitude is annihilated.

3.3. *Leading-Order Yangian*

In addition to the standard superconformal symmetries a new type of superconformal symmetry has recently been discovered for planar scattering amplitudes in $\mathcal{N}=4$.[54–57] Tree amplitudes were shown to be covariant with respect to these *dual superconformal* transformations,[58,59] and also loop amplitudes appear to be substantially constrained. The dual superconformal algebras overlaps partially with the conventional one, and therefore the two algebras must close onto a bigger one. This algebra turns out to be a Yangian.[60]

We now wish to extend the superconformal symmetry for amplitudes to a Yangian algebra. The fields transform in the superconformal representation specified in (31). It can be extended to an evaluation representation of the Yangian because the Serre relations are satisfied. Clearly, all external fields are on equal footing and should have coincident evaluation parameter. Again, its value does not play an important role because it merely multiplies the standard conformal generators; we can safely set it to zero.

The representation of the Yangian generators from the coproduct (14) then becomes, see also Fig. 8,

$$\widehat{\mathrm{J}}^A = \tfrac{1}{2} F^A_{BC} \mathrm{J}^B \wedge \mathrm{J}^C, \qquad \text{where} \qquad \mathrm{J}^B \wedge \mathrm{J}^C := \sum_{j<k=1}^{n} (\mathrm{J}^B_j \mathrm{J}^C_k - \mathrm{J}^B_k \mathrm{J}^C_j). \quad (36)$$

Invariance of tree amplitudes under Yangian symmetry $\widehat{\mathrm{J}}^A \mathcal{A} = 0$[60] follows from their conventional and dual superconformal transformation properties.[58,59]

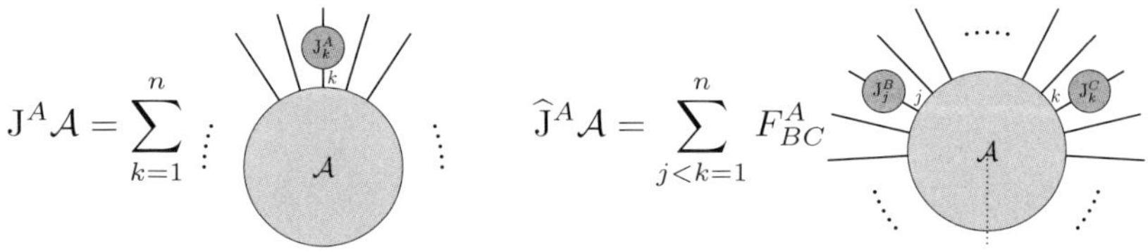

Fig. 8. Action of the free superconformal generator J and the Yangian generator $\widehat{\mathrm{J}}$ on the colour-ordered amplitude $\mathcal{A}$.

Let us demonstrate Yangian invariance of the MHV tree amplitude. The simplest of the Yangian generators is the level-one momentum generator $\widehat{\mathrm{P}}$. Due to the adjoint transformation property of the Yangian generators (12) it suffices to show invariance w.r.t. this generator in addition to superconformal invariance in order to prove complete Yangian invariance. The generator takes the explicit form

$$\widehat{\mathrm{P}}^{\beta\dot{\alpha}} = \mathrm{P}^{\beta\dot{\alpha}} \wedge \mathrm{D} + \mathrm{P}^{\delta\dot{\alpha}} \wedge \mathrm{L}^{\beta}{}_{\delta} + \mathrm{P}^{\beta\dot{\gamma}} \wedge \bar{\mathrm{L}}^{\dot{\alpha}}{}_{\dot{\gamma}} + \mathrm{Q}^{\beta c} \wedge \bar{\mathrm{Q}}^{\dot{\alpha}}{}_{c}. \quad (37)$$

This generator has one derivative which acts on the amplitude function (28). The action on delta-functions cancels straight-forwardly between the various contributions in (37). What remains is the action on the denominator terms

$$\begin{aligned} \widehat{\mathrm{P}}^{\beta\dot{\alpha}} \mathcal{A}^{\mathrm{MHV}}_n &= \sum_{j<k} \left(-\lambda^{\beta}_k \tilde{\lambda}^{\dot{\alpha}}_j \frac{\langle j, k+1\rangle}{\langle k, k+1\rangle} - \lambda^{\beta}_k \tilde{\lambda}^{\dot{\alpha}}_j \frac{\langle j, k-1\rangle}{\langle k, k-1\rangle} + \lambda^{\beta}_j \tilde{\lambda}^{\dot{\alpha}}_j \right) \mathcal{A}^{\mathrm{MHV}}_n \\ &\quad + \sum_{j<k} \left(+\lambda^{\beta}_j \tilde{\lambda}^{\dot{\alpha}}_k \frac{\langle k, j+1\rangle}{\langle j, j+1\rangle} + \lambda^{\beta}_j \tilde{\lambda}^{\dot{\alpha}}_k \frac{\langle k, j-1\rangle}{\langle j, j-1\rangle} - \lambda^{\beta}_k \tilde{\lambda}^{\dot{\alpha}}_k \right) \mathcal{A}^{\mathrm{MHV}}_n . \end{aligned} \quad (38)$$

Shifting the summation variables appropriately we can use the spinor identity

$$\lambda^{\beta}_k \langle j, k+1\rangle - \lambda^{\beta}_{k+1} \langle j, k\rangle = \lambda^{\beta}_j \langle k, k+1\rangle \quad (39)$$

to combine several terms. What remains turns out to be proportional to the overall momentum and thus vanishes proving Yangian invariance for tree MHV amplitudes

$$\widehat{\mathrm{P}}^{\beta\dot{\alpha}} \mathcal{A}_n^{\mathrm{MHV}} = \frac{\lambda_1^\delta \lambda_n^\beta + \lambda_1^\beta \lambda_n^\delta}{\langle n, 1\rangle} \varepsilon_{\delta\gamma} P_n^{\gamma\dot{\alpha}} \mathcal{A}_n^{\mathrm{MHV}} = 0. \tag{40}$$

Noting that colour-ordered amplitudes are cyclic, an important additional consideration is the cyclic behaviour of the Yangian.[60] The point is that the Yangian generators are typically not invariant under cyclic shifts: Let us compare the action on sites 1 through n with the action on sites 2 through $n+1$

$$\widehat{\mathrm{J}}_{1,n}^A = \tfrac{1}{2} F_{BC}^A \sum_{k=1}^{n} \sum_{j=1}^{k-1} \mathrm{J}_j^B \mathrm{J}_k^C, \qquad \widehat{\mathrm{J}}_{2,n+1}^A = \tfrac{1}{2} F_{BC}^A \sum_{k=2}^{n+1} \sum_{j=2}^{k-1} \mathrm{J}_j^B \mathrm{J}_k^C. \tag{41}$$

These two expressions are not equal, they differ by

$$\widehat{\mathrm{J}}_{2,n+1}^A - \widehat{\mathrm{J}}_{1,n}^A = -\tfrac{1}{2} F_{BC}^A \{\mathrm{J}_1^B, \mathrm{J}^C\} = \tfrac{1}{2} F_{BC}^A F_D^{BC} \mathrm{J}_1^D - F_{BC}^A \mathrm{J}_1^B \mathrm{J}^C. \tag{42}$$

Hence the action typically maps cyclic states to non-cyclic ones. More importantly, the action of the Yangian on periodic states is not uniquely defined; it depends on the point where the periodic chain is cut open.

For amplitudes however the situation is better because both operators on the r.h.s. are symmetries. The second term vanishes because the amplitude is invariant under conventional superconformal symmetry. The first term contains $F_{BC}^A F_D^{BC}$ which is proportional to the dual Coxeter number which equals zero for $\mathfrak{psu}(2,2|4)$.

3.4. *Collinearities and Higher Loops*

The discussion of the free superconformal symmetries in Sec. 3.2 was not entirely honest to the end that the amplitude is *not exactly* invariant under them: The special superconformal symmetries $\mathrm{S}, \bar{\mathrm{S}}, \mathrm{K}$ acting on a colour-ordered amplitude leave behind a distributional remainder supported on configurations with a pair of adjacent particles being exactly collinear, $p_k \sim p_{k+1}$. In other words, generic amplitudes are indeed annihilated by the free symmetries as discussed above, but there exist special configuration where this is not case. The extra contributions originate in the analog of (34) for $\bar{\mathrm{S}}$ from the holomorphic anomaly in the complex spinor helicity space[61,62]

$$\frac{\partial}{\partial \tilde{\lambda}^{\dot{\alpha}}} \frac{1}{\langle \lambda, \mu\rangle} = 2\pi \varepsilon_{\dot{\alpha}\dot{\gamma}} \tilde{\mu}^{\dot{\gamma}} \mathrm{sign}\big(E(\lambda)E(\mu)\big) \delta^2\big(\langle \lambda, \mu\rangle\big). \tag{43}$$

The delta-function is supported on collinear spinors λ, μ or, equivalently, when the associated momenta are collinear. Luckily these contributions can be compensated by deforming the representation of $\mathrm{S}, \bar{\mathrm{S}}, \mathrm{K}$.[63] The additional contributions map one leg of the amplitude to two or three collinear particles, cf. Fig. 9. When acting with such an operator on an amplitude with fewer legs, one can cancel the contributions from the collinear anomaly. Altogether invariance at tree level is recovered only when acting on the superposition of all amplitudes with arbitrary numbers of legs (henceforth called *the* amplitude).

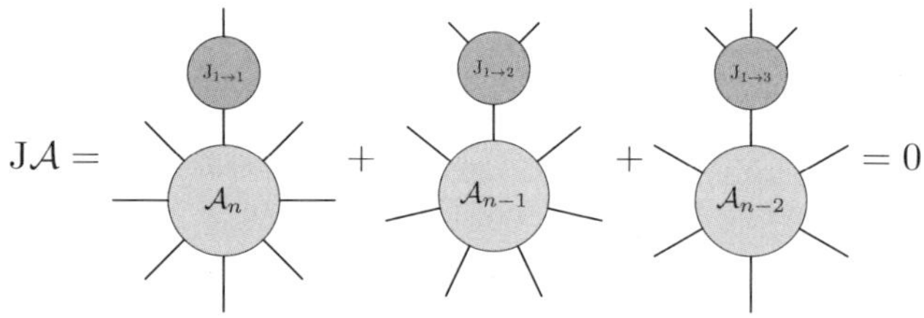

Fig. 9. Exact invariance of amplitudes under the deformed superconformal representation at tree level.

Non-invariance under the free superconformal generators turns out to be beneficial in several respects. While the free superconformal and Yangian generators only relate amplitudes with a common number of legs, the deformations introduce relations between amplitudes with different numbers of legs. Here the free and deformed generators serve two different purposes: The free superconformal generators are sensitive to the pole-like collinear singularities in the denominator of (28) through the collinear anomaly (43). The deformed generator provides the residues of the collinear singularities. Apart from those inherited from fewer-leg amplitudes, further collinear singularities are consequently prohibited by superconformal symmetry. In conclusion, conformal symmetry constrains and determines both the analytical structure and the structure of singularities of the amplitude. Together with Yangian symmetry it appears that the tree amplitude may be completely determined through symmetry arguments alone! Although there exist many ways to construct tree amplitudes conveniently, unique determination by symmetry is an interesting prospect because uniqueness is automatically inherited to all perturbative orders.[63] Nevertheless one has to bear in mind that this also requires understanding how the representation is deformed at loop level.

Concerning the latter issues, it appears that the deformations at loop level depend to a large extent on the deformation at tree level or a suitable iteration thereof.[64,65] In addition there are deformations due to infra-red divergences at loop level affecting all of the non-manifest symmetries. The divergent contributions to the one-loop planar amplitude $\mathcal{A}_n^{(1)}$ are determined by the tree amplitude $\mathcal{A}_n^{(0)}$

$$\mathcal{A}_n^{(1)} = \hat{Z}^{(1)} A_n^{(0)} + \tilde{\mathcal{A}}_n^{(1)} \quad \text{with} \quad \hat{Z}^{(1)} = -\sum_{j=1}^{n} \frac{c_\epsilon}{\epsilon^2}\left(\frac{s_{j,j+1}}{-\mu^2}\right)^{-\epsilon}, \tag{44}$$

where $s_{j,k} = (p_j + p_k)^2$ and where $\tilde{\mathcal{A}}_n^{(1)}$ is finite. The anomaly of the dilatation operator due to the presence of the scale μ is the clearest. It can be absorbed by a simple one-loop deformation $\mathrm{D}^{(1)}$ to the free dilatation generator $\mathrm{D}^{(0)}$[65]

$$\mathrm{D}^{(1)} = -\left[\mathrm{D}^{(0)}, \hat{Z}^{(1)}\right] = -2\sum_{j=1}^{n} \frac{c_\epsilon}{\epsilon}\left(\frac{s_{j,j+1}}{-\mu^2}\right)^{-\epsilon}, \tag{45}$$

so that $\mathrm{D}^{(0)}\mathcal{A}_n^{(1)} + \mathrm{D}^{(1)}\mathcal{A}_n^{(0)} = 0$ because the finite contribution $\tilde{\mathcal{A}}_n^{(1)}$ is scale invariant.

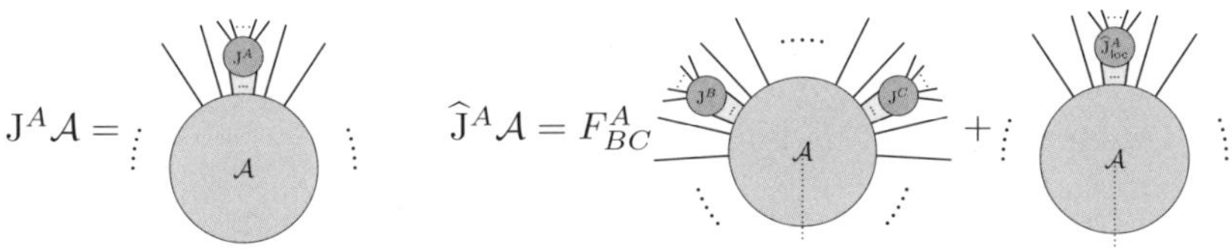

Fig. 10. Perturbative action of the superconformal generator J and the Yangian generator $\widehat{\mathrm{J}}$ on the colour-ordered amplitude $\mathcal{A}$.

The situation for the Yangian momentum generator $\widehat{\mathrm{P}}$ is similar. The anomaly due to the IR divergencies can be absorbed by a deformation analogous to the dilatation generator, but here the finite remainder is anomalous as well[57,58,66–69]

$$(\widehat{\mathrm{P}}^{(0)})^{\beta\dot{\alpha}}\,\tilde{\mathcal{A}}_n^{(1)} = 2\sum_{j=1}^{n} p_j^{\beta\dot{\alpha}} \log\left(\frac{s_{j,j+1}}{s_{j-1,j}}\right)\mathcal{A}_n^{(0)}. \tag{46}$$

Note that this anomaly depends only on neighbouring legs. Hence the total deformation of the Yangian momentum generator reads

$$\widehat{\mathrm{P}}^{(1)} = -\left[\widehat{\mathrm{P}}^{(0)}, \hat{Z}^{(1)}\right] + \widehat{\mathrm{P}}^{(1)}_{\mathrm{loc}} \qquad \text{with}$$

$$(\widehat{\mathrm{P}}^{(1)}_{\mathrm{loc}})^{\beta\dot{\alpha}} = 2\sum_{j=1}^{n}(p_j^{\beta\dot{\alpha}} - p_{j+1}^{\beta\dot{\alpha}})\,\frac{c_\epsilon}{\epsilon}\left(\frac{s_{j,j+1}}{-\mu^2}\right)^{-\epsilon}. \tag{47}$$

It follows the general structure of perturbative Yangian generators: The commutator generates the bi-local combinations of the deformed generators. In this case the dilatation generator D is the only deformed generator among the ones contributing to $\widehat{\mathrm{P}}$ in (37) because the super-Poincaré generators $\mathrm{P}, \mathrm{Q}, \bar{\mathrm{Q}}$ are manifest symmetries. In particular the deformation reads simply, see Fig. 10

$$\widehat{\mathrm{P}}^{(1)} = \mathrm{P}^{(0)} \wedge \mathrm{D}^{(1)} + \widehat{\mathrm{P}}^{(1)}_{\mathrm{loc}}. \tag{48}$$

The local contribution can be attributed to the anomaly of the finite remainder.

At higher loops we expect this general picture in Fig. 10 to remain valid. There are reasons to believe that the deformation of the dilatation D and Yangian momentum $\widehat{\mathrm{P}}$ generator at higher loops remains reasonably simple. For the other superconformal and Yangian generators, however, the deformation is already substantially more involved even at one loop.[65] And even though the deformations are known, it remains to be understood how the superconformal and Yangian algebra closes precisely.

4. Comparison and Summary

The attentive reader will have noticed that the discussions in Sec. 2 and in Sec. 3 were analogous to a large extent.

4.1. *Analogies*

Let us first concentrate on the representation of superconformal symmetry. The free representations on fields (9) and on external particles (31) are equivalent if one identifies oscillators with spinor-helicity variables

$$\mathrm{a} \sim \lambda, \quad \mathrm{b} \sim \tilde{\lambda}, \quad \mathrm{d} \sim \bar{\eta}, \qquad \mathrm{a}^\dagger \sim \frac{\partial}{\partial \lambda}, \quad \mathrm{b}^\dagger \sim \frac{\partial}{\partial \tilde{\lambda}}, \quad \mathrm{d}^\dagger \sim \frac{\partial}{\partial \bar{\eta}}. \tag{49}$$

It is clear that their algebras coincide and hence the derived representations are equivalent.

This is not surprising because both describe free on-shell fields of $\mathcal{N} = 4$ SYM: The spinor helicity superspace is designed to describe a field excitation with definite on-shell momentum. Conversely, a finite excitation of the supersymmetric oscillator describes a component of the fields expanded around a specific point in spacetime. Here, the free equations of motion are imposed through vanishing of certain components. A Fourier transformation translates between the two pictures

$$\begin{aligned}\mathcal{W}(x) &\sim \int d^4\lambda\, e^{ix\cdot p(\lambda)}\Omega(\lambda),\\ \Omega(\lambda) &\sim \int d^3x\, e^{-ix\cdot p(\lambda)}\big(\mathcal{W}(x,0) - iE(\lambda)^{-1}\dot{\mathcal{W}}(x,0)\big).\end{aligned} \tag{50}$$

For the reverse transformation it suffices to use a time slice of $\mathcal{W}$ at $t = 0$. It is however clear that for full equivalence between the two pictures one has to rely on distributions. Consequently the two representations are only equivalent in a physicist's sense or under additional assumptions.[g]

Once the equivalence of the free superconformal representations is established, it becomes straight-forward to lift the equivalence to the Yangian algebra. Indeed, the infinite-dimensional algebra appears to determine uniquely relevant structures such as the spin chain Hamiltonian as well as the scattering amplitude. There is however one crucial difference between the application of Yangians to local operators vs. scattering amplitudes. For the former it merely acts as a useful algebraic structure, cf. (16), while for the latter it is a true symmetry. This point will be discussed in more detail in the next subsection. A related issue is that Yangian symmetry is typically incompatible with cyclic symmetry. Only for the scattering amplitude it respects it due to superconformal invariance and due to vanishing dual Coxeter number, cf. (42).

We have furthermore seen that the structure of the perturbative superconformal and Yangian representation is analogous in both pictures, cf. the pairs (??), Fig. 3,8, Fig. 5,9 and Fig. 6,10. The superconformal deformations act on several adjacent fields or legs in the trace of the local operator or colour-ordered amplitude. The deformed Yangian representation is constructed as a bi-local combination of deformed superconformal representations plus local terms which can be understood as a short-distance (along the trace) regularisation of the bi-local terms. One notable difference concerns the manifest symmetries which are not deformed by radiative corrections. For local operators only the Lorentz and internal symmetries $\mathrm{L}, \bar{\mathrm{L}}, \mathrm{R}$ are manifest, while for the scattering amplitude the full super-Poincaré algebra including the (super) momentum generators $\mathrm{Q}, \bar{\mathrm{Q}}, \mathrm{P}$ are undeformed.[h]

[g]It might be interesting to Fourier transform some of the structures from one picture between position and momentum space.

[h]In fact, trying to construct a representation for local operators with manifest super-Poincaré symmetry implies vanishing anomalous dimensions.[22]

4.2. *Large-N_c Topology*

Integrability and the Yangian algebra is tightly related to the 't Hooft planar limit.[10] Let us therefore consider the large-N_c expansion. Local operators and particle configurations can be viewed as closed one-dimensional contours such that each trace corresponds to one connected component; let us refer to them as *states*, see Fig. 11. Quantum correlation functions then span two-dimensional surfaces between the contours. The topology of these surfaces determines the suppression in powers of $1/N_c$. For example, a single-trace scattering amplitude as well as two- and three-point correlators of local operators are displayed in Fig. 11. This is in line with the picture from the AdS/CFT dual string theory on $AdS_5 \times S^5$, where the surface is the string world sheet and its boundaries (or punctures) correspond to states.

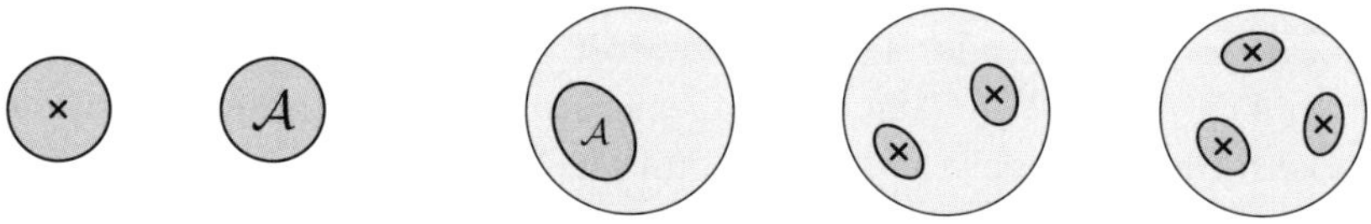

Fig. 11. Large-N_c topology representation of a single-trace local operator and a single-trace particle configuration (left two); a cross corresponds to a puncture of the surface serving as a source for charges. Quantum correlation functions insert a surface ending on the traces: planar scattering amplitude (middle) as well as two- and three-point correlators (right two); not shown are non-planar contributions where the surfaces have additional handles.

Algebra generators are represented by Wilson loops of the Lax family of flat connections on the world sheet.[70] Thus we should represent a generator by a closed loop on the surface. Due to flatness, the contour can be deformed smoothly. The action on a state corresponds to winding the loop around the trace, cf. Fig. 12.

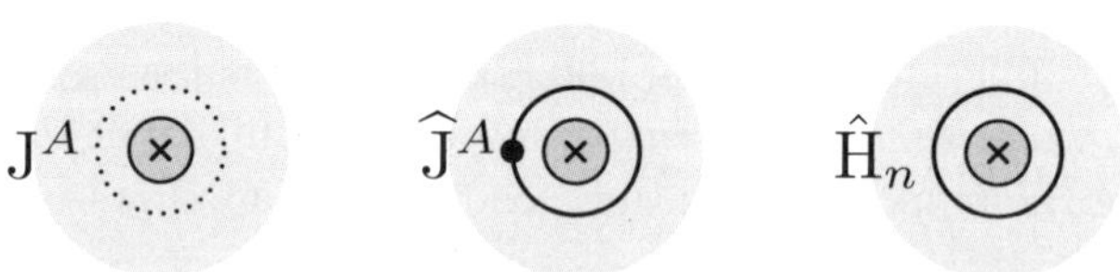

Fig. 12. Action of generators on a single-trace state: Superconformal J^A, Yangian $\widehat{\mathrm{J}}^A$ and local integrable Hamiltonians $\hat{\mathrm{H}}_n$. Fat lines represent Wilson lines; open loops have a marked base point, closed loops do not. Dotted lines are plain integrals which can be broken up.

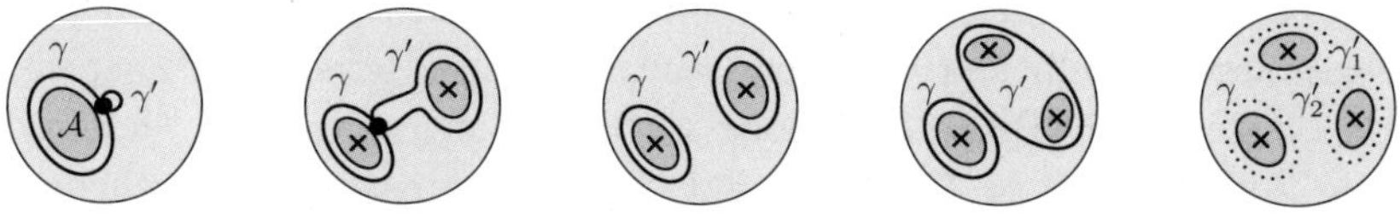

Fig. 13. Invariance conditions can be understood through deforming the contours $\gamma \to \gamma'$ associated to generators on the surface. Yangian invariance follows from shifting the contour around the surface and shrinking it to a point (leftmost). For a two-point function Yangian invariance is broken because the marked base point cannot be moved (middle left); the integrable charges are nevertheless conserved (middle). For a three-point function even the integrable charges are not conserved (middle right), but the conformal ones are (rightmost).

Now we can consider invariance conditions for particular objects, see Fig. 13: In Sec. 3 we have seen that (tree-level) planar single-trace amplitudes are invariant under Yangian symmetry. In our picture we should wind an open Wilson loop around the trace of the particle configuration. We can unwind the Wilson loop on the disc ending on the trace without having to move the base point. The Wilson loop then shrinks to a point implying invariance. The unwinding would not be possible in the non-planar case of a disc with handles.

For a correlation function of two traces, e.g. a two-point function of single-trace local operators, the Yangian action on the two states is not equivalent because the base point is different. A closed Wilson loop can, however, be deformed from one trace to the other in the planar case of an annulus connecting the traces. This implies the integrability of the problem of planar anomalous dimensions.[i]

Finally, a correlation function of three traces does not have conserved charges. Merely superconformal symmetry survives, because for these generators the loop is an abelian contour integral which can be broken up into two piccc3.

4.3. *Summary and Outlook*

In this paper we have reviewed Yangian symmetry which serves as an algebraic foundation of integrability in planar $\mathcal{N} = 4$ maximally supersymmetric gauge theory.

We have seen that the Yangian is capable of uniquely determining certain physical observables by purely algebraic means. Even more impor-

[i]For planar double-trace scattering amplitudes integrability implies the existence of a tower of conserved charges. It would be interesting to confirm them.

tantly, there exist methods to exploit the uniqueness and obtain these observables very efficiently. Among them are the Bethe ansatz, spectral curves, asymptotic Bethe equations with Lüscher corrections, thermodynamic Bethe ansatz or Y-system and Graßmannians. Applying them one can avoid highly complicated calculations in quantum field theory and arrive at the correct final result much faster.

Although these methods are already being applied reliably, it still remains to be understood *why* they work. Why is planar $\mathcal{N} = 4$ SYM governed by a Yangian algebra (technically as well as semantically)? How does it lead to the above methods? How is the algebra defined in the first place? As we have seen, at leading perturbative order the Yangian follows precisely from the established framework of quantum algebra. At higher loops the Yangian representation gets deformed, and some of the well-known rules have to be dropped in favour of new ones yet to be established.

We have discussed two subjects where the Yangian algebra makes a prominent appearance: anomalous dimensions of local operators and the spacetime scattering matrix. A third subject which was not discussed here is the worldsheet scattering matrix. The Yangian relevant to that problem is not based on $\mathfrak{psu}(2,2|4)$ but only on the subalgebra $\mathfrak{psu}(2|2)$. Although this is an exception case, several works have demonstrated that it can apparently be described by conventional quantum algebra methods, see the review.[71] Complete understanding of the smaller Yangian may eventually lead to clues for the full perturbative Yangian for $\mathcal{N} = 4$ SYM.

References

1. L. N. Lipatov, *Sov. J. Nucl. Phys.* **23**, 338 (1976).
2. E. A. Kuraev, L. N. Lipatov and V. S. Fadin, *Sov. Phys. JETP* **44**, 443 (1976).
3. E. A. Kuraev, L. N. Lipatov and V. S. Fadin, *Sov. Phys. JETP* **45**, 199 (1977).
4. V. S. Fadin, E. A. Kuraev and L. N. Lipatov, *Phys. Lett.* **B60**, 50 (1975).
5. I. I. Balitsky and L. N. Lipatov, *Sov. J. Nucl. Phys.* **28**, 822 (1978).
6. L. N. Lipatov (1993).
7. L. N. Lipatov, *JETP Lett.* **59**, 596 (1994).
8. L. D. Faddeev and G. P. Korchemsky, *Phys. Lett.* **B342**, 311 (1995).
9. L. N. Lipatov, *J. Phys.* **A42**, p. 304020 (2009).
10. G. 't Hooft, *Nucl. Phys.* **B72**, p. 461 (1974).
11. V. N. Gribov and L. N. Lipatov, *Sov. J. Nucl. Phys.* **15**, 438 (1972).
12. G. Altarelli and G. Parisi, *Nucl. Phys.* **B126**, p. 298 (1977).
13. Yu. L. Dokshitzer, *Sov. Phys. JETP* **46**, 641 (1977).
14. L. N. Lipatov (1998).

15. V. M. Braun, S. E. Derkachov and A. N. Manashov, *Phys. Rev. Lett.* **81**, 2020 (1998).
16. L. N. Lipatov, *Nucl. Phys.* **B548**, 328 (1999).
17. V. M. Braun, S. E. Derkachov, G. P. Korchemsky and A. N. Manashov, *Nucl. Phys.* **B553**, 355 (1999).
18. A. V. Belitsky, *Nucl. Phys.* **B558**, 259 (1999).
19. J. A. Minahan and K. Zarembo, *JHEP* **0303**, p. 013 (2003).
20. N. Beisert, C. Kristjansen and M. Staudacher, *Nucl. Phys.* **B664**, 131 (2003).
21. N. Beisert and M. Staudacher, *Nucl. Phys.* **B670**, 439 (2003).
22. N. Beisert, *Nucl. Phys.* **B682**, 487 (2004).
23. M. Bianchi, S. Kovacs, G. Rossi and Y. S. Stanev, *JHEP* **9908**, p. 020 (1999).
24. M. Günaydin and N. Marcus, *Class. Quant. Grav.* **2**, p. L11 (1985).
25. N. Beisert, *Nucl. Phys.* **B676**, p. 3 (2004).
26. A. V. Kotikov and L. N. Lipatov, *Nucl. Phys.* **B582**, 19 (2000).
27. A. V. Kotikov and L. N. Lipatov (2001).
28. F. A. Dolan and H. Osborn, *Nucl. Phys.* **B629**, 3 (2002).
29. A. V. Belitsky, V. M. Braun, A. S. Gorsky and G. P. Korchemsky, *Int. J. Mod. Phys.* **A19**, 4715 (2004).
30. L. D. Faddeev (1997).
31. N. Beisert, J. A. Minahan, M. Staudacher and K. Zarembo, *JHEP* **09**, p. 010 (2003).
32. N. Beisert, S. Frolov, M. Staudacher and A. A. Tseytlin, *JHEP* **10**, p. 037 (2003).
33. D. Berenstein, J. M. Maldacena and H. Nastase, *JHEP* **0204**, p. 013 (2002).
34. S. S. Gubser, I. R. Klebanov and A. M. Polyakov, *Nucl. Phys.* **B636**, 99 (2002).
35. S. Frolov and A. A. Tseytlin, *JHEP* **0206**, p. 007 (2002).
36. N. Beisert and A. A. Tseytlin, *Phys. Lett.* **B629**, 102 (2005).
37. N. Beisert, *Phys. Rept.* **405**, 1 (2004).
38. J. Plefka, *Living. Rev. Relativity* **8**, p. 9 (2005).
39. G. Arutyunov and S. Frolov, *J. Phys.* **A42**, p. 254003 (2009).
40. V. G. Drinfel'd, *Sov. Math. Dokl.* **32**, 254 (1985).
41. L. Dolan, C. R. Nappi and E. Witten (2004).
42. L. Dolan, C. R. Nappi and E. Witten, *JHEP* **10**, p. 017 (2003).
43. V. K. Dobrev and V. B. Petkova, *Phys. Lett.* **B162**, 127 (1985).
44. B. I. Zwiebel, *JHEP* **02**, p. 055 (2006).
45. D. Serban and M. Staudacher, *JHEP* **0406**, p. 001 (2004).
46. B. I. Zwiebel, *J. Phys.* **A40**, 1141 (2007).
47. N. Beisert and D. Erkal, *J. Stat. Mech.* **0803**, p. P03001 (2008).
48. T. Bargheer, N. Beisert and F. Loebbert, *J. Stat. Mech.* **0811**, p. L11001 (2008).
49. T. Bargheer, N. Beisert and F. Loebbert, *J. Phys.* **A42**, p. 285205 (2009).
50. F. A. Berends and W. T. Giele, *Nucl. Phys.* **B306**, p. 759 (1988).
51. V. P. Nair, *Phys. Lett.* **B214**, p. 215 (1988).
52. S. J. Parke and T. R. Taylor, *Phys. Rev. Lett.* **56**, p. 2459 (1986).
53. E. Witten, *Commun. Math. Phys.* **252**, 189 (2004).

54. J. M. Drummond, J. Henn, V. A. Smirnov and E. Sokatchev, *JHEP* **01**, p. 064 (2007).
55. Z. Bern, M. Czakon, L. J. Dixon, D. A. Kosower and V. A. Smirnov, *Phys. Rev.* **D75**, p. 085010 (2007).
56. Z. Bern, J. J. M. Carrasco, H. Johansson and D. A. Kosower, *Phys. Rev.* **D76**, p. 125020 (2007).
57. J. M. Drummond, J. Henn, G. P. Korchemsky and E. Sokatchev, *Nucl. Phys.* **B828**, 317 (2010).
58. A. Brandhuber, P. Heslop and G. Travaglini, *Phys. Rev.* **D78**, p. 125005 (2008).
59. J. M. Drummond and J. M. Henn, *JHEP* **04**, p. 018 (2009).
60. J. M. Drummond, J. M. Henn and J. Plefka, *JHEP* **05**, p. 046 (2009).
61. F. Cachazo, P. Svrcek and E. Witten, *JHEP* **10**, p. 077 (2004).
62. F. Cachazo (2004).
63. T. Bargheer, N. Beisert, W. Galleas, F. Loebbert and T. McLoughlin, *JHEP* **11**, p. 056 (2009).
64. A. Sever and P. Vieira (2009).
65. N. Beisert, J. Henn, T. McLoughlin and J. Plefka, *JHEP* **04**, p. 085 (2010).
66. J. M. Drummond, J. Henn, G. P. Korchemsky and E. Sokatchev (2008).
67. A. Brandhuber, P. Heslop and G. Travaglini, *JHEP* **08**, p. 095 (2009).
68. H. Elvang, D. Z. Freedman and M. Kiermaier, *JHEP* **03**, p. 075 (2010).
69. A. Brandhuber, P. Heslop and G. Travaglini, *JHEP* **10**, p. 063 (2009).
70. I. Bena, J. Polchinski and R. Roiban, *Phys. Rev.* **D69**, p. 046002 (2004).
71. A. Torrielli (2010), review of Yangians in AdS/CFT, to appear.

THE "RELATIVISTIC" MUG

L. B. OKUN

Institute of Theoretical and Experimental Physics
ul. B. Cheremushkinskaya 25
Moscow 117218, Russia
okun@itep.ru

This note is an attempt to explain in simple words why the famous relation $E = mc^2$ misrepresents the essence of Einstein's relativity theory. The note is addressed to high-school teachers, and a part of it – to those university professors who permit themselves to say that the mass of a body increases with its velocity or momentum and thus mislead the teachers and their students.

1. Introduction

The moral health of the modern society and its material well-being are unthinkable without high status of science in the country. This status in a certain degree depends on how adequate is the image of science in the mirror of mass culture. For more than twenty years I have been collecting artifacts of mass culture (from postcards and T-shirts to popular articles and books) sporting "the famous Einstein's formula".

Recently my friends added to my collection a Relativity Floxy Noxy mug. You can google (typing these four words in the search line of your computer) and see it:

In a certain sense it contains the quintessence of my collection presenting the main popular science clichès and misconceptions. As they are quite often repeated in newspapers and textbooks, I decided to reproduce the text on the mug and to explain briefly what is wrong with it. I believe that it may be useful to many people.

2. The text on the mug

There are three columns of text on the mug – to the right of the handle (**1**), to the left of the handle (**2**), opposite the handle (**3**):

Fig. 1. The Relativity Mug

Column 1.

In 1905 at the age of 26, Einstein
proposed the Special Theory
of Relativity, using the equation:
$E = mc^2$
where
E=energy,
m=mass,
c=the speed of light.
Special relativity expresses the
concept that matter and energy
are really different forms of the
same thing. Any mass has
an associated energy and vice versa.

Column 2.

Albert Einstein's
SPECIAL
Theory of
RELATIVITY
In the 1850's it was calculated that light
traveled at a fixed speed of 670 million mph.
However, whatever speed we traveled at, we
would never catch up with the speed of light.
Einstein proposed that if the speed of

light is always fixed, something else must give way, i. e. mass must change. An object must get heavier as it approaches the speed of light. He concluded that energy and mass must be interrelated.

Column 3.

His formula suggested that tiny amounts of mass can be converted into huge amounts of energy...
... which revealed the secret of how stars shine and
unlocked the key
to atomic energy.

3. My clarifications and comments

Column 1. As is well known, Maxwell in 1860-70's united optics with electricity and magnetism by establishing equations describing not only static fields, but also alternating electromagnetic fields propagating in a vacuum with velocity of light. Several physicists in the 1880-90's after realizing that Maxwell equations are incompatible with equations of Newton mechanics have attempted to preserve the latter at velocities comparable to that of light by postulating that all new physics can be reduced to the increase of mass of a body with its velocity. These attempts were continued in the 20th century; they are briefly described in the articles.[1,2] However it became clear rather soon that some of the Newton equations cannot be preserved, for instance the famous equation $\mathbf{F} = m\mathbf{a}$, where $\mathbf{F}$ is force and $\mathbf{a}$ is acceleration. It turned out that the equations themselves should be changed in such a way that mass would not depend on velocity, but there would exist an important new link between mass and energy. Let us note that relation between force and momentum remained the same: $\mathbf{F} = d\mathbf{p}/dt$, but, as we will see below, the relation between momentum and velocity $\mathbf{p} = m\mathbf{v}$ has changed. This had serious impact on the language and philosophy of physics.

In summer of 1905 Einstein published a detailed article[3] in which he presented his theory which later got the name Einstein's theory of relativity. This theory extended to electromagnetic phenomena the principle of relativity formulated by Galileo and Newton. According to it, it is impossible by any experiment to find out whether a closed space (say, a cabin of a ship) is at rest or in a uniform and rectilinear motion. Soon the theory was extended to the newly discovered nuclear phenomena and got the name Special Relativity (SR). This theory describes the motion and interaction of fast particles whose velocities are comparable with the speed of light. Such particles are called relativistic. (In 1915 Einstein proposed General Relativity (GR) to describe gravity. But in this note we will not consider it.)

In autumn of 1905 Einstein published a short note[4] in which he stated that in the framework of his theory the mass of a body is a measure of its energy content. The total energy E of a free body is equal, according to the theory of relativity, to the sum of its kinetic energy E_K (of the motion as a whole) and its energy at rest – rest energy E_0:

$$E = E_K + E_0. \tag{1}$$

Of course, the concept of a free (isolated from any external influence) body is an idealization. But idealization (abstraction) lies at the basis of scientific method and is extremely fruitful.

The realization that any body at rest possesses energy was the greatest discovery of the 20th century. The amount of this energy is given by Einstein's equation:

$$E_0 = mc^2, \tag{2}$$

where m is the mass of the body and c is the speed of light. (It was exactly in this form that Einstein had written equation (44) in 1921 in his lectures "The meaning of relativity",[5] though the notion of the rest energy E_0 appeared already in the note.[4])

The kinetic energy of ordinary bodies is given by the well known equation of Newton's mechanics $E_K = mv^2/2$. As the velocity v of an ordinary body is much less than c, the rest energy of a body is huge in comparison with its kinetic energy. But in the ordinary life the rest energy does not manifest itself. Einstein pointed out that part of it is liberated in the radioactive decays.

Unfortunately, many famous physicists during the last century have formulated the Einstein equation in a "simplified form" by omitting the index zero:

$$E = mc^2, \tag{3}$$

and treating this relation as increase of mass not only with energy but also with velocity and momentum of the body.

In 1948 Einstein warned Barnett – the author of the book "Universe of Dr. Einstein" – against using the concept of mass depending on velocity. (A copy of this handwritten letter is reproduced in Ref. 1.) But sometimes, especially in his popular writings, he himself did not care about the index zero. This semantic kink was caused by the clash of two languages – the old non-relativistic and the new, consistently relativistic one.

Column 2. The assertion that the speed of light is always fixed at a value of 670 million mph is correct, but the dating (1850's) is not quite correct. That the speed of light is finite (not infinite), was established in 1676 by Römer who deduced this from observations of Jupiter's satellite. It followed from them that the speed is around 200 000 km/s. The first and more precise measurements of c on the Earth were performed by Fizeau in 1849. But the fact that the speed about 300 000 km/s is fixed and does not depend on the velocities of the source and the observer, was discovered in 1887 by Michelson and Morely.

The statement that energy and mass are interrelated is correct: $E_0 = mc^2$, while that the mass changes with velocity is definitely wrong. In the theory of relativity (unlike the mechanics of Newton) the measure of inertia is not mass m but the total energy E of the body. The momentum $\mathbf{p}$ of a body is connected with its velocity $\mathbf{v}$ not by the Newton's relation $\mathbf{p} = m\mathbf{v}$ but by the relation

$$\mathbf{p} = (E/c^2)\mathbf{v}. \tag{4}$$

As a result it is the more difficult to change the momentum of a body, the higher its total energy E. And $E/c^2 = m$ only at zero momentum, when the total energy equals the rest energy E_0.

One can feel more deeply that the measure of inertia is energy by considering the example of the Large Electron-Positron collider LEP which operated at CERN during the last decade of the 20th century. Particles with energy 50 GeV were kept in its 27 km ring tunnel by a rather weak field of iron magnets. (Without this field particles would fly along a straight line.) Exactly the same field would maintain the circulation of protons with the same momentum (and almost the same energy), though the mass of the proton is 2000 times larger than the mass of the electron. In the year 2010 the Large Hadron Collider LHC started to operate in the same tunnel. To circulate protons with energy 3500 GeV, the magnetic field of superconducting magnets in it is 70 times stronger.

Thus, the measure of inertia of a particle is its total energy.

Column 3. Here everything is correct if one uses the equation $E_0 = mc^2$ and takes into account that in nuclear reactions in the stars, in the Sun and on the Earth a part of the rest energy of the particles which are burned is transformed into kinetic energy of the products of burning. The same is valid for any process of burning.

4. Four dimensions of the world

Now I would like to address a few words to those who are more or less familiar with the concept of the four-dimensional world (4-world) introduced in the relativity theory in 1908 by Minkowski.[6] In the 4-world the time coordinate ct of an event and its position coordinates $\mathbf{r}$ form a 4-vector. Similarly the energy E of a free (isolated) body (more precisely, E/c^2) and three components of its momentum $\mathbf{p}$ (more precisely, $\mathbf{p}/c$) form four components of the pseudo-euclidean 4-vector. The scalar length of this 4-vector is given by the mass of the body m according to the equation

$$m^2 = E^2c^{-4} - \mathbf{p}^2c^{-2}. \tag{5}$$

(The words "pseudo euclidean" indicate that the square of the length of the 4-vector is equal not to the sum but to the difference of squares of its E- and $\mathbf{p}$-components.)

Taylor and Wheeler in the book[7] put energy and momentum on the orthogonal axes, then on the hypotenuse they depict mass by a short and thick segment. But it is possible to present equation (5) simply as a right triangle if one rewrites it in the form $E^2 = m^2c^4 + \mathbf{p}^2c^2$ and puts mass and momentum on the orthogonal axes (see article[8]). Then energy is the hypotenuse, while mass and momentum are the other two legs. For any value of momentum the kinetic energy is

$$E_K = \sqrt{m^2c^4 + \mathbf{p}^2c^2} - mc^2. \tag{6}$$

The main equation (5) of relativity theory has been tested in thousands of experiments with the accuracy of up to ten digits. For a massive body whose momentum is zero it implies $E_0 = mc^2$. For a non-vanishing momentum one can rewrite it as $(Ec^{-2} - m)(Ec^{-2} + m) = \mathbf{p}^2c^{-2}$ and at $E_K \ll E_0$ derive from it the non-relativistic expression for kinetic energy $E_K = \mathbf{p}^2/2m$ without developing the square root. Similarly, for relativistic particles $E - |\mathbf{p}|c = mc^2/2E$. (This equation is essential for neutrino oscillations.) It follows also from equation (5) and from the formula (4) for velocity $\mathbf{v} = \mathbf{p}c^2/E$ that for a massless particle of light – the photon – the speed is always equal to c.

The special theory of relativity is impeccable. One cannot say the same about its image in the mass culture.

Unfortunately, the sudden illness and death of Minkowski did not allow him to persuade his contemporaries to switch to the language of the four-dimensional world, and they continued futile attempts to explain the meaning of relativity theory in terms of Newton's three-dimensional mechanics. Though Einstein used the four-dimensional mathematical apparatus in deriving the equations of his general theory of relativity for gravitational interaction, I failed to find the equation $E^2c^{-4} - \mathbf{p}^2c^{-2} = m^2$ on the pages of his writings.

It appeared first in the articles of Klein,[9] Fock,[10] Gordon[11] (1926) and especially in the works of Dirac[12] (1930) in which relativistic quantum mechanics was constructed (as is well known, Einstein, a co-founder of the concept of quantum, did not accept quantum mechanics). The equation appeared in the framework of not quantum but classical field theory much later, in the book "The classical theory of fields" by Landau and Lifshitz in 1941 (in Russian).[13]

Four-dimensional description is equally good for massive and massless particles of matter. It shows that mass and matter are not the same thing, that energy and momentum are the measures of all processes and motions in nature. As for the mass of the particles, it becomes non-essential for processes at high energies $E \gg mc^2$.

5. The speed of light as the unit of velocity

The correct equations must be correct regardless of the choice of units. The existence of the universal maximal velocity c allows one to express any velocity v in units of c as a dimensionless number $\beta = v/c$. It is evident that in these units $\beta = 1$ for $v = c$. As a result one can get rid of c in the equations of relativity theory by rewriting equations (2), (4), (6) in the form

$$E_0 = m, \qquad m^2 = E^2 - \mathbf{p}^2, \qquad \mathbf{v} = \mathbf{p}/E. \tag{7}$$

As for equation (3) $E = mc^2$, it is reduced to $E = m$, which evidently contradicts equation (1) $E = E_K + E_0 = E_K + m$ and hence is wrong.

6. Conclusion

Volodya Gribov, whose attitude concerning $E = mc^2$ was the same as mine, gave me a friendly advice in the 1980s not to struggle against the famous and false equation because this fight just cannot be won. It was with a feeling of permanent defeat that I was writing the text above for the forthcoming volume "Gribov-80" in the summer 2010, amidst the unprecedented heat and smog in Moscow, which could but adversely affect the quality of the text. In autumn, a few weeks ago, Julia Nyiri reminded me that this text is a continuation of my contribution to the volume "Gribov-75"[14] in which I compared the equation $E = mc^2$ with a virus. Indeed, the concept of relativistic mass hidden in the equation $E = mc^2$ is a semantic virus similar to computer viruses. People infected by this virus (they often call themselves relativists) believe that Relativistic Mass is the main portal to Relativity Theory because mass is the measure of inertia. They ignore the fact that mass is the measure of inertia only for very slowly moving bodies and particles for which the rest energy E_0 is much larger than the kinetic energy E_K. When velocities are not very low, mass is only an approximate measure of inertia. For fast particles for which $E_K \gg E_0$ (photons, neutrinos, protons in LHC) the measure of inertia is the total energy E.

It is well known that formulas in physics are a continuation of the ordinary language: equations are encoded sentences, while mathematical symbols in these equations are encoded words or terms. To prevent confusion, each symbol must be unambiguously connected with a corresponding term. Is it possible to introduce in Special Relativity the concept of relativistic mass? Yes, it is possible: $m_r = E/c^2$, though it will be just another symbol for energy because c is a universal constant. Is it possible then to introduce the term relativistic rest mass $m_{r0} = m$? Yes, it is possible. Of course, it is equally possible to introduce both terms and both symbols. Although they are not needed in Special Relativity as it is a complete, self consistent theory without them, their introduction is possible. What is not good is to denote the relativistic rest mass m_{r0} by m_0 and then call it simply rest mass, because this presumes that mass m depends on velocity in Special Relativity (SR), while we all know that in SR m is Lorentz invariant: it is the same at rest and in motion, and hence, there is no sense in supplying it with indices.

But the real trouble begins when m_r is called the mass, is denoted by m and at the same time the ordinary Newtonian mass m is renamed into m_0. Then the mixing of two languages ("French and Nizhegorodsky") mutilates the beautiful theory, leads to unbelievable confusion and thwarts its understanding. To top it all, some " philosophers-relativists" then allege that the mechanics of Newton is not a limiting case of mechanics of Einstein, and that these two theories are incommensurate.

As a result of the reverse action of mass culture on the scientific culture, many chapters in the best text-book on physics of the 20th century published in 1960s – "The Feynman Lectures on Physics"[15] – repeat the statement that mass changes with velocity. The little book "What is relativity?"[16] by Landau and Rumer also claims that mass increases with velocity. (The book was written in the 1930s, before the arrest of both authors, and published in the 1950s after Rumer was released from exile. (Landau remained in jail for one year.)) "The classical theory of fields" by Landau and Lifshitz published in 1940s was the first text-book in the world in which mass was velocity-independent. But even in the latest edition of this book, in the 21st century, the concept of rest energy E_0 is absent and the Einstein's formula is given in its famous but false form $E = mc^2$. Indeed, nobody is perfect.

It is impossible in this short note to refer to the articles and books of the creators of relativity theory, but it is easy to find them by clicking the hyper references,[1,2,8,14,17,18] listed below. The seventh hyper reference[19] contains slides of the talk which explained why the teaching of physics must be based on two fundamental constants of nature: c and $\hbar$. By operating with these two constants I plan to present the foundations of physics in a little book (100 pages) "The ABC of Physics".

Acknowledgments

I am grateful to Erica Gulyaeva, Marek Karliner, Elya and Vitaly Kisin, Olga Milyaeva, Boris Okun and Zurab Silagadze whose remarks helped me write this note.

The work is supported by grant of the President of RF NSh-4172.2010.2

References

1. L.B. Okun, *The concept of mass*, *Physics Today*, June 1989, 31-36. `http://www.itep.ru/theor/persons/lab180/okun/em_3.pdf`
2. L.B. Okun, *The Einstein formula: $E_0 = mc^2$. "Isn't the Lord laughing"?*, *Physics – Uspekhii* **51**, (5), 513 - 527 (2008). `http://www.itep.ru/theor/persons/lab180/okun/em_29.pdf`
3. A. Einstein, *Zur Electrodynamik bewegter Körper*, *Ann. Phys.* **17**, 891-921 (1905).
4. A. Einstein, *Ist die Tragheit eines Körpers von seinem Energieinhalt abhänging?*, *Ann. Phys.* **18**, 639-641 (1905).
5. A. Einstein, *Four lectures on the theory of relativity, held at Princeton University in May 1921. ollected papers of Albert Einstein* v **7**, doc 71, Princeton (1997).

6. H. Minkowski, *Raum und Zeit*, *Phys. Zeit.* **10**, 104-111 (1909).
7. E.F. Taylor and J.A. Wheeler, *Spacetime physics*, New York (1992) (pp 246-252 Dialog: Use and Abuse of the concept of mass).
8. L.B. Okun, *The theory of relativity and the Pythagorean theorem*, *Physics - Uspekhi* **51**, (6), 622 - 631 (2008).
   `http://www.itep.ru/theor/persons/lab180/okun/em_30.pdf`
9. O. Klein, *Quantum Theorie und fünfdimensionale Relativitätstheorie*, *Zeit. f. Physik* **37**, 895-906 (1926).
10. V. Fock, *Über die invarianten Form der Wellen- und der Bewegungsgleichungen für einen geladenen Massenpunkt*, *Zeit. f. Physik* **39**, 226-232 (1926). On the invariant form of wave and motion equations for a charged point mass, see L.B. Okun, *Physics - Uspekhi* **53**(8) (2010).
11. W. Gordon, *Der Compton Effect nach der Schrödingerschen Theorie*, *Zeit. f. Physik* **40**, 117-133 (1926).
12. P.A.M. Dirac, *The principles of quantum mechanics* (1930).
13. L.D. Landau and E.M. Lifshitz, *The classical theory of fields*, (1941) (in Russian); Reed Publishing Ltd (2000).
14. L.B. Okun, *The virus of relativistic mass in the year of physics in* Gribov memorial volume (quarks, hadrons and strong interactions), Eds. Yu. Dokshitzer, P. Lévai and J. Nyíri, World Scientific, (2006) 470-473.
   `http://www.itep.ru/theor/persons/lab180/okun/em_22.pdf`
15. R.P. Feynman, R.B. Leighton and M. Sands, *The Feynman lectures on physics*, Addison-Wesley (1963).
16. L.D. Landau and Yu.B. Rumer, *What is relativity?* (1959), (in Russian); Dover (2003).
17. L.B. Okun, *Energy and Mass in Relativity Theory*, World Scientific, (2009).
   `http://www.worldscibooks.com/physics/6833.html`
18. L.B. Okun, *Mass versus relativistic and rest masses*, *Am. J. Phys.* **77**, (5), 430-431 (2009).
   `http://www.itep.ru/theor/persons/lab180/okun/doc/AJP000430.pdf`
19. *The main concepts and laws of physics and properties of elementary particles of matter*, Talk at the Presidium of the Russian Academy of Sciences, 27. 10. 2009, presented on the site of the Division of Physical Sciences of RAS (in Russian).
   `http://www.gpad.ac.ru/info/contributions/Okun_Prez.pdf`

SUPERFLUIDITY AND ROTATION IN MERGING NEUTRON STAR BINARY SYSTEMS AND EFFECTS RELATED TO GRAVITATIONAL WAVE EMISSION

C. N. COLACINO

Dipartimento di fisica "Enrico Fermi", University of Pisa
Pisa, Italy
colacino@pi.infn.it

During the merging process of a binary system made of two neutron stars there should be a phase transition in the star, given that the superfluid is constrained by the crust to corotate on pinning sites. Two things should happen, neither of which is well-studied and that affect both the gravitational wave emission and the approach to the final structure. Even before the actual merger, the two neutron stars (or a neutron star and white dwarf or black hole, taking the former as the model system) have a tidal coupling that strongly affects the rotation. Due to tidal coupling, the interior should spin up, leading to strong glitching and dissipation. This is independent of the equation of state (for the core) but depends strongly on the superfluid and the superconductor state in the crust. So the problem is to look at how this affects the calculations of both the rate and timing of the mergers (the same issue for precession of distorted neutron stars). The same with heating and the magnetic field coupling. All of this is very schematic in the polytropic interior models now used.

Keywords: Neutron stars; gravitational waves; superfluidity.

1. Neutron Stars as sources of gravitational radiation

The merging of two neutron stars is one of the most important sources of gravitational waves (GW). Astrophysical estimates give a minimum ratio of 0.01 events per year for current interferometric detectors, and since advanced detectors will increase their sensitivity by an order of magnitude, they will be able to observe at least 10 events per year, with some more refined estimates raising this number to forty events per year.[1] Unlike black holes, which can be described as elementary objects characterized only by their mass and angular momentum, neutron stars have a very complex structure that is the subject of deep investigation and can affect the emitted gravitational radiation. For example, the still unknown equation of state (EOS) will affect predictions of the GW luminosity. Neutron stars may exhibit conditions and phenomena not observed elsewhere, such as deconfined quark matter,[2] superfluidity,[3] and superconductivity with critical temperatures near 10^{10} K, large neutrino opacities and magnetic fields in excess of 10^{13}G.[4] They are ideal astrophysical sites for testing theories of dense matter and provide connections among astrophysics, nuclear physics, and particle physics, in particular Quantum Chromodynamics (QCD). Gravitational waves are expected not only from mergers involving two neutron stars (or a neutron star and a black hole) but also from asymmetric spinning neutron stars and from gravitational collapse supernovae which eventually lead to neutron star formation.

Among these sources, we have focussed our attention primarily to binary mergers because they can yield important information, including the masses and possibly the mass-to-radius ratios of their binary components and perhaps even details of their insparaling orbits. Moreover, neutron star mergers are thought to be the central engine of short gamma-ray bursts,[5] thus it would be extremely important to combine gravitational observations with results from conventional electromagnetic telescopes, and possibly even neutrino telescopes.[6] Neutron star mergers are in this sense ideal objects for "multimessenger astronomy". They can be considered "standard sirens": 10 detections in coincidence with gamma-ray bursts would allow a very precise measurement of H_0.[7] The generic GW signal from a NS merger can be split into a chirp signal emitted by the inspiralling binary, the burst amplitude from the very final plunge when the two stars collide, and a quasi-periodic post-merger signal caused either by the rotation and internal oscillation of a newly-formed, non axisymmetric hypermassive neutron star as merger remnant or by the quasi normal ringing of a newly born black hole. Different aspects of physics enter the problem. It is not only General Relativity which describes the merger, but also nuclear and particle physics play a role in the description of the hot and dense NS fluid via an Equation of State and in the treatment of energy losses after the merging. The GW signal of the late inspiral and of the merging phase is therefore expected to contain information not only about the binary parameters such as mass and spin but also about the nuclear EOS and the internal dynamics of the star.

2. Superfluidity in Neutron Stars

We will concentrate on an issue that, to date, has found little attention in literature but that may be significant in the evolution of NS binaries: the effects of the spin-up during the merging process on the star's superfluid. For temperatures less than ~ 0.1 MeV , and for densities above the neutron drip ($\rho \geq 4 \times 10^{11}$ g/cm^3) the crustal neutron fluid forms a $^1S_0$ superfluid that alters the specific heat and the neutrino emissivity of the crust – thereby affecting how neutron stars cool. Superfluids have zero viscosity and are locally irrotational. However the superfluid is threaded by an array of quantised vortices, that lead to an effective viscosity through the so-called entrainment effect:[8] the momentum induced in one of the constituents, in this case neutrons, carries along part of the mass of the other (protons). Because of this, some of the electrons which are coupled to the protons on very short timescales, will track the entrained protons and scatter off the magnetic fields of the vortices in a dissipative process called mutual friction. In the case of individual rotating stars, one issue of utmost importance, which remains open nowadays despite a lot of work in this direction[9] concerns the role of the superfluid in the r-mode instability problem.[10] The superfluid also represents a reservoir of angular momentum that, being coupled to the crust, causes the pulsar glitch phenomena. Superfluidity quenches cooling from the direct Urca process. However, an additional cooling source from the formation and breaking of nucleonic Cooper pairs increases the cooling rate through the modified Urca process.[11] Thus, changes in the state of the superfluid will affect the evolution of the star and of the binary.

3. Pulsar glitches

Pulsar glitches are the occasional disruption of otherwise regular spindown by magnetic torquing.[4] The leading glitch model involves angular momentum transfer from the superfluid to the normal component. Both are spinning but the normal crust is decelerated by the pulsar's magnetic dipole radiation. The superfluid is very weakly coupled to normal matter and its rotation rate is not diminished. But when the difference in spin rates becomes too large, something breaks and the spin rates are brought into closer alignment.[8] While an isolated rotating neutron star spins down during its standard cooling process, a neutron star in a binary system is forced by tidal coupling and accretion torques. The neutron superfluid is constrained to rotate on pinning sites. The spin up of the interior causes strong glitching and dissipation. The superfluid's angular momentum is quantised. So, the first question is whether or not this dissipation is observable by gravitational-wave detectors such ad Advanced Virgo, even before the final moments of merging, as a change in the angular momentum and energy loss from GW emission. Another issue is whether it is associated to high-energy electromagnetic phenomena that could be a precursor alert of the coalescence of neutron stars, thus giving us a warning. We know that superfluid vorticity reaches a turbulent regime (Andersson et al[12]). It is still unknown, though, how this turbulence influences the neutron star angular momentum, and thus the dynamics of merging itself. This could be a very important effect in the calculation of the waveforms and also for the matching filter templates to be used in the data analysis. And conversely, can information about the superfluid -and the superconductor- state be extracted from the merger waveform? Although superconducting features should not affect the EOS because the condensate energy E_c is far less than the Fermi energy, we do not know whether or not the EOS have any effect on this scheme. Polytropic models have little to say on the subject, because they make very general assumptions. This picture is obviously modified in the case of a BH/NS binary system, and in the case of a violent tidal disruption of the star by a massive black hole, this will be the subject of further investigation.

4. Merger rate for neutron star binaries

Models of galactic evolution, e.g. scenario machines,[13] provide rates of production (given particular recipes and parametrisations of the relevant formation processes) for neutron stars through continued star formation. Several assumptions, e.g. the initial mass function for single and multiple stars, star formation histories, mass and angular momentum loss rates, are supplemented by particular schemes for end states of stars, e.g. mass loss rates, dynamos, magnetically-coupled systems, to produce the initial systems that merge over time. The standard picture is that double degenerates, for instance, can, through magnetically mediated spin orbit coupling, produce the initial approach fo the components to the point where the energy loss is dominated by gravitational wave emission. Inevitably, this leads to a collapse of the system and the production of a massive remnant, likely a low mass black hole. While details are still to be clarified, this general picture does not pose insuperable difficulties. The same is not true, however, if the system consists of two neutron stars (e.g. the binary pulsars). The interior structure of each component depends on the EOS, but more significantly, the evolution of the system will as well after a certain point. Depending on the radius, itself in turn dependent on the EOS, tidal coupling between the components sets in at different orbital frequencies. The normal glitch event in isolated pulsars is presumed to be a depinning of the vortices from crustal sites caused by a combination of magnetically and wind induced spindown and thermal agitation of the vortices.[14] Recent calculations also indicate that superfluid turbulence can develop, which rapidly couples the core and crust leading to a global re-adjustment of the angular momentum distribution. On the other hand, in merging systems there should be a spin-up as the orbit decays (even more extreme than what occurs for accreting systems that display anti-glitching). We would like to model the reaction of the interior to such events (with relevance as well to the production of millisecond pulsars), to investigate what happens to the vortices if they get sufficiently stressed by the tidal spin-up. It seems possible that a phase transition can occur, whose physical consequences remain unknown. If re-connection occurs, do excitations in the normal fluid produce an increased Magnus force and enhanced coupling that can lead to both massive vorticity redistribution and heating? Could such heating suffice to transform the superfluid into the normal component? We will approach this problem analytically but, given the complexity of the system itself, we will make use of numerical methods, in particular Smoothed Particle Hydrodynamics (SPH) codes.[15] Superfluid equations can be studied and solved both in a General Relativistic approach and in the Newtonian limit.

5. Summary

Neutron Star binary mergers are the most promising source of gravitational radiation, and yet the dynamics of the individual stars could drastically alter the merger process and change the picture we have and on which we base our data-analysis matched filters and algorithms. The author, together with Prof. Steven Shore of the University of Pisa, has recently started to investigate some of these issues, and perhaps the first results will be of some interests for Advanced Interferometric Gravitational Wave Detectors such as Advanced Virgo and Advanced LIGO, due to start operation in 2015.

References

1. The LIGO Scientific Collaboration and the VIRGO Collaboration, Predictions for the Rates of Compact Binary Coalescences Observable by Ground-based Gravitational-wave Detectors (2010), LSC-VIRGO internal note, to be published.
2. S. L. Shapiro and S. A. Teukolsky, *Black holes, white dwarfs, and neutron stars: The physics of compact objects* (Wiley-VCH Verlag, Weinheim, 2004).

3. T. Padmanabhan, *Theoretical Astrophysics, Volume II: Stars and Stellar Systems* (Cambridge University Press, Cambridge, U.K.; New York, U.S.A., 2001).
4. J. Lattimer and M. Prakash, *Science* **304**, 536 (2004).
5. S. Rosswog and M. Liebendörfer, *Mon. Not. R. Astron. Soc.* **342**, 673 (2003).
6. L. Dessart, C. Ott, A. Burrows, S. Rosswog and E. Livne, *ApJ* **690**, 1681 (2009).
7. N. Dalal, D. E. Holz, S. A. Hughes and B. Jain, *Phys. Rev.* **D74**, 063006 (2006).
8. N. Andersson and G. L. Comer , *Mon. Not. R. Astron. Soc.* **328**, 1129 (2001).
9. L. Lindblom and G. Mendell, *Phys. Rev.* **D61**, 104003 (2000).
10. N. Andersson and K. D. Kokkotas, *Int. J. Mod. Phys.* **D10**, 381 (2001).
11. N. Chamel and P. Haensel, *Living Reviews in Relativity* **11** (2008).
12. N. Andersson, T. Sidery and G. L. Comer, *Mon. Not. R. Astron. Soc.* **381**, 747 (2007).
13. C. Fryer and A. Hungerford, *The Electromagnetic Spectrum of Neutron Stars*, 3 (2005).
14. M. A. Alpar, H. F. Chau, K. S. Cheng and D. Pines , *ApJ* **759**, 406 (1996).
15. R. Oechslin and H.-T. Janka, *Phys. Rev. Lett.* **99**, 121102 (2007).

TWO DIMENSIONAL GRAVITY IN MATRIX MODEL, TOPOLOGICAL AND LIOUVILLE APPROACHES

A. BELAVIN

L.D. Landau Institute for Theoretical Physics, RAS
Chernogolovka, 142432 Moscow Oblast, Russia
belavin@itp.ac.ru

1. Introduction

Since the end of the eighties there exist three approaches to 2d Quantum Gravity.

The first approach is the continuous approach[1,2] in which the theory is defined as the integral over Riemannian metrics modulo diffeomorphismus. Fixing the so-called conformal gauge leads to quantum Liouville theory describing the quantum gravity coupled with some CFT describing "matter" fields. It is called the Liouville Gravity. Indeed, due to the additional Weyl (BRST) invariance the functional integrals over metrics reduces to the integrals over moduli space of curves.

The second one is the discrete approach,[3] based on the idea of approximating the fluctuating $2d$ geometry by an ensemble of planar graphs, so that the continuous theory is recovered in the scaling limit where the planar graphs of very large size dominate approximating continuous surfaces. The discrete approach is usually referred to as the Matrix Models, since technically the ensemble of the graphs is usually generated by the perturbative expansion of the integral over $N \times N$ matrices.

The third approach is the topological gravity, first of all the intersection theory on Moduli Space of Riemann Curves.[4–6] A very good earlier review on this subject is given in Ref. 7.

It was conjectured by E.Witten in the end of the eighties that the three models are equivalent. However, literally this is not correct.

The evidence that the Liouville Gravity and Matrix Models describe the same reality is the fact that operators O_k^{LG} and O_k^{MM} have identical scale dimensions and some correlation numbers coincide:

$$\langle\, O_1^{LG} \cdots O_n^{LG} \,\rangle = \langle\, O_1^{MM} \cdots O_n^{MM} \,\rangle. \tag{1}$$

However, in the general case the correlation numbers are not in agreement with the "naive" identification.

Nevertheless, there exists a nontrivial way to reach their coincidence.[8,9]

The correspondence between Liouville Gravity and Matrix Models can be reached due to the resonance relations:

$$[O_k] \;\; = [O_{k_1}] + [O_{k_2}]. \tag{2}$$

A conjecture was proposed and checked that there exists a choice of the so called "resonance" transformation of the coupling constants in MM from the so called "KdV" frame to a different one ("CFT frame"), such that the correlation functions in the two theories coincide.

The explicit form of the transformation was established[9] and checked[9,10] for a particular case, namely for the Minimal Quantum Gravity $MG(2, 2p+1)$ and the One-Matrix Model (OMM) in the p-critical point.

The correspondence between Intersection Theory and Matrix Models takes place only partly. Although the partition functions in both theories obey similar recursive relations, as well as KdV and String equations, they do not coincide, but are connected by some analytic continuation. The fact that correlation numbers obey the same recursive relations was checked in genus zero and genus one.[11] It should be mentioned that it takes place when correlation numbers in OMM are calculated in KdV frame. In this talk some relations between the three approaches are reviewed.

2. One Matrix Model

In this approach the planar graphs correspond to Feynman diagrams associated with the perturbative evaluation of the matrix integral

$$Z = \log \int dM \; e^{-N\mathrm{tr}V(M)} \tag{3}$$

$$V(M) = \frac{1}{2}\, M^2 - \sum_{n=3} \frac{a_n}{n!}\, M^n \tag{4}$$

M — Hermitian $N \times N$ matrix, N being the device for sorting out the topologies

$$Z = N^2\, Z_0 + Z_1 + \cdots + N^{2-2g}\, Z_g + \cdots \tag{5}$$

Each term Z_g generates discretized surfaces, of the topology g, made of triangles and higher polygons, with the weights determined by a_n.
We will concentrate on $g = 0$ (sphere) and $g = 1$ (torus) cases.

The one-matrix model exhibits a set of multi-critical points, labelled by the integer $p = 1, 2, 3, \ldots$ in the Σ-space of the "potentials" $V(M) = \sum_{n=2} \frac{a_n}{n!} M^n$.
Parameters a_n can be considered as coordinates on Σ.
The p - critical points belong to a subspace of Σ of codimension p.
The first step to compute the integral is the diagonalization of matrix M which reduces the integration over M to integration over its eigenvalues μ_i.

$$Z(a_k, N) = \log \int \prod_{i=1}^{N} d\mu_i \Delta^2(\mu) e^{-\sum_i V(\mu_i)} \tag{6}$$

$$\Delta(\mu) = \prod_{i<j} (\mu_i - \mu_j) \quad - \quad \text{Vandermonde determinant} \tag{7}$$

3. The Method of Orthogonal Polynomials

Introducing the set of orthogonal polynomials $P_n(\mu) = \mu^n + \cdots$,

$$\int_{-\infty}^{\infty} d\mu e^{-V(\mu)} P_n(\mu) P_m(\mu) = s_n \delta_{nm}. \tag{8}$$

one obtains for the partition function

$$Z = N \sum_{k=1}^{N-1} (1 - k/N) \log(s_k / s_{k-1}). \tag{9}$$

Using the relation

$$\lambda P_k(\mu) = P_{k+1}(\mu) + R_k P_{k-1}(\mu), \text{ one gets} \tag{10}$$

$$\int e^{-V(\mu)} P_k \mu P_{k-1} d\mu = R_k s_{k-1} = s_k \tag{11}$$

$$R_k = s_k / s_{k-1}. \tag{12}$$

Therefore

$$Z = N \sum_{k=1}^{N-1} (1 - k/N) \log R_k. \tag{13}$$

Using the integration by parts

$$k s_{k-1} = \int e^{-V} P_k' P_{k-1} = \int e^{-V} V' P_k P_{k-1}. \tag{14}$$

$$V'(\mu) = \sum_{k=1}^{p+1} 2ka_k\mu^{2k-1} \tag{15}$$

and applying $2n-1$ times the previous relation for $\mu P_k(\mu)$

$$\begin{aligned}\mu^{2n-1}P_k &= \mu^{2n-2}(P_{k+1} + R_kP_{k-1}) = \\ &= \mu^{2n-3}(P_{k+2} + R_{k+1}P_k + R_kP_k + R_kR_{k-1}P_{k-2}) = \cdots\end{aligned} \tag{16}$$

we obtain the following formula for R_k

$$\frac{k}{N} = \tilde{W}(R_k, R_{k\pm1}, \ldots, R_{k\pm p}) \tag{17}$$

where

$$\tilde{W}(R_k, R_{k\pm1}, \ldots, R_{k\pm p}) = \tag{18}$$

$$= \sum_{n=1}^{p+1} 2na_n \sum_{\{\sigma_{2n-1}\}} R_{k+m_1} \cdot \cdots \cdot R_{k+m_n}. \tag{19}$$

$\{\sigma_{2n-1}\}$ denotes all paths which consist of $2n-1$ steps, starting in k and finishing in $k-1$.

The factor $R(k+m)$ arises in the sum if the (k+m)th step goes down.

4. Evaluation of Z_0 and Z_1

Assuming the existence of a smooth function $R(\xi, N)$ of the variable $\xi \in [0,1]$, and $R(\frac{k}{N}, N) = R_k$, and Taylor expansion for $R(\xi + m/N, N)$

$$R(\xi + m/N, N) = R(\xi, N) + \frac{m}{N}R_\xi(\xi, N) + \frac{m^2}{2N^2}R_{\xi\xi}(\xi, N) + O(\frac{1}{N^3}), \tag{20}$$

we obtain

$$\tilde{W}(R(\xi, N)) = W(R(\xi, N)) + \frac{1}{N}W_1(R(\xi, N)) + \frac{1}{N^2}W_2(R(\xi, N)) + O(\frac{1}{N^3}), \tag{21}$$

A direct calculation gives

$$W(R(\xi, N)) = \sum_{n=1}^{p+1} \frac{(2n)!}{n!(n-1)!}a_nR^n(\xi, N), \tag{22}$$

$$W_1(R(\xi, N)) = 0, \tag{23}$$

$$W_2(R(\xi, N)) = \frac{RR_{\xi\xi}}{6}W''(R(\xi, N)) + \frac{RR_\xi^2}{12}W'''(R(\xi, N)). \tag{24}$$

As a result we have

$$Z = N \sum_{k=1}^{N-1} (1 - k/N) \log R(\xi, N) \tag{25}$$

where $R(\xi, N)$ is the solution of the equation

$$\xi = W(R(\xi, N)) + \frac{RR_{\xi\xi}}{6N^2} W''(R(\xi, N)) + \frac{RR_\xi^2}{12N^2} W'''(R(\xi, N)) + O(\frac{1}{N^4}). \tag{26}$$

Assuming also the expansion

$$R(\xi, N) = R(\xi) + \frac{1}{N} R_1(\xi) + \frac{1}{N^2} R_2(\xi) + \cdots , \tag{27}$$

thus

$$\begin{aligned} \xi &= W(R(\xi)), \\ R_1(\xi) &= 0, \\ R_2(\xi) &= -\frac{R(\xi)}{12W'(R)} \left(2R_{\xi\xi} W''(R(\xi)) + R_\xi^2 W'''(R(\xi))\right). \end{aligned} \tag{28}$$

Passing from sum to integral in the partition function by means of the Euler-Maclaurin formula (up to N^0 terms)

$$Z = N^2 \int_0^1 d\xi (1 - \xi) \log R(\xi, N). \tag{29}$$

Then for partition functions in genus zero and genus one we obtain

$$Z_0 = \int_0^1 d\xi (1 - \xi) \log R \tag{30}$$

$$Z_1 = -\frac{1}{12} \int_0^1 d\xi (1 - \xi) \frac{2R_{\xi\xi} W''(R) + R_\xi^2 W'''(R)}{W'(R)} \tag{31}$$

where

$$W(R) = \sum_{n=1}^{p+1} \frac{(2n)!}{n!(n-1)!} a_n R^n \tag{32}$$

and $R = R(\xi)$ is a suitable root of the equation

$$W(R(\xi)) = \xi. \tag{33}$$

5. The Vicinity of the p-Critical Point

The p-critical point is defined by the system of equations

$$W(R_c) = 1, \quad W'(R_c) = 0, \quad \cdots \quad W^{(p)}(R_c) = 0. \tag{34}$$

This is a system of equations which determine the critical values of coefficients a_k^c, $k = 1, \ldots, p$, and define the R_c. In the vicinity of this point the main contributions to the Matrix Integral are given by the surfaces of very large size. Consider small deviations $\delta a_k = a_k - a_k^c$, and let t_k be the new coordinates in the vicinity of the critical point

$$W(R_c) = 1 + t_{p+1}, \quad W'(R_c) = t_p, \quad \cdots$$
$$W^{(p-1)}(R_c) = t_0, \quad W^{(p)}(R_c) = 0. \tag{35}$$

Denoting $u = R - R_c$ one can obtain

$$\xi = W(u) = u^{p+1} + t_0 u^{p-1} + \sum_{k=1}^{p-1} t_k u^{p-k-1} + 1. \tag{36}$$

Making a substitution $\xi = 1 - y$, one can get

$$\mathcal{P}(u) + y = 0, \tag{37}$$

where the string polynomial $\mathcal{P}(u)$ is defined as

$$\mathcal{P}(u) = u^{p+1} + t_0 u^{p-1} + \sum_{k=1}^{p-1} t_k u^{p-k-1} \tag{38}$$

and $u(y)$ is its solution.

Therefore for the partition functions we obtain

$$Z_0 = \frac{1}{R_c} \int_0^1 dy\, y\, u(y), \tag{39}$$

$$Z_1 = -\frac{1}{12} \int_0^1 dy\, y \left(\frac{2\mathcal{P}''(u) u_{yy} + \mathcal{P}'''(u) u_y^2}{\mathcal{P}'(u)} \right). \tag{40}$$

These expressions can be efficiently simplified and we arrive at the final answer

$$Z_0 = \frac{1}{2} \int_0^{u^*} \mathcal{P}^2(u) du \tag{41}$$

$$Z_1 = -\frac{\log \mathcal{P}'(u^*)}{12} \tag{42}$$

where

$$u^* = u^*(t_0, t_1, \ldots, t_{p-1}) \tag{43}$$

is the suitably chosen root of the polynomial $\mathcal{P}(u)$,

$$\mathcal{P}(u) = u^{p+1} + t_0\, u^{p-1} + \sum_{k=1}^{p-1} t_k\, u^{p-k-1}\,, \tag{44}$$

i.e. $\mathcal{P}(u_*) = 0$.
It is important to remember that Z_0 and Z_1 really gives only the singular (scale invariant) part of the Matrix Model partition function.

6. Minimal Gravity

In this approach the Quantum Geometry is defined by the Polyakov integral

$$\sum_{\text{topologies}} \int D[g]\, D[\phi]\, e^{-S_0[g,\phi]}, \tag{45}$$

where $g(x)$-Riemannian metric , ϕ-"matter" fields

$$\langle\, O_{k_1} \cdots O_{k_N}\,\rangle = \int O_{k_1} \cdots O_{k_N}\, e^{-S_0[g,\phi]}\, D[g,\phi] \tag{46}$$

$$O_k = \int_{\mathbb{M}} U_k(x)\, d\mu_g(x). \tag{47}$$

$U_k(x)$ are local fields (built from ϕ and g).

It is convenient to define the generating (or partition) function

$$Z(\{\lambda\}) = \int D[g,\phi]\, e^{-S_\lambda[g,\phi]}\,, \tag{48}$$

$$S_\lambda[g,\phi] = S_0[g,\phi] + \sum_k \lambda_k\, O_k \tag{49}$$

$$\langle\, O_{k_1} \cdots O_{k_N}\,\rangle = \left.\frac{\partial^N Z(\{\lambda\})}{\partial\lambda_{k_1} \cdots \partial\lambda_{k_N}}\right|_{\lambda=0}. \tag{50}$$

The parameters $\{\lambda\}$ are the coordinates in the "theory space" Σ.

Liouville gravity is described as follows. The main defining property of any 2-d conformal field theory is the conformal anomaly relation

$$g^{\mu\nu}\, T^{\text{matter}}_{\mu\nu} = -\frac{c}{12}\, R. \tag{51}$$

In conformal gauge $g_{\mu\nu} = e^{2b\varphi}\,\hat{g}_{\mu\nu}$: $\Rightarrow$ the total action of Liouville gravity

$$S[g,\phi] \;\to\; S_{\rm L}[\varphi] + S_{\rm Ghost}[B,C] + S_{\rm Matter}[\phi] \tag{52}$$

where $S_{\rm Matter}$ is some conformal field theory

$$S_L[\phi] = \frac{1}{4\pi}\int \sqrt{\hat{g}}\left[\hat{g}^{\mu\nu}\partial_\mu\varphi\partial_\nu\varphi + Q\,\hat{R}\,\varphi + 4\pi\mu\, e^{2b\,\varphi}\right] d^2x\,, \tag{53}$$

$$S_{\rm Ghost}[B,C] = \frac{1}{2\pi}\int \sqrt{\hat{g}}\; B_{\mu\nu}\,\nabla^\mu C^\nu\, d^2x\,, \tag{54}$$

$\big(B_{\mu\nu} = B_{\nu\mu}\,,\;\; \hat{g}^{\mu\nu}B_{\mu\nu} = 0\big)$,

All three theories are conformal invariant, including $S_{\rm Matter}[\phi]$ whose central charge is c.

The relation

$$26 - c = 1 + 6\,Q^2 \qquad Q = b + 1/b\,. \tag{55}$$

together with the suitable definition the observables ensures the Weyl invariance of the theory.

The observables in Liouville gravity are defined as

$$O_k = \int V_k(x)\,\Phi_k(x)\,d^2x \tag{56}$$

$\Phi_k(x)$ – (spinless) primary fields of the matter CFT, with the conformal dimensions (Δ_k, Δ_k) and $V_k(x)$ – "gravitational dressings",

$$V_k(x) = e^{2a_k\,\varphi(x)}\,, \qquad a_k(Q - a_k) + \Delta_k = 1 \tag{57}$$

Gravitational dimensions of $\tilde{O}_k$ control the scale dependence

$$O_k \;\sim\; \mu^{\delta_k}\,, \qquad \delta_k = -\frac{a_k}{b} \tag{58}$$

of the correlation numbers

$$\langle\, O_{k_1}\cdots O_{k_n}\,\rangle = |(x_1 - x_2)(x_2 - x_3)(x_3 - x_1)|^2 \tag{59}$$

$$\times \int d^2x_4\cdots d^2x_n\;\underbrace{\langle\, U_{k_1}(x_1)U_{k_2}(x_2)U_{k_3}(x_3)U_{k_4}(x_4)\cdots U_{k_n}(x_n)\,\rangle} \tag{60}$$

$$\downarrow$$

$$\langle\, V_{k_1}(x_1)\cdots V_{k_n}(x_n)\,\rangle_{\rm Liouville}\;\langle\, \Phi_{k_1}(x_1)\cdots\Phi_{k_n}(x_n)\,\rangle_{\rm Matter} \tag{61}$$

We choose one of the Minimal Models of $(2, 2p+1)$ series as Matter Theory.

$\mathcal{M}_{2/2p+1}$ has p primary fields

$$\Phi_k \equiv \Phi_{(1,k+1)}\,, \qquad k = 0, 1, \ldots, p-1 \;\; (p, p+1, \ldots, 2p-1) \tag{62}$$

They obey the fusion rules

$$[\Phi_{k_1}][\Phi_{k_2}] = \sum_{k=|k_1-k_2|\,:\,2}^{k_1+k_2} [\Phi_k]\,, \qquad [\Phi_k] = [\Phi_{2p-k-1}] \tag{63}$$

$$\Phi_k = \Phi_{2p-k-1} \tag{64}$$

which result in the following restrictions for the correlation numbers:

$$\langle\, \Phi_k \,\rangle = \delta_{k,0}, \qquad \langle\, \Phi_k \Phi_{k'} \,\rangle \;\sim\; \delta_{k,k'} \tag{65}$$

$$\langle\, \Phi_{k_1} \Phi_{k_2} \Phi_{k_3} \,\rangle = 0$$

$$\text{if} \quad \begin{cases} k_1 + k_2 < k_3, \text{ etc,} & \text{for} \quad k_1 + k_2 + k_3 \;\; \text{even} \\ k_1 + k_2 + k_3 < 2p - 1 & \text{for} \quad k_1 + k_2 + k_3 \;\; \text{odd} \end{cases}$$

$$\langle\, \Phi_{k_1} \cdots \Phi_{k_n} \,\rangle = 0$$

$$\text{if} \quad \begin{cases} k_1 + \cdots + k_{n-1} < k_n, & \text{for} \quad k_1 + \cdots + k_n \;\; \text{even} \\ k_1 + \cdots + k_n < 2p - 1 & \text{for} \quad k_1 + \cdots + k_n \;\; \text{odd} \end{cases}$$

These restrictions are inherited by correlation numbers of the corresponding Liouville gravity, whose partition function is

$$Z_{\mathcal{MG}}(\mu, \{\lambda\}) = \left\langle \; \exp\left\{ -\sum_{i=1}^{p-1} \lambda_i\, O_i \right\} \right\rangle_{\mathcal{MG}_{2/2p+1}}. \tag{66}$$

The cosmological constant μ will be denoted below also as λ_0

$$S[\mathcal{MG}] = \cdots + \mu \underbrace{\int e^{2b\varphi(x)}\, d^2x} + \cdots \tag{67}$$

Dimensions of the variables:

$$\lambda_k \sim \mu^{\frac{k+2}{2}}\,, \quad k = 0, 1, \ldots, p-1 \tag{68}$$

Correlation numbers, i.e.

$$\langle\, O_{k_1} \cdots O_{k_n} \,\rangle = \left. \frac{\partial^n Z_{\mathcal{MG}}(\mu, \{\lambda_i\})}{\partial \lambda_{k_1} \cdots \partial \lambda_{k_n}} \right|_{\{\lambda_i\}=0}\,, \qquad \{\lambda_i\} = \{\lambda_1, \ldots, \lambda_n\} \tag{69}$$

obey the same selection rules as their constituents of $\mathcal{M}_{2/2p+1}$.

7. Comparison of the Matrix Model and Liouville Gravity

Take

$$t_0 = \mu \qquad \text{– cosmological constant} \tag{70}$$

Then

$$[u] = [\mu^{\frac{1}{2}}], \qquad [t_k] = [\mu^{\frac{k+2}{2}}], \qquad [Z] = [\mu^{\frac{2p+3}{2}}], \tag{71}$$

are exactly equal to the gravitational dimensions of $\mathcal{MG}_{2/2p+1}$,

$$t_k \sim \lambda_k\,, \quad k = 0, 1, 2, \ldots, p-1. \tag{72}$$

It is convenient to separate $t_0 = \mu$ and $\{t_i\} = \{t_1, t_2, \ldots, t_{p-1}\}$

If we define Matrix Model correlation numbers as

$$\langle\, \mathcal{O}_{k_1} \cdots \mathcal{O}_{k_n} \,\rangle_{MM} \equiv \left. \frac{\partial^n Z_{MM}(\mu, \{t_i\})}{\partial t_{k_1} \cdots \partial t_{k_n}} \right|_{\{t_i\}=0}, \qquad \{t_i\} = \{t_1, \ldots, t_n\} \tag{73}$$

then the naive identification $t_k \sim \lambda_k$ predicts

$$\langle\, \mathcal{O}_{k_1} \cdots \mathcal{O}_{k_n} \,\rangle_{MM} = \langle\, \tilde{O}_{k_1} \cdots \tilde{O}_{k_n} \,\rangle_{\mathcal{MG}} \tag{74}$$

However the checks show that this expectation fails.
Since

$$\mathcal{P}(u) = u^{p+1} + \mu\, u^{p-1} + \sum_{k=1}^{p-1} t_k\, u^{p-k-1}\,, \qquad Z = \frac{1}{2}\int_0^{u_*} \mathcal{P}^2(u)\, du \tag{75}$$

we have $u_*(\mu, 0, \ldots, 0) = \sqrt{\mu}$, and

$$\left. \frac{\partial Z}{\partial t_k} \right|_{\{t=0\}} = \left. \int_0^{u_*} \mathcal{P}(u)\, \frac{\partial \mathcal{P}(u)}{\partial t_k}\, du \right|_{\{t=0\}} = -\frac{2\,\mu^{\frac{2p-k+1}{2}}}{(2p-k-1)(2p-k+1)} \tag{76}$$

$$\left. \frac{\partial^2 Z}{\partial t_k \partial t_{k'}} \right|_{\{t=0\}} = \left. \int_0^{u_*} \frac{\partial \mathcal{P}(u)}{\partial t_k}\, \frac{\partial \mathcal{P}(u)}{\partial t_{k'}}\, du \right|_{\{t=0\}} = \frac{\mu^{\frac{2p-k-k'-1}{2}}}{2p-k-k'-1} \tag{77}$$

etc, in sharp contrast with

$$\langle\, O_k \,\rangle_{\mathcal{MG}} \;=\; 0\,, \quad k = 1, 2, \ldots, p-1 \quad (\text{since } \langle\, \Phi_k \,\rangle_{CFT} = 0) \tag{78}$$

$$\langle\, O_k \tilde{O}_{k'} \,\rangle_{\mathcal{MG}} \;\sim\; \delta_{kk'} \qquad\qquad (\text{since } \langle\, \Phi_k \Phi_{k'} \,\rangle_{CFT} \sim \delta_{kk'}). \tag{79}$$

8. Resonance Transformations

Taking into account that

$$[t_k] = [\mu^{\frac{k+2}{2}}] , \qquad [\lambda_k] = [\mu^{\frac{k+2}{2}}] \tag{80}$$

there exist so-called resonance relations, e.g.

$$[t_k] = [\lambda_{k_1}] + [\lambda_{k_2}] \qquad (k = k_1 + k_2 + 2 \geq 2) \tag{81}$$

$(k = 0, 1, 2, \ldots, p-1)$. It means that there is some freedom in the identifying of the t_k and λ_k variables

$$t_k \;=\; \lambda_k \;+\; \sum_{\substack{k_1,k_2=0\\ k_1+k_2=k+2}}^{p-1} c_k^{k_1 k_2}\, \lambda_{k_1}\lambda_{k_2} \;+\; \text{higher order terms} \tag{82}$$

For example,

$$t_0 = \lambda_0 = \mu \,, \tag{83}$$

$$t_1 = \lambda_1 \;, \qquad ([t_1] = [\mu^{3/2}]) \tag{84}$$

$$t_2 = \lambda_2 + A_2\,\mu^2 \;, \qquad ([t_2] = [\mu^2]) \tag{85}$$

$$t_3 = \lambda_3 + B_3\,\mu\,\lambda_1 \;, \qquad ([t_3] = [\mu][t_1]) \tag{86}$$

$$t_4 = \lambda_4 + A_4\,\mu^3 + B_4\,\mu\,\lambda_2 + C_4\,\lambda_1^2, \text{ etc.} \tag{87}$$

Generally we want to find such a resonance transformation

$$t_k = \lambda_k + \underbrace{A_k\,\mu^{\frac{k+2}{2}}} + \sum_{n=0}^{n\leq k/2} \underbrace{B_k^{k-2n}\;\mu^n\,\lambda_{k-2n}} + \tag{88}$$

$$\frac{1}{2}\sum_{n=0}\;\sum_{k_1+k_2=k-2-2n} \underbrace{C_k^{k_1,k_2}\;\mu^n\,\lambda_{k_1}\lambda_{k_2}} + \cdots \tag{89}$$

$$Z_{MM}(\{t\}) \;\rightarrow\; \tilde{Z}_{MM}(\{\lambda\}) \equiv Z_{MM}(\{t(\lambda)\}) \tag{90}$$

that the newly defined correlation numbers

$$\frac{\partial^N \tilde{Z}_{MM}(\{\lambda\})}{\partial\lambda_{k_1}\cdots\partial\lambda_{k_N}} = \langle\,\tilde{O}_{k_1}\cdots\tilde{O}_{k_n}\,\rangle_{\mathcal{MG}} \tag{91}$$

would satisfy the same "conformal" and fusion rules like those ones in Liouville gravity.

Thus we demand that one-point numbers:

$$\langle \tilde{O}_k \rangle_{MM} = \frac{\partial \tilde{Z}(\mu, \{\lambda\})}{\partial \lambda_k}\bigg|_{\{\lambda\}=0} = 0 \qquad \text{for} \quad k = 1, 2, \ldots, p-1, \tag{92}$$

two-point numbers:

$$\langle \tilde{O}_k \tilde{O}_{k'} \rangle_{MM} = \frac{\partial^2 \tilde{Z}(\mu, \{\lambda\})}{\partial \lambda_k} \partial \lambda_{k'}\bigg|_{\{\lambda\}=0} \sim \delta_{kk'}\,, \tag{93}$$

three-point numbers:

$$\langle \tilde{O}_{k_1} \tilde{O}_{k_2} \tilde{O}_{k_3} \rangle_{MM} = \frac{\partial^3 \tilde{Z}(\mu, \{\lambda\})}{\partial \lambda_{k_1}} \partial \lambda_{k_2} \partial \lambda_{k_3}\bigg|_{\{\lambda\}=0} \tag{94}$$

and multi-point numbers obey fusion rules, e.g.
for even $k_1 + \cdots + k_n$

$$\langle \tilde{O}_{k_1} \tilde{O}_{k_2} \cdots \tilde{O}_{k_n} \rangle_{MM} = 0 \qquad \text{if} \quad k_n > k_1 + k_2 + \cdots + k_{n-1} \tag{95}$$

for odd $k_1 + \cdots + k_n$

$$\langle \tilde{O}_{k_1} \tilde{O}_{k_2} \cdots \tilde{O}_{k_n} \rangle_{MM} = 0 \qquad \text{if} \quad k_1 + k_2 + \cdots + k_n < 2p - 1 \tag{96}$$

9. Finding the Liouville Coordinates

When one plugs $t_k(\lambda)$, the polynomial

$$\mathcal{P}(u) = u^{p+1} + t_0\, u^{p-1} + \sum_{k=1}^{p-1} t_k\, u^{p-k-1}\,, \tag{97}$$

takes the form

$$\mathcal{P}(u) = \mathcal{P}_0(u) + \sum_{k=1}^{p-1} \lambda_k\, \mathcal{P}_k(u) + \cdots + \sum_{k_i=1}^{p-1} \frac{\lambda_{k_1} \cdots \lambda_{k_n}}{n!}\, \mathcal{P}_{k_1 \ldots k_n}(u) + \cdots \tag{98}$$

where $\mathcal{P}_0(u)$ and $\mathcal{P}_{k_1 \ldots k_n}(u)$ are the polynomials of u whose coefficients involve non-negative powers of μ.

$\mathcal{P}_{k_1 \ldots k_n}(u)$ are polynomials of the degree

$$p + 1 - 2n - \sum k_i\,, \tag{99}$$

Of course, only polynomials of non-negative degree appear, so that the sum in $\mathcal{P}(u)$ is finite.

When the fusion rules are violated, the correlation numbers vanish. This requirement for the n-point numbers imposes strong conditions on the form of the polynomials $\mathcal{P}_{k_1\ldots k_{n-1}}(x)$, which fix them uniquely.

Technically, this is done by constructing the polynomial $\mathcal{P}(u)$, order by order in λ_k. We have executed this program up to the fifth order.

For higher n direct calculations become rather involved. But a quick glance at a few first orders immediately suggests the general form,

$$Q_{k_1\ldots k_n}(u) = \left(\frac{d}{du}\right)^{n-1} L_{p-\sum k-n}(u)\,, \tag{100}$$

where again $\sum k = k_1 + \cdots + k_n$.

Here $Q_{k_1\ldots k_n}(u)$ differs from $\mathcal{P}_{k_1\ldots k_{n-1}}(x)$ by slight renormalizations.

So we arrive to the conjecture:

The partition function of the one-MM is expressed through $Q(u)$

$$\mathcal{Z} = \frac{1}{2}\int_0^{u_*} Q^2(u)\,du\,, \tag{101}$$

u_* is the solution of the "string equation"

$$Q(u_*) = 0, \tag{102}$$

$$Q(u) = \sum_{n=0} \sum_{k_1,\ldots,k_n=1}^{p-1} \frac{\lambda_{k_1}\cdots\lambda_{k_n}}{n!} L^{(n-1)}_{p-\sum k-n}(u)\,. \tag{103}$$

Here we denote

$$L_k^{(n)}(u) = \left(\frac{d}{du}\right)^n L_k(u)\,. \tag{104}$$

The partition function $\mathcal{Z}$ coincides with the generating functions of the correlation numbers in $\mathcal{MG}_{2/2p+1}$

$$\mathcal{Z} = \left\langle \exp\left\{-\sum_{i=1}^{p-1}\lambda_i\,\tilde{O}_i\right\}\right\rangle_{\mathcal{MG}_{2/2p+1}}\,. \tag{105}$$

In particular, for the four-point correlation number in the even sector, assuming that $0 < k_1 < k_2 < k_3 < k_4 < p-1$, we obtain

$$\langle \tilde{O}_{k_1}\tilde{O}_{k_2}\tilde{O}_{k_3}\tilde{O}_{k_4}\rangle_{MM} = \tag{106}$$

$$\begin{cases} \frac{1}{2}(\sum k_i^2 + 3p^2 - 5p - (2p-1)k_{1234}), & p-1 < k_{12} \\ (k_1+1)(2p-k_{1234}-3) + F(k_{14}) + F(k_{13}), & k_{12} < p-1 < k_{13}, \\ (k_1+1)(2p-k_{1234}-3) + F(k_{14}), & k_{13} < p-1 < k_{14}, k_{23}, \\ (k_1+1)(2p-k_{1234}-3), & k_{14} < k_{23},\ k_{14} < p-1, \\ \frac{1}{2}(k_{123}-k_4+2)(2p-k_{1234}-3), & k_{14} < k_{23},\ k_{23} < p-1, \\ 0, & k_{123} < k_4, \end{cases}$$

where $F(k) = \frac{1}{2}(p-k-1)(p-k-2)$, $k_{ij} = k_i + k_j$, $k_{ijl} = k_i + k_j + k_l$, $k_{1234} = k_1 + k_2 + k_3 + k_4$.

10. Comparison of the Matrix Model and Topological Gravity

The Witten recursion relation for the intersection numbers of the stable cohomology classes on moduli space reads

$$\langle\sigma_{k_1}\sigma_{k_2}\cdots\sigma_{k_s}\rangle_0 = k_1 \sum_{S=X\cup Y}\langle\sigma_{k_1-1}\prod_{i\in X}\sigma_{k_i}\sigma_0\rangle_0\langle\sigma_0\prod_{j\in Y}\sigma_{k_j}\sigma_{k_{s-1}}\sigma_{k_s}\rangle_0,$$

$$\begin{aligned}\langle\sigma_{k_1}\sigma_{k_2}\cdots\sigma_{k_s}\rangle_1 &= \frac{1}{12}k_1\langle\sigma_{k_1-1}\sigma_{k_2}\cdots\sigma_{k_s}\sigma_0\sigma_0\rangle_0 \\ &\quad + k_1\sum_{S=X\cup Y}\langle\sigma_{k_1-1}\prod_{i\in X}\sigma_{k_i}\sigma_0\rangle_0\langle\sigma_0\prod_{j\in Y}\sigma_{k_j}\rangle_1,\end{aligned} \tag{107}$$

All of them follow from the basic recursion relation

$$\begin{aligned}\langle\sigma_{k_1}\sigma_{k_2}\sigma_{k_3}\rangle_0 &= k_1\langle\sigma_{k_1-1}\sigma_0\rangle_0\langle\sigma_0\sigma_{k_2}\sigma_{k_3}\rangle_0, \\ \langle\sigma_k\rangle_1 &= \frac{1}{12}k\langle\sigma_{k-1}\sigma_0\sigma_0\rangle_0 + k\langle\sigma_{k-1}\sigma_0\rangle_0\langle\sigma_0\rangle_1, \text{ and}\end{aligned} \tag{108}$$

$$\frac{\partial}{\partial a_k}\langle N\rangle = \langle\sigma_k N\rangle, \tag{109}$$

Substituting the operators O_{p-k-1} of the Matrix Model for operators σ_k of the Intersection Theory, we obtain and can try to check the relations

$$\langle O_{p-k_1-1}O_{p-k_2-1}O_{p-k_3-1}\rangle_0 = k_1\langle O_{p-k_1}O_{p-1}\rangle_0\langle O_{p-1}O_{p-k_2-1}O_{p-k_3-1}\rangle_0,$$

$$\langle O_{p-k-1}\rangle_1 = \frac{1}{12}k\langle O_{p-k}O_{p-1}O_{p-1}\rangle_0 + k\langle O_{p-k}O_{p-1}\rangle_0\langle O_{p-1}\rangle_1 \qquad (110)$$

At arbitrary $\{t_k\}$ one can get

$$\langle O_{k_1}O_{k_2}\rangle_0 = \frac{\partial^2 Z_0}{\partial t_{k_1}\partial t_{k_2}} = \frac{(u^*)^{2p-k_1-k_2-1}}{2p-k_1-k_2-1},$$

$$\langle O_{k_1}O_{k_2}O_{k_3}\rangle_0 = \frac{\partial^3 Z_0}{\partial t_{k_1}\partial t_{k_2}\partial t_{k_3}} = -\frac{(u^*)^{3p-k_1-k_2-k_3-3}}{\mathcal{P}'(u^*)}, \qquad (111)$$

$$\langle O_k\rangle_1 = \frac{\partial Z_1}{\partial t_k} = -\frac{p-k-1}{12\mathcal{P}'(u^*)}(u^*)^{p-k-2} + \frac{\mathcal{P}''(u^*)}{12(\mathcal{P}'(u^*))^2}(u^*)^{p-k-1}.$$

Using these expressions, we see that recursion relations are really fulfilled.

11. Conclusion

We have derived the torus partition function Z_1 in the p-critical One-Matrix Model. Using this expression as well as the known expression of Z_0 we can compute the correlation numbers in KdV, as well as in CFT frames. The results in CFT frame in genus one should be compared with the correlation numbers in the Minimal Liouville Gravity. We expect that the coincidence in genus one will take place similarly to that observed in genus zero. It has been already checked for the one-point case on torus.[12] The Witten recursive relation is obeyed by the correlation numbers of One Matrix Model in KdV frame in genus zero and genus one. Some open problems are:
What is the form of the resonance transformations for general (p, q) case?
How to derive the recursive KdV and string relations in Liouville Gravity?

Acknowledgments

I am grateful to Julia Nyíri and Péter Lévai for the possibility to take part in the V.N.Gribov Memorial conference. I am thankful to V.Belavin, M.Bershtein, G.Tarnopolsky and A.Zamolodchikov for useful discussions. This research was carried out in the framework of the Federal Program "Scientific and Scientific-Pedagogical personnel of innovational Russia" No 1339, RFBR initiative interdisciplinary project 09-02-12446-ofi-m and RFBR-CNRS project PICS-09-02-91064.

References

1. A. Polyakov, *Phys. Lett.* **B103**, 207 (1981).
2. V.G. Knizhnik *et al.*, *Mod. Phys. Lett.* **A3**, 819 (1988).
3. V.A. Kazakov, *Phys. Lett.* **A119**, 140 (1986);
   V. Kazakov, *Mod. Phys. Lett.* **A4**, 215 (1989);
   M.Staudacher, *Nucl. Phys.* **B336**, 349 (1990).
4. E. Witten, *Nucl. Phys.* **B340**, 281 (1990).
5. M. Kontsevich, *Comm. Math. Phys.* **147**, 1 (1992).
6. E. Verlinde and H. Verlinde, *A Solution Of Two Dimensional Topological Quantum Gravity*, preprint IASSNS-HEP-90/40, PUPT-1176 (1990).
7. P.H. Ginsparg and G.W. Moore, arXiv:hep-th/9304011;
   P. Di Francesco, P.H. Ginsparg, J. Zinn-Justin, *Phys. Rep.* **254**, 1 (1995).
8. G.W. Moore, N. Seiberg and M. Staudacher, *Nucl. Phys.* **B362**, 665 (1991).
9. A. Belavin and A. Zamolodchikov, arXiv:0811.0450v1 [hep-th] (2008).
10. G. Tarnopolsky, arXiv:0912.4971 (2009).
11. A. Belavin and G. Tarnopolsky, hep-th/1006.2056 (2010).
12. V. Belavin, to be published (2010)

CAPTURE OF DARK MATTER BY THE SOLAR SYSTEM. ANALYTICAL ESTIMATES

I. B. KHRIPLOVICH

Budker Institute of Nuclear Physics,
11 Lavrentjev pr., 630090 Novosibirsk, Russia
khriplovich@inp.nsk.su

We consider the capture of galactic dark matter by the Solar System, due to the gravitational three-body interaction of the Sun, a planet, and a dark matter particle. Closed (up to an overall numerical factor) analytical expression is derived for the capture cross-section. With this result we arrive at simple estimates for density and velocity distribution of captured dark matter particles close to the Earth.

Keywords: Restricted three-body problem; dark matter; solar system.

This year Vladimir Naumovich Gribov, a great physicist and a remarkable personality, would be 80.

Gribov was truly an epoch in high energy physics. The modern theoretical conceptions, concerning the processes at high energies, to a large extent belong to him. It is no accident that for physicists working in this field the Theoretical Department of the Petersburg Institute of Nuclear Physics was a sort of Mecca when Gribov was working there.

In fact, the interests of Gribov were much wider than just high energy processes, physics was a single whole for him. Among his well-known contributions to this single whole, there are for instance the investigations of the neutrino oscillations and of the subtle problems of quantization of non-abelian gauge theories. However, Gribov's influence was due not only to his published papers. How valuable were discussions with him! Gribov was ready to discuss any interesting problem, irrespective of the field of physics it belonged to. His criticisms were sometimes merciless, but always useful.

In discussions, Gribov was really working, again and again generating truly remarkable ideas. In particular, it was in a discussion that Gribov gave his interpretation of the instanton, the classical solution of nonlinear field equations, as an underbarrier trajectory connecting degenerate vacua. At present, this interpretation is a sort of folklore.

In 1971 or 1972, 2–3 years before the famous paper by Hawking, Gribov precisely formulated, also in a discussion, the conclusion that black holes radiate. In those, now remote times this his assertion caused a strong negative reaction of a respectable representative of the astrophysical community, who insisted that the idea of black hole radiation contradicts the very fundamentals of general relativity.[a] One can only regret that Gribov did not publish this result, apparently he considered it self-evident.

When Gribov had gone, the words were pronounced: "A great physicist lived among us". This is no exaggeration. Lev Lipatov said, at last, what many of us had known, though had not dared to say aloud these rather unconventional words.

When pondering over the subject of my talk at the meeting dedicated to Gribov, I have decided to present at it (instead of "telling old jokes") a quite recent at that moment analysis of the so-called restricted three-body problem. It results in a closed (up to an overall numerical factor) analytical expression for the capture cross-section in this problem. With thus obtained result, one arrives at the estimate for the dark matter density close to the Earth.

Possibly, with Gribov's broad interests in physics, the present investigation would be also of interest to him.

1. Introduction

The local density of dark matter (dm) in our Galaxy can be estimated, according to the most recent analysis of Ref. 1 (see also references therein), as

$$\rho_g = (8.31 \pm 0.59(\text{stat}) \pm 1.37(\text{syst})) \cdot 10^{-25}\ \text{g/cm}^3\,. \tag{1}$$

However, only upper limits on the level of 10^{-19} g/cm^3 (see, e.g., Refs. 2, 3) are known for the density of dark matter particles (dmp) in the Solar System (SS). Besides, even these upper limits are derived under the quite strong assumption that the distribution of dm density in the SS is spherically-symmetric with respect to the Sun. Meanwhile, information on dm density

[a]It is worth mentioning, to the credit of this person, that, after publication of the mentioned paper by Hawking, he said aloud: "We were just fools when did not to listen to Volodya."

in SS is very important, in particular for the experiments aimed at the detection of dark matter.

The capture of dark matter by the SS was addressed previously in Refs. 4–7. In particular, in Ref. 7 the total mass of the captured dark matter was estimated analytically. In the present note the analytical estimates are given for the capture cross-section, as well as for the density and velocity distribution of captured dm close to the Earth.

Of course, a particle cannot be captured by the Sun alone. The interaction with a planet is necessary for it, i.e. this is essentially a three-body (the Sun, planet and dmp) problem. Obviously, the capture is dominated by the particles with orbits close to parabolic ones with respect to the Sun; besides, the distances between their perihelia and the Sun should be comparable with the radius of the planet orbit r_p . Just the trajectories of these particles are most sensitive to the attractive perturbation by the planet.

The capture can be effectively described by the so-called restricted three-body problem (see, e.g. Ref. 8). In this approach the interaction between two heavy bodies (the Sun and a planet in our case) is treated exactly. As exactly is treated the motion of the third, light body (a dmp in our case) in the gravitational field of the two heavy ones. One neglects however the back reaction of a light particle upon the motion of the two heavy bodies. Obviously, this approximation is fully legitimate for our purpose.

Still, the restricted three-body problem is rather complicated, its solution requires both subtle analytical treatment and serious numerical calculations (see, e.g. Ref. 9). Under certain conditions the dynamics of light particle becomes chaotic. The "chaotic" effects are extremely important for the problem. However their quantitative investigation is quite complicated and remains beyond the scope of the present note. We confine here instead to simple estimates which could be also of a methodological interest by themselves. On the other hand, thus derived results for the total mass and density of the captured dark matter constitute at least an upper limit for their true value. As to the velocity distribution of dmp's given here, together with the mentioned result for the dark matter density, it could be possibly of some practical interest for planning the experimental searches for dm.

2. Total mass of dark matter captured by the Earth

The Solar System is immersed in the halo of dark matter and moves together with it around the center of our Galaxy. To simplify the estimates, we assume that the Sun is at rest with respect to the halo. The dark matter particles in the halo are assumed also to have in the reference frame, comoving with the halo, the Maxwell distribution:

$$f(v)\,dv = \sqrt{\frac{54}{\pi}}\,\frac{v^2 dv}{u^3}\,\exp\left(-\frac{3}{2}\frac{v^2}{u^2}\right), \tag{2}$$

with the local rms velocity $u \simeq 220$ km/s. Let us note that the velocities v discussed in this section are the asymptotic ones, they refer to large distances from the Sun, so that their values start at $v = 0$ and formally extend to ∞.

The amount of dm captured by the SS can be found by means of simple estimates.[b] The total mass captured by the Sun (its mass is M) together with a planet with the mass m_p, during the lifetime

$$T \simeq 4.5 \cdot 10^9 \text{ years} \simeq 10^{17} \text{ s} \tag{3}$$

of the SS, can be written as follows:

$$\mu_p = \rho_g T < v\,\sigma >; \tag{4}$$

here σ is the capture cross-section. The product σv is averaged over distribution (2); with all typical velocities in the SS much smaller than u, this distribution simplifies to

$$f(v)\,dv = \sqrt{\frac{54}{\pi}}\,\frac{v^2 dv}{u^3}\,. \tag{5}$$

To estimate the average value $< v\,\sigma >$, we resort to dimensional arguments, supplemented by two rather obvious physical requirements: the masses m_p and M of the two heavy components of our restricted three-body problem should enter the result symmetrically, and the mass of the dmp should not enter the result at all in virtue of the equivalence principle. Thus, we arrive at

$$< v\,\sigma > \sim \sqrt{54\pi}\;\frac{k^2\, m_p\, M}{u^3}\,, \tag{6}$$

or

$$\int_0^\infty dv\, v^3\, \sigma \;\sim \pi\, k^2\, m_p\, M\,; \tag{7}$$

here k is the Newton gravitation constant; an extra power of π, inserted into these expressions, is perhaps inherent in σ. Since the capture would be

[b]These estimates were given previously in Ref. 7. Here we repeat them, as well as results (8), (9), (20) (see below), since they are essential for the present discussions.

impossible if the planet were not bound to the Sun, it is only natural that the result is proportional to the corresponding effective "coupling constant" $k\, m_p\, M$. One more power of k corresponds to the gravitational interaction of the dark matter particle. The final estimate for the captured mass is

$$\mu_p \sim \rho_g T \sqrt{54\pi}\;\; k^2\, m_p\, M/u^3\,. \tag{8}$$

For the Earth it constitutes

$$\mu_E \sim 8 \cdot 10^{18}\, \mathrm{g}\,. \tag{9}$$

3. Capture cross-section

By the same dimensional reasons (and in the complete correspondence with formula (7)), the total capture cross-section for the Earth should look as follows:

$$\sigma \sim \pi\, k^2\, m_E\, M\, /\tilde{v}^4\,, \tag{10}$$

where m_E is the mass of the Earth, and $\tilde{v}$ is some velocity which can be estimated as follows. It is natural to assume that the capture of dm particles occurs when they are close to the Earth, i.e. at the distances $\sim r_E$ from the Sun. As natural are the following assumptions: 1) the initial velocities of the captured dmp's exceed only slightly the parabolic one $v_{\rm par}$ ($v_{\rm par}^2 = 2kM/r_E$); 2) their final velocities are only slightly less than v_{par}. To our accuracy, here we omit the factor of 2 in the definition of $v_{\rm par}^2$, and thus put $\tilde{v}^2 \sim v_E^2 = kM/r_E$ ($v_E = 30$ km/s is the velocity of the Earth). Thus, the capture cross-section is

$$\sigma \;\sim\; \pi\, k^2\, m_E\, M\, /v_E^4\,. \tag{11}$$

This formula can be also conveniently rewritten as

$$\sigma \;\sim\; \pi\, r_E^2\, (m_E/M)\,. \tag{12}$$

Let us note here that the impact parameter corresponding to formula (12), i.e. the typical distance between a dmp and the Earth crucial for the capture, is

$$r_{\rm imp} \;\sim\; r_E\, (m_E/M)^{1/2} \ll r_E\,. \tag{13}$$

In fact, this impact parameter corresponds to the distance at which the attraction to the Earth equals the attraction to the Sun, i.e. where

$$km/r^2 > kM/r_E^2 \;\;\; (r \ll r_E)\,. \tag{14}$$

Up to now, in all relevant formulae, (7), (10), (11), we dealt with the capture cross-section averaged over the directions of the dmp velocity $\mathbf{v}$. However, this cross-section depends essentially on the mutual orientation of $\mathbf{v}$ and $\mathbf{v}_E$. Certainly, it is maximum when these velocities are parallel and as close as possible by modulus. Besides, the impact parameter $r_{\rm imp}$ of the collision is much less than the radius r_E of the Earth orbit (see (13)), and thus within the distances $\simeq r_{\rm imp}$ both the Earth and dmp trajectories can be treated as rectilinear. Therefore, it looks quite natural to identify $\tilde{v}$ in (10) with the relative velocity $|\mathbf{v} - \mathbf{v}_E|$ of the dmp and the Earth, i.e. to generalize formula (11) as follows:

$$d\sigma \sim \frac{k^2\, m_p\, M}{(\mathbf{v} - \mathbf{v}_E)^4}\, \frac{1}{4}\, d\Omega \tag{15}$$

(factor 1/4 is introduced here for the correspondence with factor π in (11): $(1/4)\int d\Omega = \pi$).

Thus derived total cross-section is

$$\sigma \sim \frac{1}{4}\int d\Omega\, \frac{k^2\, m_p\, M}{(\mathbf{v} - \mathbf{v}_E)^4} = \frac{\pi\, k^2\, m_p\, M}{(v^2 - v_E^2)^2}\,. \tag{16}$$

Clearly, it is the particles moving initially with the velocities only slightly above the parabolic one $\sqrt{2}\, v_E = 42$ km/s that are captured predominantly, and thus, with $v = \sqrt{2}\, v_E$, cross-sections (11) and (16) practically coincide.

On the other hand, it follows from (16) that in the vicinity of the Earth the captured particles move with respect to it with the velocities close to $(\sqrt{2} - 1)v_p \simeq 12$ km/s.

4. Space distribution of captured dark matter

The captured dmp's had initial trajectories predominantly close to parabolas focussed at the Sun, and the velocities of these dmp's change only slightly as a result of scattering. Therefore, their trajectories become elongate ellipses with large semimajor axes, still focussed at the Sun. The ratio of their maximum $r_{\rm max}$ and minimum $r_{\rm min}$ distances from the Sun is (see Ref. 10)

$$\frac{r_{\rm max}}{r_{\rm min}} = \frac{1+e}{1-e}\,, \tag{17}$$

where e is the eccentricity of the trajectory. In the numerator of this ratio, we can safely put with our accuracy $1 + e \simeq 2$ both for the initial parabolas and final ellipses. In our case, as a result of the capture, the eccentricity changes from $1+\varepsilon_1$ to $1-\varepsilon_2$, where $\varepsilon_{1,2} \ll 1$. This loss of eccentricity is due to the gravitational perturbation by the Earth, and therefore both ε_1 and ε_2 are proportional to m_E. In particular, for the final ellipse

$$r_{\max} \sim 1/m_E \,. \tag{18}$$

On the other hand, $r_{\min}$ is close to the radius r_E of the Earth orbit. Thus, for dimensional reasons, we arrive (see Ref. 7) at

$$r_{\max} \sim r_E \, (M/m_E) \,. \tag{19}$$

Let us note here that the analogous estimate for the case of Jupiter complies qualitatively with the results of corresponding numerical calculations presented in Ref. 9.

Obviously, the semimajor axis a_{dmp} of the trajectory of a captured dmp is on the same order of magnitude as $r_{\max}$. Then, the time spent by a dmp, with the characteristic velocity close to v_E and at the distance from the Sun close to r_E, is comparable to the orbital period of the Earth $T_E = 1$ year. Besides, the orbital period T is related to the semimajor axis a as follows (see Ref. 10): $T \sim a^{3/2}$. Thus, we arrive at the following estimate for the orbital period of the captured dmp[c]:

$$T_{\mathrm{dmp}} \sim T_E \, (M/m_E)^{3/2} \,. \tag{20}$$

In other words, the relative time spent by a dmp at the distances $\sim r_E$ from the Earth can be estimated as $(m_E/M)^{3/2}$. Moreover, the typical distances from the Earth at which a dmp can be captured, should be less than the impact parameter $r_{\mathrm{imp}} \sim r_E \, (m_E/M)^{1/2}$ (see (1)). Thus, the relative time spent by a dmp sufficiently close to the Earth to be captured, can be estimated as $(m_E/M)^2$.

With the impact parameter (1), the corresponding volume V, centered at the Earth and crucial for the capture, can be estimated as

$$V \sim \frac{4\pi}{3}\, r_{\mathrm{imp}}^3 \sim \frac{4\pi}{3}\, r_E^3 \, (m_E/M)^{3/2} \ll \frac{4\pi}{3}\, r_E^3 . \tag{21}$$

Let us combine formula (9) for the total captured mass with the effective volume (21) occupied by this mass and with the estimate $(m_E/M)^2$ for the relative time spent by a dmp within the impact parameter (1) with respect to the Earth. In this way we arrive at the following estimate for the density of dark matter, captured by the SS, in the vicinity of the Earth:

$$\rho_E \sim 10^{-24} \ \mathrm{g/cm}^3 \,. \tag{22}$$

---

[c]In the case of the Earth, this orbital period is huge, $\sim 10^8$ years. Still, it is much less than the lifetime of the SS, $\sim 5 \cdot 10^9$ years.

This estimate practically coincides with the value (1) for the galactic dm density.

In fact, the result (22) for the density of the captured dm, as well as the estimates (8) and (9) for its total mass, should be considered as upper limits only, since we have neglected therein the inverse process, that of the ejection of the captured dm from the SS. The characteristic time of the inverse process is not exactly clear now. Therefore, it cannot be excluded that it is comparable to, or even larger than, the lifetime T of the SS (see Ref. 7). In this case our estimates for the total mass and the density of the captured dm are valid.

If this is the case indeed, then the dm around the Earth consists essentially of two components with comparable densities. In line with the common component with the typical velocity around $u \sim 220$ km/s, there is one more, with the velocity relative to the Earth $\gtrsim$12 km/s.

Acknowledgments

I am grateful to V.V. Sokolov, A.A. Pomeransky, and V.G. Serbo for useful discussions. The work was supported by the Russian Foundation for Basic Research through Grant No. 08-02-00960-a.

References

1. M. Pato, O. Agertz, G. Bertone, B. Moore and R. Teyssier *Phys. Rev.* **D82**, 023531 (2010).
2. I.B. Khriplovich and E.V. Pitjeva, *Int. J. Mod. Phys.* **D15**, 615 (2006).
3. I.B. Khriplovich, *Int. J. Mod. Phys.* **D16**, 1475 (2007).
4. A. Gould and S.M.K. Alam, *Astrophys. J.* **549**, 72 (2001).
5. J. Lundberg and J. Edsjo, *Phys. Rev.* **D69**, 123505 (2004).
6. A.H.G. Peter, *Phys. Rev.* **D79**, 1003531, 1003532, 1003533 (2009).
7. I.B. Khriplovich and D.L. Shepelyansky, *Int. J. Mod. Phys.* **D18**, 1903 (2009).
8. V. Szebehely, *Theory of Orbits*, Academic Press, N.Y. (1964).
9. T.Y. Petrosky, *Phys. Lett.* **A117**, 328 (1986).
10. L.D. Landau and E.M. Lifshitz, *Mechanics*, Nauka, Moscow (1988), §15.

AN OLD EINSTEIN — EDDINGTON GENERALIZED GRAVITY AND MODERN IDEAS ON BRANES AND COSMOLOGY

A.T. FILIPPOV

Joint Institute for Nuclear Research, Dubna, Moscow Region RU-141980, Russia
alexandre.filippov@jinr.ru

We briefly discuss new models of an 'affine' theory of gravity in multidimensional space-times with *symmetric connections*. We use and generalize Einstein's proposal to specify the space-time geometry by use of the Hamilton principle to determine the connection coefficients from a *geometric Lagrangian* that is an arbitrary function of the generalized Ricci curvature tensor and of other fundamental tensors. Such a theory supplements the standard Einstein gravity with dark energy (the cosmological constant, in the first approximation), a neutral massive (or tachyonic) vector field (*vecton*), and massive (or tachyonic) *scalar fields*. These fields couple only to gravity and can generate dark matter and/or inflation. The concrete choice of the geometric Lagrangian determines further details of the theory. The most natural geometric models look similar to recently proposed brane models of cosmology usually derived from string theory.

The history of science teaches us not to completely forget beautiful and logically consistent papers of the past that were not understood in time. Even though not recognized by the contemporaries (and, often, by the authors) some of them happen to become of interest many years after their publishing. One can recall sufficiently many examples of such work and here we discuss a misunderstood and forgotten model based on work of three eminent scientists (Weyl, Eddington and Einstein) that was reinterpreted and generalized in Refs. 1–3.

By the end of 1922, Einstein deeply studied and seriously reconsidered attempts of Weyl and Eddington (see Refs. 4–6) to construct an affine modification* of his general relativity. In 1923 he published three beautiful

*In 1918, Weyl introduced a special symmetric non-Riemannian connection depending on a metric tensor and on a vector field ('Weyl's connection'), which he attempted to identify with the electromagnetic potential. His theory was severely criticized and is

and concise papers[7] later summarized in Ref. 8 and soon forgotten (but see brief discussions in Refs. 9, 10). The most clear exposition of Einstein's approach is given in Ref. 8 while the most beautiful model was proposed in the first paper of the series.[7] Here we only briefly summarize general principles, which can more or less naturally restrict possible choice of the physical models. Then a simple model satisfying these principles and generalizing Einstein's first paper, which we call the Einstein–Eddington model, will be introduced and compared to some recently discussed cosmologies based on string theory.

The most important properties of the affine theory are the following. **1.** It predicts the existence of one or more vector fields with real or imaginary mass. **2.** Its D-dimensional generalization predicts (after the simplest dimensional reduction) $(D-4)$ scalar fields with the same mass. **3.** Both the vector and scalar fields couple to gravity only (being the part of the generalized gravitation). **4.** The most natural effective ('physical') Lagrangian contains Eddington–Einstein terms (nowadays often called Dirac–Born–Infeld terms).

Einstein's key idea was to derive the concrete form of the affine connection by applying the Hamilton principle to a generic Lagrangian depending on the generalized Ricci curvature. This assumption completely fixes a geometry, which does not coincide with Weyl's geometry, but belongs to the same simple class of connections introduced and discussed in Refs. 1–3. Einstein's unusual result was difficult to comprehend in the first half of the last century and it remains somewhat puzzling even these days. From the modern mathematics viewpoint, its origin could be ascribed to a sort of a mismatch between the affine connection geometry and the Lagrangian 'geometry'. At the moment, it is difficult to find a more detailed explanation. Possibly, this is an interesting mathematical problem.

In Refs. 1–3 we follow Einstein's approach and first construct a *geometric Lagrangian density* having the dimension L^{-D} (in units $c = 1$). Then we show that, without a metric, one can use scalar densities of weight two constructed of pure geometric fields (see Refs. 1–3), the square roots of which give the desirable scalar densities of weight one. The effective *physical Lagrangians* are derived from the geometric ones. A more detailed presentation of the main steps briefly discussed here can be found in Refs. 1–3.

Here we first outline basic **geometrical facts** and then concentrate on a physical model that looks most interesting for applications to cosmology.

mostly remembered because in it he first introduced a fairly general concept of the gauge symmetry.

A general exposition of the theory of non-Riemannian spaces equipped with a *symmetric connection* can be found in Refs. 9, 11, and in our previous papers. In general, the connection coefficients can be expressed in terms of a Riemannian connection Γ^i_{jk} and of an arbitrary third rank *tensor* a^i_{jk} that is symmetric in the lower indices

$$\gamma^i_{jk} = \Gamma^i_{jk}[g] + a^i_{jk} \,. \qquad (1)$$

Here g_{ij} is and arbitrary symmetric tensor and $\Gamma^i_{jk}[g]$ is its Christoffel symbol. More precisely, for any symmetric connection γ^i_{jk}, there exists a symmetric tensor g_{ij} and a tensor $a^i_{jk} = a^i_{kj}$ such that (1) is satisfied.

The curvature tensor r^i_{jkl} can be defined in terms of γ^i_{jk} by the standard general expression not using any metric,

$$r^i_{jkl} = -\gamma^i_{jk,l} + \gamma^i_{mk}\gamma^m_{jl} + \gamma^i_{jl,k} - \gamma^i_{ml}\gamma^m_{jk} \,. \qquad (2)$$

Then, the Ricci-like (but *non-symmetric*) curvature tensor can be defined by contracting the indices i, l:

$$r_{jk} \equiv r^i_{jki} = -\gamma^i_{jk,i} + \gamma^i_{mk}\gamma^i_{ji} + \gamma^i_{ji,k} - \gamma^i_{mi}\gamma^m_{jk} \qquad (3)$$

(we again stress that $\gamma^i_{jk} = \gamma^i_{kj}$ but $r_{jk} \neq r_{kj}$). Using only these tensors and the completely antisymmetric tensor density of the rank D, we can construct a quite rich geometric structure.

The *antisymmetric part* of the Ricci curvature r_{ij} can be expressed in terms of the vector field $\gamma_i \equiv \gamma^m_{im}$ or in terms of $a_i \equiv a^m_{im}$, which differ by the gradient term $\partial_i \ln\sqrt{|g|}$ ($g \equiv \det(g_{ij})$):

$$a_{ij} \equiv \frac{1}{2}(r_{ij} - r_{j\,i}) \equiv -\frac{1}{2}(a_{i,j} - a_{j,i}) \equiv -\frac{1}{2}(\gamma_{i,j} - \gamma_{j,i}) \,. \qquad (4)$$

We call this field *vecton* and will see that it can be massive or tachyonic depending on a choice of the connection. This definition of the vecton is independent of the division of the connection (1) into the metric and non-metric parts. By the way, $r^m_{mij} = 2a_{ij}$.

Introducing the covariant derivative ∇^γ_i (with respect to the connection γ) we can write the symmetric part of the curvature as

$$s_{ij} \equiv \frac{1}{2}(r_{ij} + r_{j\,i}) = -\nabla^\gamma_m \gamma^m_{ij} + \frac{1}{2}(\nabla^\gamma_i \gamma_j + \nabla^\gamma_j \gamma_i) - \gamma^m_{ni}\gamma^n_{mj} + \gamma^n_{ij}\gamma_n \,. \quad (5)$$

Using the 'metric' covariant derivative $\nabla^g_i \equiv \nabla_i$ we can rewrite s_{ij} in the form

$$s_{ij} = R_{ij}[g] - \nabla_m a^m_{ij} + \frac{1}{2}(\nabla_i\, a_j + \nabla_j\, a_i) + a^m_{ni}a^n_{mj} - a^m_{ij}a_m \,, \qquad (6)$$

where $R_{ij}[g]$ is the standard Ricci tensor of a Riemannian space with the metric g_{ij}.

Now, suppose that $\mathbf{c}^{ij} \equiv \sqrt{-g}\, c_{ij}$ is an arbitrary tensor density. Then, its covariant derivative with respect to connection (1) is defined by

$$\nabla_i^{\gamma}\, \mathbf{c}^{kl} \;=\; \partial_i\, \mathbf{c}^{kl} + \gamma_{im}^{k}\, \mathbf{c}^{ml} + \gamma_{im}^{l}\, \mathbf{c}^{km} - \gamma_{im}^{m}\, \mathbf{c}^{kl}\,. \tag{7}$$

For any *antisymmetric* density, $\mathbf{c}^{ij} \equiv \mathbf{f}^{ij} = -\mathbf{f}^{ji}$, it follows that

$$\nabla_i^{\gamma}\, \mathbf{f}^{ik} \;=\; \nabla_i^{g}\, \mathbf{f}^{ik} \;=\; \partial_i \mathbf{f}^{ik}\,. \tag{8}$$

The *symmetric* tensor density $\mathbf{g}^{ij} \equiv \sqrt{-g}\, g^{ij}$, obviously satisfies the equations

$$\nabla_i^{\gamma}\, \mathbf{g}^{ik} \;=\; \sqrt{-g}\, a_{im}^{k}\, g^{im}\,, \qquad \nabla_i^{g}\, \mathbf{g}^{ik} \;=\; 0. \tag{9}$$

Eqs. (5)–(9) will be used in what follows.

For a general symmetric connection one can introduce the concept of the geodesic curve (path), the tangent vector to which is parallel to itself at every point of the curve. The equations for the geodesic curves of any symmetric connection γ_{jk}^{i} can be written as

$$\ddot{x}^{\,i} + \gamma_{jk}^{i}\, \dot{x}^{j}\, \dot{x}^{k} = 0\,, \tag{10}$$

where the dot denotes differentiating with respect to the so called 'affine' parameter τ of the curve $x^i(\tau)$. Using the affine parameter we can compare the distances between points on the same curve.

For a particular geodesics, the affine parameter is unique up to an affine transformation $\tau \mapsto \tau' = a\tau + b$. Each connection define the unique set of paths, but all symmetric connections (with an arbitrary vector $\hat{a}_k$)

$$\hat{\gamma}_{jk}^{i} = \gamma_{jk}^{i} + \delta_j^i\, \hat{a}_k + \delta_k^i\, \hat{a}_j\,, \tag{11}$$

define the same paths. The Weyl (conformal) tensor W_{jkl}^{i} of connection (11) is independent of $\hat{a}_k$ while the Ricci tensor and its symmetric and antisymmetric parts are $\hat{a}_i$-dependent (see Ref. 11 for more details).

Therefore, an interesting class of connections is

$$\hat{\gamma}_{jk}^{i} = \Gamma_{jk}^{i}[g] + \delta_j^i\, \hat{a}_k + \delta_k^i\, \hat{a}_j\,, \tag{12}$$

where $\Gamma_{jk}^{i}[g]$ is a Riemannian connection (the Christoffel symbol of a symmetric tensor g_{ij}). The paths of the connection $\hat{\gamma}_{jk}^{i}$ coincide with the geodesics of $\Gamma_{jk}^{i}[g]$, but the Ricci tensor of $\hat{\gamma}$ is symmetric *if and only if* $\hat{a}_i = \partial_i\, \hat{a}$ with some scalar $\hat{a}$. We see that connection (12) is maximally close to the Riemannian connection $\Gamma_{jk}^{i}[g]$ and may be called a *geodesically Riemannian* ('g-Riemannian') connection. Weyl and Einstein studied more general connections that belong to the following class introduced in Refs. 1, 2:

$$\gamma^i_{jk} = \Gamma^i_{jk}[g] + \alpha(\delta^i_j\,\hat{a}_k + \delta^i_k\,\hat{a}_j) - (\alpha - 2\beta)g_{jk}\,\hat{a}^i\,, \tag{13}$$

where $\hat{a}^i = g^{im}\hat{a}_m$. The Weyl connection corresponds to $\beta = 0$ and the g-Riemannian connection, to $\alpha = 2\beta$. Einstein derived the connection for the space-time dimension $D = 4$, his result is $\alpha = -\beta = \frac{1}{6}$ (it was generalized to any dimension in Ref. 3).

Using (6) it is easy to calculate the physically important expression for the symmetric part of the Ricci curvature. The terms linear in A are equal to

$$(\alpha + \beta)(\nabla_i\hat{a}_j + \nabla_j\hat{a}_i) + (\alpha - 2\beta)\,g_{ij}\nabla_m\hat{a}^m\,, \tag{14}$$

and the quadratic terms are

$$\hat{a}_i\hat{a}_j\left[(\alpha - 2\beta)^2 - 3\alpha^2\right] + 2\,g_{ij}\hat{a}^2(\alpha - 2\beta)(\alpha + \beta)\,. \tag{15}$$

As we shall soon see the presence of the $\hat{a}_i\hat{a}_k$ term in the expression for s_{ij} signals that the vector field a_i has in general a nonzero mass and that the sign of the first term in (15) can be positive or negative (the second term in (15) and the linear terms in (6) in general do not vanish). In particular, for the Weyl, Einstein and g-Riemannian connections the quadratic terms are, respectively:

$$\text{W:}\quad -\frac{1}{2}[\hat{a}_i\hat{a}_k - 2g_{ik}\hat{a}^2]\,, \qquad \text{E:}\ \frac{1}{6}\hat{a}_i\hat{a}_k\,, \qquad \text{g-R:}\quad -\frac{3}{4}\hat{a}_i\hat{a}_k\,. \tag{16}$$

Before we leave pure mathematics and turn to more physical problems, we should mention one of the characteristic properties of symmetric connections. For applications of geometry to gravity, it is very important that at every point of the affine-connected space-time manifold there must exist a geodesic coordinate system, such that the connection coefficients are zero at this point. Using the above formulas it is easy to prove that such a coordinate system exists if and only if the connection is symmetric. For symmetric connections, the Fermi theorem about the existence of geodesic coordinates along the curves also holds (for the precise definitions and proofs see Ref. 11).

Let us turn to **dynamics**. Weyl's approach to constructing a physical theory based on the affine geometry is direct (if we discard his ideas on 'linear metric' and on lengths calibrating): he first chooses a particular geometry (13) with $\beta = 0$ and then constructs tensor equations that should generalize the Einstein equations. He tries to identify the antisymmetric part of the curvature with the electromagnetic field tensor but the $\hat{a}_i \hat{a}_k$ terms spoil this interpretation. He eventually guessed a Lagrangian which is similar to the Einstein theory coupled to a vector field with the mass term that cannot be removed and with the cosmological constant that he introduces 'by hand'. Eddington tried to find a generalization of Einstein's theory by considering the most general nonsymmetric affine connection. In a discussion of possible scalar densities he suggests the simplest one (we call it *Eddington's scalar density*),

$$\mathcal{L} \equiv \sqrt{-\det(r_{ij})} \equiv \sqrt{-r}\,, \tag{17}$$

This resembles the fundamental scalar density of the Riemannian geometry, $\sqrt{-\det(g_{ij})} \equiv \sqrt{-g}$, and Eddington tried to directly identify s_{ij} with the metric. If we in addition identify a_{ij} with the electromagnetic field we get a Born-Infeld Lagrangian. However, Eddington did not succeed in constructing consistent equations.

A consistent *Lagrangian formulation* of the generalized theory was found by Einstein. His approach is conceptually different both from Weyl's and Eddington's ones and consists of two stages. In the first stage, he assumed that the general symmetric connection should be restricted by the Hamilton principle for a general Lagrangian density depending either on r_{ij} (see the second paper†) or on s_{ij} and a_{ij} separately (in the third paper). He gave no motivation for this assumption, but it is easy to see that the resulting theory in the limit $a_{ij} = 0$ is consistent with the standard general relativity supplemented with a cosmological term. In this stage, Einstein succeeded in deriving the remarkable expression for the connection (see (13) with $\alpha = -\beta = \frac{1}{6}$) and the general expression for s_{ij} depending on a massive (tachyonic) vector field and the metric tensor density $\mathbf{g}^{ij}$.

In the next stage, a concrete Lagrangian density $\mathcal{L}(s_{ij}\,, a_{ij})$ should be chosen. Einstein did not formulate any principle for selecting a Lagrangian, and both from geometric and physical standpoint his concrete choice seems sufficiently arbitrary, especially in the third paper where he essentially reproduced one of the Weyl results. We believe that his best choice was made in the first two papers and, indeed, very similar effective Lagrangians are considered in modern applications of the superstring theory to cosmology. We may try to formulate some properties of possible geometric Lagrangian densities $\mathcal{L}$ that are consistent with the Eddington–Einstein Lagrangian but

---

†In the first paper Einstein uses as the Lagrangian the Eddington density but later he realized that in the first stage it is sufficient to suppose that the Lagrangian is an arbitrary scalar density depending on r_{ij}

allow for a more general class of them (with different mass terms, different dependence on s_{ij}, a_{ij}, a_i, in different space-time dimensions).

Naturally, the Lagrangian must depend on *tensor variables having a direct geometric meaning.* It is desirable that in the next stage they will acquire a *natural physical interpretation.* As soon as we do not wish to fix the division of the connection into the metric and non-metric parts by Eq.(1), we can take the vector γ_i (not a_i!), second-rank tensors, s_{ij}, a_{ij}, $\gamma_{ij} \equiv \gamma_i \gamma_j$ (if we used representation (1) for γ_{ij}^k we could add to this list $\bar{\gamma}_{ij} \equiv \gamma_k a_{ij}^k$, but this tensor implicitly depends on the metric and we must not use it at the first stage). We can construct higher-rank tensors, but the tensors of the second rank (especially, the first three) look more fundamental from the physics point of view.

Consider the *second-rank tensors* as building blocks of the 'geometric' Lagrangian. They all have the dimension L^{-2} (in the units $c = 1$) and we can use as Lagrangian densities some homogeneous functions of the degree $D/2$ and dimension L^{-D} that are independent on any dimensional constants. After integrating over D-dimensional volume element $dx^0 \bigwedge ... \bigwedge dx^{D-1}$ we then get a dimensionless quantity playing a role of a *geometric action.* The simplest Lagrangian density then depends on three second-rank tensors,

$$\mathcal{L} = \mathcal{L}(s_{ij}, a_{ij}, \gamma_{ij}) \,, \tag{18}$$

and a density having the correct dimension L^{-D} can easily be written:

$$\mathcal{L}_g = \sqrt{- \det(s_{ij} + \nu a_{ij} + \nu_1 \gamma_{ij})} \,. \tag{19}$$

Here we take the minus sign because $\det(s_{ij}) < 0$ (due to the local Lorentz invariance) and we naturally assume that the same is true for $\det(s_{ij} + \nu a_{ij} + \nu_1 \gamma_{ij})$ (to reproduce Einstein's general relativity in the limit $\nu, \nu_1 \to 0$). The ν-parameters are dimensionless, we mainly introduce them to disentangle the scale of the mass parameter of the vector field from the cosmological constant. If we take the original Eddington–Einstein Lagrangian (17), the mass squared will be of the order of the cosmological constant Λ (see Ref. 2). Lagrangian (19) with $\nu_1 = 0$ was proposed and studied in some detail in Ref. 2. The general Lagrangian (19) was first considered in Ref. 3, where we also discussed a more general construction that allows to write other Lagrangians having the desired properties. Unfortunately, these generalized Lagrangians are more complicated both technically and conceptually, and we do not discuss them here.

We emphasize that the Lagrangians (17) and (19) are written in the form independent of D, although the analytic expressions for the dependence of the determinants on s_{ij} and a_{ij} essentially depend on D. Accordingly, the physical equations depend on the space-time dimension as we will shortly demonstrate.

The starting point for Einstein (in his first paper of the series[7]) was the action principle with Lagrangian density (17) depending on 40 connection functions γ^i_{kl}. Varying the action with respect to these functions, he derived 40 equations that allowed him to find the expression for γ^i_{kl} given by (13) with $\alpha = -\beta = \frac{1}{6}$ (in the four-dimensional space-time).

The main steps of his proof were reproduced in Ref. 2. Here we somewhat generalize the derivation to an arbitrary dimension D and assume that the geometric Lagrangian depends also on $\gamma_i \equiv \gamma^m_{im}$. We define the new tensor densities.[‡]

$$\frac{\partial \mathcal{L}}{\partial s_{ij}} \equiv \mathbf{g}^{ij}\,, \qquad \frac{\partial \mathcal{L}}{\partial a_{ij}} \equiv \mathbf{f}^{ij}\,, \qquad \frac{\partial \mathcal{L}}{\partial \gamma_i} \equiv \mathbf{b}^i\,, \tag{20}$$

and introduce a conjugate Lagrangian density $\mathcal{L}^* = \mathcal{L}^*(\mathbf{g}^{ij}, \mathbf{f}^{ij}, \mathbf{b}^i)$ by a Legendre transformation,

$$s_{ij} = \frac{\partial \mathcal{L}^*}{\partial \mathbf{g}^{ij}}\,, \qquad a_{ij} = \frac{\partial \mathcal{L}^*}{\partial \mathbf{f}^{ij}}\,, \qquad \gamma_i = \frac{\partial \mathcal{L}^*}{\partial \mathbf{b}^i}\,. \tag{21}$$

By varying $\mathcal{L}$ in γ^i_{kl} and using the above definitions, we can then show that the conditions $\delta\mathcal{L}/\delta\gamma^i_{kl} = 0$ are equivalent to the following 40 equations

$$2\nabla^{\gamma}_i\, \mathbf{g}^{kl} = \delta^l_i\, [\nabla^{\gamma}_m\, (\mathbf{g}^{km} + \mathbf{f}^{km}) - \mathbf{b}^k] + \delta^k_i\, [\nabla^{\gamma}_m\, (\mathbf{g}^{lm} + \mathbf{f}^{lm}) - \mathbf{b}^l]\,, \tag{22}$$

where ∇^{γ}_i is the covariant derivative with respect to the affine connection γ^i_{jk}. Remembering (8) we define the vector density $\hat{\mathbf{a}}^k$ by

$$\partial_i\, \mathbf{f}^{ki} - \mathbf{b}^k \equiv \hat{a}^k, \tag{23}$$

and then easily find that

$$\nabla^{\gamma}_i\, \mathbf{g}^{ik} = -\frac{D+1}{D-1}\hat{a}^k\,, \tag{24}$$

[‡]Following Eddington's notation, we let boldface Latin letters denote tensor densities. The derivatives in (20) and (21) must be properly symmetrized, which is easy in concrete calculations. We tacitly assume that geometry has only a single dimensional constant, e.g., the cosmological constant Λ with the dimension L^{-2}. To restore the correct dimension in (20) and (21), we must then multiply the densities by $\Lambda^{(D-2)/2}$.

Now it is easy to find the equations from which the connection coefficients can be derived (as in the Riemannian case):

$$\nabla_i^{\gamma}\, \mathbf{g}^{kl} \;=\; -\frac{1}{D-1}\big(\delta_i^{k}\hat{a}^l + \delta_i^{l}\hat{a}^k\big). \tag{25}$$

Defining the Riemann metric tensor g_{ij} by the equations

$$g^{kl}\sqrt{-g} = \mathbf{g}^{kl}\,, \quad g_{kl}\, g^{lm} \;=\; \delta_k^m\,, \tag{26}$$

we can then define the corresponding Riemannian covariant derivative ∇_i, for which

$$\nabla_i\, g_{kl} = 0, \quad \nabla_i\, g^{kl} = 0. \tag{27}$$

Taking the above into account, we can now use (25) to derive the expression for γ^i_{jk} in terms of the metric tensor g_{ij} and of the vector $\hat{a}^k \equiv \hat{\mathrm{a}}^k/\sqrt{-g}$,

$$\gamma^i_{jk} \;=\; \Gamma^i_{jk}[g] \,+\, \alpha_D\left[\,\delta^i_j\,\hat{a}_k + \delta^i_k\,\hat{a}_j - (D-1)\, g_{jk}\,\hat{a}^i\right], \tag{28}$$

which corresponds to $\alpha = \alpha_D$ and $\beta = \beta_D$ in (13), with

$$\alpha_D \equiv [(D-1)(D-2)]^{-1}\,, \qquad \beta_D \equiv -[2(D-1)]^{-1}\,. \tag{29}$$

For $D = 4$, this coincides with Einstein's result for the connection. If we add γ_{ij} as an independent variable, the connection remains the same. Note also that the added variables remain non-dynamical and attempting to make them dynamical in the second stage 'by hand' destroys the beauty of the original Einstein construction.

We cannot go deeper into discussions of further relations between geometry of affine connections and dynamical principles. But the above results show that these relations are rather complex and we do not yet understand their nature. Indeed, we tried to add new natural variables into the geometric Lagrangian, but the class of connections obtained as an output of Einstein's approach did not change at all. It can be argued that there are many other, not yet explored options, but in reality, we do not even know how to obtain Weyl's or g-Riemannian connections following Einstein's approach.

One of the possibilities could be to abandon some of Einstein's assumptions. The most serious drawback (or virtue, depending on a viewpoint) of his approach is that two pairs of the basic variables of the theory, (s_{ij}, $\mathbf{g}^{ij}$) and (a_{ij}, $\mathbf{f}^{ij}$), having very different geometrical and physical meaning, are treated symmetrically. Definition (20) looks quite natural for the metric density because Einstein's Lagrangian for the pure gravity theory is simply $\mathbf{g}^{ij}R_{ij}$. But Einstein's definition of $\mathbf{f}^{ij}$ tacitly (and, as we see, wisely!) assumes that the geometric Lagrangian is independent of γ_i or γ_{ij}. This may look rather paradoxical, but, as we have seen, the mass term is dictated by the geometry because its germ, the term $\sim a_i a_j$, is already present in the expression for s_{ij}.[§] Its interpretation as the physical mass comes when we write an effective physical Lagrangian. Then the geometric Lagrangian generates only the kinetic terms and is in fact the Lagrangian of a brane.

There are many other questions, which should be carefully discussed, but we postpone the discussion to future publications. Here, we present a simple example demonstrating how to eventually pass **from geometry to physics** and to demonstrate a relation of the Einstein approach to the present-day concerns. Our discussion suggests that the geometric Lagrangian (19) with $\nu_1 = 0$ is better motivated by geometry and physics than other ones. This Lagrangian is most natural and gives the effective physical Lagrangian belonging to a class widely discussed in relation to modern problems of cosmology. We only briefly describe this model which is, possibly, the simplest generalization of Einstein's general relativity.

Pure geometry gives us equations (4) and (6). With a^i_{jk} given by (13), their right-hand sides are given by $(a_{i,j} - a_{j,i})/2$, where $a_i = (D\alpha + 2\beta)\,\hat{a}_i$, and by the sum of R_{ij} with expressions (14), (15). To derive s_{ij} and a_{ij} in terms of the 'physical' variables g_{ij} and f_{ij} we must choose a Lagrangian (e.g., (19)) and then solve equations (20) with respect to the geometric variables s_{ij} and a_{ij}. Alternatively, if we know the conjugate Lagrangian $\mathcal{L}^*(\mathbf{g}^{ij}, \mathbf{f}^{ij})$, we can directly calculate them using (21).

In Ref. 2, we reproduced Einstein's result of Refs. 7, 5 (in which it was not written explicitly but could easily be derived):

$$\mathcal{L} \equiv \sqrt{-\det(r_{ij})} = 4\sqrt{-\det(\mathbf{g}^{ij} + \mathbf{f}^{ij})} \equiv 4\sqrt{-\det(g_{ij} + f_{ij})} = \mathcal{L}^* \,. \tag{30}$$

We emphasize that these equations are valid only in the four-dimensional theory. Note that the equality $\mathcal{L}^* = \mathcal{L}$ simply follows from the fact that $\mathcal{L}$ is a homogeneous function of the degree two but, in general, the concrete expression for $\mathcal{L}^*$ must be obtained by a direct calculation. Now we can show that the relation like (30) holds also for Lagrangian (19) with $\nu_1 = 0$ and $\nu \neq 0$, which we rewrite as

$$\mathcal{L}_\nu \equiv \sqrt{-\det(s_{ij} + \nu a_{ij})} \,. \tag{31}$$

This can be done by a direct computation but it is simpler to first dimensionally reduce $\mathcal{L}_\nu$.

[§]Therefore, it would be more natural to identify the field tensor of the massive vector field directly with a_{ij}, up to a necessary dimensional multiplier.

Consider the $D = 4$ case and define a *'spherical reduction'* not using any metric. Suppose that s_{ij} and a_{ij} are functions of (x^0, x^1) and that $a_2 = a_3 = 0$ (therefore, only $a_{01} = -a_{10} \neq 0$). We then assume that the symmetric matrix has the following nonzero elements: $s_{ij} = \delta_{ij}\, s_i\,,\ s_{01} = s_{10}$ (our result will not change if also $s_{23} \neq 0$). By explicitly deriving $s_{ij} + \nu a_{ij}$, we can find $\mathbf{g}^{ij}$ and $\mathbf{f}^{ij}$ (using (20)) and hence express $\det(\mathbf{g}^{ij} + \lambda \mathbf{f}^{ij})$ in terms of s_{ij} and a_{ij} :

$$16 \det(\mathbf{g}^{ij} + \lambda \mathbf{f}^{ij}) = \det[s_{ij} + (\nu^2 \lambda)\, a_{ij}]. \tag{32}$$

It follows that choosing $\lambda = \nu^{-1}$ we have¶

$$\begin{aligned} \mathcal{L}_\nu = \sqrt{-\det(s_{ij} + \lambda^{-1}\, a_{ij})} &= 4\sqrt{-\det(\mathbf{g}^{ij} + \lambda \mathbf{f}^{ij})} = \\ &= 4\sqrt{-\det(g_{ij} + \lambda f_{ij})} = \mathcal{L}^*_\lambda\,, \end{aligned} \tag{33}$$

where the sign and normalization are arbitrary chosen in relation to the cosmological interpretation. This result is written in the form not implying the spherical reduction, and we suppose it is true in a general four-dimensional theory. In arbitrary dimension ($D \neq 2$) it must be somewhat modified as was first shown in Ref. 2.

To similarly treat the **higher dimensional case** we first reduce the D-dimensional Lagrangian to the dimension four. For simplicity, let us consider $D = 5$. Then the field a_k $(k = 0, .., 4)$ depends only on x_i $(i = 0, .., 3)$, $a_{ij} = \frac{1}{2}(\partial_j a_i - \partial_i a_j)$, and $a_{4i} = 1/2\, (\partial_i\, a_4)$. Therefore the terms containing a^2_{4i} should be interpreted in four dimensions as kinetic terms of the *scalar field* a_4.‖ Applying spherical reduction to the resulting four-dimensional Lagrangian, we can construct a two-dimensional model effectively describing spherically symmetric solutions of the four-dimensional gravity coupled to the vecton and to the scalar fields. To get the corresponding Lagrangian we derive the determinant of the matrix

$$s_{ij} + \nu a_{ij} \equiv s_i \delta_{ij} + (\delta_{0i}\delta_{1j} + \delta_{0j}\delta_{1i})(s_{ij} + \nu a_{ij}) + (\delta_{i4} + \delta_{j4}) a_{ij}\,, \tag{34}$$

¶We ignore the dimensional constants while working mainly with geometrical theory, where presumably exists just one dimensional constant Λ (with $c = 1$). Then emergence of some dimensionless parameters may signal that there exist other dimensional constants (e.g., different scales in symmetric and antisymmetric sectors of geometry may be described by introducing our parameter ν). We eventually restore dimensions in the effective physical Lagrangian.

‖It can be seen that this scalar field is massive or tachyonic. In the simplest reduction, its mass coincides with that of the vecton.

where a_{ij} are defined in terms of a_i, and all the functions in (34) depend on x^0, x^1 (thus $a_{24} = a_{34} = 0$). The determinant is

$$\det(s_{ij} + \nu a_{ij}) = \prod_{i=0}^{4} s_i \,[1 + \tilde{s}_{01}^2 - \nu^2(\tilde{a}_{01}^2 + \tilde{a}_{04}^2 - \tilde{a}_{14}^2)], \tag{35}$$

where we define $\tilde{m}_{ij} \equiv m_{ij}\,|s_i s_j|^{-1/2}$. The determinant obviously has zeroes and thus its square root is always singular. Therefore, the corresponding two-dimensional dilaton gravity describing spherically symmetric solutions is rather unusual and complex. By further reductions to static or cosmological configurations we can construct corresponding one-dimensional dynamical systems describing static states with horizons as well as cosmological models. The cosmological models look realistic enough because they incorporate a natural sources of the dark energy, inflation, and, possibly, some candidates for the dark matter (for a more detailed discussion see Refs. 1, 2).

Before presenting a simplest cosmological model, we write the general D-dimensional theory. In addition to predicting scalar fields, the higher-dimensional Lagrangians differ from the ones usually considered in modern brane cosmology. In fact, while the square-root Lagrangian $\mathcal{L}$ produces the square-root Lagrangian $\mathcal{L}^*$, which gives the so-called DBI-like term in the effective physical Lagrangian (see many examples in Refs. 12–22), our higher-dimensional Lagrangian essentially depends on D:

$$\begin{aligned}\sqrt{-\det(s_{ij} + \nu a_{ij})} &= [-2^D \det(\mathbf{g}^{ij} + \lambda \mathbf{f}^{ij})]^{1/(D-2)} \\ &= \sqrt{-g}\,[-2^D \det(\delta_i^j + \lambda f_i^j)]^{1/(D-2)}\,, \end{aligned} \tag{36}$$

which coincides with (33) for $D = 4$. Following,[2] we may write the corresponding **physical Lagrangian**

$$\mathcal{L}_{eff} = \sqrt{-g}\left[-2\Lambda\,[\det(\delta_i^j + \lambda f_i^j)]^{1/(D-2)} + R(g) + c_a\, g^{ij} a_i a_j\right], \tag{37}$$

which should be varied with respect to the metric and the vector field; c_a is a parameter depending on D (Einstein's first model is obtained for $D = 4$ and $c_a = 1/6$). When the vecton field is zero, we have the standard Einstein gravity with the cosmological constant. Making the dimensional reduction from $D = 5$ to $D = 4$, we obtain the Lagrangian describing the vecton a_i, $f_{ij} \sim \partial_i a_j - \partial_j a_i$ and $(D - 4)$ scalar fields a_k, $k = 4, .., D$.

The theory (37) is very complex, even at the classical level. Its spherically symmetric sector is described by a (1+1)-dimensional dilaton gravity coupled to one massive vector and to several scalar fields. If the mass of the vector field is zero and the scalar fields vanish,** the *dilaton gravity is classically integrable with a rather general dependence of the Lagrangian on the massless Abelian gauge fields*, $X(\phi, F^2)$, where $F^2 = F_{ij}F^{ij}$ and ϕ is the dilaton field, see Refs. 23–25. If $\mu^2 \neq 0$, the theory is certainly not integrable even with vanishing scalar fields. It is also not easy to analytically construct its physically interesting approximate solutions.††

Further dimensional reductions to one-dimensional static or cosmological theories also give non-integrable dynamical systems although some approximate solutions can possibly be derived. The naive cosmological reduction of four-dimensional theory (37) can be written using the metric

$$ds_4^2 = e^{2\alpha}dr^2 + e^{2\beta}d\Omega^2(\theta, \phi) - e^{2\gamma}dt^2\,, \tag{38}$$

where α, β, γ depend on t and $d\Omega^2$ is the metric on the two-dimensional sphere.‡‡ Now the effective cosmological (one-dimensional) Lagrangian corresponding to theory (37) in the $D = 4$ case is

$$\mathcal{L}_c = -2e^{2\beta}\left[e^{\alpha-\gamma}(\dot{\beta}^2 + 2\dot{\beta}\dot{\alpha}) + \Lambda\sqrt{e^{2(\alpha+\gamma)} - \lambda^2\dot{A}^2} + \frac{1}{2}\,\mu^2 A^2 e^{-\alpha+\gamma}\right]. \tag{39}$$

As γ is obviously a Lagrange multiplier we can fix the remaining gauge freedom by choosing $\gamma = -\alpha$. (The standard gauge fixings are $\gamma = 0$ or $\gamma = \alpha$; in Ref. 2 we also used the gauge fixing $\gamma = 3\rho \equiv \alpha + 2\beta$. Varying the Lagrangian multiplier γ gives the *energy constraint*, i.e. vanishing of the Hamiltonian.) Using this gauge and denoting the anisotropy function by $3\sigma \equiv \beta - \alpha$, we have the gauge fixed Lagrangian

$$\mathcal{L}_c = -2e^{2\beta}\left[3e^{2\alpha}(\dot{\rho}^2 - \dot{\sigma}^2) + \Lambda\sqrt{1 - \lambda^2\dot{A}^2} + \frac{1}{2}\,\mu^2 A^2 e^{-2\alpha}\right], \tag{40}$$

where $\alpha = \rho - 2\sigma$ and $\gamma = \rho + \sigma$.

Up to the dilaton multiplier $e^{2\beta}$, the second term in (40) is the DBI (or, 0-brane) Lagrangian. If we consider constant metric functions α, β,

**Such models can be derived by dimensional reductions of some higher-dimensional gravity and supergravity theories, see, e.g, Refs. 23–26 and references therein.

††At first sight, a perturbation theory in μ^2 seems to be a viable alternative to numeric approximations but, when $\mu^2 = 0$, an additional gauge symmetry emerges that makes it more difficult to estimate the validity of the approximation, especially, in the physically important asymptotic regions. We consider this approximation and its relation to DBI-models in a forthcoming publication.

‡‡The function $\beta(t)$ is the two-dimensional dilaton field and, usually, it is supposed that $\alpha = \beta$ (isotropy condition). With the massive vector field $A_i(t)$, this is not possible because the equations of motion require $A_0 \equiv A_t = 0$ and $A_1 \equiv A_r \neq 0$, which obviously gives an anisotropic configuration, see Refs. 1, 2.

and denote $M_A \equiv 2\lambda^2 \Lambda e^{2\beta}$, we see that the 0-brane term is the relativistic Lagrangian of a particle with the mass M_A (the analog of the velocity of light c is $\lambda^{-1} \equiv \bar{c}$). Introducing the canonical momenta p_ρ, p_σ, p_A we find the Hamiltonian (one should not forget that M_A depends on $\beta(t)$):

$$\mathcal{H} = \bar{c}\sqrt{p_A^2 + M_A^2\,\bar{c}^2} + \mu^2 A^2 e^{2(\beta-\alpha)} + \frac{1}{24}\,e^{2(\beta+\alpha)}\,(p_\sigma^2 - p_\rho^2) = 0. \qquad (41)$$

If $\mu^2 = 0$, the momentum P_A is the integral of motion and we get an integrable 1-dimensional dilaton gravity. (with $\mu^2 \neq 0$, it is not integrable and rather unusual theory). If α and β vary much slower than $A(t)$ this is a more tractable model of a relativistic 'particle' with the slowly varying time dependent mass M_A in a simple potential having time dependent parameters. A simpler effective 'particle' model was used by Gribov for discovering the famous *Gribov copies.* One may hope that a similar interpretation of the theory (41) will help to understand some unusual qualitative features of our generalized gravity.

For small A and slowly varying gravitational fields, one can also use the small-field approximation (see Refs. 1, 2), which is formally equivalent to expanding (37) in powers of λ^2. Keeping only the first-order correction we then obtain a nice-looking field theory:

$$\mathcal{L}_{eff} \cong \sqrt{-g}\left[R[g] - 2\Lambda - \kappa\left(\frac{1}{2}F_{ij}F^{ij} + \mu^2 A_i A^i + g^{ij}\partial_i\psi\,\partial_j\psi + m^2\psi^2\right)\right], \qquad (42)$$

where $A_i \sim a_i$, $F_{ij} \sim f_{ij}$, $\kappa \equiv G/c^4$ and we use the CGS dimensions. Note that here we choose the standard normalization of the fields and thus the dimensionless parameters of the theory (D, λ) are hidden in the masses μ and m. Note also that for Einstein's geometry the masses are imaginary, but we should study the general case when they may also be real.

This simplified theory still keeps traces of its geometric origin: the simplest form of the dark energy (the cosmological constant Λ), massive (or tachyonic) vector and scalar fields, which can describe inflation and/or imitate dark matter. The most popular inflationary models require a few massive scalar particles usually called inflatons (see, e.g., Refs. 27–31). Without massive scalar fields, there is no simple inflation mechanism with one massive vecton. However, with the tachyonic vecton (see Ref. 32) or with several massive vector particles, it is probably easier to find more realistic inflation models (see Refs. 33–37; some of these papers also discuss a possible role of massive vector particles in dark energy and dark matter mechanisms).

Conclusion

We note that the geometrical and dynamical models discussed in this paper are not well understood, both conceptually and technically. Much work on them should be done before a realistic cosmological model could be constructed. In particular, one should study the relation between the geometry and dynamics discovered by Einstein. Possibly, one shall find behind it some symmetry principles which are not yet understood. One should also study more general theories. For example, why we not add to the geometric Lagrangian the terms quadratic in the curvature tensor that can be constructed not using any metric? Of course, the Eddington–Einstein Lagrangian and its simplest generalizations discussed here are most beautiful and are closely related to the modern theory of branes, but this is not a good enough argument for restricting alternative geometric proposals. The new part of the connection a^i_{jk} is a tensor that can generate some higher spin fields and we must have some serious arguments for excluding this possibility from the very beginning.

Finally, we must clearly state once more that the generalization of gravity considered here has nothing to do with other matter fields. It is not suggesting any unification of gravity with other forces of nature and with the standard matter. The true meaning of it and its unexpected relation to recent discoveries and ideas in cosmology is a real puzzle. Possibly, a role of this theory is to replace the standard gravity inside the string theory which did not yet completely succeed in giving a simple and natural explanation of dark energy, inflation, and dark matter.

It is a great sorrow to dedicate this article to the dear memory of Volodya Gribov and not to hear his sharp critical and highly stimulating remarks on its content. I realize that the ideas treated here might look to Volodya a bit far from physics he liked, but his incredible ability to penetrate deep to the heart of any problem would certainly help to solve a puzzle left to us by three great scientists of the last century.

Acknowledgments

This work was supported in part by the Russian Foundation for Basic Research (Grant No. 09-02-12417 ofi-M). The author is also thankful to CERN-TH members and, especially, to L. Alvarez-Gaume for kind hospitality at CERN, where this paper was completed.

References

1. A.T. Filippov, arXiv:0812.2616v2 [gr-qc] (2008).
2. A.T. Filippov, *Theor. Math. Phys.* **163** (3), 753-767 (2010).
3. A.T. Filippov, arXiv:1008.2333v2 [gr-qc] (2010).
4. H. Weyl, *Raum-Zeit-Materie*, Springer, Berlin, 1923 (1st ed. 1918; English translation 1950).
5. A.S. Eddington, *Proc. Roy. Soc. London A*, **99**, 104 (1919).
6. A.S. Eddington, *The mathematical theory of relativity*, Cambridge Univ. Press, New York, 1923 (German translation of the 2-nd ed. 1925).
7. A. Einstein, *Sitzungsber. Preuss. Akad. Wiss., Phys.-Math.*, 32-38 (1923); *ibid.* 76-77 (1923); *ibid.* 137-140 (1923).
8. A. Einstein, *Nature*, **112**, 448-449 (1923); 'Eddington's Theorie und Hamiltonisches Prinzip', Appendix to the book: A.Eddington, *Relativitäts theorie in mathematischer Behandlung*, Springer, Berlin, 1925.
9. E. Schrödinger, *Space–time structure*, Cambridge Univ. Press, New York, 1950.
10. W. Pauli, *'Relativitätstheorie'* in: *Enzykl. d. Math. Wiss.*, Vol. 5, Teubner, Leipzig (1921), 539-775; W. Pauli, *Theory of Relativity*, Pergamon Press, Cambridge, New York, 1958.
11. L.P. Eisenhart, *Nonriemannian geometry*, *Amer. Math. Soc. Publ.* New York, 1927.
12. M. Born, *Proc. Roy. Soc. London*, **A143**, 410-437 (1933/34);
    M. Born and L. Infeld, *Proc. Roy. Soc. London*, **A144** 425-451 (1934); **A147** 522-546 (1934); **A150** 141-166 (1935).
13. P.A.M. Dirac, *Proc. Roy. Soc. London*, **A268**, 57-67 (1960).
14. C. Callan and J. Maldacena, arXiv: hep-th/9708147 (1997).
15. G.W. Gibbons, arXiv: hep-th/9709027 (1997).
16. S. Deser and G.W. Gibbons, *Class. Quant. Grav.* **15**, L35-L39 (1998); arXiv: hep-th/9803049v1 (1998).
17. A. Tseylin, arXiv: hep-th/9908105v5 (1999).
18. P.K. Townsend, arXiv: hep-th/0004039 (2000).
19. J. Schwarz, arXiv: hep-th/0103165 (2001).
20. G.W. Gibbons, arXiv: hep-th/010659 (2001).
21. M. Bañados, *Phys. Rev.* **D77**, 123534 (2008).
22. D. Langlois, S. Renaux-Petel and D.A. Steer, *J. Cosmol. Astropart. Phys.*, **0904**, 021 (2009).
23. A.T. Filippov, *Modern Phys. Lett.* **A11**, 1691 (1996);
    Internat. J. Mod. Phys. **A12**, 13 (1997).
24. V. de Alfaro and A.T. Filippov, arXiv: hep-th/0307269 (2003).
25. V. de Alfaro and A.T. Filippov, arXiv: hep-th/0504101 (2005).
26. A.T. Filippov, arXiv: hep-th/0605276 (2006).
27. A. Linde, arXiv: hep-th/0503203v1 (2005).
28. V. Sahni and A. Starobinsky, *Internat. J. Mod. Phys.* **D15**, 2105-2132 (2006).
29. V. Mukhanov, *Physical foundations of cosmology*, Cambridge Univ. Press, New York, 2005.

30. V. Rubakov and D. Gorbunov, *Introduction into the theory of early Universe*, Vols. 1 and 2, Moscow, 2008-2009 [in Russian].
31. S. Weinberg, *Cosmology*, Oxford Univ. Press, Oxford, 2008.
32. L.H. Ford, *Phys. Rev.* **D40**, 967 (1989).
33. M.C. Bento, O. Bertolami, P.V. Moniz, J.M. Mourão and P.M. Sá, *Class. Quant. Grav.* **10**, 285 (1993).
34. C. Armendariz-Picón, *J. Cosmol. Astropart. Phys.* **0407**, 007 (2004).
35. A. Golovnev, V. Mukhanov and V. Vanchurin, *J. Cosmol. Astropart. Phys.* **0806**, 009 (2008).
36. A. Golovnev and V. Vanchurin, *Phys. Rev.* **D 79**, 103524 (2009).
37. T.S. Koivisto and D.F. Mota, arXiv: 0801.3776v.2 [astro-phys] (2008).

Bonn, 1994.

COULD A WEAK COUPLING MASSLESS SU(5) THEORY UNDERLY THE STANDARD MODEL S-MATRIX?

ALAN R. WHITE

Argonne National Laboratory, Il 60439, USA

The unitary Critical Pomeron connects to a unique massless left-handed SU(5) theory that, remarkably, might provide an unconventional underlying unification for the Standard Model. Multi-regge theory suggests the existence of a *bound-state high-energy S-Matrix* that replicates Standard Model states and interactions via massless fermion anomaly dynamics. Configurations of anomalous wee gauge boson reggeons play a vacuum-like role. All particles, including neutrinos, are bound-states with dynamical masses *(there is no Higgs field)* that are formed (in part) by anomaly poles. The contributing zero-momentum chirality transitions break the SU(5) symmetry to vector SU(3)⊗U(1) in the S-Matrix. The high-energy interactions are vector reggeon exchanges accompanied by wee boson sums *(odd-signature for the strong interaction and even-signature for the electroweak interaction)* that strongly enhance couplings. The very small SU(5) coupling, $\alpha_{QUD} \lesssim 1/120$, should be reflected in small (Majorana) neutrino masses. A color sextet quark sector, still to be discovered, produces both Dark Matter and Electroweak Symmetry Breaking. Anomaly color factors imply this sector could be produced at the LHC with *large cross-sections*, and would be definitively identified in double pomeron processes.

1. The Critical Pomeron Leads First to Electroweak Symmetry Breaking and Dark Matter, Then to Neutrino Masses

The Reggeon Field Theory Critical Pomeron (alone) satisfies all high-energy unitarity constraints.[1] Supercritical RFT implies[1] it occurs via reggeon anomaly interactions in " QCD_S " – QCD with six color triplet quarks plus two color sextet quarks. SU(3) color produces the right superconducting phase, while adding the sextet sector provides the infra-red scaling interactions and asymptotic freedom saturation needed for the critical behavior.

Strikingly, the sextet sector could also solve what are currently regarded as the two most fundamental problems of particle physics. Sextet "pions" provide an effective Higgs sector that produces Electroweak Symmetry Breaking, while stable sextet "neutrons" provide Dark Matter. Anomaly color factors imply that the sextet sector, including dark matter and multiple electroweak bosons, will dominate high-energy cross-sections and so could be responsible for the dominance of dark matter production in the early universe. These cross-sections could also be responsible for the Cosmic Ray spectrum knee. If so, they should be seen at the LHC.

Most extraordinarily, though, consistently adding the electroweak interaction to QCD_S requires a unique massless SU(5) theory that might provide an unexpected and novel origin for the Standard Model. I will suggest that the Standard Model may actually be reproducing an anomaly-driven bound-state S-Matrix that contains both the Critical Pomeron and massive neutrinos and which sits within the SU(5) theory. Small neutrino masses could be direct evidence for the, *perforce very small,* SU(5) coupling.

2. QUD – a Bound-State S-Matrix Theory?

Kyungsik Kang and I discovered[1] some years ago that, *uniquely,*

QUD* ≡ SU(5) gauge theory with left-handed couplings to $\mathbf{5} \oplus \mathbf{15} \oplus \mathbf{40} \oplus \mathbf{45}^*$ massless fermions

contains a potential electroweak symmetry-breaking color sextet doublet, is anomaly free, and is asymptotically free. Under $SU(3)_C \otimes SU(2)_L \otimes U(1)$

$$\mathbf{5} = [\mathbf{3},\mathbf{1},-\tfrac{1}{3}]^{\mathbf{3}} + [\mathbf{1},\mathbf{2},\tfrac{1}{2}]^{\mathbf{2}}\ , \qquad \mathbf{15} = [\mathbf{1},\mathbf{3},\mathbf{1}] + [\mathbf{3},\mathbf{2},\tfrac{1}{6}]^{\mathbf{1}} + \{\mathbf{6},\mathbf{1},-\tfrac{2}{3}\}\ ,$$

$$\mathbf{40} = [\mathbf{1},\mathbf{2},-\tfrac{3}{2}]^{\mathbf{3}} + [\mathbf{3},\mathbf{2},\tfrac{1}{6}]^{\mathbf{2}} + [\mathbf{3}^*,\mathbf{1},-\tfrac{2}{3}]\ +\ [\mathbf{3}^*,\mathbf{3},-\tfrac{2}{3}] + \{\mathbf{6}^*,\mathbf{2},\tfrac{1}{6}\} + [\mathbf{8},\mathbf{1},\mathbf{1}]\ ,$$

$$\mathbf{45}^* = [\mathbf{1},\mathbf{2},-\tfrac{1}{2}]^{\mathbf{1}} + [\mathbf{3}^*,\mathbf{1},\tfrac{1}{3}] + [\mathbf{3}^*,\mathbf{3},\tfrac{1}{3}] + [\mathbf{3},\mathbf{1},-\tfrac{4}{3}] + [\mathbf{3},\mathbf{2},\tfrac{7}{6}]^{\mathbf{3}} + \{\mathbf{6},\mathbf{1},\tfrac{1}{3}\} + [\mathbf{8},\mathbf{2},-\tfrac{1}{2}]$$

It was a welcome surprise that, in addition to the sextet quarks {...} having the right quantum numbers for sextet "pions" to provide the longitudinal components of the electroweak vector bosons, QUD also contains QCD_S. In fact, both the triplet quark and lepton sectors, neither of which were asked for, are amazingly close to the Standard Model. There are "almost" three generations – denoted by superscripts **1,2,3**.

*Quantum Uno/Unification/Unique/Unitary/Underlying Dynamics

Very importantly, QUD is real {*vector-like*} with respect to $SU(3)_C \otimes U(1)_{em}$. Obviously, the $SU(2)_L \otimes U(1)$ quantum numbers are not quite right for the Standard Model, but (also very importantly) the lepton anomaly is correct. Only after I fully understood the reggeon anomaly dynamics of QCD_S, did I realise that QUD could be physically realistic if the same dynamics is present. In this case, *all elementary leptons and quarks would be confined and remain massless, with infra-red anomalies dominating the dynamics!!* The Standard Model would have to be an effective theory obtained (in principle) by integrating out the elementary leptons.

A priori, the bound-states of a massless field theory are prohibitively difficult to access. Fortunately, multi-regge theory provides a key! As is well-known, infinite momentum wee partons can, in principle, play a vacuum role. The multi-regge region involves multiple infinite-momenta that, specifically for QCD_S and QUD, allow "universal wee partons" to play this role in the reggeon diagram construction of bound-state amplitudes. Crucially, the infinite-momenta also introduce anomaly pole bound-states.

3. The QUD S-Matrix

Because of it's uniqueness, QUD is either right or wrong, in it's entirety. If it fails to reproduce the Standard Model S-Matrix, it is necessarily wrong. Although much further development is obviously needed and very many details are missing, the arguments outlined in the following imply that

(1) ***All elementary fermions are confined.*** *Infinite-momentum bound-states contain anomaly poles involving zero-momentum chirality transitions that produce* $\boldsymbol{SU(5) \rightarrow SU(3) \otimes U(1)_{em}}$ *symmetry breaking.*

(2) Infinite-momentum ***interactions are vector boson reggeons,*** *accompanied by sums of anomalous wee gauge bosons – odd-signature for the pomeron and even-signature for the electroweak interactions.*

(3) The S-Matrix is a ***massless fermion anomaly phenomenon,*** *without corresponding off-shell amplitudes. Although QUD lies in the SU(5) "conformal window" and the symmetry is unbroken at large* $k_\perp$

the S-Matrix has only Standard Model interactions and a spectrum of Standard Model form.

(4) ***There is no Higgs!!*** *All particles, including (Majorana) neutrinos, are bound-states with dynamical masses.*

Consequently, QUD could, perhaps, provide an amazingly economic underlying unification for the Standard Model, while also producing neutrino masses. Beyond the known generations and the sextet quark sector that, potentially, solves the other outstanding mysteries of dark matter and electroweak symmetry breaking, there is only a lepton-like octet quark sector and a pair of exotically charged quarks. My construction implies that the octet quark sector is buried in all states in an infinite-momentum (light-cone) subtraction role that produces leptons and hadrons in Standard Model generations. Although the physics is both novel and radical, it is consistent with all established Standard Model physics and explains many puzzles. Unfortunately, the multi-regge theory that I use to uncover it is so erudite that general interest may well require, what would surely be,[2,3]

> *A MAJOR EXPERIMENTAL DISCOVERY, i.e. the LHC observes BIG x-sections for multiple Z's and W's,* N_6 *and* P_6 *pairs and, distinctively,* $\gamma\,\mathbb{P}\!\to Z$ *together with* $\mathbb{P}\,\mathbb{P}\!\to ZZ,\ WW$ *pairs.*

An immediate major issue is whether, and how, physical scales can be produced by QUD. Unfortunately, I do not yet have the calculational tools to address this issue directly. An infra-red fixed-point implies a very small coupling, with the second-order β-function giving $\alpha_{QUD} \lesssim 1/120$. While this has the great attraction that it might be an immediate explanation for small neutrino masses, it also implies that Standard Model couplings can not be obtained via QUD evolution. It is essential, therefore, that all particles are anomaly-produced bound-states without the off-shell amplitudes needed to invoke evolution. In this case, α_{QUD} has no direct physical meaning, Moreover, the elementary interaction strengths are enhanced, in the high-energy S-Matrix, by infinite sums of wee gauge bosons involving anomaly color factors. Indeed, the larger color factors for sextet states imply that, at high-energy, this new sector will be produced with cross-sections that are even larger than normal hadronic cross-sections.

4. Cosmic Rays and Dark Matter

Cosmic rays already suggest that new large x-section physics including dark matter could appear at the LHC! As shown in Figure 1, the spectrum knee occurs between Tevatron and LHC energies. It is remarkably well-established, yet not understood. Although dark matter was unknown, a major threshold for neutral particles, unobserved in detectors, was initially suggested ($\sim$ 40 years ago). Underestimation of the energy would pile-up events as a "knee". Unbelievably, perhaps, neutral particles must also dominate the x-section far above the knee. But, if the dark matter x-section is large at the highest LHC energy, a link to the knee is surely inevitable.

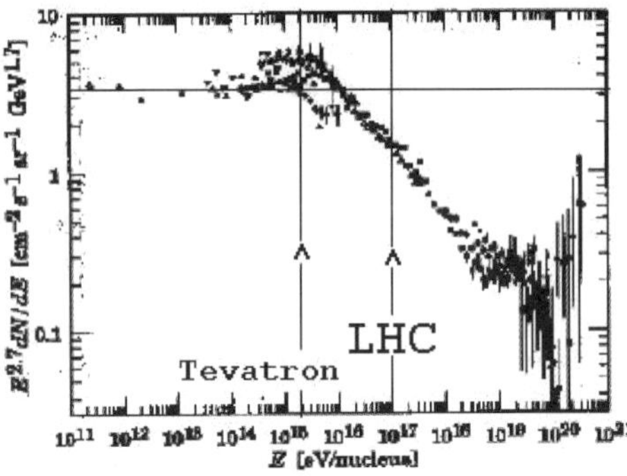

Fig. 1. The Cosmic Ray Knee

For the sextet sector, three related effects could produce a knee.

(1) *Prolific production of electroweak bosons increases* $< p_\perp >$ *dramatically and increases neutrino production – leading to energy underestimation.*
(2) *Direct production of sextet neutron dark matter.*
(3) *Sextet neutrons as incoming cosmic rays (including UHE?) with a threshold for atmospheric interaction not far below the knee.*

But, can sextet x-sections be large enough to dominate the total x-section at the highest energies? If so, a natural explanation for the existence and dominance of stable dark matter in the universe would be provided.

5. Wee Partons in the Multi-Regge Region

Multiple regge poles appear in multi-regge limits of multiparticle amplitudes. Most familiar, perhaps, is the triple-regge limit, illustrated in Figure 2(a), in which $P_1, P_2, P_3 \to \infty$ along distinct light-cones. In the P_3 rest-frame, the regge pole pions have ∞-momentum and continuation to $Q_1^2 = Q_2^2 = m_\pi^2$ gives the on-shell pion coupling to the pomeron. More important, for our purposes, is the "di-triple regge" (DTR) limit in which two triple-regge limits are separated by a further ∞-momentum. Now, as illustrated in Figure 2(b), regge-pole pions can scatter via the pomeron. All the pions and the pomeron have ∞-momentum in some frame, suggesting that both bound-states ($\pi's$) and interactions ($\mathbb{P}$) could appear as reggeon states in which "universal wee partons" play a vacuum role. If this can be shown, it amounts to a derivation of (much more than) the parton model.

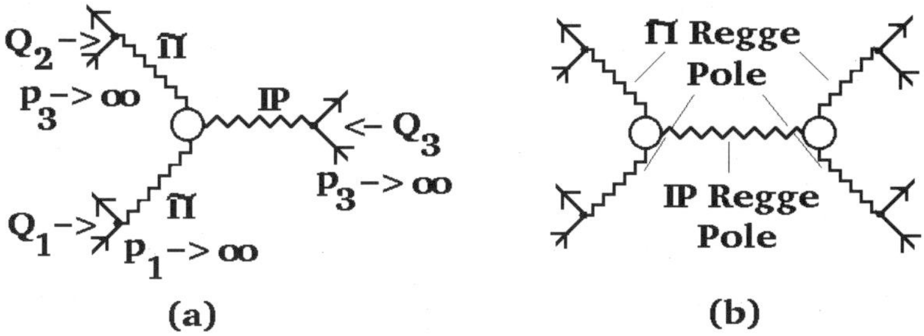

Fig. 2. (a) The Triple-Regge Limit (b) The Di-Triple-Regge Limit.

In my construction of DTR amplitudes an initial, cut-off induced, $k_\perp$ infra-red divergence produces universal wee gluon reggeons in both QCD_S and QUD, as illustrated in Figure 3. The wee gluon reggeons have opposite sign color and space parities and so are "anomalous". They couple via

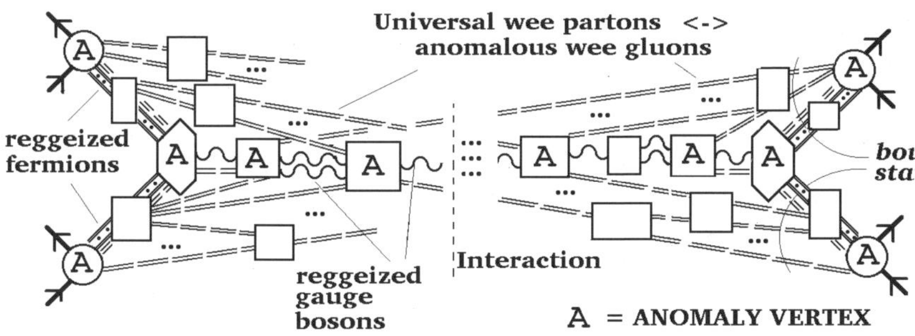

Fig. 3. A Typical Initial QCD_S/QUD DTR Amplitude.

anomaly vertices that are chirality violating and contain *infra-red chirality transitions* of massless quarks. Such vertices, *necessarily*, involve more than one DTR reggeon channel. Moreover, in combined ∞-momentum and small $k_\perp$ limits, they reduce to **anomaly poles** that provide both **bound-state particles** and the **vertex factorization** of the wee gluons. After outlining the origin of reggeon anomaly vertices, I will discuss QCD_S briefly and then discuss QUD in more detail.

That Standard Model states and interactions emerge in QUD (in my construction) as the complexity of the wee partons increases, is what has to be demonstrated. The wide range of scales and interactions in the Standard Model has to be a consequence of the build-up of the (infinite) variety of wee parton anomaly vertices that couple the interactions in the distinct DTR channels. The arguments that follow will uncover, at best, only the simplest components of what is surely a truly complicated, if beautifully elaborate, phenomenon.

6. Anomalies and Anomaly Poles

Reggeon diagrams are generated when large light-cone momenta are routed through feynman diagrams so that internal particles are maximally close to mass-shell, while also having large relative rapidities. Internal particles with finite relative rapidity generate reggeon interaction vertices and, in a gauge theory, fermion loop reggeon vertices include triangle anomalies. Because a four-dimensional interaction is involved, the anomalies occur only in special vertices coupling reggeon channels with distinct light-cone momenta (such as appear in the DTR limit). Included are axial-vector/vector/vector triangle diagrams T^{AVV} that, in both QUD and massless QCD_S, must be defined as the zero fermion mass limit of massive reggeon diagrams.

At first sight, chirality is conserved in zero mass triangle diagrams, implying $T^{AAA} = T^{AVV} = T^{RRR}+T^{LLL}$ - producing a conflict between the axial-vector anomaly and vector current conservation. However, the regularization of γ_5 amplitudes is a major problem. Fortunately, for our purposes, it can be shown that vector current conservation plus the axial anomaly implies unique massless infra-red anomaly pole chiral amplitudes.

In infra-red limits, pseudoscalar anomaly poles appear similiarly in chirality-violating and non chirality-violating amplitudes via the triangle singularity. When there is a chirality transition, as illustrated in Figure 4, the pseudoscalar pole can be a chiral Goldstone boson. ∞-momentum chirality transitions can also occur as part of a Pauli-Villars subtraction.

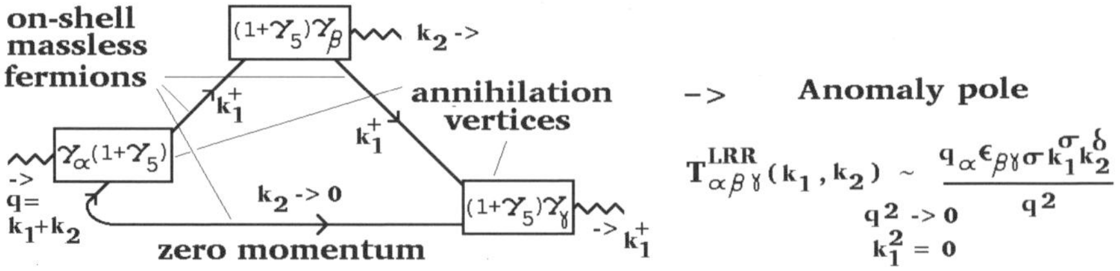

Fig. 4. Zero Momentum Chirality Violation in T^{LRR}.

7. Massless QCD_S

The structure of DTR amplitudes in massless QCD_S is summarized, in a first approximation, in Figure 5. Anomalous wee gluon reggeons appear

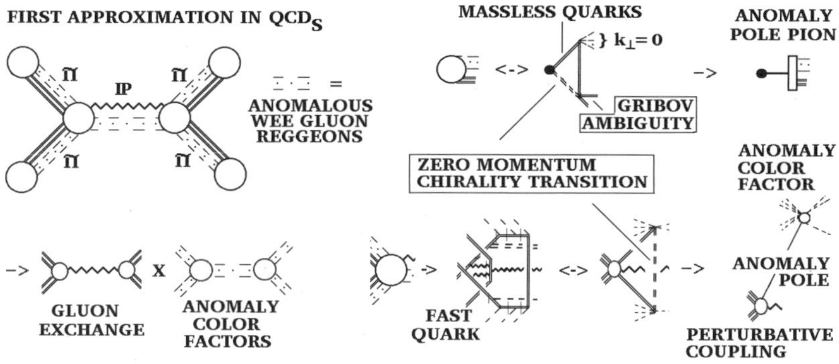

Fig. 5. A DTR Amplitude for Pion Scattering in Massless QCD_S

in both the pions and the pomeron via anomaly vertices involving zero-momentum quark chirality transitions (and, for the pion, a longitudinal gluon exchange allowed by the Gribov ambiguity). As illustrated, the full pion coupling to the pomeron also contains a perturbative coupling of the dynamical quark and gluon reggeons that are involved. The wee gluons lie in an SU(2) color subgroup, but their combination with the dynamical reggeons produces a color zero projection in each channel. This results in SU(3) color zero states when the SU(2) subgroup is randomized (averaged over) within SU(3), as we briefly discuss later, via the Critical Pomeron.

- *The bound-states are triplet and sextet (pseudoscalar) mesons and (with an extra quark reggeon) triplet and sextet baryons. There are*
- ***NO** pseudoscalar anomaly poles producing hybrid sextet/triplet states,*
- ***NO** glueballs, **NO** BFKL pomeron, and **NO** odderon.*
- *The Critical Pomeron occurs as a factorized regge pole, plus triple pomeron interactions – consistent with the parton model.*

If sextet pions become the longitudinal components of massive electroweak bosons, sextet nucleons are the only new states. (The η_6 aquires an electroweak scale mass by mixing with the pomeron.) That sextet quarks have zero current mass implies the N_6 neutron is stable. (Electric charge makes the P_6 proton heavier – in contrast to the triplet sector.) The very strong, very short range, QCD self-interaction implies the N_6's could form

DARK MATTER "clumps". They will only interact with normal matter at ultra high energies. N_6 production will dominate the high-energy cross-sections responsible for early universe stable matter formation, and will also explain the Cosmic Ray knee!

These results are at variance with conventional expections for high-energy QCD. There are fewer states (than requiring just confinement and chiral symmetry breaking) and the interaction is simpler. Both features are strongly suggested by experiment! Although the anomaly dynamics appears to require the quarks to be massless, effective quark masses that do not disturb the dynamics are produced by embedding QCD_S in QUD.

8. QUD Reggeon Diagrams – the Massless Limit

The chirality transitions in QCD_S do not conflict with the vector gauge symmetry. In QUD, more fundamentally, they produce a dynamical breaking of the non-vector part of the gauge symmetry. In the following construction they are a direct consequence of the zero fermion mass limit and so they retain the initial mass symmetry breaking. In the physical S-Matrix, they should be dynamical and randomized via the Critical Pomeron.

I start with masses for all reggeons and a $k_\perp$ cut-off $\lambda_\perp$. How the masses and cut-off are removed is crucial in resolving the (light-cone) Gribov ambiguity and, in effect, produces the wee partons of the massless theory. A combination of **24** and $\mathbf{5}\oplus\mathbf{5}^*$ scalar VeV's is needed to give masses to all the fermions. This also identifies particle/antiparticle pairs and so determines possible chirality transitions. For the gauge bosons, using only $\mathbf{5}\oplus\mathbf{5}^*$ VeVs ensures a smooth massless limit (via complementarity).

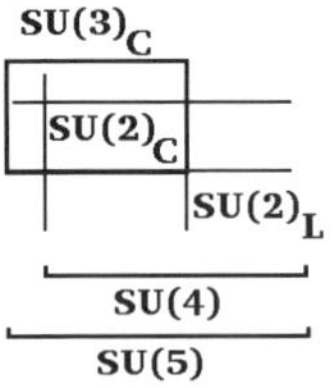

(1) *I decouple fermion mass scalars first, leaving chirality transitions that break SU(5) to* $\mathbf{SU(3)}_C\otimes\mathbf{U(1)}_{em}$ **in anomaly vertices only.**

(2) *I decouple gauge boson scalars successively, giving global symmetries*

$$\rightarrow\ \mathbf{SU(2)_C}\ ,\quad \rightarrow\ \mathbf{SU(4)}\ ,\quad \lambda_\perp\rightarrow\infty\ ,\quad \rightarrow\ \mathbf{SU(5)}$$

Closely related to the weak coupling of QUD, the last scalar to be removed is asymptotically free, allowing $\lambda_\perp\rightarrow\infty$ before the SU(5) limit. This is essential for obtaining the Critical Pomeron and for the emergence of Standard Model generations via ∞-momentum color octet contributions.

In non-anomaly reggeon diagrams, the exponentiation of reggeization divergences leaves only infra-red finite, global color zero, interaction kernels. **Crucially**, $\lambda_\perp$ *implies fermion loop vertices,* ***including anomaly vertices,*** *do not satisfy Ward identities producing* $k_\perp = 0$ *zeroes.* Consequently, the exponentiation of divergences goes well beyond reggeization, *particularly for left-handed bosons.*

Parity non-conservation allows fermion loop vertices that exponentiate all left-handed bosons contributions that would be part of, or are accompanied by, the anomalous wee boson divergences that we discuss next.

9. SU(2) Color Restoration and Anomalous Wee Gluons

With a $k_\perp$ cut-off, **vector** SU$(2)_C$ produces a *non-exponentiating* divergence involving the **anomalous wee gluons** of Figure 3, which are

$I = 0$ sets of massless gluon reggeons, with $k_\perp$'s
scaled to zero and color parity $C \neq \mathcal{T} =$ signature.

The anomalous color parity implies that the *vector* wee gluons couple only via anomaly vertices with chirality transitions. For SU(2), only $\mathcal{T} = -C = -1$ is possible ($\leftrightarrow 3, 5, ...\infty$ reggeons). Infra-red fixed-point scaling implies that the iteration of $I = 0$ kernels reproduces the basic divergence, with a factorized residue. Also, as for QCD$_S$, anomaly poles in vertices connect the divergence in different channels. Factoring off the overall divergence leaves a universal wee gluon component in all reggeon states.

As for the pion anomaly pole in Figure 5, bound-state anomaly poles are also present in external vertices. To extract residues, it is necessary to go to an ∞-momentum frame in which the wee gluons carry vanishingly small light-cone momenta orthogonal to the ∞-momentum of the fermions and for the polarizations of the fermions and the wee gluons to be (additionally) orthogonal. In the process generating the pole a reggeon state containing a same chirality physical fermion pair and an anomalous wee gluon component, is coupled to a state containing only two opposite chirality fermions, one of which is unphysical and has zero momentum. In effect, there is a zero-momentum shift of the Dirac sea. By absorbing anomalous wee gluons, a physical fermion makes a symmetry-breaking chirality transition to an unphysical "hole state" and so produces a pseudoscalar Goldstone boson[†]. The chiral symmetries we discuss next do not conflict with the SU$(2)_C$

[†]At ∞-momentum, an anomaly pole has physical Goldstone boson couplings.

gauge symmetry. Later, when the gauge symmetry is broken, it will be important that the reggeon state involved has a projection on a (color zero) unbroken symmetry state.

Via 5⊕5* chirality transitions, reggeon states containing $SU(2)_C$ anomalous wee gluons produce chiral Goldstones (π_C's),that are $q\bar{q}$ "mesons" or qq and $\bar{q}\bar{q}$ "nucleons". The q's are **3's, 6's,** and **8's** under $SU(3)_C$. Because the **8's** are real representations, they can not give an $SU(3)_C$ anomaly. However, they contain complex $SU(2)_C$ chiral doublets that produce anomaly poles when only $SU(2)_C$ is restored. Other reggeon states containing a π_C are also selected by the wee gluon divergence and will, ultimately, give leptons and SU(5) symmetric reggeon states. To avoid fermion loop exponentiation of the anomaly divergence, the massive gauge boson reggeons in such states must all be vectors (i.e gluons or photons).

The leading interaction exchanges are even signature and contain an $SU(2)_C$ singlet massive vector boson accompanied by anomalous wee gluons. As for the pomeron/pion coupling in Figure 5, the coupling to bound-states contains both an anomaly vertex involving wee gluons and a perturbative coupling of dynamical fermions to the exchanged boson. $SU(3)_C$ massive gluon exchange straightforwardly gives a supercritical pomeron. The gluon can also be replaced by a massive γ or, after inclusion of the vertices involving bound-state wee gluons that we discuss next, a $W^{\pm}$ or Z^0 (using $SU(3)_C \otimes SU(2)_L \otimes U(1)$ quantum numbers). There are also non-leading odd-signature interactions $\gamma\,\mathbb{P}$, $W^{\pm}\,\mathbb{P}$ and $Z^0\,\mathbb{P}$, that will actually provide the physical γ, $W^{\pm}$, and Z^0, after $SU(3)_C$ restoration.

10. Bound-State Anomalous Wee Gluon Vertices

Elementary left-handed $W^{\pm}$ and Z^0 exchanges, accompanied by wee gluons, are exponentiated to zero via fermion loop interactions, but 5⊕5* chirality transitions provide crucial wee-gluon vertex couplings to the $\pi_C's$ ($\sim \pi_6's$). In an appropriate ∞-momentum frame, the vertices can be evaluated via anomaly pole contributions, as illustrated in Figure 6. They provide a mass M_W^2 ($\sim g_W^2 \int dkk$ – a wee gluon integral multiplied by an electroweak scale determining sextet anomaly factor) that survives the SU(5) symmetry restoration, while also providing a chirality transition and sextet flavor quantum number that prevents the exponentiation. The perturbative coupling of the $W^{\pm}$ and Z^0 is retained, however.

The **24** chirality transitions also provide very important wee gluon vertices, as illustrated in Figure 7. The SU(3) d-tensor shown gives a triple pomeron vertex with the symmetry properties needed to produce the

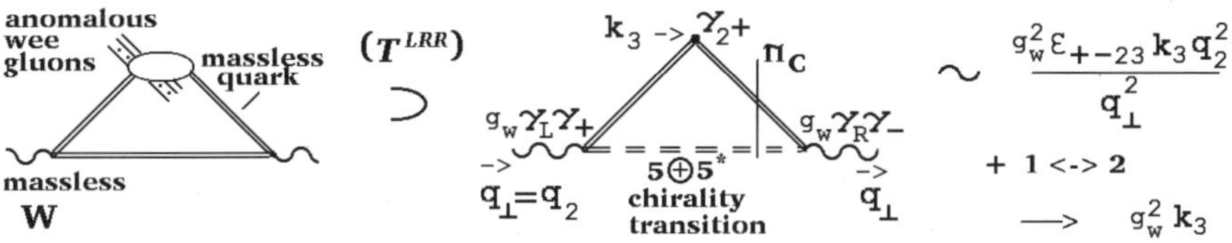

Fig. 6. Vector Boson Mass Generation

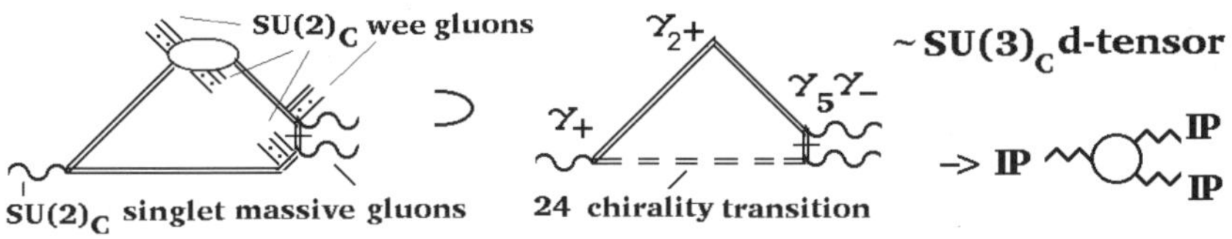

Fig. 7. A Triple Pomeron Vertex

Critical Pomeron. The wee gluons in the scattering states provide the orthogonal γ-matrices needed to generate a γ_5.

11. SU(4) Color Restoration

As SU(4) color is restored, a vector-coupling boson γ_4 that is a singlet under both SU(2)$_C$ and SU(2)$_L$(a linear combination of the γ and a massive gluon), becomes massless. The $W^{\pm}$ and Z^0 survive and retain the mass discussed above, while other gauge bosons that become massless are left-handed and their reggeon amplitudes are amongst those exponentiated to zero by fermion loop interactions. (They survive, however, in interaction kernels.) γ_4 pairs are even signature and have a scaling interaction that also exponentiates amplitudes to zero via fermion loop interactions, *except* when the pairs couple via anomaly vertices ($e^{-\infty} \to 1 - e^{-\infty}$) and combine with the anomalous wee gluons. Consequently, as reggeon states, bound-states now contain two fermion pairs produced by anomaly poles and accompanied by odd-signature **anomalous wee bosons**, which are

SU(2)$_C$ anomalous wee gluons plus γ_4 pairs (each with $k_\perp = 0$).

The anomaly vertices coupling to γ_4 pairs are produced by

(1) **lepton pairs -** *(1, 2, $\frac{1}{2}$)/(1, 2, -$\frac{1}{2}$) chiral symmetry plus* **24** *chirality transitions gives anomaly pole* **pseudoscalars** $\pi_L^{\pm,0}$

(2) **SU(2)$_C$ singlet octets -** *(8, 1, 1)/(8, 2, -$\frac{1}{2}$) chiral symmetry plus* ***5***$\oplus$***5**** *chirality transitions gives anomaly pole* **pseudoscalars** $\eta_8^{\pm,0}$

The new Goldstone boson states explicitly break the SU(4) gauge symmetry; but this is already broken by the fermion loop removal of left-handed reggeons. In place of the full symmetry, reggeon states must have SU(4) singlet projections that, after the randomizing of the symmetry breaking within SU(5) (via the Critical Pomeron), will give finite amplitudes. The fermion states corresponding to SU(4) singlet reggeon states are then

LEPTONS - $\pi_L + \pi_8 +$ elementary lepton $\rightarrow$ 3 generations.

MESONS - $\pi_{3,6} + \eta_8$, **BARYONS** $\leftrightarrow$ additional quark.

Note that both leptons and hadrons have *octet quark* components.

The leading interaction is the even signature supercritical pomeron – the remaining massive SU(4) singlet vector reggeon plus *odd-signature wee bosons.* Interactions involving electroweak exchanges are present, but do not have the SU(3)$_C$ singlet projection needed to survive the SU(5) restoration. The odd-signature exchanges $\gamma\,\mathbb{P}$, $W^{\pm}\,\mathbb{P}$, and $Z^0\,\mathbb{P}$, accompanied by *even-signature wee bosons,* will provide the physical interactions.

12. SU(5) Color Restoration

As the remaining SU(4) singlet vector becomes massless, the supercritical pomeron becomes Critical and, simultaneously, the photon becomes massless. The massless photon is, therefore, an odd-signature partner of the even signature Critical Pomeron. There is no "triple-photon" vertex and the photon does not have the anomaly couplings to hadrons that make the pomeron interaction so much stronger. We assume that the Critical Pomeron is analagous to a second-order phase transition phenomenon in that the transition randomizes the symmetry breaking. Critical amplitudes can, nevertheless, be obtained from supercritical amplitudes in which order parameters explicitly breaking the symmetry are introduced and then removed. Most likely, the symmetry breaking is randomized in all sets of reggeon diagrams connecting anomaly vertices. As discussed elsewhere,[1] the $k_\perp$ cut-off is an order parameter that must be removed first, if the restoration of SU(3)$_C$ is to be sufficient to give the critical behavior.

The new massless vector (carrying zero $k_\perp$ via a divergence) also joins odd-signature wee bosons to give **even-signature anomalous wee bosons** that have SU(3)$_C$ color zero and combine with γ , $W^\pm$, and Z^0 exchanges to give the SU(5) singlet projections that are randomized to provide the physical electroweak interaction. Because the π_8 and η_8 combine to form a real octet SU(3)$_C$ representation **(Π_8)**, the **infra-red octet anomaly poles cancel in all amplitudes.** However, the $\lambda_\perp \to \infty$ limit, taken before the SU(5) limit, introduces an ∞-momentum octet anomaly contribution (as a companion to the canceling infra-red anomaly pole) in vertices coupling the SU(3)$_C$ component of the *even-signature wee bosons*. Via a Ward identity, the anomaly couples like an ∞-momentum Π_8 anomaly pole so that, in effect, octet quark anomaly poles coupled to SU(3)$_C$ wee gluons provide an ∞-momentum contribution in all bound-states and interaction vertices.

In reggeon bound-states the infra-red dynamical fermion reggeons, that couple perturbatively to the dynamical exchanged vector bosons, must combine with an ∞-momentum Π_8 pair and adjoint representation anomalous wee bosons to give an SU(5) singlet projection. In the corresponding fermion states, this requires that three elementary fermion reggeons, two of which are produced by an anomaly pole, combine to provide the $SU(2) \otimes U(1)$ representations

$$(\mathbf{2}, -\tfrac{\mathbf{1}}{\mathbf{2}})_\mathbf{L} \ , \quad or \ (\mathbf{2}, \tfrac{\mathbf{1}}{\mathbf{2}})_\mathbf{R} \ , \quad or \quad (\mathbf{1}, \mathbf{1})_\mathbf{L} \ , \quad or \ (\mathbf{1}, -\mathbf{1})_\mathbf{R}$$

Consequently, leptons and hadrons form Standard Model generations.

13. Physical Leptons

The SU(3)×SU(2)$_L$×U(1) fermion reggeons that combine with SU(5) adjoint wee bosons to form physical leptons are

- $(e^-, \nu) \ \leftrightarrow \ (1,2,-\tfrac{1}{2}) \times \pi_L^0 \times [\Pi_8]_\infty$
  $\leftrightarrow \ (1,2,-\tfrac{1}{2}) \times (1,2,-\tfrac{1}{2})(1,2,\tfrac{1}{2}) \times [(8,1,1)(8,2,-\tfrac{1}{2})]_\infty$
  $\leftrightarrow \ \mathbf{SU(5)\ adjoint} \ - \ \mathbf{45^*} \times \mathbf{45^*} \times \mathbf{5} \times [\mathbf{40} \times \mathbf{45^*}]$
- $(\mu^-, \nu) \ \leftrightarrow \ (1,2,\tfrac{1}{2}) \times \pi_L^- \times [\Pi_8]_\infty$
  $\leftrightarrow \ (1,2,\tfrac{1}{2}) \times (1,2,-\tfrac{1}{2})(1,2,-\tfrac{1}{2}) \times [(8,1,1)(8,2,-\tfrac{1}{2})]_\infty$
  $\leftrightarrow \ \mathbf{SU(5)\ adjoint} \ - \ \mathbf{5} \times \mathbf{45^*} \times \mathbf{45^*} \times [\mathbf{40} \times \mathbf{45^*}]$

- $(\tau^-,\nu) \leftrightarrow (1,2,-\frac{3}{2}) \times \pi_L^+ \times [\Pi_8]_\infty$

$$\leftrightarrow (1,2,-\tfrac{3}{2}) \times (1,2,\tfrac{1}{2})(1,2,\tfrac{1}{2}) \times [(8,1,1)(8,2,-\tfrac{1}{2})]_\infty$$

$$\leftrightarrow \mathbf{SU(5)\ adjoint} \ - \ \mathbf{40 \times 5 \times 5 \times [40 \times 45^*]}$$

In principle, e^+, μ^+ and τ^+ can be obtained via charge conjugation, once elementary antiparticles are explicitly identified. Neutrinos will necessarily be Majorana fermions. Chirality transitions connect the constituents of neutrino and anti-neutrino candidates and there is no quantum number that would prevent anomalous wee boson vertices from generating mass terms involving left-handed neutrinos and right-handed antineutrinos.

14. The Bound-State Mass Spectrum

I have outlined my understanding of only a small part of the structure of DTR amplitudes in QUD. It is obviously a major challenge to develop my outline into a calculable reggeon diagram formalism that would explicitly provide scales and masses. Wee gluon anomaly vertices need to be catalogued, as a starting point. They will mix reggeon states and introduce color factors, with related wee gluon distributions needed to determine how many, if any, parameters are involved. At this stage, we can only say

(1) *Since α_{QUD} is so small, perturbative reggeization is a small effect, reflected only in small masses for the, zero color and charge, neutrinos.*

(2) *$SU(3)_C$ interactions and masses will be enhanced: by anomaly color factors, by the triple pomeron interaction, and by the high mass sector.*

(3) *Assuming bound-state fermions have constituent masses, connecting the η_6 to top production suggests*[2]

$$m_{q6} \sim m_{top} \implies m_{N_6} \sim 500\ GeV$$

(4) *Electromagnetic anomaly factors will enhance charged particle masses, but less strongly. There is no triple photon interaction.*

(5) *There is no symmetry conflicting with the Standard Model spectrum.*

(6) *CP violation can be introduced via the anomalies, but is it essential?*

15. Potential QUD Virtues – Beyond QCD

Of course, the scientific and aesthetic importance of an underlying massless field theory for the Standard Model can not be exagerated. In addition,

- QUD is self-contained, with only Standard Model Interactions. It has to be completely right – or else it is completely wrong!

- The massless photon partners the "massless" Critical Pomeron.
- The only new physics still to be discovered is a high mass sector of the strong interaction: giving **electroweak symmetry breaking, dark matter, and unification,** without supersymmetry!
- **Parity properties** of the strong, electromagnetic, and weak interactions are naturally explained.
- Anomaly vertices mix the reggeon states. Color factors could produce the wide range of Standard Model scales and masses, with **small Majorana neutrino masses** due to the very small QUD coupling.
- Despite the underlying SU(5) symmetry, there is **no proton decay.**
- Particles and fields are truly distinct, with physical hadrons and leptons having equal status. Symmetries and masses are S-Matrix properties. There are no off-shell amplitudes and there is **no Higgs field.**
- The QUD S-Matrix is the only "non-perturbative" part of quantum field theory needed – with **infinite momentum** physics retaining a **diagrammatic "parton model" description.**
- Perturbatively, QUD is a massless, asymptotically free, fixed-point theory that has no renormalons, and so, no vacuum energy. Therefore,[4] it would induce Einstein gravity with **zero cosmological constant.**

References

1. A. R. White, "*High-Energy Unitarity and the Standard Model*", Proceedings of Gribov-75: hep-ph/0508125. This paper provides a good introduction to the present paper. The path from high-energy unitarity to the Critical Pomeron, then to QCD_S via the Supercritical Pomeron, and finally to QUD, is outlined (with references). The underlying influence of Gribov's work is emphasized.

   The following two papers provide background perspective and further comments on the general significance of QUD, as well as additional references. There is also more discussion of the physics that should be seen at the LHC.

2. A. R. White, "*The LHC Pomeron and Unification of the Standard Model: A Bound-State S-Matrix Within a Fixed-Point Field Theory?*", 6th Workshop on Small-x and Diffractive Physics: arXiv:0708.1306.

3. A. R. White, "*The 'Crisis in Fundamental Physics": Will the LHC Pomeron End it?*", 5th Manchester Forward Physics Workshop: arXiv:0803.1151.

4. B. Holdom, arXiv:0708.1057 (2007).

NEARLY CONFORMAL GAUGE THEORIES ON THE LATTICE

ZOLTÁN FODOR

Department of Physics, University of Wuppertal,
Gauss Strasse 20, D-42119, Germany
and
Institute for Theoretical Physics, Eotvos University,
1/A Pazmany P. Sétány, H-1117, Hungary
fodor@bodri.elte.hu

KIERAN HOLLAND

Department of Physics, University of the Pacific
3601 Pacific Ave, Stockton CA 95211, USA
kholland@pacific.edu

JULIUS KUTI*

Department of Physics 0319, University of California, San Diego
9500 Gilman Drive, La Jolla CA 92093, USA
jkuti@ucsd.edu

DÁNIEL NÓGRÁDI

Institute for Theoretical Physics, Eotvos University,
1/A Pazmany P. Sétány, H-1117, Hungary
nogradi@bodri.elte.hu

CHRIS SCHROEDER

Department of Physics, University of Wuppertal,
Gauss Strasse 20, D-42119, Germany
crs@physics.ucsd.edu

We present selected new results on chiral symmetry breaking in nearly conformal gauge theories with fermions in the fundamental representation of the SU(3) color gauge group. We found chiral symmetry breaking (χSB) for all flavors between $N_f = 4$ and $N_f = 12$ with most of the results discussed here for $N_f = 4, 8, 12$ as we approach the conformal window. To identify χSB we

*Contributor to the Proceedings.

apply several methods which include, within the framework of chiral perturbation theory, the analysis of the Goldstone spectrum in the p-regime and the spectrum of the fermion Dirac operator with eigenvalue distributions of random matrix theory in the ϵ-regime. Chiral condensate enhancement is observed with increasing N_f when the electroweak symmetry breaking scale F is held fixed in technicolor language. Important finite-volume consistency checks from the theoretical understanding of the $SU(N_f)$ rotator spectrum of the δ-regime are discussed. We also consider these gauge theories at $N_f = 16$ inside the conformal window.

Remembering Volodya Gribov

When the organizers kindly invited me to the Gribov-80 Memorial Workshop, I was very much hoping to participate and enjoy again the shared experience in paying tribute to Volodya, his life, his work, and his great humanity. Unfortunately, I was not able to attend the meeting which I deeply regretted. Julia Nyíri kindly suggested to send a summary of my recent work as my token participation. I took the opportunity and prepared this note with my collaborators to summarize our recent studies of nearly conformal gauge theories which is a rapidly growing research field with the audacity of hope to realize the composite Higgs mechanism beyond the Standard Model. Gribov maintained keen interest in the Higgs mechanism over the years and frequently expressed his views how the Higgs sector of the Standard Model has to be an interim and incomplete solution yet to be surpassed with some unknown new physics. I have known Volodya for more than two decades before his untimely death. He was a great man with towering intellect, intense energy and curiosity. I continue to miss him dreaming about time travel where I can explain to him what we would like to do today with the composite Higgs mechanism. I would be listening to his insight and wise counsel. It would not be an easy and smooth discussion but I would treasure every minute of it.

1. Introduction

It is an intriguing possibility that new physics beyond the Standard Model might take the form of some new strongly-interacting gauge theory building on the original technicolor idea.[1–3] This approach has lately been revived by new explorations of the multi-dimensional theory space of nearly conformal gauge theories.[4–7] Model building of a strongly interacting electroweak sector requires the knowledge of the phase diagram of nearly conformal gauge theories as the number of colors N_c, number of fermion flavors N_f, and the fermion representation R of the technicolor group are varied in theory space. For fixed N_c and R the theory is in the chirally broken phase for low N_f, and asymptotic freedom is maintained with a negative β function. On the other hand, if N_f is large enough, the β function is positive for all couplings, and the theory is trivial. There is some range of N_f for which the β function might have a non-trivial zero, an infrared fixed point, where the theory is in fact conformal.[10,11] This method has been refined by estimating the critical value of N_f, above which spontaneous chiral symmetry breaking no longer occurs.[12–14]

Interesting models require the theory to be very close to, but below, the conformal window, with a running coupling which is almost constant over a large energy range .[15–20] The nonperturbative knowledge of the critical N_f^{crit} separating the two phases is essential and this has generated much interest and many new lattice studies.[21–54] To provide theoretical framework for the analysis of simulation results, we review first a series of tests expected to hold in the setting of χPT in finite volume and in the infinite volume limit.

2. Chiral symmetry breaking below the conformal window

We will identify in lattice simulations the chirally broken phases with $N_f = 4, 8, 12$ flavors of staggered fermions in the fundamental SU(3) color representation using finite volume analysis. We deploy staggered fermions with exponential (stout) smearing[55] in the lattice action to reduce well-known cutoff effects with taste breaking in the Goldstone spectrum.[56] The presence of taste breaking requires careful analysis of staggered χPT following the important work of Lee, Sharpe, Aubin and Bernard.[57–59]

2.1. *Finite volume analysis in the p-regime*

Three different regimes can be selected in simulations to identify the chirally broken phase from finite volume spectra and correlators. For a lattice size $L_s^3 \times L_t$ in euclidean space and in the limit $L_t \gg L_s$, the conditions $F_\pi L_s > 1$ and $M_\pi L_s > 1$ select the the p-regime, in analogy with low momentum counting.[60,61]

For arbitrary N_f, in the continuum and in infinite volume, the one-loop chiral corrections to M_π and F_π of the degenerate Goldstone pions are given by

$$M_\pi^2 = M^2\left[1 - \frac{M^2}{8\pi^2 N_f F^2} ln\left(\frac{\Lambda_3}{M}\right)\right], \tag{1}$$

$$F_\pi = F\left[1 + \frac{N_f M^2}{16\pi^2 F^2} ln\left(\frac{\Lambda_4}{M}\right)\right], \quad (2)$$

where $M^2 = 2B \cdot m_q$ and $F, B, \Lambda_3, \Lambda_4$ are four fundamental parameters of the chiral Lagrangian, and the small quark mass m_q explicitly breaks the symmetry.[62] The chiral parameters F, B appear in the leading part of the chiral Lagrangian, while Λ_3, Λ_4 enter in next order. There is the well-known GMOR relation $\Sigma_{cond} = BF^2$ in the $m_q \to 0$ limit for the chiral condensate per unit flavor.[63] It is important to note that the one-loop correction to the pion coupling constant F_π is enhanced by a factor N_f^2 compared to M_π^2. The chiral expansion for large N_f will break down for F_π much faster for a given M_π/F_π ratio. The NNLO terms have been recently calculated[64] showing potentially dangerous N_f^2 corrections to Eqs. (1), (2).

The finite volume corrections to M_π and F_π are given in the p-regime by

$$M_\pi(L_s, \eta) = M_\pi\left[1 + \frac{1}{2N_f}\frac{M^2}{16\pi^2 F^2} \cdot \widetilde{g}_1(\lambda, \eta)\right], \quad (3)$$

$$F_\pi(L_s, \eta) = F_\pi\left[1 - \frac{N_f}{2}\frac{M^2}{16\pi^2 F^2} \cdot \widetilde{g}_1(\lambda, \eta)\right], \quad (4)$$

where $\widetilde{g}_1(\lambda, \eta)$ describes the finite volume corrections with $\lambda = M \cdot L_s$ and aspect ratio $\eta = L_t/L_s$. The form of $\widetilde{g}_1(\lambda, \eta)$ is a complicated infinite sum which contains Bessel functions and requires numerical evaluation.[61] Eqs. (1-4) provide the foundation of the p-regime fits in simulations.

2.2. *δ-regime and ϵ-regime*

At fixed L_s and in cylindrical geometry $L_t/L_s \gg 1$, a crossover occurs from the p-regime to the δ-regime when $m_q \to 0$, as shown in Fig. 1. The dynamics is dominated by the rotator states of the chiral condensate in this limit[65] which is characterized by the conditions $FL_s > 1$ and $ML_s \ll 1$. The densely spaced rotator spectrum scales with gaps of the order $\sim 1/F^2L_s^3$, and at $m_q = 0$ the chiral symmetry is apparently restored. However, the rotator spectrum, even at $m_q = 0$ in the finite volume, will signal that the infinite system is in the chirally broken phase for the particular parameter set of the Lagrangian. This is often misunderstood in the interpretation of lattice simulations. Measuring finite energy levels with pion quantum numbers at fixed L_s in the $m_q \to 0$ limit is not a signal for chiral symmetry restoration of the infinite system.[39]

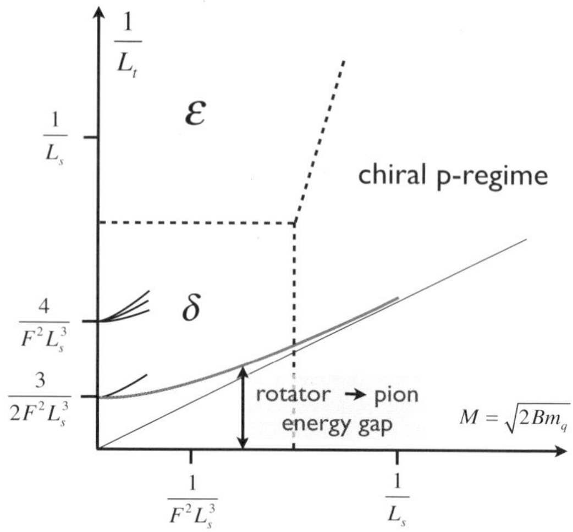

Fig. 1. Schematic plot of the regions in which the three low energy chiral expansions are valid. The vertical axis shows the finite temperature scale (euclidean time in the path integral) which probes the rotator dynamics of the δ-regime and the ϵ-regime. The first two low lying rotator levels are also shown on the vertical axis for the simple case of $N_f = 2$. The fourfold degenerate lowest rotator excitation at $m_q = 0$ will split into an isotriplet state (lowest energy level), which evolves into the p-regime pion as m_q increases, and into an isosinglet state representing a multi-pion state in the p-regime. Higher rotator excitations have similar interpretations.

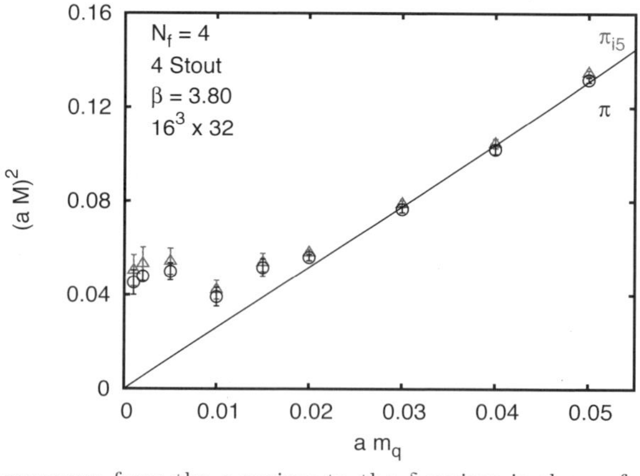

Fig. 2. The crossover from the p-regime to the δ-regime is shown for the π and π_{i5} states at $N_f = 4$.

If $L_t \sim L_s$ under the conditions $FL_s > 1$ and $ML_s \ll 1$, the system will be driven into the ϵ-regime which can be viewed as the high temperature limit of the δ-regime quantum rotator. Although the δ-regime and ϵ-regime have an overlapping region, there is an important difference in their dynamics. In the δ-regime of the quantum rotator, the mode of the pion field $U(x)$ with zero spatial momentum dominates with time-dependent quantum dynamics. The ϵ-regime is dominated by the four-dimensional zero momentum mode of the chiral Lagrangian.

We report simulation results of all three regimes in the chirally broken phase of the technicolor models we investigate. The analysis of the three regimes complement each other and provide cross-checks for the correct identification of the phases. First, we will probe Eqs. (1-4) in the p-regime, and follow with the study of Dirac spectra and RMT eigenvalue distributions in the ϵ-regime. The spectrum in the δ-regime is used as a signal to monitor p-regime spectra as m_q decreases. Fig. 2 is an illustrative example of this crossover in our simulations. It is important to note that the energy levels in the chiral limit do not always match the rotator spectrum at the small $F \cdot L_s$ values of the simulations. This squeezing of insufficiently large enough $F \cdot L_s$ for undistorted, finite volume chiral behavior in the p-regime, ϵ-regime, and δ-regime is a serious limitation of all recent simulations.

3. Goldstone spectrum and χSB at $\mathrm{N_f} = 12$

We find the chiral symmetry breaking pattern for the controversial $N_f = 12$ case similar to the $N_f = 8, 9$ cases. The Goldstone spectrum remains separated from the technicolor scale of the ρ-meson. The true Goldstone pion and two additional split pseudo-Goldstone states are shown again in Fig. 3 with different slopes as $a \cdot m_q$ increases. The trends and the underlying explanation are similar to the $N_f = 8, 9$ cases. The chiral fit to M_π^2/m_q shown at the top right side of Fig. 3 is based on Eq. (1) only since the F_π data points are outside the convergence range of the chiral expansion. At $\beta = 2.2$ the fitted value of B is $a \cdot B = 2.7(2)$ in lattice units with $a \cdot F = 0.0120(1)$ and $a \cdot \Lambda_3 = 0.50(3)$ also fitted. The fitted ρ-mass in the chiral limit is $a \cdot M_\rho = 0.115(15)$ from $a \cdot m_q = 0.025 - 0.045$ with $M_\rho/F = 10(1)$. The fitted value of $B/F = 223(17)$ is not very reliable but consistent with the enhancement of the chiral condensate found at $N_f = 8, 9$ without including renormalization scale effects. Again, at fixed lattice spacing, the small chiral condensate $\langle\overline{\psi}\psi\rangle$ summed over all flavors is dominated by the linear term in m_q from UV contributions. The linear fit gives $\langle\overline{\psi}\psi\rangle = 0.0033(13)$ in the chiral limit which came out unexpectedly close the GMOR relation of $\langle\overline{\psi}\psi\rangle = 12F^2B$ with $12F^2B = 0.0046(4)$ fitted. Issues and concerns in the systematics are similar to the $N_f = 8, 9$ cases

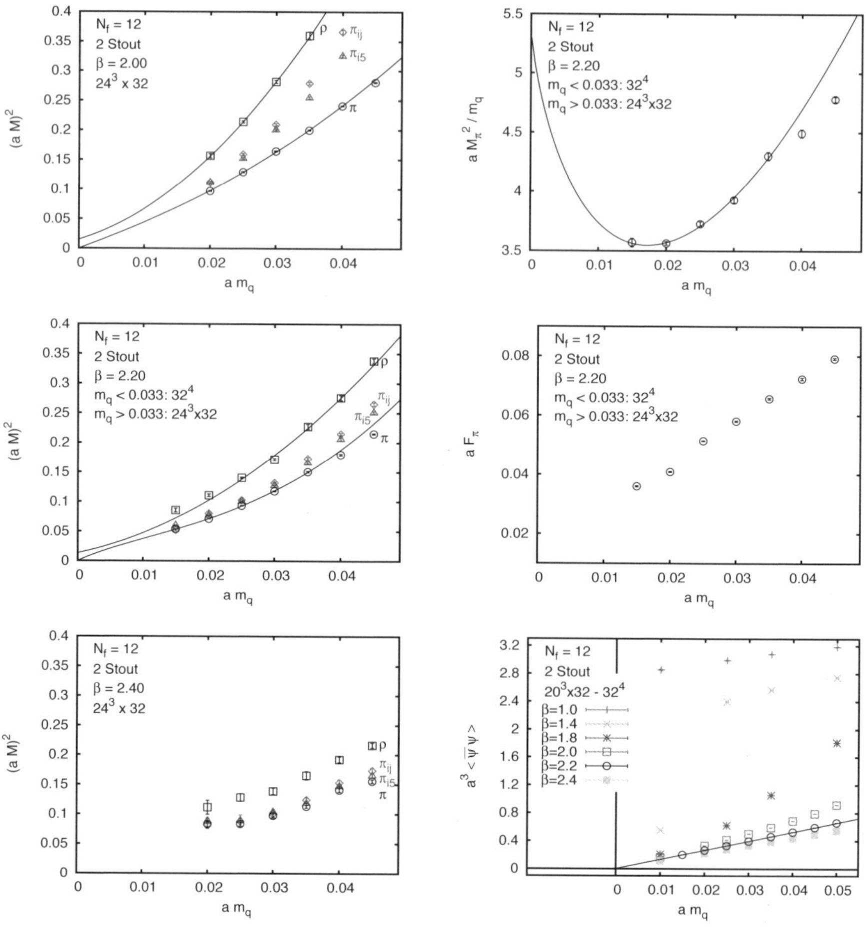

Fig. 3. The pseudo-Goldstone spectrum and chiral fits are shown for $N_f = 12$ simulations with lattice size $24^3 \times 32$ and 32^4. The left column shows the pseudo-Goldstone spectrum with decreasing taste breaking as the gauge coupling is varied from $\beta = 2.0$ to $\beta = 2.4$. Although the bottom figure on the left at $\beta = 2.4$ illustrates the continued restoration of taste symmetry, the volume is too small for the Goldstone spectrum. The middle value at $\beta = 2.2$ was chosen in the top right figure with fitting range $a \cdot m_q = 0.015 - 0.035$ of the NLO chiral fit to M_π^2/m_q which approaches $2B$ in the chiral limit. The middle figure on the right shows the F_π data with no NLO fit far away from the chiral limit. The bottom right figure, with its additional features discussed in the text, is the linear fit to the chiral condensate with fitting range $a \cdot m_q = 0.02 - 0.04$. The physical fit parameters B, F, Λ_3 are discussed in the text. Two stout steps were used in all $N_f = 12$ simulations.

but finite volume limitations and the convergence of the chiral expansion are more problematic. Currently we are investigating the important $N_f = 12$ model on larger lattices to probe the possible influence of unwanted squeezing effects on the spectra. This should also clarify the mass splitting pattern of the ρ and A_1 states we are seeing in the chiral limit as N_f is varied.

Our findings at $N_f = 12$ are not consistent with some earlier work[30,31] where the runing coupling was studied. Lessons from the Dirac spectra and RMT to complement p-regime tests are discussed in the next section.

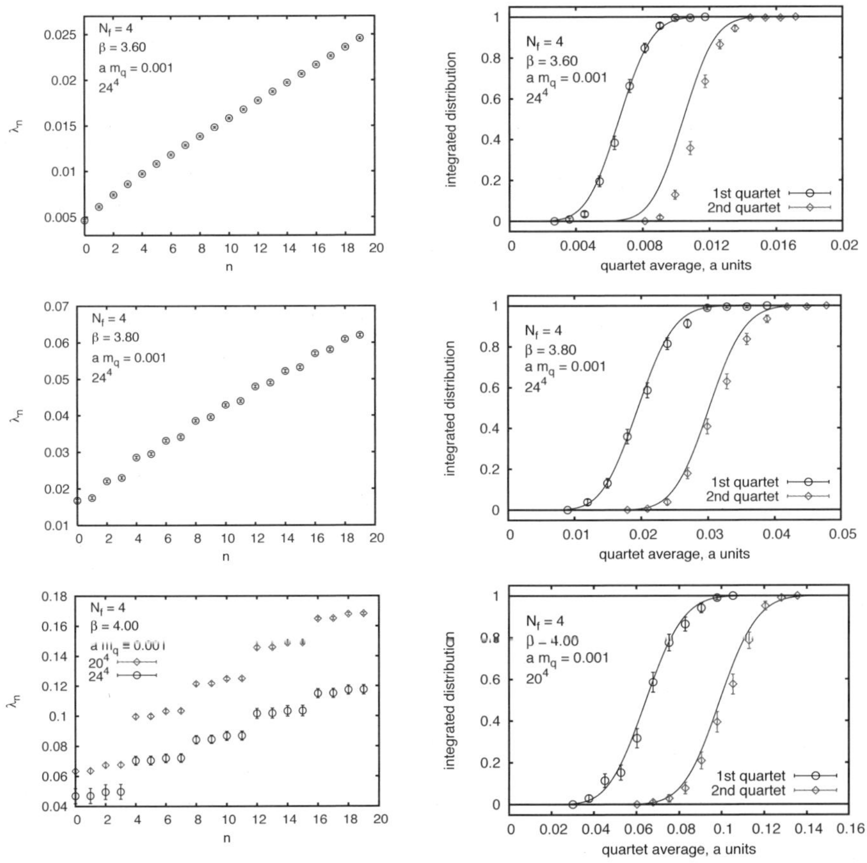

Fig. 4. From simulations at $N_f = 4$ the first column shows the approach to quartet degeneracy of the spectrum as β increases. The second column shows the integrated distribution of the two lowest quartets averaged. The solid line compares this procedure to RMT with $N_f = 4$.

4. Epsilon regime, Dirac spectrum and RMT

If the bare parameters of a gauge theory are tuned to the ϵ-regime in the chirally broken phase, the low-lying Dirac spectrum follows the predictions of random matrix theory. The corresponding random matrix model is only sensitive to the pattern of chiral symmetry breaking, the topological charge and the rescaled fermion mass once the eigenvalues are also rescaled by the same factor $\Sigma_{cond}V$. This idea has been confirmed in various settings both in quenched and fully dynamical simulations. The same method is applied here to nearly conformal gauge models.

The connection between the eigenvalues λ of the Dirac operator and chiral symmetry breaking is given in the Banks-Casher relation,[69]

$$\Sigma_{cond} = -\langle \overline{\Psi}\Psi \rangle = \lim_{\lambda \to 0} \lim_{m \to 0} \lim_{V \to \infty} \frac{\pi \rho(\lambda)}{V},$$

where Σ_{cond} designates the quark condensate normalized to a single flavor. To generate a non-zero density $\rho(0)$, the smallest eigenvalues must become densely packed as the volume increases, with an eigenvalue spacing $\Delta\lambda \approx 1/\rho(0) = \pi/(\Sigma_{cond}V)$. This allows a crude estimate of the quark condensate Σ_{cond}. One can do better by exploring the ϵ-regime: If chiral symmetry is spontaneously broken, tune the volume and quark mass such that $\frac{1}{F_\pi} \ll L \ll \frac{1}{M_\pi}$, so that the Goldstone pion is much lighter than the physical value, and finite volume effects are dominant as we discussed in Section 2. The chiral Lagrangian is dominated by the zero-momentum mode from the mass term and all kinetic terms are suppressed. In this limit, the distributions of the lowest eigenvalues are identical to those of random matrix theory, a theory of large matrices obeying certain symmetries.[70–72] To connect with RMT, the eigenvalues and quark mass are rescaled as $z = \lambda\Sigma_{cond}V$ and $\mu = m_q\Sigma_{cond}V$, and the eigenvalue distributions also depend on the topological charge ν and the number of quark flavors N_f. RMT is a very useful tool to calculate analytically all of the eigenvalue distributions.[73] The eigenvalue distributions in various topological sectors are measured via lattice simulations, and via comparison with RMT, the value of the condensate Σ_{cond} can be extracted.

After we generate large thermalized ensembles, we calculate the lowest twenty eigenvalues of the Dirac operator using the PRIMME package.[74] In the continuum limit, the staggered eigenvalues form degenerate quartets, with restored taste symmetry. The first column of Fig. 4 shows the change in the eigenvalue structure for $N_f = 4$ as the coupling constant is varied. At $\beta = 3.6$ grouping into quartets is not seen, the Goldstone pions are somewhat still split, and staggered perturbation theory is just beginning to kick in. At $\beta = 3.8$ doublet pairing appears and at $\beta = 4.0$ the quartets are nearly degenerate. The Dirac spectrum is collapsed as required by the Banks-Casher relation. In the second column we show the integrated distributions of the two lowest eigenvalue quartet averages,

$$\int_0^\lambda p_k(\lambda')d\lambda', \qquad k = 1, 2 \tag{5}$$

which is only justified close to quartet degeneracy. All low eigenvalues are selected with zero topology. To compare with RMT, we vary $\mu = m_q \Sigma_{cond} V$ until we satisfy

$$\frac{\langle \lambda_1 \rangle_{\rm sim}}{m} = \frac{\langle z_1 \rangle_{\rm RMT}}{\mu}, \tag{6}$$

where $\langle \lambda_1 \rangle_{\rm sim}$ is the lowest quartet average from simulations and the RMT average $\langle z \rangle_{\rm RMT}$ depends implicitly on μ and N_f. With this optimal value of μ, we can predict the shapes of $p_k(\lambda)$ and their integrated distributions, and compare to the simulations. The agreement with the two lowest integrated RMT eigenvalue shapes is excellent for the larger β values.

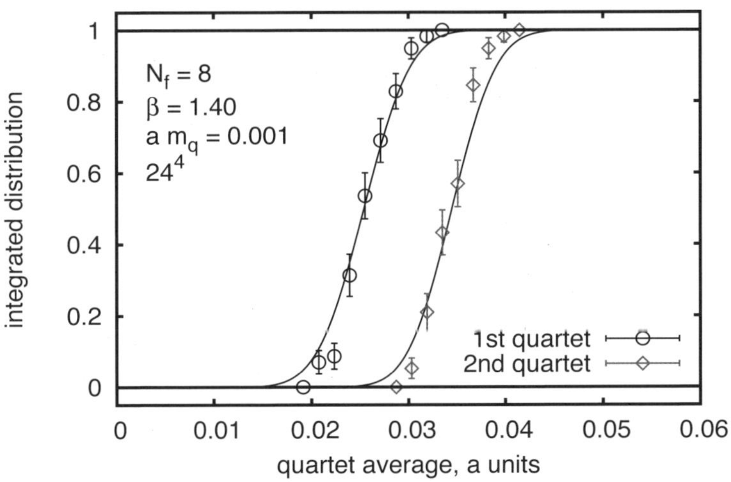

Fig. 5. The solid lines compare the integrated distribution of the two lowest quartet averages to RMT predictions with $N_f = 8$.

The main qualitative features of the RMT spectrum are very similar in our $N_f = 8$ simulations as shown in Fig. 5 . One marked quantitative difference is a noticeable slowdown in response to change in the coupling constant. As β grows the recovery of the quartet degeneracy is considerably delayed in comparison with the onset of p-regime Goldstone dynamics. Overall, for the $N_f = 4, 8$ models we find consistency between the p-regime analysis and the RMT tests. Earlier, using Asqtad fermions at a particular β value, we found agreement with RMT even at $N_f = 12$ which indicated a chirally broken phase.[22] Strong taste breaking with Asqtad fermions leaves the quartet averaging in question and the bulk pronounced crossover of the Asqtad action as β grows is also an issue. Currently we are investigating the RMT picture for $N_f = 9, 10, 11, 12$ with our much improved action with stout smearing. This action shows no artifact transitions and handles taste breaking much more effectively. Firm conclusions on the $N_f = 12$ model to support our findings of χSB in the p-regime will require continued investigations.

5. Inside the conformal window

A distinguished feature of the $N_f = 16$ conformal model is how the renormalized coupling $g^2(L)$ runs with L, the linear size of the spatial volume in a Hamiltonian or Transfer Matrix description. On very small scales the running coupling $g^2(L)$ grows with L as in any other asymptotically free theory. However, $g^2(L)$ will not grow large, and in the $L \to \infty$ limit it will converge to the fixed point g^{*2} which is rather weak,[75] within the reach of perturbation theory. There is non-trivial, small-volume dynamics which is illustrated first in the pure gauge sector.

At small g^2, without fermions, the zero-momentum components of the gauge field are known to dominate the dynamics.[76–78] With $SU(3)$ gauge group, there are twenty-seven degenerate vacuum states, separated by energy barriers which are generated by the integrated effects of the non-zero momentum components of the gauge field in the Born-Oppenheimer approximation. The lowest-energy excitations of the gauge field Hamiltonian scale as $\sim g^{2/3}(L)/L$ evolving into glueball states and becoming independent of the volume as the coupling constant grows with L. Non-trivial dynamics evolves through three stages as L grows. In the first regime, in very small boxes, tunneling is suppressed between vacua which remain isolated. In the second regime, for larger L, tunneling sets in and electric flux states will not be exponentially suppressed. Both regimes represent small worlds with zero-momentum spectra separated from higher momentum modes of the theory with energies on the scale of $2\pi/L$. At large enough L the gauge dynamics overcomes the energy barrier, and wave functions spread over the vacuum valley. This third regime is the crossover to confinement where the electric fluxes collapse into thin string states wrapping around the box.

It is likely that a conformal theory with a weak coupling fixed point at $N_f = 16$ will have only the first two regimes which are common with QCD. Now the calculations have to include fermion loops.[79,80] The vacuum structure in small enough volumes, for which the wave functional is sufficiently localized around the vacuum configuration, remains calculable by adding in one-loop order the quantum effects of the fermion field fluctuations. The spatially constant abelian gauge fields parametrizing the vacuum valley are given by $A_i(\mathbf{x}) = T^a C_i^a/L$ where T_a are the (N-1) generators for the Cartan subalgebra of $SU(N)$. For $SU(3)$, $T_1 = \lambda_3/2$ and $T_2 = \lambda_8/2$. With N_f flavors of massless fermion fields the effective potential of the constant mode is given by

$$V_{\text{eff}}^{\mathbf{k}}(\mathbf{C}^b) = \sum_{i>j} V(\mathbf{C}^b[\mu_b^{(i)} - \mu_b^{(j)}]) - N_f \sum_i V(\mathbf{C}^b \mu_b^{(i)} + \pi\mathbf{k}), \tag{7}$$

with $\mathbf{k} = \mathbf{0}$ for periodic, or $\mathbf{k} = (1,1,1)$, for antiperiodic boundary conditions on the fermion fields. The function $V(\mathbf{C})$ is the one-loop effective potential for $N_f = 0$ and the weight vectors $\mu^{(i)}$ are determined by the eigenvalues of the abelian generators. For SU(3) $\mu^{(1)} = (1,1,-2)/\sqrt{12}$ and $\mu^{(2)} = \frac{1}{2}(1,-1,0)$. The correct quantum vacuum is found at the minimum of this effective potential which is dramatically changed by the fermion loop contributions. The Polyakov loop observables remain center elements at the

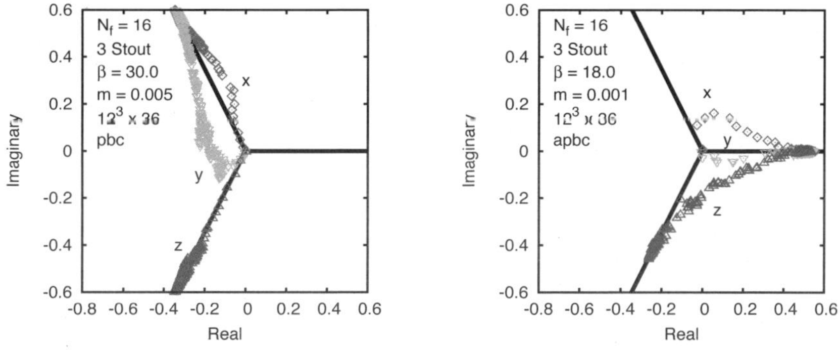

Fig. 6. The time evolution of complex Polyakov loop distributions are shown from our $N_f = 16$ simulations with $12^3 \times 36$ lattice volume. Tree-level Symanzik-improved gauge action is used in the simulations and staggered fermions with three stout steps and very small fermion masses.

new vacuum configurations with complex values; for $SU(N)$

$$P_j = \frac{1}{N}\,\mathrm{tr}\,\left(\exp(iC_j^b T_b)\right) = \frac{1}{N}\sum_n \exp(i\mu_b^{(n)} C_j^b) = \exp(2\pi i l_j/N). \quad (8)$$

This implies $\mu_b^{(n)}\mathbf{C}^b = 2\pi\mathbf{l}/N$ (mod 2π), independent of n, and $V_{\mathrm{eff}}^{\mathbf{k}} = -N_f N V(2\pi\mathbf{l}/N + \pi\mathbf{k})$. In the case of antiperiodic boundary conditions, $\mathbf{k} = (1,1,1)$, this is minimal only when $\mathbf{l} = \mathbf{0}$ (mod 2π). The quantum vacuum in this case is the naive one, $A = 0$ ($P_j = 1$). In the case of periodic boundary conditions, $\mathbf{k} = \mathbf{0}$, the vacua have $\mathbf{l} \neq \mathbf{0}$, so that P_j correspond to non-trivial center elements. For SU(3), there are now 8 degenerate vacua characterized by eight different Polyakov loops, $P_j = \exp(\pm 2\pi i/3)$. Since they are related by coordinate reflections, in a small volume parity (P) and charge conjugation (C) are spontaneously broken, although CP is still a good symmetry.[79]

Our simulations of the $N_f = 16$ model below the conformal fixed point g^{*2} confirm the theoretical vacuum structure. Fig. 6 shows the time evolution of Polyakov loop distributions monitored along the three separate spatial directions. On the left side, with periodic spatial boundary conditions, the time evolution is shown starting from randomized gauge configuration with the Polyakov loop at the origin. The system evolves into one of the eight degenerate vacua selected by the positive imaginary part of the complex Polyakov loop along the x and y direction and negative imaginary part along the z direction. On the right, with antiperiodic spatial boundary conditions, the vacuum is unique and trivial with real Polyakov loop in all three directions. The time evolution is particularly interesting in the z direction with a swing first from the randomized gauge configuration to a complex metastable minimum first, and eventually tunneling back to the trivial vacuum and staying there, as expected. The measured fermion-antifermion spectra and the spectrum of the Dirac operator further confirm this vacuum structure.

Acknowledgments

We thank Sándor Katz and Kálmán Szabó for the Wuppertal RHMC code. For some calculations, we used the publicly available MILC code. We performed simulations on the Wuppertal GPU cluster, Fermilab clusters under the auspices of USQCD and SciDAC, and the Ranger cluster of the Teragrid organization. This research is supported by the NSF under grant 0704171, by the DOE under grants DOE-FG03-97ER40546, DOE-FG-02-97ER25308, by the DFG under grant FO 502/1 and by SFB-TR/55.

References

1. S. Weinberg, *Phys. Rev.* **D19**, 1277 (1979).
2. L. Susskind, *Phys. Rev.* **D20**, 2619 (1979).
3. E. Farhi and L. Susskind, *Phys. Rept.* **74**, 277 (1981).
4. F. Sannino, arXiv:0902.3494 [hep-ph].
5. T. A. Ryttov and F. Sannino, *Phys. Rev.* **D76**, 105004 (2007).
6. D. D. Dietrich and F. Sannino, *Phys. Rev.* **D75**, 085018 (2007).
7. D. K. Hong *et al.*, *Phys. Lett.* **B597**, 89 (2004).
8. H. Georgi, *Phys. Rev. Lett.* **98**, 221601 (2007).
9. M. A. Luty and T. Okui, *JHEP* **0609**, 070 (2006).
10. W. E. Caswell, *Phys. Rev. Lett.* **33**, 244 (1974).
11. T. Banks and A. Zaks, *Nucl. Phys.* **B196**, 189 (1982).
12. T. Appelquist *et al.* , *Phys. Rev. Lett.* **61**, 1553 (1988).
13. A. G. Cohen and H. Georgi, *Nucl. Phys.* **B314**, 7 (1989).
14. T. Appelquist *et al.*, *Phys. Rev. Lett.* **77**, 1214 (1996).
15. B. Holdom, *Phys. Rev.* **D24**, 1441 (1981).
16. K. Yamawaki *et al.*, *Phys. Rev. Lett.* **56**, 1335 (1986).
17. T. W. Appelquist *et al.*, *Phys. Rev. Lett.* **57**, 957 (1986).
18. V. A. Miransky and K. Yamawaki, *Phys. Rev.* **D55**, 5051 (1997).
19. M. Kurachi and R. Shrock, *JHEP* **0612**, 034 (2006).
20. E. Eichten and K. D. Lane, *Phys. Lett.* **B90**, 125 (1980).
21. Z. Fodor, K. Holland, J. Kuti, D. Nogradi and C. Schroeder, *Phys. Lett.* **B681**, 353 (2009).
22. Z. Fodor *et al.*, *PoS* **LATTICE2008**, 066 (2008).
23. Z. Fodor *et al.*, *PoS* **LATTICE2008**, 058 (2008).
24. Z. Fodor, K. Holland, J. Kuti, D. Nogradi and C. Schroeder, arXiv:0908.2466 [hep-lat].
25. Z. Fodor, K. Holland, J. Kuti, D. Nogradi and C. Schroeder, *JHEP* **0908**, 084 (2009) [arXiv:0905.3586 [hep-lat]].
26. T. DeGrand *et al.*, *Phys. Rev.* **D79**, 034501 (2009).
27. T. DeGrand *et al.*, arXiv:0809.2953 [hep-lat].
28. B. Svetitsky et al., arXiv:0809.2885 [hep-lat].
29. Y. Shamir *et al.*, *Phys. Rev.* **D78**, 031502 (2008).
30. T. Appelquist *et al.*, *Phys. Rev. Lett.* **100**, 171607 (2008).
31. T. Appelquist *et al.*, arXiv:0901.3766 [hep-ph].
32. G. T. Fleming, *PoS* **LATTICE2008**, 021 (2008).
33. L. Del Debbio *et al.*, arXiv:0812.0570 [hep-lat].
34. L. Del Debbio *et al.*, arXiv:0805.2058 [hep-lat].
35. L. Del Debbio *et al.*, *JHEP* **0806**, 007 (2008).
36. A. J. Hietanen *et al.*, arXiv:0904.0864 [hep-lat].
37. A. J. Hietanen *et al.*, arXiv:0812.1467 [hep-lat].
38. A. Hietanen *et al.*, *PoS* **LATTICE2008**, 065 (2008).
39. A. Deuzeman *et al.*, arXiv:0904.4662 [hep-ph].
40. A. Deuzeman *et al.*, arXiv:0810.3117 [hep-lat].
41. A. Deuzeman *et al.*, *PoS* **LATTICE2008**, 060 (2008).
42. A. Deuzeman *et al.*, *Phys. Lett.* **B670**, 41 (2008).

43. S. Catterall and F. Sannino, *Phys. Rev.* **D76**, 034504 (2007).
44. S. Catterall *et al.*, *JHEP* **0811**, 009 (2008).
45. X. Y. Jin and R. D. Mawhinney, *PoS* **LATTICE2008**, 059 (2008).
46. A. Hasenfratz, arXiv:0907.0919 [hep-lat].
47. T. DeGrand and A. Hasenfratz, arXiv:0906.1976 [hep-lat].
48. T. DeGrand, arXiv:0906.4543 [hep-lat].
49. L. Del Debbio *et al.*, arXiv:0907.3896 [hep-lat].
50. T. DeGrand, arXiv:0910.3072 [hep-lat].
51. T. Appelquist *et al.*, arXiv:0910.2224 [hep-ph].
52. X. Y. Jin and R. D. Mawhinney, *PoS* **LAT2009**, 049 (2009).
53. A. Hasenfratz, arXiv:0911.0646 [hep-lat].
54. D. K. Sinclair and J. B. Kogut, arXiv:0909.2019 [hep-lat].
55. C. Morningstar and M. J. Peardon, *Phys. Rev.* **D69**, 054501 (2004).
56. Y. Aoki *et al.*, it JHEP **0601**, 089 (2006); *Phys. Lett.* **B643**, 46 (2006).
57. W. J. Lee and S. R. Sharpe, Phys. Rev. **D60**, 114503 (1999).
58. C. Aubin and C. Bernard, *Phys. Rev.* **D68**, 034014 (2003).
59. C. Aubin and C. Bernard, *Phys. Rev.* **D68**, 074011 (2003).
60. J. Gasser and H. Leutwyler, *Nucl. Phys.* **B307**, 763 (1988).
61. F. C. Hansen and H. Leutwyler, *Nucl. Phys.* **B350**, 201 (1991).
62. J. Gasser and H. Leutwyler, *Annals Phys.* **158**, 142 (1984).
63. M. Gell-Mann et al., *Phys. Rev.* **175**, 2195 (1968).
64. J. Bijnens and J. Lu, arXiv:0910.5424 [hep-ph].
65. H. Leutwyler, *Phys. Lett.* **B189**, 197 (1987).
66. J. Noaki *et al.* [JLQCD and TWQCD Collaborations], *Phys. Rev. Lett.* **101**, 202004 (2008).
67. P. Hasenfratz and F. Niedermayer, *Z. Phys.* **B92**, 91 (1993), arXiv:hep-lat/9212022.
68. P. Hasenfratz, arXiv:0909.3419 [hep-th].
69. T. Banks and A. Casher, *Nucl. Phys.* **B169**, 103 (1980).
70. E. V. Shuryak and J. J. M. Verbaarschot, *Nucl. Phys.* **A560**, 306 (1993).
71. P. H. Damgaard, *Nucl. Phys. Proc. Suppl.* **128**, 47 (2004).
72. J. J. M. Verbaarschot and T. Wettig, *Ann. Rev. Nucl. Part. Sci.* **50**, 343 (2000).
73. P. H. Damgaard and S. M. Nishigaki, *Phys. Rev.* **D63**, 045012 (2001).
74. A. Stathopoulos and J. R. McCombs, *SIAM J. Sci. Comput.*, Vol. **29, No. 5**, 2162 (2007).
75. U. M. Heller, *Nucl. Phys. Proc. Suppl.* **63**, 248 (1998).
76. G. 't Hooft, *Nucl. Phys.* **B153**, 141 (1979).
77. M. Luscher, *Nucl. Phys.* **B219**, 233 (1983).
78. P. van Baal and J. Koller, *Annals Phys.* **174**, 299 (1987).
79. P. van Baal, *Nucl. Phys.* **B307**, 274 (1988).
80. J. Kripfganz and C. Michael, *Nucl. Phys.* **B314**, 25 (1989).

KINETICS OF CHIRAL PHASE TRANSITIONS IN QUARK MATTER

AWANEESH SINGH, SANJAY PURI

School of Physical Sciences, Jawaharlal Nehru University, New Delhi – 110067, India

HIRANMAYA MISHRA

Theory Division, Physical Research Laboratory,
Navrangpura, Ahmedabad – 380009, India
hm@prl.res.in

We study the kinetics of chiral transitions in quark matter using a microscopic framework (Nambu-Jona-Lasinio model) and a phenomenological model (Ginzburg-Landau free energy). We focus on the coarsening dynamics subsequent to a quench from the massless quark phase to the massive quark phase. The morphology of the ordering system is characterized by the scaling of the order-parameter correlation function. The domain growth process obeys the Allen-Cahn growth law, $L(t) \sim t^{1/2}$. We also study the growth of bubbles of the stable massive phase from the metastable massless phase.

The nature of the QCD phase diagram in the plane of temperature (T) and baryon chemical potential (μ) has been studied extensively over the last few decades.[1,2] For $\mu = 0$, finite-temperature perturbative QCD calculations have been complemented by first-principle calculations like lattice QCD simulations.[3–5] However, for $\mu \neq 0$, lattice QCD calculations are limited to small values of μ.[6] In QCD with two massless quarks, the equilibrium chiral phase transition is expected to be a second-order transition at zero baryon densities. However, in nature, the light quarks are not exactly massless and the second-order phase transition is replaced by an analytical crossover. This picture is consistent with lattice QCD simulations with a transition temperature $T_c \sim 140 - 190$ MeV.[7] Calculations based on different effective potentials, on the other hand, indicate the phase transition to be first-order at large μ and small T. This means that the phase diagram will have a tri-critical point, where the first-order chiral transition becomes second-order (for vanishing quark masses) or ends (for non-vanishing quark masses) at a critical end point in the phase diagram. The location of the tri-critical point (TCP)(μ^E, T^E) in the phase digram has been estimated by lattice QCD simulations as $\mu^E/T_c \simeq 0.94$ and $\mu^E/T^E \simeq 1.8$.[8]

Heavy-ion collision experiments at high energies produce hot and dense strongly-interacting matter, and provide the opportunity to explore the phase diagram of QCD. While the high-T and small-μ region of the QCD phase diagram has been explored in recent experiments. Future heavy-ion collision experiments plan to explore the high baryon density regime, particularly the region around the tricritical point.[9] The experiments at Relativistic Heavy Ion Collider (RHIC) provide clear signals that the nuclear matter undergoes a phase transition to partonic phases at sufficiently large value of the energy density. However, the nature of the phase transition still remains an open question. We might mention here that lattice QCD thermodynamics has an built in equilibrium assumption while the heavy ion experiments are essentially nonequilibrium processes. It is worthwhile to mention here that, in a phase transition process, information about which equilibrium phase has lowest free energy is not sufficient to discuss all possible structures that the system can have. One has to understand the kinetics of the process by which the phase ordering or disordering proceeds and the nature of nonequilibrium structures that the system must go through on its way to reach equilibrium.

It is not surprising, therefore that much attention has been focused on dynamics of chiral transition in dense quark matter, particularly near the critical point. The critical behavior and the fluctuations of conserved charges in the presence of spinodal decomposition in the context of chiral transition have been explored in Nambu-Jona-Lasinio (NJL) model.[10] There have also been preliminary studies[11] of the kinetics of the chiral phase transition in a nonlinear sigma model coupled to quarks. In this context, the effect of dissipation in slowing down spinodal decomposition has also been studied in this model.[12] The dynamics of first-order phase transitions has been considered recently in Ref.[13] Simulated by the findings of lattice QCD, few studies have also been considered for the case of smooth crossover, its dynamics as compared to a strongly first order transition.[14] Further, a Langevin equation has been derived within NJL model describing the behavior of the fluctuation above the critical temperature at finite chemical potential within a linear approximation.[15]

On the otherhand, there has been intense research interest in the kinetics of phase transitions, and the phase ordering process that occurs after a rapid quench in system parameters e.g., temperature, pressure[16,17] in various condensed matter systems. During the transition, the system develops a spatial structure of randomly-distributed domains which coarsen with time. This domain growth process has been extensively studied in many condensed matter systems like ferromagnets, binary fluids, liquid crystals, etc.

In the present work, we focus on the kinetics of the chiral phase transition subsequent to a quench from the disordered phase (with vanishing quark condensate) to the ordered phase for quark matter. This means while the equilibrium configuration at zero temperature is an ordered state, the system finds itself in a configuration from the ensemble appropriate to a high temperature. We are thus interested in the far-from-equilibrium evolution of the system and pattern dynamics associated with it.

To model chiral symmetry breaking in QCD, we use the two-flavor Nambu-Jona-Lasinio (NJL) model[18,19] with the Hamiltonian

$$\mathcal{H} = \sum_{i,a} \psi^{ia\dagger} \left(-i\vec{\alpha} \cdot \vec{\nabla} + \gamma^0 m_i \right) \psi^{ia} - G \left[(\bar{\psi}\psi)^2 - (\bar{\psi}\gamma^5 \vec{\tau}\psi)^2 \right]. \tag{1}$$

Here, m_i is the current quark mass – we take this to be the same ($m_i = m$) for both u and d quarks. The parameter G denotes the quark-quark interaction strength. Further, τ is the Pauli matrix acting in flavor space. The quark operator ψ has two indices i and a, denoting theflavor and color indices, respectively. This model exhibits a second-order chiral phase transition for massless quarks at small μ and high T, and a first-order transition at large μ and small T.

To describe the ground state, we take an ansatz with quark-antiquark condensates:[20]

$$|vac\rangle = \exp\left[\int d\vec{k}\ q_I^{0i}(\vec{k})^\dagger (\vec{\sigma} \cdot \vec{k}) h_i(\vec{k}) \tilde{q}_I^{0i}(-\vec{k}) - \text{h.c.} \right] |0\rangle. \tag{2}$$

Here, $q^\dagger$, $\tilde{q}$ are two-component quark and antiquark creation operators, and $|0\rangle$ is the perturbative chiral vacuum. Further, $h_i(\vec{k})$ is a variational function related to the quark-antiquark condensate as

$$\langle \bar{\psi}\psi \rangle = -\frac{3}{\pi^3} \sum_{i=1}^{2} \int d\vec{k}\ \sin[2h_i(\vec{k})]. \tag{3}$$

This flavor-dependent function can be determined by minimizing the energy at $T = 0$, or the thermodynamic potential at nonzero T and density. In the mean-field approximation and, near the chiral phase transition, we write down the expression for the thermodynamic potential as

$$\tilde{\Omega}(M,\beta,\mu) = -\frac{12}{(2\pi)^3\beta}\int d\vec{k}\left\{\ln\left[1+e^{-\beta\left(\sqrt{k^2+M^2}-\mu\right)}\right] + \ln\left[1+e^{-\beta\left(\sqrt{k^2+M^2}+\mu\right)}\right]\right\} - \frac{12}{(2\pi)^3}\int d\vec{k}\left(\sqrt{\vec{k}^2+M^2}-k\right)+\frac{M^2}{4G}. \tag{4}$$

Here, we have taken vanishing current quark mass, and introduce $M = -2g\rho_s$ with $\rho_s = \langle\bar{\psi}\psi\rangle$ being the scalar density and $g = G[1 + 1/(4N_c)]$. The details of the mean-field approximation are reported elsewhere.[21]

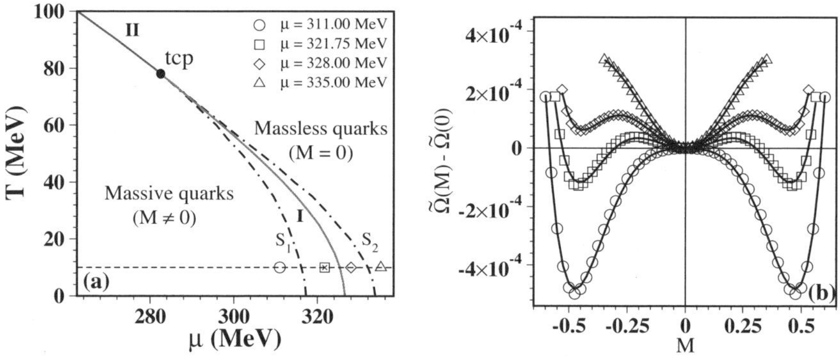

Fig. 1. (a) Phase diagram of the Nambu-Jona-Lasinio (NJL) model in the (μ, T)-plane for zero current quark mass. A line of first-order transitions (I, green online) meets a line of second-order transitions (II, blue online) at the tricritical point (tcp). We have $(\mu_{\rm tcp}, T_{\rm tcp}) \simeq (282.58, 78)$ MeV. The dot-dashed lines S_1 and S_2 denote the spinodals or metastability limits for the first-order transitions. The open symbols denote 4 combinations of (μ, T), chosen to represent qualitatively different shapes of the NJL potential. The cross denotes the point at which we quench the system for $b < 0$. (b) Plot of $\tilde{\Omega}(M, \beta, \mu)$ from Eq. (4) as a function of M. The (μ, T)-values are marked in (a). The solid lines superposed on the potentials correspond to the GL potential in Eq. (5) with a from Eq. (6), and b, d being fit parameters (cf. Table 1).

The phase diagram for the chiral transition in the (μ, T)-plane resulting from Eq. (4) is shown in Fig. 1(a). For the numerical calculation of the thermodynamic potential, we have taken here a three-momentum ultraviolet cutoff $\Lambda = 653.30$ MeV, and the four-fermion coupling $G = 5.0163\times 10^{-6}$ MeV^{-2}.[22] With these parameters, the vacuum mass of quarks is $M \simeq 312$ MeV. At $T = 0$, a first-order transition takes place at $\mu \simeq 326.321$ MeV. For $\mu = 0$, a second-order transition takes place at $T \simeq 190$ MeV. The first-order line (I) meets the second-order line (II) at the tricritical point

$(\mu_{\rm tcp}, T_{\rm tcp}) \simeq (282.58, 78)$ MeV. The first-order transition is characterized by the existence of metastable phases. The limit of metastability is denoted by the dashed lines S_1, S_2 in Fig. 1(a), referred to as spinodal lines.

Close to the phase boundary, the potential in Eq. (4) may be expanded as a Ginzburg-Landau (GL) potential in the order parameter M:

$$\tilde{\Omega}\,(M) = \tilde{\Omega}\,(0) + \frac{a}{2}M^2 + \frac{b}{4}M^4 + \frac{d}{6}M^6 + O(M^8), \tag{5}$$

correct upto logarithmic corrections.[10,19] In the following, we consider the expansion of potential $\tilde{\Omega}\,(M)$ upto the M^6-term. This will prove adequate to recover the phase diagram in Fig. 1(a), as we see shortly. The first two coefficients in Eq. (5) can be obtained by comparison with Eq. (4) as

$$\tilde{\Omega}(0) = -\frac{6}{\pi^2\beta}\int_0^\Lambda dk\; k^2 \left\{\ln\left[1+e^{-\beta(k-\mu)}\right] + \ln\left[1+e^{-\beta(k+\mu)}\right]\right\},$$

$$a = \frac{1}{2G} - \frac{3\Lambda^2}{\pi^2} + \frac{6}{\pi^2}\int_0^\Lambda dk\; k\left[\frac{1}{1+e^{\beta(k-\mu)}} + \frac{1}{1+e^{\beta(k+\mu)}}\right]. \tag{6}$$

We treat the higher coefficients as phenomenological parameters, which are obtained by fitting $\tilde{\Omega}\,(M)$ in Eq. (5) to the integral expression for $\tilde{\Omega}$ in Eq. (4). There are two free parameters in the microscopic theory (μ and T), so we consider the M^6-GL potential with parameters b and d. For stability, we require $d > 0$.

In Fig. 1(b), we plot $\tilde{\Omega}\,(M) - \tilde{\Omega}\,(0)$ from Eq. (4) as a function of M. We show 4 combinations of (μ, T) as marked in Fig. 1(a), chosen to represent qualitatively different shapes of the potential. The solid lines superposed on the data sets in Fig. 1(b) correspond to the GL potential in Eq. (5) with a from Eq. (6), and b, d being fit parameters (see Table 1).

Table 1. The coefficients (a, b, d) of the GL-potential for 4 different values of μ at $T = 10$ MeV. The coefficients a, b and d are measured in units of Λ^2, Λ^0 and Λ^{-2} respectively, where $\Lambda = 653.30$ MeV.

T=10 MeV				
μ (MeV)	a/Λ^2	b	$d\Lambda^2$	$\lambda = \lvert a\rvert d/b^2$
311.00	-1.30552×10^{-3}	0.0923644	0.4389108	0.067
321.75	3.53885×10^{-3}	−0.1005344	0.4015734	0.140
328.00	6.43101×10^{-3}	−0.1109344	0.3957498	0.206
335.00	9.73628×10^{-3}	−0.1005564	0.2652330	0.255

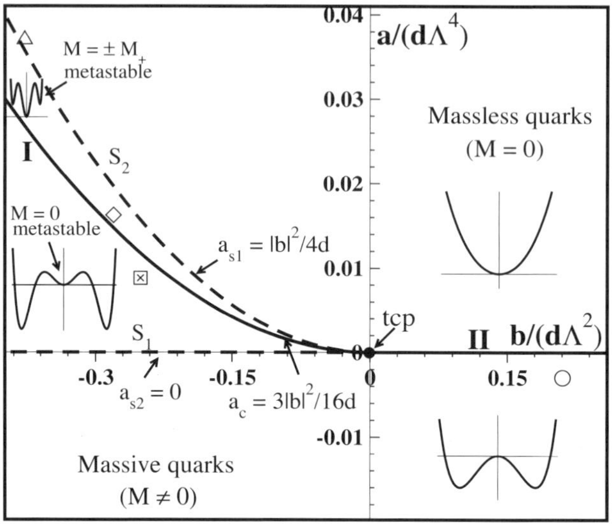

Fig. 2. Phase diagram in $(b/(d\Lambda^2), a/(d\Lambda^4))$-space for the GL-free energy in Eq. (5). The typical forms of the GL potential in various regions are shown in the figure. The open symbols denote the (μ, T)-values marked in Fig. 1(a). The cross denotes the point where we quench the system for $b < 0$, shows all different possible quenches.

The extrema of the potential in Eq. (5) are determined by the gap equation $f'(M) = aM + bM^3 + dM^5 = 0$. The solutions are $M = 0$, and $M_\pm^2 = (-b \pm \sqrt{b^2 - 4ad})/(2d)$. For $b > 0$, the transition is second-order, analogous to an M^4-potential – the stationary points are $M = 0$ (for $a > 0$) or $M = 0, \pm M_+$ (for $a < 0$). For $a < 0$, the preferred equilibrium state is the one with massive quarks. For $b < 0$, the solutions of the gap equation are as follows: (i) $M = 0$ for $a > b^2/(4d)$, (ii) $M = 0, \pm M_+$, $\pm M_-$ for $b^2/(4d) > a > 0$, and (iii) $M = 0, \pm M_+$ for $a < 0$. A first-order transition takes place at $a_c = 3b^2/(16d)$ with the order parameter jumping discontinuously from $M = 0$ to $M = \pm M_+ = \pm(3|b|/4d)^{1/2}$. The phase diagram for the GL potential is shown in Fig. 2. The tricritical point is located at $b_{\rm tcp} = 0$, $a_{\rm tcp} = 0$ [cf. Fig. 1(a)]. The 4 combinations of (μ, T)-values marked in Fig. 1(a) are identified with the same symbols in Fig. 2.

Next, we study dynamical problems in the context of the above free energy. Consider the dynamical environment of a heavy-ion collision. As long as the evolution is slow compared to the typical re-equilibration time, the order parameter field will be in local equilibrium. We consider a system which is rendered thermodynamically unstable by a rapid quench from the massless phase to the massive phase in Figs. 1 or 2. The unstable disordered state evolves via the emergence and growth of domains rich in the preferred phase.[16,17] The coarsening system is inhomogeneous, and we account for this by including a surface tension term in the GL free energy:

$$\Omega[M] = \int d\vec{r} \left[\frac{a}{2} M^2 + \frac{b}{4} M^4 + \frac{d}{6} M^6 + \frac{K}{2} \left(\vec{\nabla} M \right)^2 \right]. \tag{7}$$

In Eq. (7), $\Omega[M]$ is a functional of the spatially-dependent order parameter $M(\vec{r})$, and K measures the energy cost of spatial inhomogeneities, i.e., surface tension.

The evolution of the system is described by the time-dependent Ginzburg-Landau (TDGL) equation:

$$\frac{\partial}{\partial t} M(\vec{r}, t) = -\Gamma \frac{\delta \Omega [M]}{\delta M} + \theta(\vec{r}, t), \tag{8}$$

which models the over-damped relaxational dynamics of $M(\vec{r}, t)$ to the minimum of $\Omega[M]$.[23] Here, Γ is the inverse damping coefficient, and $\theta(\vec{r}, t)$ is the noise term satisfying the fluctuation-dissipation relation: $\langle \theta(\vec{r}, t) \rangle = 0$ and $\left\langle \theta(\vec{r'}, t')\theta(\vec{r''}, t'') \right\rangle = 2\Gamma T \delta(\vec{r'} - \vec{r''}) \delta(t' - t'')$. We use the natural scales of order parameter, space and time to introduce dimensionless variables: $M = M_0 M'$ ($M_0 = \sqrt{|a|/|b|}$); $\vec{r} = \xi \vec{r'}$ ($\xi = \sqrt{K/|a|}$); $t = \tau t'$ ($\tau = (\Gamma |a|)^{-1}$); $\theta = (\Gamma |a|^{3/2} T^{1/2} / |b|^{1/2})\, \theta'$. Dropping the primes, we obtain the dimensionless TDGL equation:

$$\frac{\partial}{\partial t} M(\vec{r}, t) = -\mathrm{sgn}(a) M - \mathrm{sgn}(b) M^3 - \lambda M^5 + \nabla^2 M + \theta(\vec{r}, t), \tag{9}$$

where $\lambda = |a| d / b^2 > 0$. For $T = 10$ MeV, as μ takes values (in MeV) 311, 321.75, 328 and 335, the corresponding values of λ are 0.067, 0.14, 0.206 and 0.255 from Table 1.

The following evolutions are performed in the dimensionless units of length and time. To get them in physical units, one has to multiply them with the appropriate dimensional parameters. Two unknown parameters are K, the strength of the surface tension and the inverse damping coefficient Γ. These parameters are obtained from the GL coefficients (see Table 1). Let us note that the surface energy can be calculated to be $\sigma = \sqrt{K}(a^{3/2}/b) \int dz (dM/dz)^2$. The surface energy ($\sigma$) for quark matter is poorly known and varies from 10-100 MeV/fm^2 at small temperatures.[24] To get an idea about the length scale we can take $\sigma = 50$ MeV/fm^2. For $T = 10$ MeV, we can estimate $\sqrt{K/a} = 1.6$ fm. This factor has to be multiplied with the dimensionless length scale to get the physical spatial length scale in units of fermi. Similarly for the inverse damping coefficient Γ, we take it to be of the order $2T/s$, where s is a quantity of order 1.[25] This leads to $t = 2.6\ t'$ fm$/s$.

In the following, we consider the phase transition kinetics for two different quench possibilities. The first case corresponds to high T and low baryon density (μ), where the quenching is done through the second-order line (II) in Fig. 1(a) or Fig. 2. The second case corresponds to low T and high baryon density (μ), where the phase conversion process can probe the metastable region of the phase diagram. This can be achieved by quenching through the first-order line (I) in Fig. 1(a) or Fig. 2. Note that both these scenarios can be studied using Eq. (9) by choosing a fixed value of λ and appropriate values of a, b. In our simulation, we have used $\lambda = 0.14$. This value of λ corresponds to, e.g., (μ, T)= (231.6 MeV, 85 MeV) or (321.75 MeV, 10 MeV) in Fig. 1(a).

First, we focus our attention to the ordering dynamics in the context of the phase diagram of Fig. 2 for the case of $b > 0$. This corresponds to the second order transition case relevant for low chemical potential regime in the context of the phase diagram of Fig. 1(a). For $b > 0$, the chiral transition occurs when $a < 0$. We solve Eq. (9) numerically using an Euler-discretization scheme with an isotropic Laplacian. We have implemented it on a d=3 lattice of size N^3 ($N = 256$), with periodic boundary conditions in all directions. The dimensionless mesh sizes are $\Delta x = 1.0$ and $\Delta t = 0.1$, which satisfy the numerical stability condition. We have further confirmed that the spatial mesh size is sufficiently small to resolve the interface region.[26]

Figure 3 shows the evolution of a disordered initial condition for Eq. (9) with $b > 0$ and $a < 0$. It corresponds to a temperature quench through the second-order line (II) in Figs. 1(a) or 2. The initial state consists of small-amplitude thermal fluctuations about the massless phase $M = 0$. The system rapidly evolves into domains of the massive phase with $M \simeq M_+$ and $M \simeq -M_+$. The interfaces between these massive domains correspond to $M = 0$. The interface evolution is shown in the snapshots (frames on left) of Fig. 3. The frames on the right show the interface structure in a cross-section of the snapshots.

The characteristic length scale $L(t)$ of the domains grows with time. The growth process is analogous to coarsening in the TDGL equation with an M^4-potential, where the coarsening is driven by kinks with the equilibrium profile $M(z) = \tanh(\pm z/\sqrt{2})$.[16,17] The order-parameter correlation function $C(r,t)$ shows *dynamical scaling* $C(r,t) = f(r/L)$. The scaling function $f(x) = (2/\pi)\sin^{-1}(e^{-x^2})$ has been calculated by Ohta *et al.* (OJK)[27]

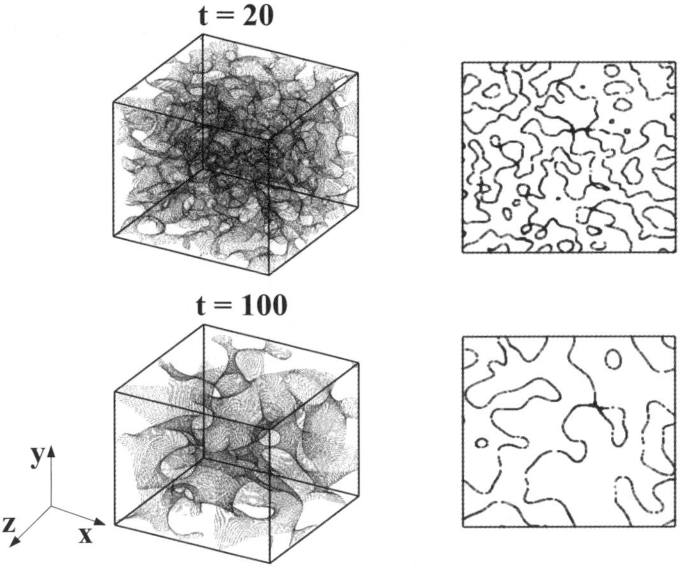

Fig. 3. Interface evolution after temperature quench through second-order line (II) in Figs. 1(a) or 2. The d=3 snapshots on the left show the interfaces ($M = 0$) at $t = 20$, 100 (in units of τ). They were obtained by numerically solving Eq. (9) as described in the text with $a < 0$, $b > 0$, $\lambda = 0.14$. The noise amplitude was $\epsilon = 0.008$. The frames on the right show a cross-section of the snapshots at $z = N/2$.

in the context of an ordering ferromagnet. In Fig. 4(a), we demonstrate that $C(r,t)$ shows dynamical scaling for the evolution in Fig. 3 shows the scaling property. Further, the domain scale obeys the Allen-Cahn (AC) growth law, $L(t) \sim t^{1/2}$[17] [see Fig. 4(b)]. Typically, the interface velocity $v \sim dL/dt \sim 1/L$, where L^{-1} measures the local curvature of the interface. This yields the AC growth law. The same growth law has also been obtained via a closed time path formalism of relativistic finite-temperature field theory applied to the NJL model.[28]

Next, let us consider the case with $b < 0$ in Fig. 2. In this case, a first-order chiral transition occurs for $a < a_c = 3b^2/(16d)$ (or $\lambda < \lambda_c = 3/16$). For $a < 0$, the potential has a double-well structure and the ordering dynamics is equivalent to M^4-theory, i.e., the domain growth scenario is similar to Figs. 3 and 4. We focus on a quench from the disordered state (with $M = 0$) to $0 < a < a_c$ or $0 < \lambda < \lambda_c$, corresponding to a quench between the first-order line (I) and S_1 in Figs. 1(a) or 2 – the corresponding points are denoted by crosses. The massless state ($M = 0$) is now a metastable

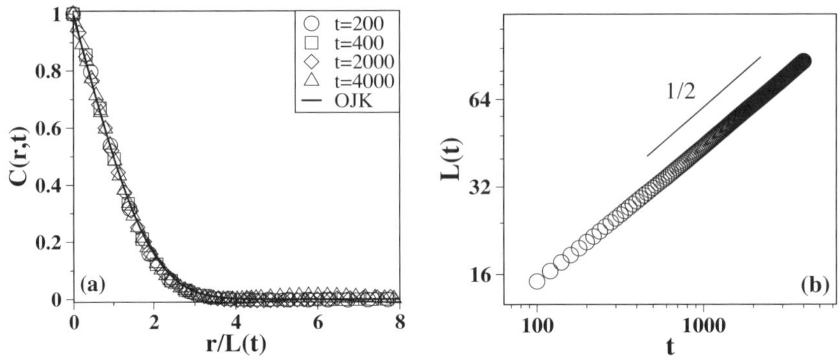

Fig. 4. (a) Scaling of order-parameter correlation function, $C(r,t)$ vs. r/L, for $\lambda = 0.14$ at four different times. The different data sets collaps onto a master curve. The statistical data is obtained as an average over 10 independent runs on 4096^2 lattices. The length scale $L(t)$ is defined as the distance over which $C(r,t)$ decays from $1 \rightarrow 1/2$. The solid line denotes the OJK result.[27] (b) Time-dependence of domain size, $L(t)$ vs. t, for $\lambda = 0.14$. The coarsening process obeys the Allen-Cahn (AC) growth law, $L(t) \sim t^{1/2}$.

state of the M^6-potential. The chiral transition proceeds via the nucleation and growth of droplets of the preferred phase ($M = \pm M_+$). The nucleation results from large fluctuations in the initial condition or thermal fluctuations during the evolution.

In Fig. 5, we show the nucleation and growth process. At early times ($t = 400$), the system is primarily in the $M = 0$ phase with small droplets of the preferred phase. These droplets grow in time and coalesce into domains. The subsequent coarsening of these domains is analogous to that in Figs. 3-4, through the interfacial kinks are slightly flatter in the $M \simeq 0$ region. In the late stages of growth, there is no memory of the nucleation dynamics which characterized growth during the early stages.

For $b < 0$ and $0 < a < a_c$, we have also studied the growth of single droplets of the preferred phase ($M = +M_+$) in a background of the metastable phase ($M = 0$) for $\lambda = 0.14$ [see Fig. 6(a)]. We start with an initial configuration of a bubble of radius $R_0 > R_c$ such that $M(r) = M_+$ for $r < R_0$ and $M(r) = 0$ for $r > R_0$, where R_c is the critical size of the droplet. This configuration is evolved as per Eq. (9) with the noise term. The noise term is taken sufficiently large to overcome the metastable barrier. The droplets have a unique growth velocity $v(\lambda)$, which depends on the degree of undercooling (λ): $v \rightarrow 0$ as $\lambda \rightarrow \lambda_c^-$. We have obtained $v(\lambda)$ by undertaking a phase-plane analysis of the traveling-wave solutions of Eq. (9) with $\theta = 0$. The droplet interface corresponds to a saddle

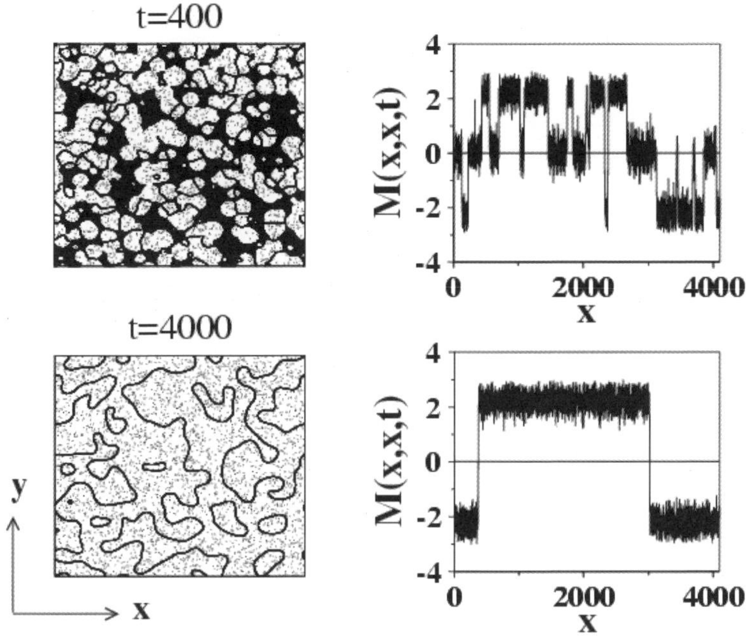

Fig. 5. Interface evolution after shallow temperature quench through first-order line (I) in Figs. 1(a) or 2. The d=2 snapshots on the left show the interfaces ($M = 0$) at $t = 400$, 4000. They were obtained by solving Eq. (9) with $b < 0$, $a_c > a > 0$ and $\lambda = 0.14$. The frames on the right show the variation of the order parameter along the diagonal. Notice that the metastable patches ($M = 0$) at $t = 400$ are absent at later times.

connection between the fixed points $+M_+ \to 0$ or $-M_+ \to 0$. The details of this analysis will be presented later.[21] In Fig. 6(b), we plot numerical results for $v(\lambda)$ vs. λ along with our theoretical result.

Before concluding, we briefly discuss the inertial counterpart of the overdamped TDGL equation (8). The TDGL equation contains first-order time-derivatives and is not Lorentz invariant. This is reasonable as we are considering finite-temperature field theory, where there is no Lorentz invariance. However, in some situations, the inertial terms can also play an important role in the evolution dynamics. The inertial TDGL equation has the following dimensionless form:

$$\frac{\partial^2 M}{\partial t^2} + \gamma \frac{\partial M}{\partial t} = -\text{sgn}\,(a)\, M - \text{sgn}\,(b)\, M^3 - \lambda M^5 + \nabla^2 M + \theta\,(\vec{r}, t)\,, \quad (10)$$

where γ measures the relative strengths of the damping and inertial terms. In principle, Eq. (10) can be obtained from a microscopic field-theoretic description of the nonequilibrium dynamics of the scalar field at finite

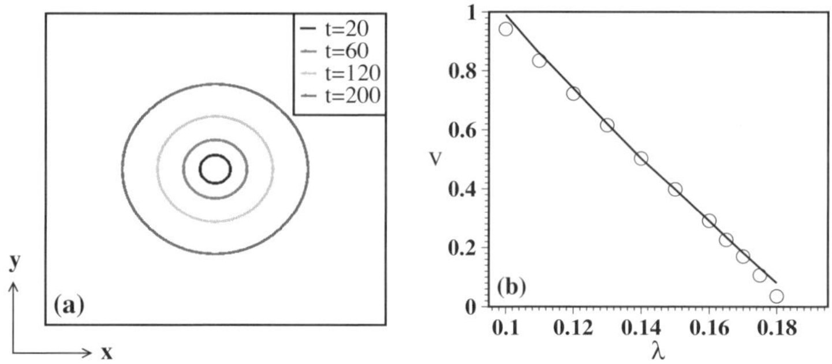

Fig. 6. (a) Growth of droplet of the preferred phase ($M = +M_+$) in a background of the metastable phase ($M = 0$) for $b < 0$, $a_c > a > 0$ and $\lambda = 0.14$. We show the boundary of the droplet at four different times. The inner circle corresponds to a droplet at time $t = 20$. (b) Droplet velocity v vs. λ. The circles refer to numerical data, while the solid line corresponds to the result from a phase-plane analysis.[21]

temperatures.[29] We have also studied ordering dynamics for the inertial case, and will present details of our results later.[21] Here, we mention the main results of our study. The ordering dynamics in the inertial case is analogous to that in the overdamped case, except that nucleation does not have a significant effect even during the early stages of evolution for quenches to $b < 0$ and $0 < a < a_c$. The droplets grow very rapidly and merge to form a bicontinuous domain structure characteristic of late-stage domain growth. The domain growth law is again the AC law, $L(t) \sim t^{1/2}$.

To summarize: we have studied the kinetics of chiral phase transitions in QCD. In terms of the quark degrees of freedom, the phase diagram is obtained using the NJL model. An equivalent coarse-grained description is obtained from a M^6-Ginzburg-Landau (GL) free energy. The chiral kinetics is modeled via the TDGL equation, and we consider both the overdamped and inertial cases. We study the ordering dynamics resulting from a sudden temperature quench through the first-order (I) or second-order (II) transition lines in Figs. 1(a) or 2. For quenches through II and deep quenches through I, the massless phase is spontaneously unstable and evolves to the massive phase via spinodal decomposition. For shallow quenches through I, the massless phase is metastable and the phase transition proceeds via the nucleation and growth of droplets of the massive phase. The merger of these droplets results in late-stage domain growth analogous to that for the unstable case. In all cases, the asymptotic growth process exhibits dynamical scaling, and the growth law is $L(t) \sim t^{1/2}$. Given the dynamical universality of the processes involved, our results are of much under applicability than the simple NJL Hamiltonian considered here. We hope that our results will motivate fresh experimental interest in hot and dense quark matter, and that our predictions will be subjected to experimental verification.

Acknowledgments

A.S. would like to thank CSIR (India) for financial support. H.M. would like to thank the School of Physical Sciences, Jawaharlal Nehru University, New Delhi for hospitality.

References

1. D.H. Rischke, *Prog. Part. Nucl. Phys.* **52**, 197 (2004).
2. K. Rajagopal and F. Wilczek, arXiv:hep-ph/0011333, in *At the frontier of particle physics*, Vol. **3**, M. Shifman(Ed.) World Scientific, 2061-2151.
3. F. Karsch, *Lect. Notes. Phys.* **583** 209, (2002).
4. E. Laerman and O. Philipsen, *Ann. Rev. Nucl. Part. Sci.* **53**, 163 (2003).
5. M. Cheng *et al.*, *Phys. Rev.* **D77**, 014511 (2008).
6. Z. Fodor and S. Katz, *JHEP* **203**, 014 (2002);
   Ph. de Forcrand and O. Phillipsen, *Nucl. Phys.* **B642**, 290 (2002);
   M.P. Lombardo and M. D'Elia, *Phys. Rev.* **D67**, 014505 (2003);
   C. Alton *et al.*, *Phys. Rev.* **D66**, 074507 (2002);
   ibid. Phys. Rev. **D68**, 014507 (2002).
7. S. Gottlieb *et al.*, *Phys. Rev.* **D35**, 2531 (1987);
   M. Cheng *et al.*, *Phys. Rev.* **D74**, 054507 (2006).
8. R.V. Gavai and S. Gupta, *Phys. Rev.* **D68**, 034506 (2003);
   ibid. Phys. Rev. **D78**, 114503 (2008).
9. B. Mohanty, *Nucl. Phys.* **A830**, 899 (2009); T. Schuster, *PoS* **CPOD2009**, 029 (2009); G. Stefanek, *PoS* **CPOD2009**, 049 (2009).
10. C. Sasaki, B. Friman and K. Redlich, *Phys. Rev.* **D77**, 034024 (2008).
11. O. Scavenius, A. Dumitru, E.S. Fraga, J. Lenaghan and A.D. Jackson, *Phys. Rev.* **D63**, 116003 (2001).
12. E.S. Fraga and G. Krein, *Phys. Lett.* **B614**, 181 (2005).
13. A. Bessa, E.S. Fraga and B.W. Mintz, *Phys. Rev.* **D79**, 034012 (2009).
14. K. Paech, H. Stoecker and A. Dumitru, *Phys. Rev.* **C68**, 044907, (2003);
    K. Paech and A. Dumitru, *Phys. Lett.* **B623**, 1200 (2005);
    A. Dumitru, L. Portugal and D. Zschiesche, *Phys. Rev.* **C73**, 024902 (2006).
15. T. Koide and M. Maruyama, *Nucl. Phys.* **A742**, 95 (2004).
16. S. Puri and V.K. Wadhawan (eds.), *Kinetics of Phase Transitions*, CRC Press, Boca Raton, Florida (2009).
17. A.J. Bray, *Adv. Phys.* **43**, 357 (1994).
18. S.P. Klevansky, *Rev. Mod. Phys.* **64**, 649 (1992).
19. M. Iwasaki, *Phys. Rev.* **D70**, 114031 (2004); H. Fujii and M. Ohtani, *Phys. Rev.* **D70**, 014016 (2004).
20. H. Mishra and S.P. Misra, *Phys. Rev.* **D48**, 5376 (1993).

21. A. Singh, S. Puri and H. Mishra (*in preparation*).
22. M. Asakawa and K. Yazaki, *Nucl. Phys.* **A504**, 668 (1989).
23. P.C. Hohenberg and B.I. Halperin, *Rev. Mod. Phys.* **49**, 435 (1977).
24. H. Heiselberg, C.J. Pethick and E.F. Staubo, *Phys. Rev. Lett.* **70**, 1355 (1993).
25. K. Kajantie, *Phys. Lett.* **B285**, 331 (1992).
26. Y. Oono and S. Puri, *Phys. Rev. Lett.* **58**, 836 (1987); *Phys. Rev.* **A38**, 434 (1988); S. Puri and Y. Oono, *Phys. Rev.* **A38**, 1542 (1988).
27. T. Ohta, D. Jasnow and K. Kawasaki, *Phys. Rev. Lett.* **49**, 1223 (1982).
28. P.F. Bedaque and A.K. Das, *Mod. Phys. Lett.* **A8**, 3151 (1993).
29. D.H. Rischke, *Phys. Rev.* **C58**, 2331 (1998); M. Gleiser and R.O. Ramos, *Phys. Rev.* **D50**, 2441 (1994).

PROGRESS ON CALORONS, CONTINUED

PIERRE VAN BAAL

Institute Lorentz for Theoretical Physics
University of Leiden, P.O. Box 9506
NL-2300 RA Leiden, The Netherlands

The progress on calorons (finite temperature instantons) is sketched. In particular there is some interest for confining temperatures, where the holonomy (the asymptotic value of the Polyakov loop) is non-trivial. In the last section I give more recent results by others.

1. Introduction

This is what I said in a special talk[1] given at the Minneapolis workshop "Continuous Advances in QCD" in April, 1998:

Volodya Gribov was a lightning and thundering example of scientific integrity and honesty. The first time I heard him speak was at a meeting in Eger in August 1988. My first interactions with him stem from a visit to my friends here in Minneapolis in May of 1995. At that time I gave two talks, both on the global issue of gauge fixing and its applications to the dynamics in gauge theories. I did get to the end of those talks and had a few private discussions with Volodya. It is not that I believe to have convinced him this was the way to go - our results are furthermore only very modest. Later I learned[2] that he considered having a "picture" to be very important for approaching a problem. In February 1996 I managed to convince Volodya to come and present a colloquium Ehrenfestii, a visit none of us in Leiden and his wife Julia are likely to forget. With sadness we have to bear the loss of a great and unique physicist who relentlessly looked for the truth and forcefully reminded us that we have the duty to do the same.

Today I am happy I can repeat this, because it remains true, no matter how long ago this was said.

During the memorial in Budapest, called "Gribov75", I decided to talk about "Cosmic Rays: a bridge between high schools and real research". You can still find the slides at www.kfki.hu/~gribov75/programt.html. Alan

White remembered it and I would have loved to talk about this also at the "Gribov80" memorial, here in Trieste, but I couldn't. Jan-Willem van Holten who is now an extraordinary Professor in Leiden has taken it over, so it is doing fine. For myself, I doubted if I could talk again, but it finally came back in time (I thank Dmitri Kharzeev who has convinced me to try at BNL, where I did give a talk on April 28). My condition is apparently now finally going up, instead of down, but Julia Nyiri knew this all the time and scheduled me for a talk on May 28. But I still was afraid, so the talk is similar to the one I gave at BNL, and also similar to the one I gave in Mainz on September 6, 2008 (which I have written up, arXiv:0901.2853).

2. The setting

There has been a revised interest in studying instantons at finite temperature T, so-called calorons,[3,4] because new explicit solutions could be obtained where the Polyakov loop at spatial infinity (the so-called holonomy) is non-trivial. They reveal more clearly the monopole constituent nature of these calorons.[5] Non-trivial holonomy is therefore expected to play a role in the confined phase (i.e. for $T < T_c$) where the trace of the Polyakov loop fluctuates around small values. The properties of instantons are therefore directly coupled to the order parameter for the deconfining phase transition.

At finite temperature A_0 plays in some sense the role of a Higgs field in the adjoint representation, which explains why magnetic monopoles occur as constituents of calorons. Since A_0 is not necessarily static it is better to consider the Polyakov loop as the analog of the Higgs field, $P(t,\vec{x}) = \mathrm{Pexp}\left(\int_0^\beta A_0(t+s,\vec{x})ds\right)$, which transforms under a periodic gauge transformation $g(x)$ to $g(x)P(x)g^{-1}(x)$, like an adjoint Higgs field. Here $\beta = 1/kT$ is the period in the imaginary time direction, under which the gauge field is assumed to be periodic. Finite action requires the Polyakov loop at spatial infinity to be constant. For SU(n) gauge theory this gives $\mathcal{P}_\infty = \lim_{|\vec{x}|\to\infty} P(0,\vec{x}) = g^\dagger \exp(2\pi i \mathrm{diag}(\mu_1,\mu_2,\ldots,\mu_n))g$, where g brings $\mathcal{P}_\infty$ to its diagonal form, with n eigenvalues being ordered according to $\sum_{i=1}^n \mu_i = 0$ and $\mu_1 \le \mu_2 \le \ldots \le \mu_n \le \mu_{n+1} \equiv 1 + \mu_1$. In the algebraic gauge, where $A_0(x)$ is transformed to zero at spatial infinity, the gauge fields satisfy the boundary condition $A_\mu(t+\beta,\vec{x}) = \mathcal{P}_\infty A_\mu(t,\vec{x})\mathcal{P}_\infty^{-1}$.

Caloron solutions are such that the total magnetic charge vanishes. A single caloron with topological charge one contains $n-1$ monopoles with a unit magnetic charge in the i-th U(1) subgroup, which are compensated by the n-th monopole of so-called type $(1,1,\ldots,1)$, having a magnetic charge in each of these subgroups.[6] At topological charge k there are kn constituents, k monopoles of each of the n types. Monopoles of type j have a mass $8\pi^2\nu_j/\beta$, with $\nu_j \equiv \mu_{j+1} - \mu_j$. The sum rule $\sum_{j=1}^{n}\nu_j = 1$ guarantees the correct action, $8\pi^2 k$.

Prior to their explicit construction, calorons with non-trivial holonomy were considered irrelevant,[4] because the one-loop correction gives rise to an infinite action barrier. However, the infinity simply arises due to the integration over the finite energy density induced by the perturbative fluctuations in the background of a non-trivial Polyakov loop.[7] The calculation of the non-perturbative contribution was performed in.[8] When added to this perturbative contribution, with minima at center elements, these minima turn unstable for decreasing temperature right around the expected value of T_c. This lends some support to monopole constituents being the relevant degrees of freedom which drive the transition from a phase in which the center symmetry is broken at high temperatures to one in which the center symmetry is restored at low temperatures. Lattice studies, both using cooling[9] and chiral fermion zero-modes[10] as filters, have also conclusively confirmed that monopole constituents do dynamically occur in the confined phase.

3. Some Properties of Caloron Solutions

Using the classical scale invariance we can always arrange $\beta = 1$, as will be assumed throughout. A remarkably simple formula for the SU(n) action density exists,[6]

$$\mathrm{Tr}F_{\alpha\beta}^2(x) = \partial_\alpha^2\partial_\beta^2 \log\psi(x), \quad \psi(x) = \tfrac{1}{2}\mathrm{tr}(\mathcal{A}_n\cdots\mathcal{A}_1) - \cos(2\pi t),$$

$$\mathcal{A}_m \equiv \frac{1}{r_m}\begin{pmatrix} r_m & |\vec{\rho}_{m+1}| \\ 0 & r_{m+1}\end{pmatrix}\begin{pmatrix}\cosh(2\pi\nu_m r_m) & \sinh(2\pi\nu_m r_m) \\ \sinh(2\pi\nu_m r_m) & \cosh(2\pi\nu_m r_m)\end{pmatrix},$$

with $r_m \equiv |\vec{x} - \vec{y}_m|$ and $\vec{\rho}_m \equiv \vec{y}_m - \vec{y}_{m-1}$, where $\vec{y}_m$ is the location of the mth constituent monopole with a mass $8\pi^2\nu_m$. Note that the index m should be considered mod n, such that e.g. $r_{n+1} = r_1$ and $\vec{y}_{n+1} = \vec{y}_1$ (there is one exception, $\mu_{n+1} = 1 + \mu_1$). It is sufficient that only one constituent location is far separated from the others, to show that one can neglect the $\cos(2\pi t)$ term in $\psi(x)$, giving rise to a static action density in this limit.[6]

In Fig. 1 we show how for SU(2) there are two lumps, except that the second lump is absent for trivial holonomy. Fig. 2 demonstrates for SU(2) and SU(3) that there are indeed n lumps (for SU(n)) which can be put anywhere. These lumps are constituent monopoles, where one of them has a winding in the temporal direction (which cannot be seen from the action density).

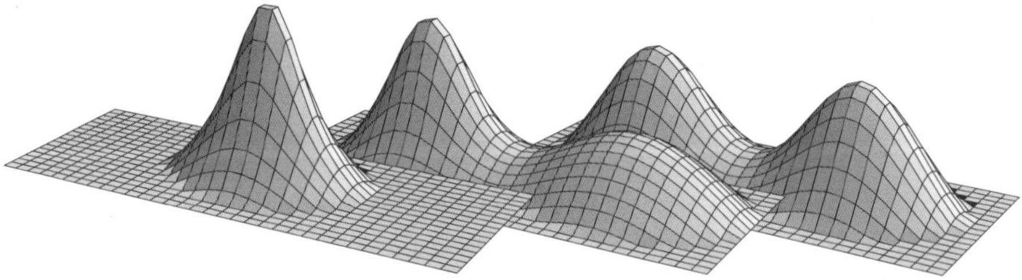

Fig. 1. Shown are three charge one SU(2) caloron profiles at $t = 0$ with $\beta = 1$ and $\rho = 1$. From left to right for $\mu_2 = -\mu_1 = 0$ ($\nu_1 = 0, \nu_2 = 1$), $\mu_2 = -\mu_1 = 0.125$ ($\nu_1 = 1/4, \nu_2 = 3/4$) and $\mu_2 = -\mu_1 = 0.25$ ($\nu_1 = \nu_2 = 1/2$) on equal logarithmic scales, cutoff below an action density of $1/(2e)$.

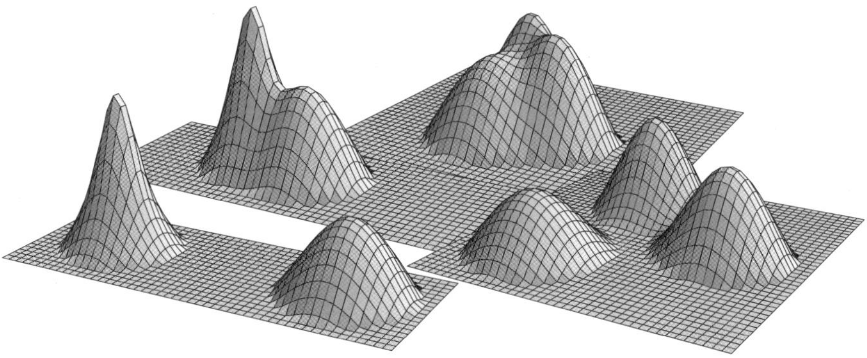

Fig. 2. On the left are shown two charge one SU(2) caloron profiles at $t = 0$ with $\beta = 1$ and $\mu_2 = -\mu_1 = 0.125$, for $\rho = 1.6$ (bottom) and 0.8 (top) on equal logarithmic scales, cutoff below an action density of $1/(2e^2)$. On the right are shown two charge one SU(3) caloron profiles at $t = 0$ and $(\nu_1, \nu_2, \nu_3) = (1/4, 7/20, 2/5)$, implemented by $(\mu_1, \mu_2, \mu_3) = (-17/60, -1/30, 19/60)$. The bottom configuration has the location of the lumps scaled by 8/3. They are cutoff at $1/(2e)$.

3.1. *Fermion Zero-Modes*

An essential property of calorons is that the chiral fermion zero-modes are localized to constituents of a certain charge only. The latter depends on the choice of boundary condition for the fermions in the imaginary time direction (allowing for an arbitrary U(1) phase $\exp(2\pi i z)$).[11] This provides an important signature for the dynamical lattice studies, using chiral fermion zero-modes as a filter.[10] To be precise, the zero-modes are localized to the monopoles of type m provided $\mu_m < z < \mu_{m+1}$. Denoting the zero-modes by $\hat{\Psi}_z(x)$, we can write $\hat{\Psi}_z^\dagger(x)\hat{\Psi}_z(x) = -(2\pi)^{-2}\partial_\mu^2 \hat{f}_x(z,z)$, where $\hat{f}_x(z,z')$ is a Green's function which for $z \in [\mu_m, \mu_{m+1}]$ satisfies $\hat{f}_z(z,z) = \pi < v_m(z)|\mathcal{A}_{m-1}\cdots\mathcal{A}_1\mathcal{A}_n\cdots\mathcal{A}_m|w_m(z) > /(r_m\psi)$, where the spinors v_m and w_m are defined by $v_m^1(z) = -w_m^2(z) = \sinh(2\pi(z-\mu_m)r_m)$, and $v_m^2(z) = w_m^1(z) = \cosh(2\pi(z-\mu_m)r_m)$.

To obtain the finite temperature fermion zero-mode one puts $z = \frac{1}{2}$, whereas for the fermion zero-mode with periodic boundary conditions one takes $z = 0$. From this it is easily seen that in case of well separated constituents the zero-mode is localized only at $\vec{y}_m$ for which $z \in [\mu_m, \mu_{m+1}]$. To be specific, in this limit $\hat{f}_x(z,z) = \pi \tanh(\pi r_m \nu_m)/r_m$ for SU(2), and more generally $\hat{f}_x(z,z) = 2\pi \sinh[2\pi(z-\mu_m)r_m]\sinh[2\pi(\mu_{m+1}-z)r_m]/(r_m \sinh[2\pi\nu_m r_m])$. We illustrate in Fig. 3 the localization of the fermion zero-modes for the case of SU(3).

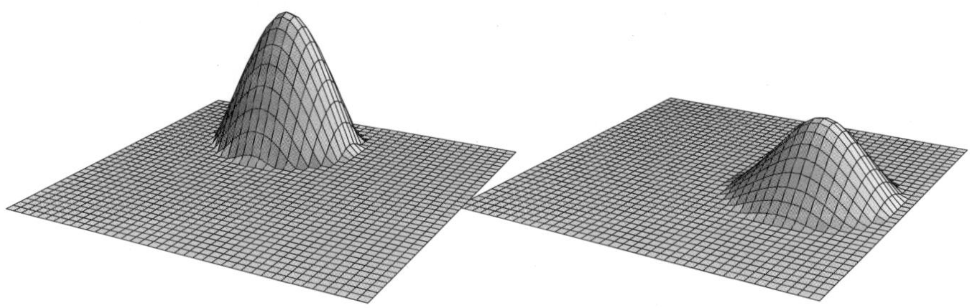

Fig. 3. For the SU(3) configuration in the lower right corner of Fig. 2 we have determined on the left the zero-mode density for fermions with anti-periodic boundary conditions in time and on the right for periodic boundary conditions. They are plotted at equal logarithmic scales, cut off below $1/e^5$.

3.2. *Calorons of Higher Charge*

We have been able to use a "mix" of the ADHM and Nahm formalism,[12] both in making powerful approximations, like in the far field limit (based on our ability to identify the exponentially rising and falling terms), and for finding exact solutions through solving the homogeneous Green's function.[13] We found axially symmetric solutions for arbitrary k, as well as for $k = 2$ two sets of non-trivial solutions for the matching conditions that interpolate between overlapping and well-separated constituents. For this task we could make use of an existing analytic result for charge-2 monopoles,[14] adapting it to the case of carolons. An example is shown in Fig. 4.

4. More recent results

There are more recent lectures by Bruckmann[15] and Diakonov.[16] Also, Diakonov and Petrov made some progress on constructing the hyperKähler

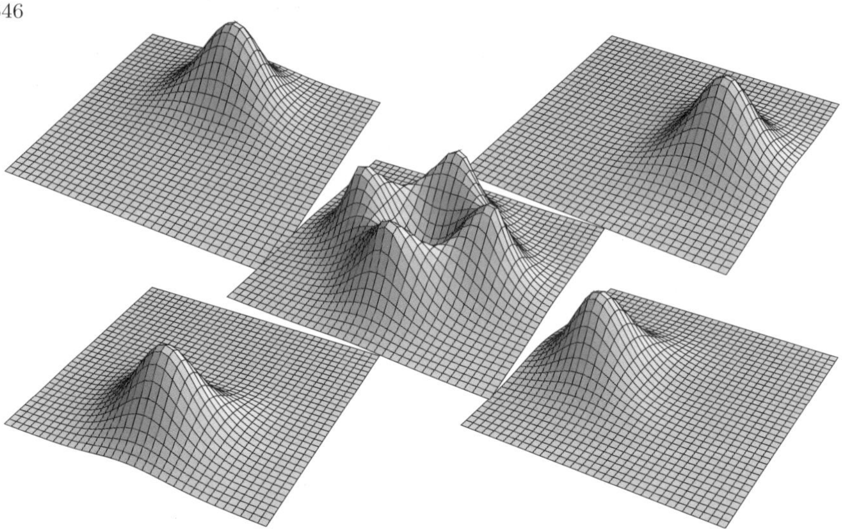

Fig. 4. In the middle is shown the action density in the plane of the constituents at $t = 0$ for an SU(2) charge 2 caloron with $\mathrm{tr}\,\mathcal{P}_\infty = 0$, where all constituents strongly overlap. On a scale enhanced by a factor $10\pi^2$ are shown the densities for the two zero-modes, using either periodic (left) or anti-periodic (right) boundary conditions in the time direction.

metric which approximates the metric for an arbitrary number of calorons. They claim that this already gives confinement.[16,17] But some cautionary remarks can be made.[18] Also multi-calorons were revisited,[19] and the authors claim to have the full SU(2) moduli space for $k = 2$.

The calorons have also adjoint fermionic zero-modes, and they are now known in analytical form.[20] Finally, Ünsal has published a paper concerning the mechanism of confinement in QCD-like theories,[21] for example SU(2) with $1 \leq n_f \leq 4$ adjoint Majorana fermions. He argues that there are BPS and KK monopoles (precisely the constituents of the caloron), which have zero-modes under the adjoint fermions. They then make BPS-$\overline{\text{KK}}$ bound states (instead of BPS-KK).

Acknowledgments

Thanks to everybody, and especially to Arkady and Nelly Vainshtein for a warm welcome. And the real person behind the Gribov memorials is of course Julia Nyiri, and to her I say "till Gribov85". Also I would like to thank FOM for financial support.

References

1. P. van Baal, *"Gribov's Horizon - a Vista on Non-perturbative Gauge Theory"*, in: Continuous Advances in QCD 1998, ed. A.V. Smilga (World Scientific, Singapore, 1999, pp. 3-18).
2. Yu.L. Dokshitzer, *"V.N. Gribov (1930-1997)"*, in: Continuous Advances in QCD 1998, ed. A.V. Smilga (World Scientific, Singapore, 1999, pp. xv-xxi); physics/9801025.
3. B.J. Harrington and H.K. Shepard, *Phys. Rev.* **D17**, 2122 (1978); *Phys. Rev.* **D18**, 2990 (1978).
4. D.J. Gross, R.D. Pisarski and L.G. Yaffe, *Rev. Mod. Phys.* **53**, 43 (1981).
5. T.C. Kraan and P. van Baal, *Phys. Lett.* **B428**, 268 (1998); *Nucl. Phys.* **B533**, 627 (1998).
   K. Lee, *Phys. Lett.* **B426**, 323 (1998); K. Lee and C. Lu, *Phys. Rev.* **D58**, 025011 (1998).
6. T.C. Kraan and P. van Baal, *Phys. Lett.* **B435**, 389 (1998).
7. N. Weiss, *Phys. Rev.* **D24**, 475 (1981).
8. D. Diakonov, N. Gromov, V. Petrov and S. Slizovskiy, *Phys. Rev.* **D70**, 036003 (2004);
   D. Diakonov and N. Gromov, *Phys. Rev.* **D72**, 025003 (2005).
9. E.-M. Ilgenfritz, B.V. Martemyanov, M. Müller-Preussker, S. Shcheredin and A.I. Veselov, *Phys. Rev.* **D66**, 074503 (2002);
   F. Bruckmann, E.-M. Ilgenfritz, B.V. Martemyanov and P. van Baal, *Phys. Rev.* **D70**, 105013 (2004);
   P. Gerhold, E.-M. Ilgenfritz and M. Müller-Preussker, *Nucl. Phys.* **B760**, 1 (2007).
10. C. Gattringer and S. Schaefer, *Nucl. Phys.* **B654**, 30 (2003);
    C. Gattringer and R. Pullirsch, *Phys. Rev.* **D69**, 094510 (2004).
11. M. Garcia Pérez, A. González-Arroyo, C. Pena and P. van Baal, *Phys. Rev.* **D60**, 031901 (1999);
    M. N. Chernodub, T.C. Kraan and P. van Baal, *Nucl. Phys.* **B** *(Proc.Suppl.)* **83-84**, 556 (2000).
12. M.F. Atiyah, N.J. Hitchin, V. Drinfeld and Yu.I. Manin, *Phys. Lett.* **A65**, 185 (1978); W. Nahm, *"Self-dual monopoles and calorons,"* in: *Lecture Notes in Physics*, **201**, 189 (1984).
13. F. Bruckmann and P. van Baal, *Nucl. Phys.* **B645**, 105 (2002);
    F. Bruckmann, D. Nógrádi and P. van Baal, *Nucl. Phys.* **B666**, 197 (2003); *Nucl. Phys.* **B698**, 233 (2004).
14. H. Panagopoulos, *Phys. Rev.* **D28**, 380 (1983).
15. F. Bruckmann, *Eur. Phys. J.* **ST152**, 61 (2007).
16. D. Diakonov, *Acta Phys. Polon.* **B39**, 3365 (2008); *Nucl. Phys.* **B** *(Proc.Suppl.)* **195**, 5 (2009).
17. D. Diakonov and V. Petrov, *Phys. Rev.* **D76**, 056001 (2007); *AIP Conf. Proc.* **1134**, 190 (2009).
18. F. Bruckmann, S. Dinter, E.-M. Ilgenfritz, M. Müller-Preussker and M. Wagner, *Phys. Rev.* **D79**, 116007 (2009).
19. A. Nakamula and J. Sakaguchi, *J. Math. Phys.* **51**, 043503 (2010).

20. M. Garcia Perez and A. González-Arroyo, *JHEP* **0611**, 091 (2006); M. Garcia Perez, A. González-Arroyo and A. Sastre, *Phys. Lett.* **B668**, 340 (2008); *JHEP* **0906**, 065 (2009).
21. M. Ünsal, *Phys. Rev.* **D80**, 065001 (2009).

Minneapolis, 1995.

ON ANOMALOUS QUARK TRIANGLES

ARKADY VAINSHTEIN

William I. Fine Theoretical Physics Institute, University of Minnesota, 116 Church St. SE, Minneapolis, MN 55455, USA

Anomalous quark triangles with one axial and two vector currents are studied in special kinematics when one of the vector currents carries a soft momentum. According to the Adler-Bardeen theorem the anomalous longitudinal part of the triangle is not renormalized in the chiral limit. We show that perturbative corrections the transversal part of the triangle is also absent. This nonrenormalization, in difference with the longitudinal part, holds on only perturbatively.

1. Introduction

Study of fermion triangle diagrams with one axial and two vector currents represents a remarkable story. The Adler-Bell-Jackiw anomaly in the divergence of axial current, the Adler-Bardeen nonrenormalization theorem, the Wess-Zumino effective action, calculation of the $\pi^0 \to \gamma\gamma$ amplitude, the 't Hooft consistency condition, and the solution of the U(1) problem give an incomplete list of acts where these triangles played a major role.

For Vladimir Gribov the axial anomaly was an important element in his overall picture of the quantum field world. This is visible already from the titles of his two publications[1,2] on the topic. "Anomalies, as a Manifestation of the High Momentum Collective Motion in the Vacuum" and "Anomalies and a Possible Solution of Problems of Zero Charge and Infrared Instability." Besides having a privilege of numerous discussions with Volodya I often observed directly his attitude to the axial anomaly phenomenon.

About 7 years ago quark triangles show up in our study[3] of hadronic effects for the muon anomalous magnetic moment, $g-2$, in which I have collaborated with Andrzej Czarnecki and Bill Marciano. As depicted in Fig. 1 such triangles appear in the two-loop order electroweak corrections. The axial current enters Z boson fermionic vertex.

The famous Adler-Bardeen theorem[4] proves nonrenormalization for the longitudinal part of triangles associated with the anomalous divergence

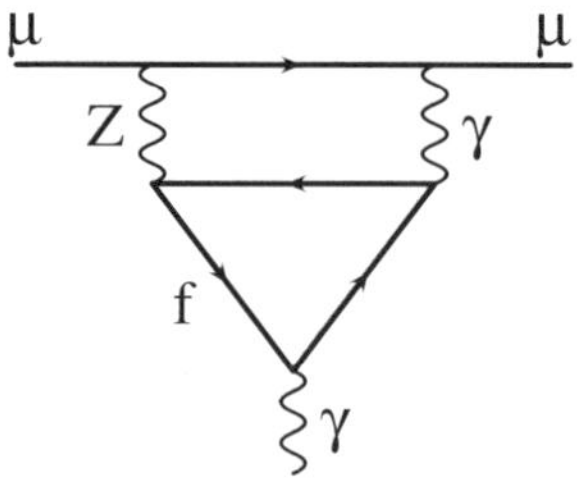

Fig. 1. Fermion triangle in electroweak correction to $g-2$.

of the axial current. There is no general statement about the transversal part of the triangle. This part, even its existence, depends on the choice of external momenta. In case of the two-loop corrections to $g-2$ both, longitudinal and transversal parts of the triangle contribute.

Then, an immediate question is what kind of theoretical control one has for the transversal part of the triangle amplitude. In particular, whether gluon corrections to triangles with light quarks are present in the transversal part in contrast to the longitudinal one. In the special kinematics characteristic for $g-2$ when one of the vector currents (associated with the magnetic field) carries a soft momentum the transversal part is unambiguously fixed by the longitudinal one in the chiral limit of perturbation theory.[5] Such relation immediately proves an absence of perturbative corrections to the transversal part of fermion triangles in the kinematics considered.

The difference between longitudinal and transversal parts shows up at a nonperturbative level. The Operator Product Expansion (OPE) demonstrates this explicitly: in the chiral limit only the transversal part in the OPE contains nonleading operators.[3,6] Nonrenormalization of the longitudinal part, both perturbatively and nonperturbatively, constitutes the 't Hooft consistency condition,[7] i.e. the exact quark-hadron duality. In QCD this duality is realized as a correspondence between the infrared singularity of the quark triangle and the massless pion pole in terms of hadrons.* It is clear that for the transversal part this kind of duality cannot be exact in QCD: there is no massless particle contributing to the transversal part. Thus, the transversal part of the triangle with a soft momentum in one of the vector currents provides us with an interesting object: no perturbative corrections but nonperturbatively it is modified.

*A pioneering effort to analyze the axial current anomaly in terms of infrared singularity was made by Dolgov and Zakharov.[8]

2. Perturbative Calculations

Let us start with a definition of vector, j_μ, and axial, j^5_ν, currents,

$$j_\mu = \bar{q}\, V \gamma_\mu q, \qquad j^5_\nu = \bar{q}\, A \gamma_\nu \gamma_5 q\,, \tag{1}$$

where the quark field q^i_f has color (i) and flavor (f) indices and the matrices V and A are diagonal matrices of vector and axial couplings acting on flavor indexes. To avoid dealing with the U(1) anomaly in respect to gluon interactions we assume that $\mathrm{Tr}\, A = 0$. In the case of electroweak corrections one can view the vector current as an electromagnetic one with V being the matrix of electric charges and j^5_ν as the axial part of the Z boson current with matrix A given by the weak isospin projection.

The amplitude for the triangle diagram in Fig. 2 involving the axial

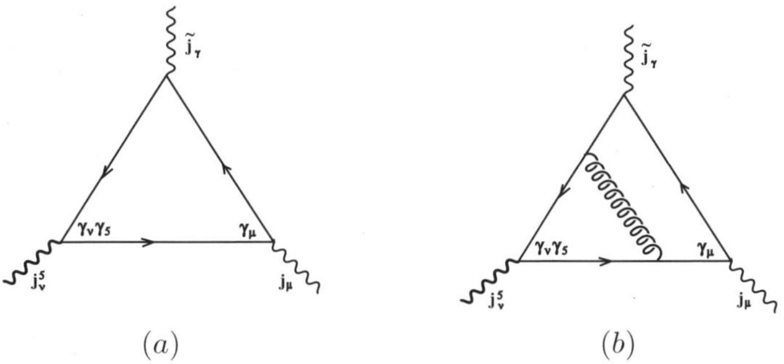

Fig. 2. Quark triangle, diagram (a), and a gluon correction to it, diagram (b).

current j^5_ν and two vector currents j_μ and $\tilde{j}_\gamma = \bar{q}\, \widetilde{V} \gamma_\mu q$ (for generality we use a different matrix $\widetilde{V}$ for the soft momentum current) can be written as

$$T_{\mu\gamma\nu} = -\int \mathrm{d}^4x \mathrm{d}^4y\, \mathrm{e}^{iqx - iky}\, \langle 0|\, T\{j_\mu(x)\, j_\gamma(y)\, j^5_\nu(0)\}|0\rangle\,. \tag{2}$$

We can view the current $\tilde{j}_\gamma$ as a source of a soft photon with the momentum k. Introducing a polarization vector of a soft photon $e^\gamma(k)$ we come to the amplitude $T_{\mu\nu}$

$$T_{\mu\nu} = T_{\mu\gamma\nu} e^\gamma(k) = i\int \mathrm{d}^4x\, \mathrm{e}^{iqx}\, \langle 0|\, T\{j_\mu(x)\, j^5_\nu(0)\}|\, \gamma(k)\rangle\,, \tag{3}$$

which can be viewed as a mixing between the axial and vector currents in the external electromagnetic field.

It is clear that the expansion of $T_{\mu\nu}$ in the small momentum k starts with linear terms and we neglect quadratic and higher powers of k. There are only two Lorentz structures for $T_{\mu\nu}$ which are linear in k and consistent with the conservation of electromagnetic current,

$$T_{\mu\nu} = -\frac{i}{4\pi^2}\left[w_T(q^2)\left(-q^2\tilde{f}_{\mu\nu} + q_\mu q^\sigma \tilde{f}_{\sigma\nu} - q_\nu q^\sigma \tilde{f}_{\sigma\mu}\right) + w_L(q^2)\, q_\nu q^\sigma \tilde{f}_{\sigma\mu}\right], \tag{4}$$
$$\tilde{f}_{\mu\nu} = \frac{1}{2}\,\epsilon_{\mu\nu\gamma\delta} f^{\gamma\delta}\,, \qquad f_{\mu\nu} = k_\mu e_\nu - k_\nu e_\mu\,.$$

Both structures are transversal with respect to vector current, $q^\mu T_{\mu\nu} = 0$. As for the axial current, the first structure is transversal with respect to q^ν while the second is longitudinal.

The one-loop result for the invariant functions w_T and w_L can be taken from the classic papers by Bell and Jackiw,[9] Adler[10] and Rosenberg[11] (it simplifies considerably in the limit of the small photon momentum[12]),

$$w_L^{1-\text{loop}} = 2\, w_T^{1-\text{loop}} = 2N_c \,\text{Tr}\, A\, V\, \widetilde{V} \int_0^1 \frac{d\alpha\, \alpha(1-\alpha)}{\alpha(1-\alpha)Q^2 + m^2}\,, \tag{5}$$

where $Q^2 = q^2$, the factor N_c accounts for the color of quarks and m is the diagonal quark mass matrix, $m = \text{diag}\{m_{q_1}, m_{q_2}, \ldots\}$. In the chiral limit, $m = 0$, the invariant functions $w_{T,L}$ are

$$w_L^{1-\text{loop}}[m=0] = 2\, w_T^{1-\text{loop}}[m=0] = \frac{2N_c\, \text{Tr}\,(A\, V\, \widetilde{V})}{Q^2}\,. \tag{6}$$

Nonvanishing in the chiral limit, $m = 0$, the longitudinal part $q^\nu T_{\mu\nu}$ represents the axial anomaly,[9,10]

$$q^\nu T_{\mu\nu} = \frac{i}{4\pi^2}\, Q^2 w_L\, q^\sigma \tilde{f}_{\sigma\mu} = \frac{i}{2\pi^2}\, N_c\, \text{Tr}\,(A\, V\, \widetilde{V})\, q^\sigma \tilde{f}_{\sigma\mu}\,, \tag{7}$$

and its nonrenormalization implies that the one-loop result (6) for w_L stays intact when interaction with gluons is switched on.

3. Nonrenormalization theorem for the transversal part of the triangle

Our claim is that the relation

$$w_L[m=0] = 2\, w_T[m=0] \tag{8}$$

which holds at the one-loop level, see Eq. (6), gets no perturbative corrections from gluon exchanges. This follows from the following line of argumentation.

In the chosen kinematics the fermion triangle with $m = 0$ possesses a special feature: namely, a symmetry under permutation of indexes of axial and vector currents, $\mu \leftrightarrow \nu$. Indeed, in the triangle diagrams (a) and (b) in Fig. 2 one can move γ_5 from the axial vertex $\gamma_\nu\gamma_5$ to the vector vertex γ_μ. In the chiral limit it moves via even number of gamma matrices in any order of perturbation theory. Together with the change of the momentum $q \to -q$ (which does not affect $T_{\mu\nu}$) it shows the symmetry of the amplitude $T_{\mu\nu}$.

At first glance the symmetry under the $\mu \leftrightarrow \nu$ permutation seems to be in contradiction with the general decomposition (4): the transversal part of $T_{\mu\nu}$ is antisymmetric, the longitudinal part has no symmetry, and there is no way to choose w_T and w_L which makes the $T_{\mu\nu}$ symmetric. Note, however, that the term $q^2 \tilde{f}_{\mu\nu}$ in the transversal structure in Eq. (4) actually produces a term in $T_{\mu\nu}$ which does not depend on q. It is because $w_T \propto 1/q^2$. The $\mu \leftrightarrow \nu$ symmetry holds for a singular in q part of $T_{\mu\nu}$ when the condition (8) relating w_T to w_L is fulfilled. The constant in q part is then fixed by the conservation of the vector current, $q^\mu T_{\mu\nu} = 0$. An independence on q for the antisymmetric part provides, in fact, an alternative proof of the Adler-Bardeen theorem. Indeed, gluon corrections would lead to logarithmic dependence on q instead of the constant.

Another way to be automatically consistent with the vector current conservation is to use the Pauli-Vilars regulators. Technically it reduces to subtraction from the triangles with massless quarks similar triangles with the heavy regulator fermions propagating on the loops. The regulator triangles produce terms which are polynomial in momenta, in our case terms linear in k and independent on q. Moreover, it is simple to see that these terms are antisymmetric under the $\mu \leftrightarrow \nu$ permutation. Indeed, in the propagator of the heavy regulator the leading term contains no gamma-matrix that leads to the sign change when γ_5 from the axial vertex $\gamma_\nu\gamma_5$ is moved to the vector vertex γ_μ.

Thus, we see that the crossing symmetry of the singular part in the triangle amplitude $T_{\mu\nu}$ leads to the relation (8) in perturbation theory. Nonrenormalization of w_L implies the same for w_T.

For a general kinematics the relation was found in Ref. [13] and checked by explicit two-loop calculation in Ref. [14].

4. Nonperturbative effects and OPE

What about nonperturbative corrections? None in the longitudinal part ('t Hooft consistency condition [7]) should present in the transversal part – there is no massless spin one states. To study a nonperturbative effect in the triangle amplitude $T_{\mu\nu}$ one can use the OPE methods.

The OPE for the T-product of electromagnetic and axial currents at large Euclidean q^2 has the form

$$\hat{T}_{\mu\nu} \equiv i \int \mathrm{d}^4x\, \mathrm{e}^{iqx}\, T\{j_\mu(x)\, j^5_\nu(0)\} = \sum_i c^i_{\mu\nu\gamma_1\ldots\gamma_i}(q)\, \mathcal{O}_i^{\gamma_1\ldots\gamma_i}\,, \qquad (9)$$

where the local operators $\mathcal{O}_i^{\gamma_1\ldots\gamma_i}$ are constructed from the light fields and supplied by a normalization point μ separating short distances (accounted in the coefficients c^i) and large distances (in matrix elements of $\mathcal{O}_i$). The field can be viewed as light if its mass is less than μ. In the problem under consideration besides quark and gluon fields this includes also the soft electromagnetic field A_μ. The field A_μ could enter local operators in a form of the gauge invariant field strength $F_{\mu\nu} = \partial_\mu A_\nu - \partial_\nu A_\mu$.

The amplitude $T_{\mu\nu}$ is given by the matrix element of the operator $\hat{T}_{\mu\nu}$ between the photon and vacuum states,

$$T_{\mu\nu} = \langle 0|\, \hat{T}_{\mu\nu}\, |\gamma(k)\rangle = \sum_i c^i_{\mu\nu\alpha_1\ldots\alpha_i}(q)\, \langle 0|\, \mathcal{O}_i^{\alpha_1\ldots\alpha_i}\, |\gamma(k)\rangle\,. \qquad (10)$$

In our approximation, when the matrix elements are linear in $f_{\alpha\beta} = k_\alpha e_\beta - k_\beta e_\alpha$, they are nonvanishing only for operators with a pair of antisymmetric indexes,

$$\langle 0|\, \mathcal{O}_i^{\alpha\beta}\, |\gamma(k)\rangle = -\frac{i}{4\pi^2}\, \kappa_i\, \tilde{f}^{\alpha\beta}\,, \qquad (11)$$

where constants κ_i depend on the normalization point μ. With only contributing operators the OPE takes the form

$$\hat{T}_{\mu\nu} = \sum_i \left\{ c^i_T(q^2)\left(-q^2 \mathcal{O}^i_{\mu\nu} + q_\mu q^\sigma \mathcal{O}^i_{\sigma\nu} - q_\nu q^\sigma \mathcal{O}^i_{\sigma\mu}\right) + c^i_L(q^2)\, q_\nu q^\sigma \mathcal{O}^i_{\sigma\mu} \right\}, \qquad (12)$$

and the invariant functions $w_{T,L}$ can be presented as

$$w_{T,L}(q^2) = \sum c^i_{T,L}(q^2)\, \kappa_i\,. \qquad (13)$$

The leading (by a minimal dimension) is the $d = 2$ operator

$$\mathcal{O}_F^{\alpha\beta} = \frac{1}{4\pi^2}\, \tilde{F}^{\alpha\beta} = \frac{1}{4\pi^2}\, \epsilon^{\alpha\beta\rho\delta}\partial_\rho A_\delta\,, \qquad (14)$$

where $\tilde{F}^{\alpha\beta}$ is the dual of the electromagnetic field strength. The numerical factor in (14) is such that $\kappa_F = 1$. The OPE coefficients for $\mathcal{O}_F^{\alpha\beta}$ follow from one-loop expressions (5) for $w_{L,T}$,

$$c^F_L[\text{1-loop}] = 2c^F_T[\text{1-loop}] = \frac{2\,N_c}{Q^2}\,\mathrm{Tr}\, A\, V\tilde{V}\left[1 + \mathcal{O}\left(\frac{m^2}{Q^2}\right)\right], \qquad (15)$$

where we imply that $m \ll \mu \ll Q$, see[3] for a more detailed discussion.

The next, by dimension, are $d = 3$ operators

$$\mathcal{O}_f^{\alpha\beta} = -i\, \bar{q}_f\, \sigma^{\alpha\beta}\gamma_5\, q^f \equiv \frac{1}{2}\, \epsilon^{\alpha\beta\gamma\delta}\bar{q}_f\, \sigma_{\gamma\delta}\, q^f\,, \tag{16}$$

where the index denotes the quark flavor. The OPE coefficients follow from tree diagrams of the Compton scattering type,

$$c_L^f = 2c_T^f = \frac{4\, A_f V_f\, m_f}{Q^4}\,. \tag{17}$$

Proportionality to m_f is in correspondence with chirality arguments. Taking matrix element of $\mathcal{O}_f^{\alpha\beta}$ between the soft photon and vacuum states we produce the following terms in the invariant functions $w_{T,L}(q^2)$:

$$\Delta^{(d=3)} w_L = 2\,\Delta^{(d=3)} w_T = \frac{4}{Q^4}\sum_f A_f V_f m_f \kappa_f\,. \tag{18}$$

In perturbation theory the matrix element κ_f of the chirality-flip operator O_f is proportional to m_f. Nonperturbatively, however, κ_f does not vanish at $m_f = 0$. Due to spontaneous breaking of the chiral symmetry in QCD the matrix elements of quark operators (16) are instead proportional to the quark condensate $\langle\bar{q}q\rangle_0 = -(240\,\text{MeV})^3$. It leads to the representation of κ_f in the form

$$\kappa_f = -4\pi^2 \widetilde{V}_f\, \langle\bar{q}q\rangle_0\, \chi\,, \tag{19}$$

introduced by Ioffe and Smilga[15] with χ as the magnetic susceptibility of quark vacuum.

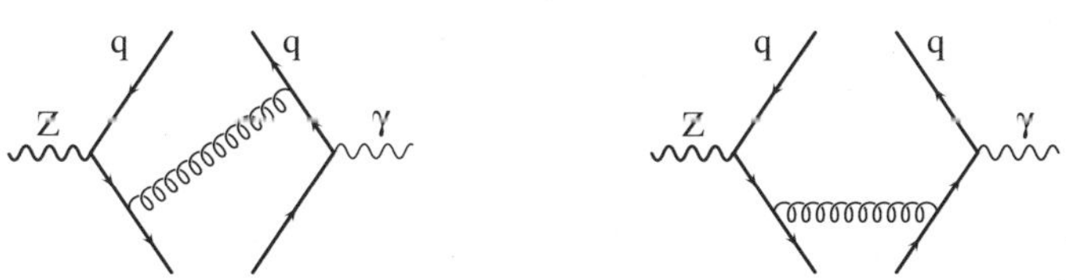

Fig. 3. Diagrams for four-fermion operator.

The distinction between longitudinal and transversal parts in the chiral limit shows up at the $d = 6$ level of four-fermion operators.[6] A particular example is the four-fermion operator, see Fig. 3,

$$-\frac{16\pi\alpha_s(Q)}{Q^6}\left[\bar{q}\gamma^\alpha\gamma_5\, V\, t^a q\; \bar{q}\gamma^\beta\, A\, t^a q - (\alpha \leftrightarrow \beta)\right]. \tag{20}$$

Arising due to these operators terms $1/Q^6$ in w_T reflect nonvanishing masses of meson resonances contributing to the transversal part. This was used in Ref. [3] for construction of a resonance model for w_T consistent with the OPE constraints,

$$w_T[u,d] = \frac{1}{m_{a_1}^2 - m_\rho^2}\left[\frac{m_{a_1}^2 - m_\pi^2}{Q^2 + m_\rho^2} - \frac{m_\rho^2 - m_\pi^2}{Q^2 + m_{a_1}^2}\right]. \tag{21}$$

5. Applications

- Hadrons in the electroweak corrections to $g-2$. The analysis [3] allowed for a quite accurate account of hadronic part in the electroweak corrections to the muon anomalous magnetic moment. The total result for $a = (g-2)/2$ is

$$a^{\mathrm{EW}} = 154(2)(1) \times 10^{-11}. \tag{22}$$

- Magnetic susceptibility of quark condensate. With an additional assumption about pion dominance this susceptibility is given by [5]

$$\chi = -\frac{N_c}{4\pi^2 F_\pi^2} = -\frac{1}{(335\ \mathrm{MeV})^2}. \tag{23}$$

  This expression, similar to the Gell-Mann, Oakes, Renner relation,[16] was recently obtained within holographic approach to QCD.[17,18]
- Hadronic light-by-light in the muon $g-2$. This contribution is quite

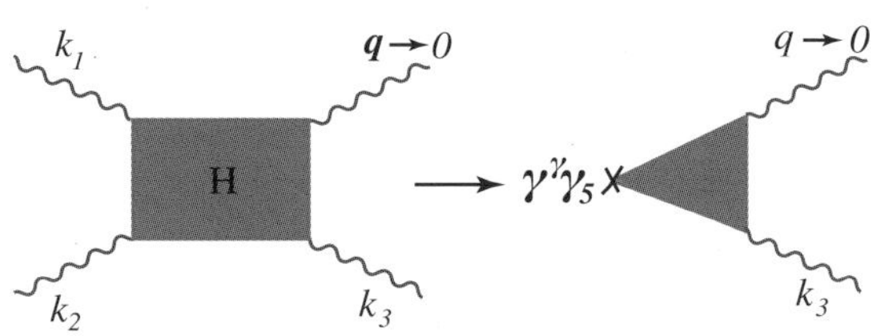

Fig. 4. OPE relation between box and triangle amplitudes.

  difficult to control theoretically and results depend on low-energy hadron dynamics. Certain constraints on the models come from OPE. When two photon virtualities k_1^2, k_2^2 are much larger than the third one, see Fig. 4, and also much larger than the hadronic scale Λ_{QCD} the quark box diagram is related by OPE to to the triangle amplitude which we discussed above. In the model [19] satisfying this constraint the result for the hadronic light-by-light in the muon $g-2$ is

$$a^{\mathrm{HLbL}} = (105 \pm 26) \times 10^{-11}\,. \qquad (24)$$

Acknowledgments

I am grateful to Andrzej Czarnecki, Bill Marciano and Kirill Melnikov for collaboration and to the organizers of the Workshop for invitation. The work is supported in part by DOE grant DE-FG02-94ER408.

References

1. V. N. Gribov, *"Anomalies, as a manifestation of the high momentum collective motion in the vacuum,"* Preprint KFKI-1981-66, Aug 1981. Published in Balaton Conf/Workshop 1986:0065.
2. V. N. Gribov, *Phys. Lett.* **B194**, 119 (1987).
3. A. Czarnecki, W. J. Marciano and A. Vainshtein, *Phys. Rev.* **D67**, 073006 (2003). [*Erratum-ibid.* **D73**, 119901 (2006)].
4. S. L. Adler and W. A. Bardeen, *Phys. Rev.* **182**, 1517 (1969).
5. A. Vainshtein, *Phys. Lett.* **B569**, 187 (2003).
6. M. Knecht, S. Peris, M. Perrottet and E. De Rafael, *JHEP* **0211**, 003 (2002).
7. G. 't Hooft, in *Recent Developments In Gauge Theories,* Eds. G. 't Hooft *et al.*, (Plenum Press, New York, 1980).
8. A.D. Dolgov and V.I. Zakharov, *Nucl. Phys.* **B27**, 525 (1971).
9. J. S. Bell and R. Jackiw, *Nuovo Cim.* **A60**, 47 (1969).
10. S. L. Adler, *Phys. Rev.* **177**, 2426 (1969).
11. L. Rosenberg, *Phys. Rev.* **129**, 2786 (1963).
12. T. V. Kukhto, E. A. Kuraev, A. Schiller and Z. K. Silagadze, *Nucl. Phys.* **B371**, 567 (1992).
13. M. Knecht, S. Peris, M. Perrottet and E. de Rafael, *JHEP* **0403**, 035 (2004).
14. F. Jegerlehner and O. V. Tarasov, *Phys. Lett.* **B639**, 299 (2006).
15. B. L. Ioffe and A. V. Smilga, *Nucl. Phys.* **B232**, 109 (1984).
16. M. Gell-Mann, R. J. Oakes and B. Renner, *Phys. Rev.* **175**, 2195 (1968).
17. A. Gorsky and A. Krikun, *Phys. Rev.* **D79**, 086015 (2009).
18. D. T. Son and N. Yamamoto, *"Holography and Anomaly Matching for Resonances,"* arXiv:1010.0718 [hep-ph].
19. K. Melnikov and A. Vainshtein, *Phys. Rev.* **D70**, 113006 (2004).

Bonn, 1995.

CLOSING ADDRESS

CONCLUDING REMARKS

YU.L. DOKSHITZER*.

Laboratory of Theoretical High Energy Physics (LPTHE),
University Paris–6, Paris, France
yuri@lpthe.jussieu.fr

The nature of the actual presentation that I gave in the "Concluding remarks" slot of the Gribov–80 Workshop does not make it suitable to be included into Proceedings. So, I use this opportunity to present you another essential part of the Gribov Memorial: his lectures on *Strong Interactions of Hadrons at High Energies.*

1. Gribov lectures

In the academic year 1972/73, Vladimir Naumovich Gribov gave a year-long lecture course on strong interactions for 5th-year students of the physics faculty of the Leningrad University. The majority of students who followed that course were future theorists who belonged to the "field theory", "quantum mechanics" and "mathematical physics" specialization groups. I had a chance to be among them.

These lectures were about hadrons and strong interactions in the "old" — pre-QCD, even pre-quark — sense.

The "old approach" to strong interactions took off in the early 60s when it was realised that the general properties of the relativistic S-matrix theory — crossing symmetry, unitarity, causality — put severe restrictions on the possible high energy behaviour of hadron interactions (elastic, inelastic, total cross sections).

The effective theory describing the high energy asymptotics of σ_{tot}, as well as fluctuations in multi-particle production — the Gribov Reggeon Field Theory (RFT) of interacting pomerons — was constructed. It was a turning point when it has been found to be *intrinsically unstable* in the specific, but the only practically relevant, case of (nearly) constant total

*On leave of absence from PNPI, 188350, St. Petersburg, Russia

cross sections.

Looking into possible solutions of the corresponding *infrared unstable dynamics* has helped to develop the theory of *second order phase transitions* in condensed matter physics (the "scaling" solution). However, the original pomeron problem remained unsolved.

By the time when V. Gribov started teaching strong interactions to graduate students†, it became clear that the pomeron instability is a serious problem that could not be resolved without an input from outside the S-matrix theory, namely, without understanding the structure of the hadronic vacuum.

2. Challenging experience

For students that was a challenging time. Each week Gribov was giving a four-hour lecture (with one short break), and the six days — till the next Thursday — students spent desperately trying to digest the material. Not that we got caught totally unprepared, having been already taught quantum mechanics (including scattering theory) and special relativity. Moreover, we were familiar with quantum field theory too, since the previous year we have followed Gribov's course on Quantum Electrodynamics.[1]

His way of teaching Quantum Electrodynamics had a bizarre flavor to it: Gribov managed to construct and fully explore QED without mentioning the word "Lagrangian" a single time! He did it on purpose. Apparently, he already had in mind the future *strong interactions* course, and wanted to demonstrate the way of introducing relativistic quantum field theory based on the language of Feynman diagrams — "*the laboratory of theoretical physics*", in his words.

The strong interactions course was anything but a familiar type of fact-flowing lecturing: "I know things, you don't, so listen and remember". Obviously, there was a general introduction part about relativistic description of particle propagation, S-matrix, causality and analyticity, unitarity. However, already at the very beginning of the course the air of adiabatic "fact consumption" has evaporated. For example, one topic — *Singularities of Feynman diagrams* — that the unsuspecting audience took for a technical exercise, all over sudden turned into a powerful physical message. Having shown that the position and the nature of each singularity of a Feynman diagrams are driven exclusively by the unitarity — physical mass spec-

†He has later formulated his motivation: "I wanted to tell everything I ever learnt about hadron interactions".

trum[‡], Gribov has established the basis for using Feynman diagrams beyond perturbation theory, or even in the situation when the interaction Lagrangian is not known.

Another fascinating lesson of how to approach a difficult theoretical issue was taught in relation with the reggeon branching singularities, where Gribov's "stereoscopic" approach to a problem shone in full light. By then we already knew about resonances and Regge poles in quantum mechanics, learned how to analytically continue partial wave amplitudes to complex angular momenta in relativistic theory. We became familiar with the Regge pole picture and, in particular, with the hypothesis of the rightmost one in the angular momentum plane — the "vacuum pole" (pomeron).

The picture that possesses only poles in angular momentum is incomplete. To show this, Gribov could repeat Mandelstam's derivation of the generation of branch cut singularities from the initial poles. However, he chose a much longer road to the destination point, systematically hunting down various inconsistencies of the pole picture. This allowed students to deeper appreciate the power of the analyticity and unitarity in the crossing channels.

The first signal that *something is rotten in the state of* the pole picture, came with the story of a "kinematical" pole of the partial wave in the unphysical angular momentum point $\ell = -1$. It turned, by the power of the t-channel unitarity, into the strong statement that in relativistic theory the scattering amplitude cannot fall with energy faster than $1/s$ (though its *imaginary part* may). This has led to the concept of moving branch cut singularities in the complex J-plane as the natural means of rescuing the unitarity relation for partial waves in the physical region of the t channel.

Another problem transpired directly from physics in the s-channel. First of all, contributions to the total cross section from exclusive processes (like elastic scattering) turned out to decrease with energy *logarithmically*, the behaviour that looked unnatural from the point of view of the model where only poles were present. Moreover, the analysis of specific multiparticle processes (such as production of particles with large rapidity gaps, high mass inelastic diffraction (3–Pomeron limit), etc.) resulted in a dramatic observation, namely that the cross sections of some particular subprocesses increase with s faster than the full $\sigma_{\rm tot}$! Catastrophically large fluctuations in multiparticle production signaled incompleteness of the Regge pole approximation in relativistic theory.

[‡]and the "magnitude" of the singularity — by physical on-mass-shell amplitudes

To elucidate the nature of reggeon branching singularities, Gribov turned to quantum mechanical scattering and to the *shadowing* phenomenon.

He returned to the physics of shadowing (screening) in high energy scattering later, closer to the end of the course, when the theory of interacting reggeons had been already constructed and the dilemma arose of two possible solutions of the "reggeon field theory": the so-called *weak* and *strong coupling* regimes of the RFT. Discussion of the general physical phenomenon of the *vanishing of forward inelastic diffraction* came a climax. Strangely, Gribov has never published this study which is, unquestionably, one of the marvels of high energy theoretical physics.

3. The Book

For many years I kept student lecture notes that we have collected together with my friend and collaborator Sergey Troian, in a hope to edit and make them public one day[§]. The task proved to be difficult and took many years to accomplish. Problems were many.

First (and most important) of all, those who were taking notes barely understood the material that they have been penning down. Sufficient understanding came later when we found ourselves capable to appreciate the depth and the nature of lecturer's arguments and grasp his goals, which helped to decipher desperate abbreviations, half-sentences and telling lacunas in the notes. Gribov–the–lecturer was improvising and never had any written preparatory material, even in theses form, which we might possibly refer to. His lectures should have been audio-recorded (if one had such means in those days).

An additional complication we had to deal with was that actually Gribov delivered his strong interaction lectures *twice*: he repeated the course two years later for the next generation of particle theory students. The verb "repeated" does not suit here: it was not in his nature to literally *repeat* a story once told. Moreover, those were intensive years that spanned between the two courses. During these few years 1) the famous s-channel understanding of the nature of the pomeron has been developed (impact parameter diffusion, Mueller–Kancheli analysis of inclusive spectra and the pattern of multiplicity fluctuations, AGK cutting rules, screening, etc.) and 2) Gribov's attitude to quarks has evolved (from moderate skepticism to

[§]Working on the book, we have also consulted lecture notes taken by V. Petrov and V. Vechernin moderate interest). So, in preparing a coherent lecture course, we had to examine the fabric of two presentations and knit the most shining pieces into a consistent text.

Then, there was a standard problem of ideological incongruence of live lectures (especially Gribov's lectures) and a publishable text. The latter calls for using = signs in formulae, while Vladimir Naumovich was employing some sort of a "wiggle" in his blackboard creations which only seldom meant =. Hundreds of formulae had to be checked, re-derived, and notations made consistent through the body of the book.

By the year 1990 the first five or six lectures (or, rather, future chapters) were ready, and Gribov had a chance to read them, make remarks and approve. Unfortunately, about the rest — the most difficult part — of the book we did not have chance to consult the author. The Gribov Strong Interactions course started taking final shape about ten years after his death in 1997. Working together with Julia Nyiri, we have assembled the complete set of lectures, debugged and translated them into English. No attempts have been made to update the content of the lectures. We considered it more important to preserve Gribov's lines of reasoning, his manner of attacking, modeling, and taking about difficult theoretical problems (which high energy hadron physics has supplied in abundance).

The book "Strong Interactions at High Energies" was published by Cambridge University Press in December 2008.[3] It contains 16 chapters

(1) Introduction
(2) Analyticity and unitarity
(3) Resonances
(4) Electromagnetic interaction of hadrons
(5) Strong interactions at high energies
(6) t-channel unitarity and growing interaction radius
(7) Theory of complex angular momenta
(8) Reggeon exchange
(9) Regge poles in perturbation theory
(10) Regge pole beyond perturbation theory
(11) Reggeon branchings
(12) Branchings in the s channel and shadowing
(13) Interacting reggeons
(14) Reggeon Field Theory
(15) Particle density fluctuations and RFT
(16) Strong interactions and field theory

The book is a merger of two lecture courses delivered by V.N. Gribov in 1972/73 and 1974/75.

Its last chapter is devoted to a search for a field theory that could be responsible for microscopic dynamics of hadrons. Here QCD is introduced, and the course ends with Gribov's original derivation and analysis of the asymptotic freedom. It was very typical of Vladimir Naumovich Gribov to share with students the thoughts that preoccupied him and works that he had been doing that very time. It is this new interest in QCD that led, practically *on-line* with his lecturing, to the major discovery know today as "Gribov copies".

But this is another story.

Actually, V. Gribov turned to QCD searching for means of solving the pomeron puzzle. He came to the firm belief that the pomeron instability got to be intimately related to the infrared instability of the QCD, i.e. to the physics of quark confinement.

4. Why bother?

What sense does it make to look back into the concepts, ideas and techniques of the theory of hadrons and their interactions of the "naive", pre-QCD, epoch?

The Lagrangian theory of strong interactions — QCD as we know it today — proved to be spectacularly successful in describing small distance quark–gluon dynamics (*hard* processes). In particular, it became a working tool in the search for new physics beyond the Standard Model.

However, as far as internal problems of hadron dynamics are concerned, our understanding of even the most general characteristics of *soft* hadron processes — like total hadron interaction cross sections in the first place — remains where the "old theory" left it about thirty years ago.

The fact that the old theory had very limited means and, being devoid of microscopic dynamics, had to rely on the most general properties of the relativistic S-matrix theory, turns to its advantage nowadays.

There exists a deep relation between the "old" and "new" strong interaction theories, which can not be appreciated without studying both.

The new theory — QCD — has adopted many a notion of its predecessor: the Froissart regime; reggeization of quarks and gluons, and multi-regge kinematics; [QCD, BFKL, "hard"] pomeron, and the reggeon field theory; impact parameter diffusion; parton screening and saturation; Gribov–Glauber multiple scattering theory, and the Abramovsky–Gribov–Kancheli

(AGK) cutting rules. Many important issues of the old theory, its techniques, problems and achievements are, however, not exposed in existing textbooks. Gribov lectures fill the gap.

Topically, the new book overlaps with Gribov's *Theory of Complex Angular Momenta* (TCAM).[2] There is, however, an essential difference. The TCAM volume is based on lectures given by V.N. Gribov in 1969 to the audience of *professional theorists*, while the present course targeted *university students*. It does not presume, therefore, any additional knowledge beyond quantum mechanics (including the non-relativistic quantum scattering theory), relativistic kinematics and the QFT basics covered by Gribov's lectures on QED.[1] That is why many general topics not present in the TCAM volume are extensively discussed in the new book: analytic properties of Feynman amplitudes, resonances, electromagnetic interaction of hadrons, etc.

On the other hand, in the TCAM book some technically involved themes are elaborated in greater detail. This puts it in a position of an important supplement to the present, more pedagogical, lecture course.

So, the two books complement each other in many ways, aiming at a comprehensive exposition of the theory of hadron interactions that preceded Quantum Chromodynamics.

So, make sure you have it on your library shelf

ISBN-13: 9780521856096

References

1. V.N. Gribov and J. Nyiri, *Quantum Electrodynamics.* Cambridge University Press, 2001.
2. V.N. Gribov, *The Theory of Complex Angular Momenta.* Cambridge University Press, 2003.
3. Vladimir Gribov, *Strong Interactions of Hadrons at High Energies* Cambridge University Press, 2008.

Como, 1994.

Trieste, 2010.

LIST OF PARTICIPANTS AND CONTRIBUTORS

Victor ABRAMOVSKY
Novgorod State University,
ul. B. Peterburgskaya 41
173003 Veliky Novgorod, RUSSIA
e-mail: Victor.Abramovsky@novsu.ru

Rachid AHL LAAMARA
Mohammed V University,
Faculty of Sciences, Department of Physics,
LAB UFR PHE
Avenue Ibn Battouta, BP 1014 Agdal
10090 Rabat, MOROCCO
e-mail: ahllaamara@gmail.com

Evgeny AKHMEDOV
Max Planck Institut fuer Kernphysik,
Postfach 103960, Saufercheckweg
69117 Heidelberg, GERMANY
e-mail: akhmedov@mpi-hd.mpg.de

Massimiliano ALVIOLI
Universitá degli Studi di Perugia,
Dipartimento di Fisica
Via A. Pascoli
I-06123 Perugia, ITALY
e-mail: alvioli@pg.infn.it

Daniele AMATI
S.I.S.S.A. International School for Advanced Studies
Via Beirut 2-4
34014 Trieste, ITALY
e-mail: amati@sissa.it

Vladimir V. ANISOVICH
Petersburg Nuclear Physics Institute
Gatchina,
188300 St. Petersburg, RUSSIA
e-mail: anisovich@thd.pnpi.spb.ru

Manuel ASOREY
Universidad de Zaragoza, Facultad de Ciencias,
Departamento de Fisica Teorica,
C/ Pedro Cerbuna 12
E-50009 Zaragoza, SPAIN
e-mail: asorey@saturno.unizar.es

Pierre VAN BAAL
Instituut-Lorentz for Theoretical Physics,
P.O. Box 9506, NL-2300 RA Leiden, The Netherlands
e-mail: vanbaal@lorentz.leidenuniv.nl

Gergely Gábor BARNAFÖLDI
MTA KFKI RMKI Research Institute for Particle and Nuclear Physics,
Department of Theoretical Physics
29-33 Konkoly Thege út
P.O. Box 49, 1525 Budapest 114, HUNGARY
e-mail: bgergely@rmki.kfki.hu

Jochen BARTELS
II. Institut für Theoretische Physik, Universität Hamburg,
Luruper Chaussee 149, 22761 Hamburg, GERMANY
e-mail: jochen.bartels@desy.de

Niklas BEISERT
Albert Einstein Institut
Max Planck Institut für Gravitationsphysik
Am Muhlenberg 1
D-14476 Potsdam, GERMANY
e-mail: nbeisert@aei.mpg.de

Alexander BELAVIN
Russian Academy of Sciences
L.D. Landau Institute for Theoretical Physics
ul. Akad. Semenova 1A
Chernogolovka, 142432 RUSSIA
e-mail: belavin@itp.ac.ru

Zouina BELGHOBSI
Department of Physics
University of Jijel
B.P. 98, Ouled Aissa
Jijel DZ18000, ALGERIA
e-mail: belghob@yahoo.fr

Luciano BERTOCCHI
Dipartimento di Fisica Teorica
Universitá degli Studi di Trieste
and
The Abdus Salam International Centre for Theoretical Physics
Strada Costiera 11
34014 Trieste, ITALY
e-mail: bertocchi@ictp.it

Aparajita BHATTACHARYA
Department of Physics
Jadavpur University
Kolkata 700032, INDIA
pampa@phys.jdvu.ac.in

Stanley BRODSKY
SLAC National Accelerator Laboratory
Stanford University
2575 Sand Hill Road
Menlo Park CA 94025, USA
e-mail: sjbth@slac.stanford.edu

Alejandro Genaro CABO MONTES DE OCA
Grupo de Fisica Teorica
Instituto de Cibernetica, Matematica y Fisica (ICIMAF)
Calle e Esquina 13 y 15, N. 309
Vedado, 10400 Havana, CUBA
e-mail: cabo@icmf.inf.cu

Freddy CACHAZO
Perimeter Institute for Theoretical Physics
31 Caroline Str. N.
Waterloo ON N2L 2Y5, CANADA
e-mail: fcachazo@perimeterinstitute.ca

Giorgio CALUCCI
Dipartimento di Fisica Teorica
Universitá delli Studi di Trieste
Strada Costiera 11
34014 Trieste, ITALY
e-mail: giorgio.calucci@ts.infn.it

Stefano CATANI
INFN, Sezione di Firenze
Via G. Sansone 1, Sesto Fiorentino
I-50019 Firenze, ITALY
e-mail: catani@fi.infn.it

Carlo Nicola COLACINO
Dipartimento di Fisica & INFN Sezione di Pisa
Universitá degli Studi di Pisa
Largo Pontecorvo 3
56127 Pisa, ITALY
e-mail: carlo.nicola.colacino@pi.infn.it

Máté CSANÁD
Department of Atomic Physics
Eötvös University
Pázmány P. sétány 1A
1117 Budapest, HUNGARY
e-mail: csanad@elte.hu

Tamás CSÖRGŐ
Department of Theoretical Physics
MTA KFKI RMKI Research Institute for Particle and Nuclear Physics
29-33 Konkoly Thege Str.
P.O.B. 49, Budapest 114, 1525, HUNGARY
and
Department of Physics
Harvard University
17 Oxford Street
Cambridge MA 02138, USA
e-mail: csorgo@rmki.kfki.hu

Harleen DAHIYA
Department of Physics
Dr. B.R. Ambedkar National Institute of Technology
GT Road Bye pass
Jalandhar 144011 Punjab, INDIA
e-mail: harleen_dahiya@yahoo.com

Antonio DOBADO GONZALEZ
Departamento Fisica Teorica I
Universidad Compiutense de Madrid
28040 Madrid, SPAIN
e-mail: malcon@fis.ucm.es

Yuri L. DOKSHITZER
Laboratoire de Physique Théorique et Hautes Energies (LPTHE)
Université Paris VI
Tour 24-25, 5ème etage, Boite 126
4 Place Jussieu
76252 Paris Cedex 5, FRANCE
e-mail: yuri@lpthe.jussieu.fr

Alexander DOLGOV
Polo Scientifico e Tecnologico - Edificio C
INFN, Dipartimento di Fisica
Universitá di Ferrara
via Saragat 1
44100 Ferrara, ITALY
e-mail: dolgov@fe.infn.it

Gabor DOMOKOS
Department of Physics & Astronomy
Johns Hopkins University
3400 North Charles Street
MD 21218-2695 Baltimore, USA
e-mail: skd@pha.jhu.edu

Sidney DRELL
SLAC, Stanford University
Stanford, California 94309, USA
e-mail: drell@slac.stanford.edu

Lalla Btissam DRISSI
Departement de Physique
LPHE Laboratory
Université Mohammed V
Avenue Ibn Battouta
B.P. 1014 Rabat, MOROCCO
e-mail: drissilb@gmail.com

Michael EIDES
Dept. of Physics and Astronomy
University of Kentucky
177 Chemistry-Physics Bldg.
600 Rose Street
KY 40506-0055 Lexington, USA
e-mail: eides@ps.uky.edu

Mohamed Saad Mohamed EL ASHKAR
Radiation Protection Department
Nuclear Research Center
Atomic Energy Authority
13759 Inshas, EGYPT
e-mail: m.elashkar@gmail.com

Carlo EWERZ
ExtreMe Matter Institute EMMI
GSI Helmholtzzentrum für Schwerionenforschung
Planckstrasse 1
64291 Darmstadt
and
Institut für Theoretische Physik
Universität Heidelberg
Philosophenweg 16
D-69120 Heidelberg, GERMANY
e-mail: c.ewerz@gsi.de

Alexandre FILIPPOV
Joint Institute for Nuclear Research
Dubna
Moscow region 141980, RUSSIA
e-mail: filippov@theor.jinr.ru

Leonid FRANKFURT
School of Physics and Astronomy
Tel-Aviv University
Ramat Aviv
Tel Aviv 69978, ISRAEL
e-mail: frankfu@lev.tau.ac.il

Andor FRENKEL
Department of Theoretical Physics,
MTA KFKI Research Institute for Particle and Nuclear Physics,
29-33 Konkoly Thege Str., H-1121 Budapest, HUNGARY
e-mail: frenkel@rmki.kfki.hu

Semen S. GERSHTEIN
Institute for High Energy Physics
Protvino, Moscow region 142281, RUSSIA
e-mail: Semen.Gershtein@ihep.ru

Vahtang GOGOHIA
Department of Theoretical Physics
MTA KFKI Research Institute for Particle and Nuclear Physics
29-33 Konkoly Thege Str.
Budapest, 1121 HUNGARY
e-mail: gogohia@rmki.kfki.hu

Gösta GUSTAFSON
Department of Theoretical Physics
Lund University
Sölvegatan 14 A
Lund, 22362 SWEDEN
e-mail: Gosta.Gustafson@thep.lu.se

Peter HASENFRATZ
Institute for Theoretical Physics
University of Bern
Sidlerstr. 5., Bern, 3012 SWITZERLAND
e-mail: hasenfra@itp.unibe.ch

Dezső HORVÁTH
Department of Experimental Particle Physics
MTA KFKI Research Institute for Particle and Nuclear Physics
29-33 Konkoly Thege Street
Budapest, 1121 HUNGARY
and
Institute of Nuclear Research (ATOMKI),
Debrecen, HUNGARY
e-mail: horvath@rmki.kfki.hu

Mohammed Yousif HUSSEIN
Department of Physics
University of Bahrein
Villa 2608A, Road 2966, Area 729
Isa Town 32038, BAHRAIN
e-mail: mhussein@sci.uob.bh

Boris Lazarevich IOFFE
Institute of Theoretical and Experimental Physics
ul. B. Cheremushkinskaya 25
Moscow, 117218 RUSSIA
e-mail: ioffe@ihep.ru

Dmitri KHARZEEV
Department of Physics
Brookhaven National Laboratory
Building 510
P.O.Box 5000, Upton, NY 11973-5000 USA
e-mail: kharzeev@bnl.gov

Alinaghi KHORRAMIAN
School of Particles and Accelerators
Institute for Studies in Theoretical Physics and Mathematics
Opposite the ARAJ
Artesh Highway
P.O.Box 19395-5531, Tehran, IRAN
e-mail: khorramiana@theory.ipm.ac.ir

Valery KHOZE
Institute for Particle Physics Phenomenology
University of Durham
Durham, DH1 3LE, UK
e-mail: V.A.Khoze@durham.ac.uk

Iosif B. KHRIPLOVICH
Budker Institute of Nuclear Physics,
Lavrentiev Prospekt 11
Novosibirsk, 630090 RUSSIA
e-mail: I.B.Khriplovich@inp.nsk.su

Dubravko KLABUCAR
Department of Physics
Faculty of Science
University of Zagreb
Bijenicka Cesta 32
P.O.Box 162, Zagreb 10000 CROATIA
e-mail: klabucar@phy.hr

Susan KOVESI-DOMOKOS
Dept. of Physics & Astronomy
Johns Hopkins University
3400 North Charles Street
Baltimore, MD 21218-2695 USA
e-mail: skd@pha.jhu.edu

Julius KUTI
Department of Physics
University of California, San Diego
La Jolla, CA 92093 USA
e-mail: jkuti@ucsd.edu

Péter LÉVAI
Department of Theoretical Physics
KFKI Research Institute for Particle and Nuclear Physics
29-33 Konkoly Thege Str.
Budapest, 1121 HUNGARY
e-mail: plevai@rmki.kfki.hu

Lev N. LIPATOV
Petersburg Nuclear Physics Institute
Gatchina
188300 St. Petersburg, RUSSIA
e-mail: lipatov@thd.pnpi.spb.ru

Uri MAOR
Tel-Aviv University
School of Physics and Astronomy
Ramat Aviv, Tel-Aviv 69978 ISRAEL
e-mail: maor@post.tau.ac.il

André MARTIN
Theory Division, CERN
1211 Geneva 23, SWITZERLAND
e-mail: andre.martin@cern.ch

Larry MCLERRAN
Department of Physics
Building 510, P.O. Box 5000
Upton NY 11973-5000 USA
e-mail: MCLERRAN@QUARK.PHY.BNL.GOV

Hiranmaya MISHRA
Physical Research Laboratory
Theory Group
Navrangpura, 380 009 Ahmedabad, INDIA
e-mail: hm@prl.res.in

Márton NAGY
Department of Theoretical Physics
KFKI RMKI Research Institute for Particle and Nuclear Physics
Konkoly Thege út 29-33, P.O. Box 49
1525 Budapest, HUNGARY
e-mail: nagymarci@rmki.kfki.hu

Nikolai N. NIKOLAEV
Institut für Kernphysik
Forschungszentrum Jülich
Postfach 1913, Jülich, 52425 GERMANY
and
L.D. Landau Institute for Theoretical Physics,
Chernogolovka, 142432 RUSSIA
e-mail: n.nikolaev@fz-juelich.de

Júlia NYIRI
Department of Theoretical Physics
KFKI Research Institute for Particle and Nuclear Physics
29-33 Konkoly Thege Str.
Budapest, 1121 HUNGARY
e-mail: nyiri@rmki.kfki.hu

Lev OKUN
Institute of Theoretical and Experimental Physics
ul. B. Cheremushkinskaya 25
Moscow 117218, RUSSIA
e-mail: okun@itep.ru

András PATKÓS
Department of Atomic Physics
Eötvös University
1/A Pázmány Péter sétány
Budapest, 1117 HUNGARY
e-mail: patkos@galaxy.elte.hu

János POLÓNYi
Laboratoire de Physique Theorique
Université Louis Pasteur
3 rue de l'Université
Strasbourg Cedex, 67084 FRANCE
e-mail: polonyi@physique.u-strasbg.fr

Mikhail RYSKIN
St. Petersburg Nuclear Physics Institute
Gatchina
St. Petersburg, 188300 RUSSIA
e-mail: ryskin@thd.pnpi.spb.ru

Augustin SABIO VERA
Instituto de Fisica Teorica, UAM CSIC
Universidad Autonoma
Madrid, 28046 SPAIN
e-mail: sabio@cern.ch

Mikhail SHIFMAN
William I. Fine Theoretical Physics Institute
School of Physics and Astronomy
University of Minnesota
116 Church Street S.E.
Minneapolis, MN 55455 USA
e-mail: shifman@umn.edu

Mark STRIKMAN
Department of Physics
Pennsylvania State University
104 Davey Laboratory
University Park, PA 16802-6300 USA
e-mail: strikman@phys.psu.edu

Daniele TRELEANI
Universitá degli Studi di Trieste
Dipartimento di Fisica Teorica
Strada Costiera 11
Trieste, 34014 ITALY
e-mail: dtrelean@ictp.it

Luca TRENTADUE
Dipartimento di Fisica
Universitá di Parma
Viale delle Scienze, 7A
Parma, 43100 ITALY
e-mail: luca.trentadue@fis.unipr.it

Arkady VAINSHTEIN
William I. Fine Theoretical Physics Institute
School of Physics and Astronomy
University of Minnesota
116 Church Street S.E.
Minneapolis, MN 55455 USA
e-mail: vainshte@umn.edu

Victor Manuel VILLANUEVA
Universidad Michoacana de San Nicolas de Hidalgo
Inst. de Fisica y Matematicas
Francisco J. Mujica S.N.
Edificio C-3, Ciudad Universitaria
Col. Felicitas del Rio
Morella, 58040 MEXICO
e-mail: vvillanu@ifm.umich.mex

Bryan WEBBER
Cavendish Laboratory
University of Cambridge
Madingley Road
Cambridge CB3 OHE, UK
e-mail: webber@hep.phy.cam.ac.uk

Alan WHITE
Division of High Energy Physics
Argonne National Laboratory
9700 Southcass Avenue
Argonne, IL 60439-4815 USA
e-mail: arw@hep.anl.gov

Valentin I. ZAKHAROV
Institute of Theoretical and Nuclear Physics
ul. B. Cheremushkinskaya 25
Moscow 117125, RUSSIA
e-mail: vzakharov@itep.ru